PLANE & SPHERICAL TRIGONOMETRY

The Authors

Kaj L. Nielsen received the degrees of B.S. from the University of Michigan, M.A. from Syracuse University, and Ph.D. from the University of Illinois. He has held teaching positions at Syracuse, Illinois, Brown, and Louisiana State University. He has been associated with a number of industries as a research engineer and scientist. His present position is that of Chief Mathematician at the U.S. Naval Avionics Facility, Indianapolis. Dr. Nielsen has written numerous articles based upon original research in mathematics and published in leading mathematical and engineering journals. He is the author or coauthor of *Logarithmic and Trigonometric Tables* and of *Problems in Plane Geometry*, in the College Outline Series, and of three other mathematical books.

John H. Vanlonkhuyzen received the A.B. degree from Calvin College and the A.M. degree from the University of Michigan. He taught mathematics at various colleges until 1942, when he joined the Bell Aircraft Company as a research engineer. His present position there is that of Head of the Technical Department.

COLLEGE OUTLINE SERIES

PLANE & SPHERICAL TRIGONOMETRY

KAJ L. NIELSEN

JOHN H. VANLONKHUYZEN

BARNES & NOBLE, Inc. NEW YORK
Publishers • Booksellers • Founded 1874

Ninth Printing, 1957

L. C. catalogue card number: 54-8921

PRINTED IN THE UNITED STATES OF AMERICA

PLANE & SPHERICAL TRIGONOMETRY

PREFACE

This outline has been written with two purposes in mind: (a) to furnish a rapid review for the reader who has completed a course in plane and spherical trigonometry, and (b) to furnish supplementary material for the reader who is enrolled in a course in this subject. However, it is complete in itself and, although it is not intended to be used as a textbook, it may be read with profit by those who are beginning the study of trigonometry. The outline should provide any reader with a thorough knowledge of the usual topics studied in a college course in both plane and spherical trigonometry.

Considerable attention has been paid to definitions and explanations of the terms and symbols. Each concept has been illustrated by many figures and solved problems. The application of trigonometry to surveying, vectors, and navigation has been fully explained and illustrated. Great stress has been placed on a neat and systematic manner of solving the problems. Many errors in computation are due to carelessness and nonsystematic presentation of the solutions to problems. The reader can overcome these errors by paying strict attention to the forms of solutions given in the illustrations. The formulas have been focused so that they may readily be found.

Five-place tables are furnished with this outline, and the principle of handling any table is explained. If the reader is using a four-place table in his classwork, he may still benefit by using the accompanying tables and can check four-place answers by rounding off the numbers.

It is the authors' sincere belief that a thorough knowledge of any branch of mathematics cannot be obtained without solving problems. Consequently, a few typical exercises have been placed at various intervals in the outline. The student is urged to work all the exercises. Answers are furnished to the exercises so that the reader may check his work. A few copies of examinations have also been included with the hope that they may aid the student in his preparation for any examination.

The authors gratefully acknowledge their indebtedness to Mr. J. C. Stewart for his many criticisms and to Miss Anne Dorman for the answers to the exercises.

K. L. N.
J. V. L.

Table of Contents

PART ONE. PLANE TRIGONOMETRY

CHAPTER | PAGE

I. FUNCTIONS OF ACUTE ANGLES
1. Introduction 3
2. The Trigonometric Functions 3
3. Properties of a Right Triangle 5
4. Exercise 1 6
5. The Functions of 30°, 45°, 60° 7
6. Four-Place Trigonometric Tables 8
7. Exercise 2 12
8. Solution of a Right Triangle 13
9. Exercise 3 15

II. THE GENERAL ANGLE
10. Angles 16
11. Angular Measurements 16
12. Conversion Units 17
13. Properties of Radian Measure 19
14. Exercise 4 21

III. TRIGONOMETRIC FUNCTIONS OF ANY ANGLE
15. Rectangular Coordinates 22
16. The Trigonometric Functions of Any Angle θ 23
17. The Signs of the Trigonometric Functions in the Various Quadrants 24
18. Functions of 0°, 90°, 180°, 270° 24
19. Reduction to Acute Angles 26
20. Exercise 5 27
21. Functions of the Negative Angle $-\theta$ . . . 28
22. Simple Equations 29
23. Exercise 6 29
24. Polar Coordinates 29
25. Graphs in Polar Coordinates 30
26. Exercise 7 32

CHAPTER PAGE

IV. VARIATIONS AND GRAPHS OF THE TRIGONOMETRIC FUNCTIONS
27. Definitions 33
28. Periodicity of the Trigonometric Functions . 34
29. Variation of Tan θ 34
30. Graphs of the Functions 35
31. Graphs of Other Functions 38

V. THE FUNDAMENTAL IDENTITIES
32. Definitions 40
33. The Fundamental Identities 40
34. Proofs of General Identities 41
35. Exercise 8 43
36. Trigonometric Equations in One Unknown . 43
37. Exercise 9 45

VI. ADDITION FORMULAS
38. The Sine and Cosine of the Sum of Two Angles 46
39. Addition Formulas for Tangent and Cotangent 48
40. Functions of the Difference of Two Angles . 49
41. Double Angle Formulas 49
42. Half-Angle Formulas 50
43. Product Formulas 51
44. Sums and Differences of Sines and Cosines . 51
45. Identities Involving Addition Formulas . . 52
46. Equations Involving Multiple Angles . . . 54
47. Exercise 10 55

VII. LOGARITHMS
48. Definitions 56
49. Common Logarithms 56
50. Use of Tables 59
51. Interpolation 61
52. Exercise 11 62
53. Properties of Logarithms 62
54. Computations Using Logarithms . . . 64
55. Cologarithms 66
56. Exercise 12 66
57. Logarithms of Trigonometric Functions . . 67
58. The Small Angles 70
59. Exercise 13 72

CHAPTER PAGE

VIII. THE RIGHT TRIANGLE AND APPLICATIONS

60. Logarithmic Solution of a Right Triangle . 74
61. Exercise 14 76
62. Angle of Elevation and Depression . . . 76
63. Right Angle Vectors 78
64. Exercise 15 79

IX. OBLIQUE TRIANGLES

65. Cases 81
66. The Law of Cosines 81
67. The Use of the Law of Cosines in Solving Triangles 83
68. The Law of Sines 84
69. The Solution of Case I 85
70. The Solution of Case II, the Ambiguous Case 86
71. The Law of Tangents 89
72. The Solution of Case III by Using the Law of Tangents 89
73. Half-Angle Formulas 90
74. The Solution of Case IV by Half-Angle Formulas 91
75. The Area of the Oblique Triangle 92
76. Summary 93
77. Exercise 16 94
78. Some Applications 94

X. INVERSE TRIGONOMETRIC FUNCTIONS

79. Definitions 97
80. Exercise 17 98
81. Principal Values 98
82. Exercise 18 100

PART TWO. SPHERICAL TRIGONOMETRY

XI. SPHERICAL GEOMETRY

83. Introduction 103
84. Theorems of Solid Geometry 103
85. The Sphere 107

XII. THE SPHERICAL TRIANGLE

86. Introduction 110
87. Polar Triangles 112

CHAPTER PAGE

88. The Sine Law 114
89. The Cosine Law for Sides 115
90. The Cosine Law for Angles 117
91. The Haversine Law 118
92. The Haversine Nomogram. Exercise 19 . . 119

XIII. RIGHT SPHERICAL TRIANGLES
93. Spherical Triangles with More than One Right Angle 122
94. Formulas of the Right Spherical Triangle . 123
95. Napier's Rules 127
96. Suggestions for the Solution of the Right Spherical Triangle 129
97. Rules for Quadrants 129
98. Model Solutions 132
99. The Isosceles and Quadrantal Triangles . . 134
100. Exercise 20 135

XIV. THE OBLIQUE SPHERICAL TRIANGLE
101. Introduction 136
102. The Half-Angle Formulas 136
103. Napier's Analogies 139
104. Gauss' Formulas. Rule of Quadrants . . 141
105. The Cases 143
106. Exercise 21 151

XV. THE TERRESTRIAL AND ASTRONOMICAL TRIANGLES
107. The Terrestrial Triangle 152
108. The Celestial Sphere 155
109. Problems of Navigation and Astronomy . . 158
110. Exercise 22 163

XVI. PROJECTIONS
111. Introduction 164
112. Perspective Maps 165
113. Mercator Maps 166
114. Orthographic Maps 168
115. Stereographic Maps 168
116. Graphical Constructions 169

FINAL EXAMINATIONS 177
ANSWERS TO EXERCISES, EXAMINATIONS 182
INDEX 187

Following the Index

LOGARITHMIC AND OTHER TRIGONOMETRIC TABLES

COMMON LOGARITHMS OF NUMBERS with the Auxiliaries S and T 1– 21

NATURAL LOGARITHMS OF WHOLE NUMBERS from 1 to 200 22

LOGARITHMS OF THE TRIGONOMETRIC FUNCTIONS from 0° to 1° and 89° to 90° for Every Second, and from 1° to 6° and 84° to 89° for Every Ten Seconds 23– 46

LOGARITHMS OF THE TRIGONOMETRIC FUNCTIONS from Minute to Minute 47– 92

NATURAL TRIGONOMETRIC FUNCTIONS from Minute to Minute 93–116

VALUES AND LOGARITHMS OF HAVERSINES . 117–120

POWERS AND ROOTS 121–122

DEGREES, MINUTES AND SECONDS TO RADIANS 123–124

RADIANS TO DEGREES 125

TABULATED BIBLIOGRAPHY OF STANDARD TEXTBOOKS

The following list gives the author, title, publisher, and date of the standard textbooks referred to in the table on the two succeeding pages.

(1) Ballou and Steen. *Plane Trigonometry.* 2nd ed. Ginn, 1953.
Ballou and Steen. *Plane and Spherical Trigonometry.* 2nd ed. Ginn, 1953.
Sections on plane trigonometry in these two books are identical.)

(2) Bell and Thomas. *Essentials of Plane and Spherical Trigonometry.* Holt, 1946.

(3) Brixey and Andree. *Modern Trigonometry.* Holt, 1955.

(4) Brooks and Schock. *Trigonometry for Today.* Harper, 1951.

(5) Butler and Wren. *Trigonometry for Secondary Schools.* Heath, 1948.

(6) Cameron. *Brief Trigonometry.* Rev. ed. Holt, 1952.

(7) Corliss and Berglund. *Plane Trigonometry.* Houghton Mifflin, 1950.

(8) Crathorne and Moore. *Brief Trigonometry.* Holt, 1941.

(9) Dadourian. *Plane Trigonometry.* Addison-Wesley, 1950.

(10) Dubisch. *Trigonometry.* Ronald, 1955.

(11) Fuller. *Plane Trigonometry.* McGraw-Hill, 1950.

(12) Hart. *College Trigonometry.* Heath, 1951.

(13) Hart. *Trigonometry.* Heath, 1954.

(14) Heineman. *Plane Trigonometry.* 2nd ed. McGraw-Hill, 1956.

(15) Heineman. *Plane Trigonometry.* Alternate ed. McGraw-Hill, 1950.

(16) Holmes. *Trigonometry.* McGraw-Hill, 1951.

(17) Kells, Kern, and Bland. *Plane Trigonometry.* 3rd ed. McGraw-Hill, 1951.
Kells, Kern, and Bland. *Plane and Spherical Trigonometry.* 3rd ed. McGraw-Hill, 1951.
(Sections on plane trigonometry in these two books are identical.)

(18) Miller. *College Algebra and Trigonometry.* Rev. ed. Wiley, 1956.

(19) Morgan. *Plane and Spherical Trigonometry.* American Book, 1945.

(20) Nelson and Folley. *Plane Trigonometry.* 3rd ed. Harper, 1956.

(21) Northcott. *Plane and Spherical Trigonometry.* Rev. ed. Rinehart, 1950.

(22) Palmer, Leigh, and Kimball. *Plane and Spherical Trigonometry.* 5th ed. McGraw-Hill, 1950.

(23) Perlin. *Trigonometry.* International Textbook, 1955.

(24) Persich. *Plane Trigonometry Analyzed.* Jesuit Educational Association, 1954.

(25) Randolph. *Trigonometry.* Macmillan, 1953.

(26) Richardson. *Plane and Spherical Trigonometry.* Macmillan, 1950.

(27) Rider. *Plane Trigonometry.* Macmillan, 1949.

(28) Rietz, Reilly, and Woods. *Plane and Spherical Trigonometry.* 3rd ed. Macmillan, 1950.

(29) Rosenbach, Whitman, and Moskovitz. *Essentials of Plane Trigonometry.* 3rd ed. Ginn, 1950.

(30) Rosenbach, Whitman, and Moskovitz. *Plane and Spherical Trigonometry.* Ginn, 1943.

(31) Rutledge and Pond. *Modern Trigonometry.* Prentice-Hall, 1956.

(32) Smail. *Trigonometry, Plane and Spherical.* McGraw-Hill, 1952.

(33) Sparks and Rees. *Plane Trigonometry.* 3rd ed. Prentice-Hall, 1952.

(34) Spitzbart and Bardell. *Plane Trigonometry.* Addison-Wesley, 1955.

(35) Thompson and Cowles. *Trigonometry.* 2nd ed. Van Nostrand, 1949.

(36) Vance. *Trigonometry.* Addison-Wesley, 1954.

(37) Wentworth and Smith. *Plane and Spherical Trigonometry.* Ginn, 1943.

(38) Wylie. *Plane Trigonometry.* McGraw-Hill, 1955.

(39) Zant. *College Algebra and Plane Trigonometry.* Ginn, 1953.

QUICK REFERENCE TABLE TO STANDARD TEXTBOOKS

All figures refer to pages.

CH.	TOPIC	(1)	(2)	(3)	(4)	(5)	(6)	(7)	(8)	(9)	(10)	(11)	(12)	(13)
I	Functions of Acute Angles	13	20	37	76	11	22	25	7	5	178	19	18	1
II	General Angle	3	1	31	10, 43	105, 117	4,48	1	3	1, 37	153	1	7, 84	56, 84
III	Functions of Any Angle	13, 24	11	33, 117	13	121	6	12	23	2, 34	42, 157	12	9, 67	59
IV	Variations and Graphs	72	70	81	49	138	59	50	33	42	69	82, 160	76	76, 89
V	Fundamental Identities	56	75	93	29	153	67	159	13	57	45, 123	48	11, 93	93
VI	Addition Formulas	84	79	99	130	158	78	196	38	63	50	112	104	104
VII	Logarithms	135	45	49	107	62	93	120	48	89	218	59	33	22
VIII	Right Triangles and Applications	30	50		89, 123	32, 83	26, 38	66	18, 57	17	182, 256	28, 75	28, 53	19, 42
IX	Oblique Triangles	100	93	69	146	175	113	225	60	72	186, 258	93, 132	119	119
X	Inverse Functions	69	126	112	181	148	65	185	20, 36	52	97	151	144	144
XI	Spherical Geometry		169			211							177	
XII	Spherical Triangle	149	172			212							178	177
XIII	Right Spherical Triangle	152	178			213							179	179
XIV	Oblique Spherical Triangle	162	211			229							187	187
XV	Terrestrial & Astronomical Triangle	180	229										201	201
XVI	Projections													

QUICK REFERENCE TABLE TO STANDARD TEXTBOOKS

All figures refer to pages.

CH.	TOPIC	(14)	(15)	16)	(17)	(18)	(19)	(20)	(21)	(22)	(23)	(24)	(25)	(26)
I	Functions of Acute Angles	13	14		1	52	10	28		29	65	4	7	23
II	General Angle	4, 47	1	7	54	37	36	1	1	2, 10	5	248	1	75
III	Functions of Any Angle	5	6	10	57	44	40	8	6	16, 67	11	80	17	78
IV	Variations and Graphs	55	58	33	80	223, 239	98	107	34	76	21	103	106	179
V	Fundamental Identities	29	31	43	18	75	53	21	12	35	39	6, 113	74	120
VI	Addition Formulas	63	67	49	95	123	117	122	46	109	135	131	29	138
VII	Logarithms	83	89	62	169	246	231	71	100	223	74	15	166	213
VIII	Right Triangles and Applications	99	24, 107	71	32	51	22	83	113	49	95	55	42	39, 234
IX	Oblique Triangles	107	115	83	113	141, 262	67, 84	85	118	134	159	179	56	104, 242
X	Inverse Functions	129	138	37	141	222	110	139	74	44	47	243	117	201
XI	Spherical Geometry				183		155		165	192				287
XII	Spherical Triangle			120	185		160		165	191				287
XIII	Right Spherical Triangle			123	206		178		167	193				297
XIV	Oblique Spherical Triangle			131	201, 239		198, 217		178	200				310
XV	Terrestrial & Astronomical Triangle			138	196, 222		175		214	217				332
XVI	Projections				189									

QUICK REFERENCE TABLE TO STANDARD TEXTBOOKS

All figures refer to pages.

CH.	TOPIC	(27)	(28)	(29)	(30)	(31)	(32)	(33)	(34)	(35)	(36)	(37)	(38)	(39)
I	Functions of Acute Angles	1	1	41	21, 28	114	24	7	31	39	35	1	55	70
II	General Angle	115	34, 143	7	10, 83	8	1	57	1	2	16	77	11	52
III	Functions of Any Angle	54	34	13	13	54	11	7, 50	13	23	22	82	21, 37	59
IV	Variations and Graphs	127	60	123	232	161	81, 115	66	49	115	28, 81	151	89	100
V	Fundamental Identities	99	54	21	17, 96	74	94	39	79	96	87	13	33	86
VI	Addition Formulas	102	83	64	111	127	126	81	92	123	41	97	114	122
VII	Logarithms	29	101	135	139	23	274	99	116	A-i	125	30	169	145
VIII	Right Triangles and Applications	48	15, 120	41	55, 187	114	37	19, 112	31	53	51, 58	63, 133	55	169
IX	Oblique Triangles	66	68, 122	85	76, 195	85	152	121	132	155	51, 66	107, 137	195	170
X	Inverse Functions	139	149	113	253	185	197	155	71		74	156	135	103
XI	Spherical Geometry		164											
XII	Spherical Triangle		165		289		236					187		
XIII	Right Spherical Triangle		166		292		239					190		
XIV	Oblique Spherical Triangle		175		304		245					205		
XV	Terrestrial & Astronomical Triangle		187, 188		328		268							
XVI	Projections													

Part One
PLANE TRIGONOMETRY

GREEK ALPHABET

Letters		Names	Letters		Names	Letters		Names
A	α	Alpha	I	ι	Iota	P	ρ	Rho
B	β	Beta	K	κ	Kappa	Σ	σ	Sigma
Γ	γ	Gamma	Λ	λ	Lambda	T	τ	Tau
Δ	δ	Delta	M	μ	Mu	Υ	υ	Upsilon
E	ϵ	Epsilon	N	ν	Nu	Φ	ϕ	Phi
Z	ζ	Zeta	Ξ	ξ	Xi	X	χ	Chi
H	η	Eta	O	o	Omicron	Ψ	ψ	Psi
Θ	θ	Theta	Π	π	Pi	Ω	ω	Omega

SYMBOLS

$=$, is equal to;

$\neq$, is not equal to;

$<$, is less than;

$>$, is greater than;

$\leqq$, is less than or equal to;

$\geqq$, is greater than or equal to;

P_1, P sub 1;

$n \to \infty$, n approaches infinity

$\sqrt{n}$, square root of n;

$\angle ABC$, angle with vertex at B;

$\triangle ABC$, triangle ABC;

$\cong$, is congruent to.

Chapter I

FUNCTIONS OF ACUTE ANGLES

1. Introduction.

Trigonometry has been defined as *that branch of mathematics which deals with the solutions of triangles;* however, it is more inclusive than that, and we shall say that trigonometry is *that part of mathematical investigations which is performed by means of trigonometric functions.* The *trigonometric functions* are certain *ratios* depending upon an angle; thus they are functions of an angle. We shall assume that the student is familiar with the number system and the fundamental operations of algebra which he has met in his high school algebra course. In order that he may make use of what he has already learned in his high school mathematics courses concerning the right triangle, we shall begin by defining the trigonometric functions of the acute angles in terms of the sides of the right triangle. This is, of course, a special case of the general angle, as we shall point out later. However, this special case has many applications and forms an easy introduction to what is to follow.

2. The Trigonometric Functions.

There are six trigonometric functions of any angle; these six functions are called

sine of θ;	abbreviated: **sin** θ;
cosine of θ;	abbreviated: **cos** θ;
tangent of θ;	abbreviated: **tan** θ;
cotangent of θ;	abbreviated: **cot** θ [also **ctn** θ];
secant of θ;	abbreviated: **sec** θ;
cosecant of θ;	abbreviated: **csc** θ;

where θ (Greek theta) represents the angle.

If θ is an acute angle of a right triangle, these six trigonometric functions are defined to be

$$(1)\quad \begin{cases} \sin\theta = \dfrac{\text{opposite side}}{\text{hypotenuse}}; & \cot\theta = \dfrac{\text{adjacent side}}{\text{opposite side}}; \\ \cos\theta = \dfrac{\text{adjacent side}}{\text{hypotenuse}}; & \sec\theta = \dfrac{\text{hypotenuse}}{\text{adjacent side}}; \\ \tan\theta = \dfrac{\text{opposite side}}{\text{adjacent side}}; & \csc\theta = \dfrac{\text{hypotenuse}}{\text{opposite side}}. \end{cases}$$

HINT: These definitions should be committed to memory.

Let us consider the right triangle of Fig. 1. By the above definitions, it is easily seen that

$$\sin\alpha = \frac{a}{c};\quad \cot\alpha = \frac{b}{a};$$
$$\cos\alpha = \frac{b}{c};\quad \sec\alpha = \frac{c}{b};$$
$$\tan\alpha = \frac{a}{b};\quad \csc\alpha = \frac{c}{a};$$

and

$$\sin\beta = \frac{b}{c};\quad \cot\beta = \frac{a}{b};$$
$$\cos\beta = \frac{a}{c};\quad \sec\beta = \frac{c}{a};$$
$$\tan\beta = \frac{b}{a};\quad \csc\beta = \frac{c}{b}.$$

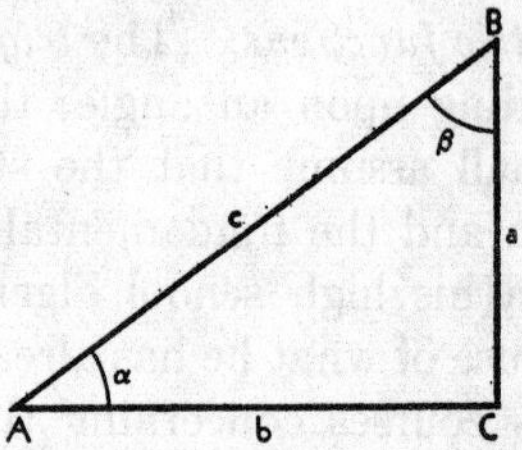

Fig. 1

It is recommended that the student learn the definitions in terms of the words rather than in terms of the symbols.

Illustration.

Consider the right triangle of Fig. 2; here we see that

$$\sin\alpha = \tfrac{4}{5};\quad \cot\alpha = \tfrac{3}{4};$$
$$\cos\alpha = \tfrac{3}{5};\quad \sec\alpha = \tfrac{5}{3};$$
$$\tan\alpha = \tfrac{4}{3};\quad \csc\alpha = \tfrac{5}{4}.$$

What are the functions of β?

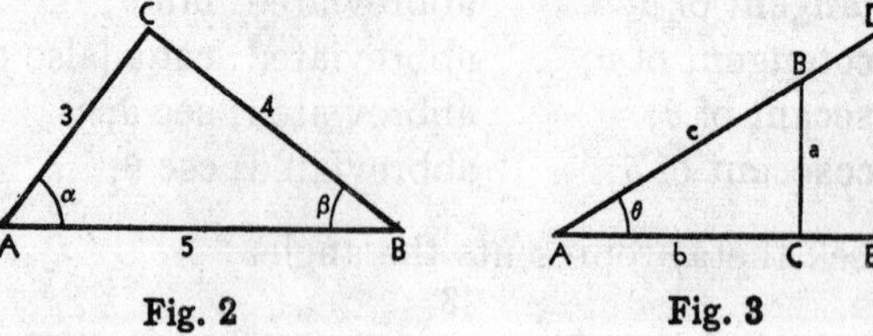

Fig. 2

Fig. 3

3. Properties of a Right Triangle.

I. Construct the right triangles ACB and AED as shown in Fig. 3. The two triangles are similar, and from elementary geometry we have $\frac{BC}{AB} = \frac{DE}{AD}$. Likewise, the ratios of any pair of corresponding sides in the two triangles are equal. Therefore the ratio of any two sides of a right triangle depends only upon the size of the angle θ and not upon the lengths of the sides of the triangle. *The trigonometric functions of any given angle are therefore constants.*

II. The Pythagorean Theorem.

The square on the hypotenuse of a right triangle is equal to the sum of the squares on the other two sides;

$$c^2 = a^2 + b^2.$$

From this theorem we can find any side of a right triangle if we know the other two sides; thus

$$c = \sqrt{a^2 + b^2}, \quad a = \sqrt{c^2 - b^2}, \quad \text{and} \quad b = \sqrt{c^2 - a^2}.$$

III. Complementary Angles.

We may recall from elementary geometry that if two acute angles are such that their sum is 90°, they are said to be **complementary angles.** Thus the two acute angles of a right triangle are complementary. In the right triangle of Fig. 1 we see that

$$\sin \alpha = \frac{a}{c} = \cos \beta; \qquad \cot \alpha = \frac{b}{a} = \tan \beta;$$

$$\cos \alpha = \frac{b}{c} = \sin \beta; \qquad \sec \alpha = \frac{c}{b} = \csc \beta;$$

$$\tan \alpha = \frac{a}{b} = \cot \beta; \qquad \csc \alpha = \frac{c}{a} = \sec \beta.$$

If we look closely at the names of the trigonometric functions we see that they may be paired: sine and cosine, tangent and cotangent, secant and cosecant. In each pair we may say that one function is the *cofunction* of the other. We thus have

Any function of α equals the cofunction of the complement of α. This theorem can be stated as follows:

$$\sin\alpha = \cos(90^\circ - \alpha); \qquad \cot\alpha = \tan(90^\circ - \alpha);$$
$$\cos\alpha = \sin(90^\circ - \alpha); \qquad \sec\alpha = \csc(90^\circ - \alpha);$$
$$\tan\alpha = \cot(90^\circ - \alpha); \qquad \csc\alpha = \sec(90^\circ - \alpha).$$

Illustration 1.

In the right triangle ABC of Fig. 1, suppose we have given c = 13 and a = 5. Find the side b and the functions of α and of β.

Solution: By the Pythagorean theorem we have

$$b = \sqrt{c^2 - a^2} = \sqrt{169 - 25} = \sqrt{144} = 12.$$

By definition we then have

$$\sin\alpha = \tfrac{5}{13}; \quad \tan\alpha = \tfrac{5}{12}; \quad \sec\alpha = \tfrac{13}{12};$$
$$\cos\alpha = \tfrac{12}{13}; \quad \cot\alpha = \tfrac{12}{5}; \quad \csc\alpha = \tfrac{13}{5};$$
$$\sin\beta = \tfrac{12}{13}; \quad \tan\beta = \tfrac{12}{5}; \quad \sec\beta = \tfrac{13}{5};$$
$$\cos\beta = \tfrac{5}{13}; \quad \cot\beta = \tfrac{5}{12}; \quad \csc\beta = \tfrac{13}{12}.$$

Illustration 2.

Given $\cot\alpha = \frac{3}{8}$; find the other trigonometric functions of α.

Solution: Since $\cot\alpha = \dfrac{3}{8} = \dfrac{b}{a}$ we may take b = 3 and a = 8.

Then $c = \sqrt{a^2 + b^2} = \sqrt{64 + 9} = \sqrt{73}$. We thus get

$$\sin\alpha = \frac{8}{\sqrt{73}} = \frac{8}{73}\sqrt{73}; \tan\alpha = \frac{8}{3}; \csc\alpha = \frac{\sqrt{73}}{8};$$
$$\cos\alpha = \frac{3}{\sqrt{73}} = \frac{3}{73}\sqrt{73}; \sec\alpha = \frac{\sqrt{73}}{3}.$$

Illustration 3.

Express tan 43°47′ as a function of another acute angle.

Solution: $\tan 43^\circ 47' = \cot(90^\circ - 43^\circ 47') = \cot 46^\circ 13'$.

4. Exercise 1.

1. The following problems refer to a right triangle ABC lettered as in Fig. 1. Find the unknown side and all functions of α and of β.

a) a = 5; b = 12.
b) $b = \sqrt{3}$; a = 1.
c) $c = \sqrt{2}$; b = 1.
d) a = 1; b = 1.
e) a = 3; b = 4.
f) $c = \sqrt{65}$; a = 4.

2. Construct the corresponding triangle and find the other functions of the given angle.

a) $\cot \alpha = \frac{4}{3}$. c) $\tan \alpha = 1$. e) $\csc \alpha = \sqrt{2}$.
b) $\sin \alpha = \frac{2}{3}$. d) $\sec \alpha = \frac{13}{5}$. f) $\cos \alpha = \frac{4}{5}$.

3. Express each function as a function of the complementary angle.

a) sin 45°. c) tan 36°18′. e) sec 37°.
b) cos 68°. d) cot 79°53′. f) csc 12°.

5. The Functions of 30°, 45°, 60°.

1. Construct an isosceles right triangle ABC and let the length of the equal legs be 1 unit each. The acute angles are each 45°, and by the Pythagorean theorem the hypotenuse is $\sqrt{2}$.

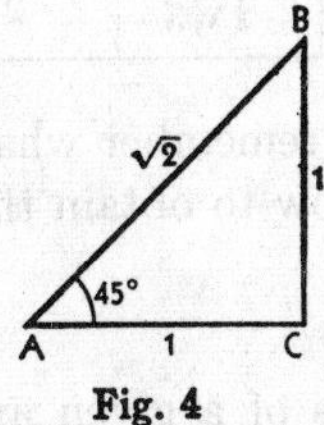

Fig. 4

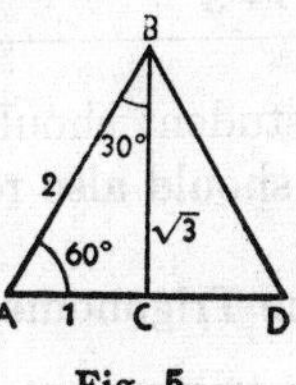

Fig. 5

By the definitions of the trigonometric functions we then have

$$\sin 45° = \frac{1}{\sqrt{2}} = \tfrac{1}{2}\sqrt{2}; \quad \cot 45° = \frac{1}{1} = 1;$$
$$\cos 45° = \frac{1}{\sqrt{2}} = \tfrac{1}{2}\sqrt{2}; \quad \sec 45° = \frac{\sqrt{2}}{1} = \sqrt{2};$$
$$\tan 45° = \frac{1}{1} = 1; \quad \csc 45° = \frac{\sqrt{2}}{1} = \sqrt{2}.$$

2. Construct an equilateral triangle and let the length of each side be 2 units. The angles are 60° each, and the bisector of any angle will also be the perpendicular bisector of the opposite side. We thus have Fig. 5. By the definitions we thus have

$$\sin 30° = \frac{1}{2}; \quad \cot 30° = \frac{\sqrt{3}}{1} = \sqrt{3};$$
$$\cos 30° = \frac{\sqrt{3}}{2}; \quad \sec 30° = \frac{2}{\sqrt{3}} = \tfrac{2}{3}\sqrt{3};$$

$$\tan 30° = \frac{1}{\sqrt{3}} = \tfrac{1}{3}\sqrt{3}; \quad \csc 30° = \frac{2}{1} = 2;$$

$$\sin 60° = \frac{\sqrt{3}}{2}; \quad \cot 60° = \frac{1}{\sqrt{3}} = \tfrac{1}{3}\sqrt{3};$$

$$\cos 60° = \frac{1}{2}; \quad \sec 60° = \frac{2}{1} = 2;$$

$$\tan 60° = \frac{\sqrt{3}}{1} = \sqrt{3}; \quad \csc 60° = \frac{2}{\sqrt{3}} = \tfrac{2}{3}\sqrt{3}.$$

These values may be put into the following table:

	sin	cos	tan	cot	sec	csc
30°	$\frac{1}{2}$	$\frac{1}{2}\sqrt{3}$	$\frac{1}{3}\sqrt{3}$	$\sqrt{3}$	$\frac{2}{3}\sqrt{3}$	2
45°	$\frac{1}{2}\sqrt{2}$	$\frac{1}{2}\sqrt{2}$	1	1	$\sqrt{2}$	$\sqrt{2}$
60°	$\frac{1}{2}\sqrt{3}$	$\frac{1}{2}$	$\sqrt{3}$	$\frac{1}{3}\sqrt{3}$	2	$\frac{2}{3}\sqrt{3}$

HINT: The student should not only remember what these functions are but should also remember how to obtain them.

6. Four-Place Trigonometric Tables.

Since the trigonometric functions of a given angle are constants, it is only necessary to compute these constants once and for all. These values are expressed in decimal form, and in general it is an *endless* decimal. By use of advanced mathematics these values can be computed to as many decimal places as desired, and the computed values have been placed in tables. Four-place, five-place, seven-place, and even higher-place tables are available. However, it is sufficient to consider a four-place table in order to learn the theory and use of such tables.

Before we begin the study of the table, we shall consider the question of "rounding off" a number. In using a four-place table, always "round off" the numbers to four significant digits. The *significant digits are those commencing with the first one, reading from left to right, which is not zero and ending with the last one definitely specified.* To "round off" a number to four places is to write the approximate value of that number to four significant digits. The error is not more than one-half of a unit in the last place.

Illustrations.

1. Express 36.747 to 4 places. Answer 36.75.
2. Express 37.743 to 4 places. Answer 37.74.
3. Express 0.00937825 to 4 places. Answer 0.009378.

If the last digit is exactly 5 we use the convention of statistics, which says that if the preceding number is even we simply drop the 5 and if the preceding number is odd we add one and make the last digit even.

Illustrations.

1. Express 2.5735 to 4 places. Answer 2.574.
2. Express 2.5765 to 4 places. Answer 2.576.

The Table of the Natural Trigonometric Functions* is a four-place table of the trigonometric functions of the acute angles. It gives the natural trigonometric functions from minute to minute. The angles from 0° to 45° are labelled at the *top* of the pages (two angles to one page except for 44°). The minutes for these angles are given in the extreme *left*-hand column under *each* degree and the titles of the columns are at the *top* of the page. The angles from 45° to 90° are labelled at the *bottom* of the pages and the minutes for these angles are given in the extreme *right*-hand column for *each* degree. For these degrees we read the titles of the columns at the *bottom* of the page. Thus each entry in the columns of the functions is a function of some angle and, moreover, is the cofunction of the complementary angle. This double representation is possible because of the relations of §3, III.

A. *Given the angle; to find the trigonometric function.*

Illustrations.

1. To find tan 17°20′:
   Find 17° at the *top* of the page.
   Find 20′ in the *left* column under 17°.
   Opposite 20′ in the column headed by "Tan" at the *top* of the page read 0.3121.
   Thus tan 17°20′ = 0.3121.

*See *Tables* in back of this book, pp. 93-116.

2. To find cos 67°43′:
   Find 67° at the *bottom* of the page.
   Find 43′ in the *right* column above 67°.
   Opposite 43′ in the column headed by "Cos" at the *bottom* of the page read 0.3792.
   Thus $\cos 67°43' = 0.3792$

B. *Given the trigonometric function; to find the angle.*

Illustrations.

1. Given $\sin\theta = 0.2840$. Find θ.
   In the columns headed by "Sin" search for 0.2840. This is found in the column which is headed by "Sin" at the *top* under 16°; therefore, look in the *left* column for the minutes and read 30. Thus $\theta = 16°30'$.
2. Given $\cot\theta = 0.1210$. Find θ.
   In the columns headed by "Cot" search for 0.1210. This is found in the column headed by "Cot" at the *bottom* of the page above 83°; therefore, look in the *right*-hand column for the minutes and read 6. Thus $\theta = 83°06'$.

Note: In the *left* column the minutes increase as one reads *downward;* in the *right* column, as one reads *upward.*

This table does not give the values for the secant and cosecant. A larger book of tables has these values given. However, since they do not occur too frequently in practice, the student can usually obtain the desired results by using only the four functions given in this table. For further relations see §33.

C. *Interpolation.*

At times it is desirable to obtain the trigonometric functions of an angle to the nearest second or tenth of a minute. To accomplish this by the table it becomes necessary to interpolate for the values of the trigonometric functions of the angles. This can be done either by linear interpolation or by using tables of proportional parts, if such tables are given. We shall discuss linear interpolation, which can be employed in most cases.

Illustrations.

1. Find tan 59°27.3′.

   It is easily seen that tan 59°27.3′ is between tan 59°27′ and tan 59°28′. From the table we get tan 59°27′ = 1.6943 and tan 59°28′ = 1.6954. The **tabular difference** is 1.6954 − 1.6943 = 0.0011. Since we desire the value for 59°27.3′, multiply the tabular difference by .3: (0.0011)(.3) = 0.00033. Add this amount to the tangent of the *smaller* angle (59°27′) and round off to four decimal places; the result is 1.6946.

   Thus tan 59°27.3 = 1.6946.

   This computation can be written as follows:

$$\left.\begin{aligned} \tan 59^\circ 27' &= 1.6943 \\ \tan 59^\circ 27.3' &= 1.6946 \\ \tan 59^\circ 28' &= 1.6954 \end{aligned}\right]\ (0.0011)(.3) = 0.00033.$$

2. Find cot 21°5.7′.

   It is between cot 21°5′ and cot 21°6′.

$$\left.\begin{aligned} \text{From table: } \cot 21^\circ 5' &= 2.5938 \\ \cot 21^\circ 5.7' &= 2.5923 \\ \text{From table: } \cot 21^\circ 6' &= 2.5916 \end{aligned}\right]\ \begin{aligned} &(-.0022)(.7) \\ &= -.00154. \end{aligned}$$

   Thus cot 21°5.7′ = 2.5923.

Hint: The student will find it advantageous to be consistent in his procedure. We therefore suggest that he always work from the *smaller* angle. Thus, to obtain the tabular difference, always subtract the value of the trigonometric function of the smaller angle from the value of the trigonometric function of the larger angle; then always *add* the computed proportional part to the function of the *smaller* angle. In case the trigonometric function is increasing, i.e., if the value of the function becomes *larger* as the angle becomes *larger* (as in sin θ, tan θ, sec θ), then the tabular difference is positive. In case the trigonometric function is decreasing, i.e., if the value of the trigonometric function becomes *smaller* as the angle becomes *larger* (as in cos θ, cot θ, csc θ), then the tabular difference is negative and, as we add the proportional part to the value of the trigonometric function of the smaller angle, we are in reality subtracting.

3. Find tan 76°58′36″.

$$\left.\begin{array}{ll} \text{From table: } \tan 76°58' & = 4.3200 \\ \qquad\qquad \tan 76°58'36'' & = 4.3234 \\ \text{From table: } \tan 76°59' & = 4.3257 \end{array}\right] \begin{array}{l} (.0057)(\tfrac{36}{60}) \\ \quad = .00342. \end{array}$$

Thus tan 76°58′36″ = 4.3234.

Note: In this problem we multiplied the tabular difference by $\frac{36}{60}$ since there are 60 seconds in 1 minute and we were given 36 seconds.

4. Given cot θ = 3.2392. Find θ.

In the column headed by "Cot," search for the value closest to 3.2392.

$$\left.\begin{array}{ll} \text{We find: } \cot 17°9' & = 3.2405 \\ \qquad \cot \theta & = 3.2392 \\ \qquad \cot 17°10' & = 3.2371 \end{array}\right] \begin{array}{l} (\tfrac{13}{34})(10) = 3.82 \\ \quad = 4. \end{array}$$

Subtract the value of the trigonometric function of the smaller angle from that of the larger angle and also from the given value. Form the fraction indicated and multiply by 10 (the angular difference). The result rounded off to the nearest integer will be the tenth of a minute to be added to the *smaller* angle; we thus have θ = 17°9.4′.

If we want the answer to be given in seconds, we multiply the above fraction by 60 and round off; that is,

$$(\tfrac{13}{34})(60) = 22.92 = 23$$

and θ = 17°9′23″.

7. Exercise 2.

I. Find each function by use of Table III.

1. tan 37°20′.	6. cot 18°8.4′.	11. tan 73°27.8′.
2. cos 65°40′.	7. sin 62°13′.	12. cot 8°2′43″.
3. sin 78°10′.	8. cos 37°24′.	13. cos 21°12′.
4. cos 14°.	9. tan 42°35′.	14. sin 37°29′.
5. cot 28°50′.	10. cot 51°46′.	15. tan 40°33′.

II. Find the acute angle α by use of Table III.

1. $\sin \alpha = .2306$.
2. $\cos \alpha = .3201$.
3. $\tan \alpha = .4279$.
4. $\cot \alpha = .6916$.
5. $\sin \alpha = .7325$.
6. $\cos \alpha = .8652$.
7. $\tan \alpha = 1.4972$.
8. $\cot \alpha = 1.3333$.

8. Solution of a Right Triangle.

A triangle has six parts; namely, three angles and three sides. In a right triangle one angle is 90°; thus it is only necessary to have given two sides or an acute angle and a side in order to compute the remaining parts of the right triangle. The finding of the unknown parts is called the *solution* of the triangle.

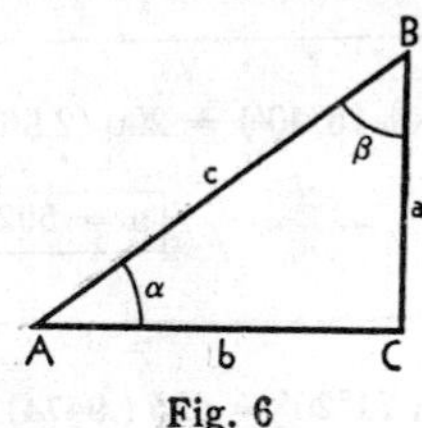

Fig. 6

We recall the formulas:

(1) $\sin \alpha = \frac{a}{c} = \cos \beta$. (4) $\cot \alpha = \frac{b}{a} = \tan \beta$.

(2) $\cos \alpha = \frac{b}{c} = \sin \beta$. (5) $\sec \alpha = \frac{c}{b} = \csc \beta$.

(3) $\tan \alpha = \frac{a}{b} = \cot \beta$. (6) $\csc \alpha = \frac{c}{a} = \sec \beta$.

and

(7) $a^2 + b^2 = c^2$. (8) $\alpha + \beta = 90°$.

We note that from (1) and (2) we can solve for a and b, getting

$$a = c \sin \alpha; \quad b = c \cos \alpha.$$

The other formulas can be solved in a similar manner.

Procedure in solving a triangle:

I. *Always draw the figure and label it.*

II. *To find any unknown part, use a formula which involves that part but no other unknown. To simplify the arithmetic select, when possible, a formula which leads to multiplication instead of division.*

III. *As a check, substitute the results in any one of the formulas which were not used in the solution.*

Illustration 1.

Solve the right triangle ABC, if b = 200 and c = 625.

Solution:

Formulas	Computation	Figure
$\sin\beta = \frac{b}{c}$	$\sin\beta = \frac{200}{625} = .3200;$ $\boxed{\beta = 18°40'.}$	
$\alpha = 90° - \beta$	$\boxed{\alpha = 90° - 18°40' = 71°20'.}$	
$\cot\beta = \frac{a}{b}$ $a = b\cot\beta$	$a = 200\ (\cot 18°40') = 200\ (2.960);$ $\boxed{a = 592.0}$	Fig. 7
check: $a = c\sin\alpha$	$a = 625\sin 71°20' = 625\ (.9474) = 592.1;$ checks.	

Illustration 2.

Solve the right triangle ABC, if b = 25 and $\alpha = 42°40'$.

Solution:

Formulas	Computation	Figure
$\beta = 90° - \alpha$	$\boxed{\beta = 90° - 42°40' = 47°20'}$	
$\tan\alpha = \frac{a}{b}$ $a = b\tan\alpha$	$a = 25\tan 42°40' = 25\ (.9217);$ $\boxed{a = 23.04}$	
$\cos\alpha = \frac{b}{c}$ $c = \frac{b}{\cos\alpha}$	$c = \frac{25}{\cos 42°40'} = \frac{25}{.7353}$ $\boxed{c = 34.00}$	Fig. 8
check: $a = c\cos\beta$	$a = 34\ (\cos 47°20') = 34\ (.6777) = 23.03;$ checks.	

Note: We can not stress too strongly the importance of having a systematic and neat form in solving triangles. Even though it may take a little longer, the student should use the above form in working his problems.

9. Exercise 3.

Solve the right triangle ABC, given the following parts:

1. $a = 20$; $\alpha = 26°10'$.
2. $a = 50$; $\beta = 63°40'$.
3. $a = 125$; $b = 180$.
4. $a = 2.5$; $c = 5.8$.
5. $b = .20$; $c = .60$.
6. $c = 30$; $\beta = 43°17'$.

CHAPTER II

THE GENERAL ANGLE

10. Angles.

We shall now define an angle in its most general sense. Let any line segment coinciding with OA (Fig. 9) be revolved about one of its end points, say O, until it takes the position OB.

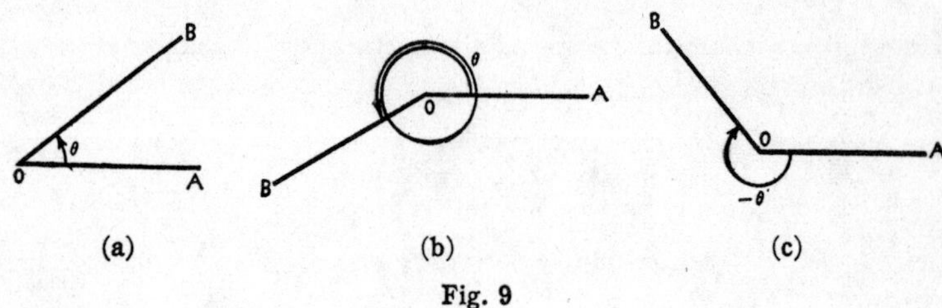

Fig. 9

This revolution of the line has generated the angle AOB. There is no limit to the number of times the line may revolve about O, and thus there is no limit to the size of the angle. We use the following terminology:

O is called the *origin;*

the revolving line is called the *radius vector;*

OA is called the *initial side;*

OB is called the *terminal side.*

If the radius vector moves counter-clockwise, the angle generated is said to be *positive.* If the radius vector moves clockwise, the angle generated is said to be *negative.*

11. Angular Measurements.

There are three systems of angular measurements in common usage, the sexagesimal measure, the radian (or circular) measure, and the mil, which is used by the United States armed forces.

1. **Sexagesimal Measure:**
the degree $= \frac{1}{360}$ of one revolution $= 1°$;
the minute $= \frac{1}{60}$ of one degree $= 1'$;
the second $= \frac{1}{60}$ of one minute $= 1''$.
2. **Radian Measure:**
A *radian* is defined to be the central angle subtended by an arc of a circle equal in length to the radius of the circle.
3. **The Mil:**
The mil is defined to be an angle equal to $\frac{1}{6400}$ of one revolution; i.e., 6400 mils = 360°. Roughly, an object one yard wide subtends an angle of approximately 1 mil at a distance of 1000 yards.

12. Conversion Units.

To change from one system of measurement to another we use the following relations, which are obtained directly from the definitions.

$$\textbf{1 degree} = \frac{\pi}{180} \textbf{ radians};$$

$$\textbf{1 radian} = \frac{180}{\pi} \textbf{ degrees.}$$

Thus *to change from degrees to radians:* multiply the number ot degrees by $\frac{\pi}{180}$;

to change from radians to degrees: multiply the number of radians by $\frac{180}{\pi}$.

$$\textbf{1 degree} = \frac{6400}{360} \textbf{ mils} = \frac{160}{9} \textbf{ mils};$$

$$\textbf{1 mil} = \frac{360}{6400} \textbf{ degrees} = \frac{9}{160} \textbf{ degrees.}$$

To decimal point approximations we have ($\pi = 3.1416$),

1° = 0.017453 radians = 17.778 mils;
1 rdn. = 57.2958°+ = 1018.6 mils;
1 mil = 0.05625° = 0.0009817 radians.

Illustrations.

1. Change 30° to radian measure.

Solution: $30\left(\frac{\pi}{180}\right) = \frac{\pi}{6}$ radians.

2. Change $5\pi/3$ radians to degrees.

Solution: $\left(\frac{5\pi}{3}\right)\left(\frac{180}{\pi}\right) = 300°.$

3. Change 36° to mils.

Solution: $36\left(\frac{160}{9}\right) = 640$ mils.

The conversion from degrees to radians and from radians to degrees for any angle has been done accurately to many decimal places and the values have been put into tabular form. Pages 123–125 in back of this book contain such tables.

That on pages 123–124 converts degrees, minutes, and seconds to radians.

Illustration.

Express 36°47′12″ in radian measure.

Solution: From the table:

$$\begin{aligned} 36° &= 0.6283185 \text{ rdns.} \\ 47' &= 0.0136717 \text{ rdns.} \\ 12'' &= 0.0000582 \text{ rdns.} \\ \text{Therefore } 36°47'12'' &= 0.6420484 \text{ rdns.} \end{aligned}$$

The table on page 125 converts radians to degrees, minutes, and seconds.

Illustration.

Express 2.5793 radians in degrees, minutes, and seconds.

Solution: From the table:

$$\begin{aligned} 2 \text{ rdns.} &= 114°\ 35'\ 29.6'' \\ .5 \text{ rdns.} &= 28°\ 38'\ 52.4'' \\ .07 \text{ rdns.} &= 4°\ 00'\ 38.5'' \\ .009 \text{ rdns.} &= 0°\ 30'\ 56.4'' \\ .0003 \text{ rdns.} &= 0°\ \ 1'\ 01.9'' \\ \text{Therefore } 2.5793 \text{ rdns.} &= 146°104'178.8'' \\ &= 147°\ 46'\ 58.8'' \end{aligned}$$

It is recommended that the student learn the following table of equivalences between degrees and radians:

Degrees	30°	45°	60°	90°	120°	135°	150°	180°	270°	360°
Radians	$\frac{\pi}{6}$	$\frac{\pi}{4}$	$\frac{\pi}{3}$	$\frac{\pi}{2}$	$\frac{2\pi}{3}$	$\frac{3\pi}{4}$	$\frac{5\pi}{6}$	π	$\frac{3\pi}{2}$	2π

13. Properties of Radian Measure.

1. By definition, the number of radians in an angle at the center of a circle is the number of times the radius will divide into the arc subtended by the angle. Thus

$$\text{angle in radians} = \frac{\text{arc}}{\text{radius}}; \ \theta = \frac{s}{r}, \ \theta \text{ in radians.}$$

Therefore $s = r\theta$;
the length of an arc of a circle is equal to the product of the radius and the number of radians in the central angle subtended by the arc.

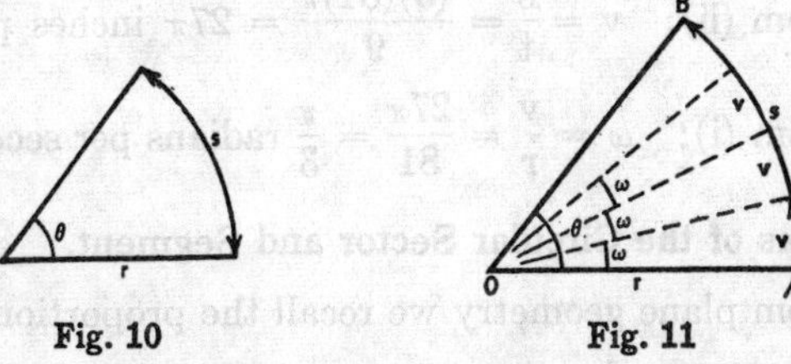

Fig. 10 Fig. 11

Since one mil is approximately .001 radians, for angles less than 5°, we can use the approximation

$$s = \frac{r\theta}{1000} \ (\theta, \text{ in mils}).$$

Illustration.

Through what angle in radians would a runner turn in going 100 yards on a circular track 160 yards in diameter?

Solution: Given $s = 100$, $r = 80$.
Substituting in the formula, we have
$100 = 80\,\theta$ or $\theta = \frac{5}{4}$ radians.

2. Linear and Angular Velocity.

If an object A is moving with uniform velocity on the circumference of a circle with center at O and radius r, then the linear velocity v is the length of arc passed over by A in one unit of time. Let ω be the angular velocity of A in one unit of time. Let t be the time. Then we have the following formulas:

(i) $v = r\omega$; since arc = radius X angle in radians;
(ii) $s = vt$; since distance = rate X time;

(iii) $\theta = \omega t$; since OA moves through ω radians per unit of time for t units of time;

where ω and θ are measured in radians. See Fig. 11.

Illustration.

Find the linear and angular velocities of an object which in 9 seconds moves 3/2 times around the circumference of a circle with a radius of 81 inches

Solution: Given $r = 81$, $t = 9$.

The circumference of a circle is $2\pi r$; we thus have $s = 2\pi r(\frac{3}{2}) = (3)(81)\pi$.

From (ii): $v = \frac{s}{t} = \frac{(3)(81)\pi}{9} = 27\pi$ inches per second.

From (i): $\omega = \frac{v}{r} = \frac{27\pi}{81} = \frac{\pi}{3}$ radians per second.

3. Areas of the Circular Sector and Segment.

From plane geometry we recall the proportion

$$\frac{\text{area of sector}}{\text{angle of sector}} = \frac{\text{area of the circle}}{\text{four right angles}}.$$

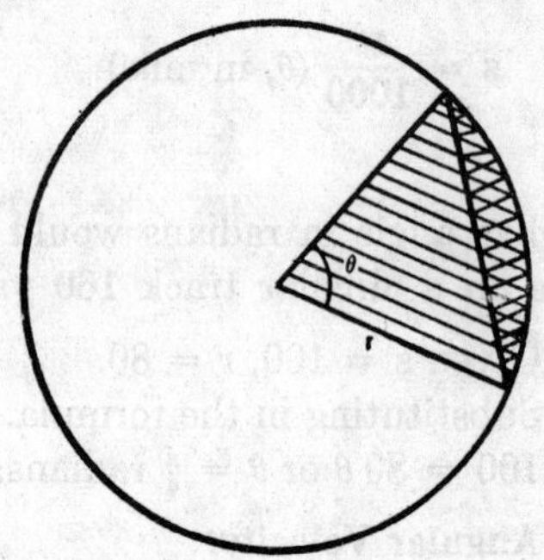

Fig. 12

Thus we have

$$\frac{K}{\theta} = \frac{\pi r^2}{2\pi}$$

or the area of a sector, K, is given by

$\mathbf{K} = \frac{1}{2} r^2\theta$ (θ, in radians).

The area of a segment, which is the area between the arc and the chord subtended by a given angle, is given by
$K' = \frac{1}{2} r^2 (\theta - \sin \theta)$ (θ, in radians).

Illustration.

Find the areas of the circular sector and segment having a central angle of $\frac{\pi}{3}$ radians and a radius of 8 feet.

Solution:

$$K = \tfrac{1}{2} r^2\theta = \tfrac{1}{2} (64) \frac{\pi}{3} = \frac{32\pi}{3} = \frac{32}{3} (3.1416)$$

$$= 33.5104 \text{ sq. ft.}$$

$$K' = \tfrac{1}{2} r^2 (\theta - \sin \theta) = \tfrac{1}{2} (64)\left(\frac{\pi}{3} - \sin \frac{\pi}{3}\right)$$

$$= 32[1.0472 - .8660];$$

$$K' = 5.7984 \text{ sq. ft.}$$

14. Exercise 4.

1. Change to radian measure:
   10°, 75°, 125°12′13″, 256°43′37″
2. Change to sexagesimal measure:
   $\frac{\pi}{2}$, $\frac{11\pi}{6}$, 2.3572 radians, 0.0163 radians.
3. A pendulum 40 inches long vibrates 2°30′ on each side of its mean position. Find the length of the arc through which it swings.
4. A wheel 6 feet in diameter makes 300 revolutions per minute. Find the linear velocity of a point on the rim in feet per second.
5. Find the area of the sector and the segment of a circle of radius 9 inches if the central angle is 40°.

Chapter III

TRIGONOMETRIC FUNCTIONS OF ANY ANGLE

We shall now discuss the trigonometric functions of any angle, using the definition of the general angle given in the last chapter.

15. Rectangular Coordinates.

The positive and negative numbers may be represented on a number scale such as is shown in Fig. 13.

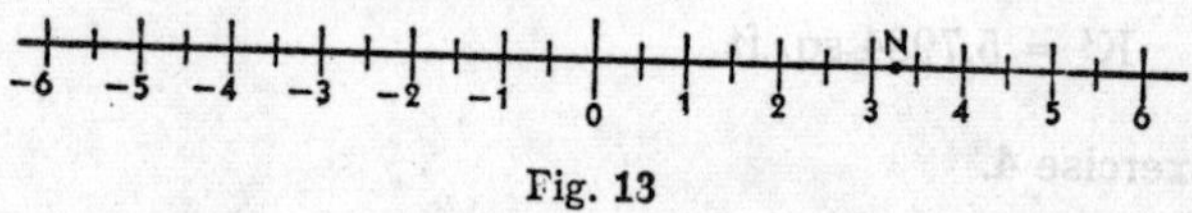

Fig. 13

Thus any real number N is represented on this line.
The notation $N < O$ means that N is negative.
The notation $N > O$ means that N is positive.

If we take two such number scales to intersect each other at right angles at the point O, we form a *rectangular coordinate system* and the two lines are called the *axes,* denoted by OX and OY. Let P be any point in the plane of OX and OY. We then have the following terminology. The *horizontal coordinate* of P is the perpendicular distance, x, from OY to P. It is considered *positive* when P is to the *right* of OY and *negative* when P is to the *left* of OY. It is called the *abscissa.* The *vertical coordinate* of P is the perpendicular distance, y, from OX to P. It is considered *positive* when P is *above* OX and *negative* when P is *below* OX. It is called the *ordinate.* The abscissa and ordinate together are called the *coordinates* of P and are denoted by (x, y). The point O is called the *origin.* OP is called the *radius vector* and is always positive when it is not zero. The coordinate axes divide the plane into four *quadrants,* which will be numbered counter-clockwise I, II, III, IV; see Fig. 14.

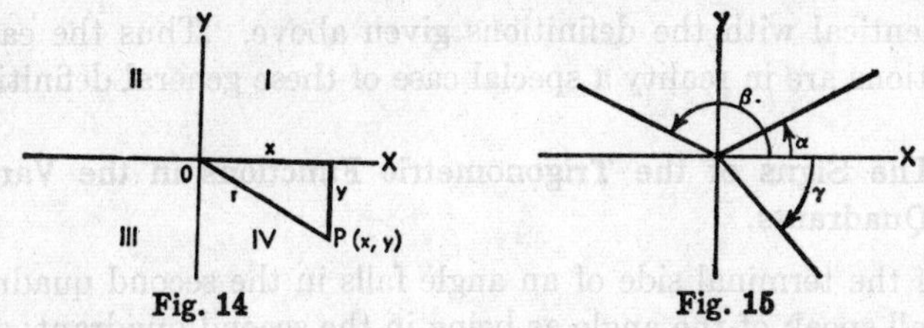

Fig. 14 Fig. 15

If we take any angle as defined in §10 and placed so that O is at the origin of the rectangular coordinate plane and its initial side coincident with the positive X-axis, this angle is said to be in its *standard position.* Fig. 15 shows the three angles:

$$\alpha = 30°; \beta = 150°; \gamma = -60°.$$

16. The Trigonometric Functions of Any Angle θ.

Definition: *Let θ be any angle in its standard position. Let* P *be any point (not the origin) having coordinates* (x, y) *and lying on the terminal side of θ. Let* r *be the radius vector of* P. *Then*

$$\sin\theta = \frac{\text{ordinate}}{\text{radius vector}} = \frac{y}{r};$$

$$\cos\theta = \frac{\text{abscissa}}{\text{radius vector}} = \frac{x}{r};$$

$$\tan\theta = \frac{\text{ordinate}}{\text{abscissa}} = \frac{y}{x};$$

$$\cot\theta = \frac{\text{abscissa}}{\text{ordinate}} = \frac{x}{y};$$

$$\sec\theta = \frac{\text{radius vector}}{\text{abscissa}} = \frac{r}{x};$$

$$\csc\theta = \frac{\text{radius vector}}{\text{ordinate}} = \frac{r}{y}.$$

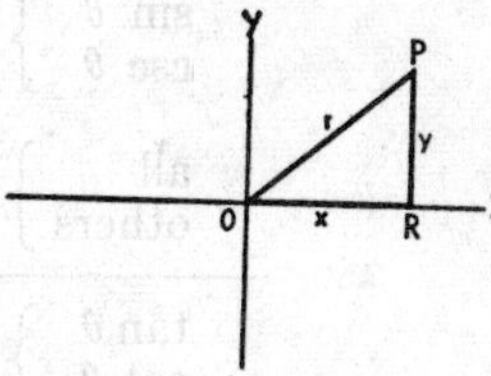

Fig. 16

The student should commit these definitions to memory.

If the angle θ is acute, the terminal side would fall in the first quadrant. We then have a right triangle ORP, with PR = y, OR = x, and OP = r. It is easily seen that the definitions of the trigonometric functions in terms of a right triangle (see §2) are identical with the definitions given above. Thus the earlier definitions are in reality a special case of these general definitions.

17. The Signs of the Trigonometric Functions in the Various Quadrants.

If the terminal side of an angle falls in the second quadrant, we shall speak of the angle as being in the second quadrant; similarly for the other quadrants. Since x and y can be both negative and positive depending upon the quadrant in which they fall (r is always positive), we notice that the trigonometric functions may be positive or negative. We proceed to determine an exact formulation of this fact.

In the first quadrant x and y are both positive, and therefore all the functions are positive.

In the second quadrant x is negative, y is positive; thus we have all the functions negative except $\sin\theta$ and $\csc\theta$. For example

$$\sin\theta = \frac{y}{r} = \frac{+}{+} = +;\ \tan\theta = \frac{y}{x} = \frac{+}{-} = -.$$

After examining the other quadrants similarly, we may summarize the results in the following diagram:

$\left.\begin{matrix}\sin\theta\\ \csc\theta\end{matrix}\right\}+$ $\left.\begin{matrix}\text{all}\\ \text{others}\end{matrix}\right\}-$	all } +
$\left.\begin{matrix}\tan\theta\\ \cot\theta\end{matrix}\right\}+$ $\left.\begin{matrix}\text{all}\\ \text{others}\end{matrix}\right\}-$	$\left.\begin{matrix}\cos\theta\\ \sec\theta\end{matrix}\right\}+$ all others −

The student should commit this diagram to memory.

18. Functions of 0°, 90°, 180°, 270°.

The functions of the quadrantal angles are determined directly from the definitions of §16. We recall that division by zero is not a permissible operation so that, for example a/0 is not a defined number.

Illustration 1.

Find the functions of 0°.

Solution: Place the angle 0° in its standard position. The initial side and terminal side will both coincide with the X-axis so that the coordinates of any point P on the terminal side will be $(x_1, 0)$ and by the Pythagorean theorem, $r = \sqrt{x_1^2 + 0} = x_1$. Thus we have $x = x_1 = r$; $y = 0$, and by the definitions

$$\sin 0° = \frac{y}{r} = \frac{0}{r} = 0; \quad \cot 0° = \frac{x}{y} = \frac{x_1}{0}, \text{ which is undefined;}$$

$$\cos 0° = \frac{x}{r} = \frac{x_1}{x_1} = 1; \quad \sec 0° = \frac{r}{x} = \frac{x_1}{x_1} = 1;$$

$$\tan 0° = \frac{y}{x} = \frac{0}{x_1} = 0; \quad \csc 0° = \frac{r}{y} = \frac{x_1}{0}, \text{ which is undefined.}$$

Illustration 2.

Find the functions of 270°.

Solution: Place 270° in its standard position. A point P on the terminal side will have $(0, -y_1)$ for its coordinates and $r = \sqrt{0 + y_1^2} = y_1$. Therefore we have

$$x = 0, \ y = -y_1, \text{ and } r = y_1,$$

and by the definitions

$$\sin 270° = \frac{y}{r} = \frac{-y_1}{y_1} = -1;$$

$$\cos 270° = \frac{x}{r} = \frac{0}{r} = 0;$$

$$\tan 270° = \frac{y}{x} = \frac{y_1}{0}, \text{ undefined;}$$

$$\cot 270° = \frac{x}{y} = \frac{0}{-y_1} = 0;$$

$$\sec 270° = \frac{r}{x} = \frac{y_1}{0}, \text{ undefined}$$

$$\csc 270° = \frac{r}{y} = \frac{y_1}{-y_1} = -1.$$

We thus get the following table. It is suggested that the student work out each function for himself before he commits the table to memory.

Angle	sin	cos	tan	cot	sec	csc
0°	0	1	0	—	1	—
90°	1	0	—	0	—	1
180°	0	−1	0	—	−1	—
270°	−1	0	—	0	—	−1

19. Reduction to Acute Angles.

In the tables, values of the trigonometric functions are given only for angles from 0° to 90°; i.e., the acute angles. Thus it becomes necessary to obtain the functions of any angle by means of the functions of an acute angle. This is easily done by the use of a reference angle; *the acute angle α between the terminal side of θ and the* X-axis *is called the reference angle of θ.*

Theorem.

Any function of an angle θ, in any quadrant, is numerically equal to the same function of the reference angle for θ; i.e.,

(any function of θ) = $\pm$(same function of α).

The "+" or "−" is determined by the quadrant in which the angle falls.

Proof.

Let θ be any angle in its standard position. On the terminal side of θ pick a point P, and drop a perpendicular from P to the X-axis. (See Fig. 17.) The reference angle α is

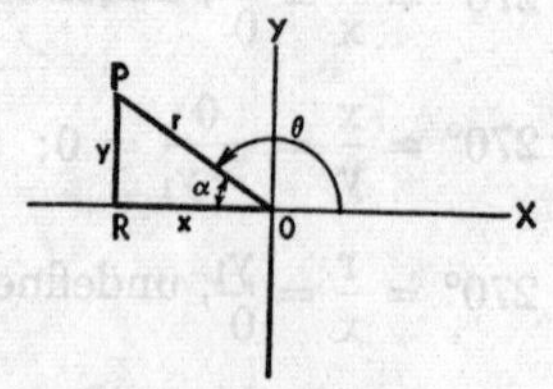

Fig. 17

then the acute angle of the right triangle ORP. Furthermore, ±OR = x, ±RP = y, and OP = r, the "+" or "−" depending upon the quadrant in which P falls. Using both the right triangle definitions and the general definitions, we have

$$\sin \alpha = \frac{RP}{OP}, \text{ and } \sin \theta = \frac{y}{r} = \pm\frac{RP}{OP} = \pm\sin \alpha;$$

$$\cot \alpha = \frac{OR}{RP}, \text{ and } \cot \theta = \frac{x}{y} = \pm\frac{OR}{RP} = \pm\cot \alpha,$$

with similar results for the other functions of θ and α. The determination of the reference angle α can be summarized as follows:

If θ is between 90° and 180°, $\alpha = 180° - \theta$.
If θ is between 180° and 270°, $\alpha = \theta - 180°$.
If θ is between 270° and 360°, $\alpha = 360° - \theta$.

If $\theta > 360°$, we have

any function of θ = same function of (θ − n 360°),

where n is an integer.

Illustrations.

1. Find sin 237°.

Solution: $\sin 237° = -\sin(237° - 180°) = -\sin 57°$
$= -.8387.$

2. Find cos 343°.

Solution: $\cos 343° = \cos(360° - 343°) = \cos 17°$
$= .9563.$

3. Find tan 517°.

Solution: $\tan 517° = \tan(517° - 360°) = \tan 157°$
$= -\tan(180° - 157°)$
$= -\tan 23° = -.4245.$

20. Exercise 5.

Find the following functions, using the Table of Natural Functions.

1. cos 167°.	4. cot 313°.	7. tan 303°09′.
2. sin 112°.	5. sin 213°12′.	8. cot 216°34′.
3. tan 256°.	6. cos 136°43′.	9. cos 736°15′.

21. Functions of the Negative Angle $-\theta$.

If θ is any angle, then

$$\sin(-\theta) = -\sin\theta; \qquad \cot(-\theta) = -\cot\theta;$$
$$\cos(-\theta) = \cos\theta; \qquad \sec(-\theta) = \sec\theta;$$
$$\tan(-\theta) = -\tan\theta; \qquad \csc(-\theta) = -\csc\theta.$$

Proof.

Place the angle $(-\theta)$ in its standard position. Construct the angle θ numerically equal to $-\theta$ in its standard position. On the terminal sides of θ and $-\theta$, choose P and P_1 so that $OP = OP_1$. Then, since angles θ and $-\theta$ are numerically equal, PP_1 is perpendicular to OX and is bisected by OX. We therefore have (see Fig. 18)

$r_1 = r;\ x_1 = x;\ y_1 = -y.$

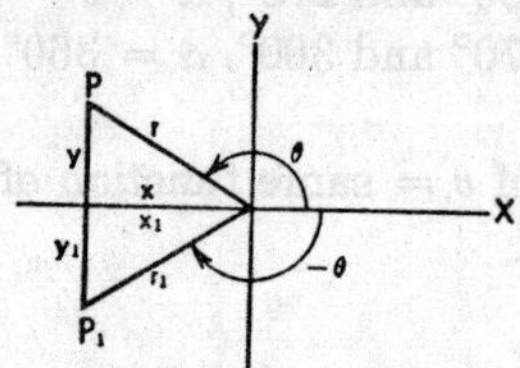

Fig. 18

From the definitions we have

$$\sin(-\theta) = \frac{y_1}{r_1} = \frac{-y}{r} = -\sin\theta,$$

$$\cos(-\theta) = \frac{x_1}{r_1} = \frac{x}{r} = \cos\theta,$$

and similarly for the other functions.

The student should complete the proof for the other four functions.

To find the functions of a negative angle: **first** *change to the function of the corresponding positive angle by using the above formulas; then find the function of the positive angle.*

Illustrations.

1. Find $\sin(-210°)$.

Solution: $\sin(-210°) = -\sin 210°$
$= -(-\sin 30°) = \frac{1}{2}.$

2. Find $\tan(-237°)$.
Solution: $\tan(-237°) = -\tan 237°$
$= -\tan 57° = -1.5399.$

22. Simple Equations.

It is now quite clear that more than one angle has the same trigonometric function. Therefore if we attempt to find an angle which has a given trigonometric function, we find that we may name more than one such angle. Usually we desire only the positive angles less than 360° that have the given function. At this point we shall limit our discussion to finding these angles.*

Illustrations.

1. Find the positive angle $\theta < 360°$, if $\tan \theta = 1.5399$.
Solution: From Table III we get $\theta = 57°$. Since $\tan \theta$ is positive in the first and third quadrants, we have $\theta = 57°$ and $\theta = 237°$.

2. Find the positive angle $\theta < 360°$, if $\cos \theta = -.3467$.
Solution: From Table III we find that $\cos 69°43' = .3467$. Since $\cos \theta$ is negative in the second and third quadrants, we get $\theta = 180° - 69°43' = 110°17'$, and $\theta = 180° + 69°43' = 249°43'$.

23. Exercise 6.

Find the following functions, using Table of Natural Functions.

1. $\sin(-113°43')$.
2. $\cos(-212°39')$.
3. $\tan(-139°12')$.
4. $\cot(-316°10')$.
5. $\sin(-69°)$.
6. $\tan(-219°27')$.

Find the positive angles $\theta < 360°$, given:

7. $\tan \theta = -.3793$.
8. $\cot \theta = 1.6932$.
9. $\sin \theta = -.3567$.
10. $\cos \theta = .6345$.

24. Polar Coordinates.

We have already discussed the concept of rectangular coordinates in §15. We saw there how any point P can be located in a plane by two coordinates, x and y. A point P can also be located in a plane by means of the **polar coordinates** (ρ, θ), where ρ is the

* A general discussion is given in Chapter X.

distance from the origin O to P and θ is the angle formed at 0 by a fixed initial line and the line OP. The number ρ may be positive or negative. *Positive values* of ρ are defined as *distances from the origin measured along the terminal side of* θ. *Negative values* of ρ are defined as *distances from the origin measured along the extension of the terminal side of* θ *through the origin.*

To plot a point P (ρ, θ):

From OI, the initial side, measure off the given angle θ and on its terminal side lay off the length ρ.

Illustration.

The following points are located on Fig. 19:

$$P_1\left(4, \frac{\pi}{6}\right) = P_1(4, 30°), \qquad P_2\left(3, \frac{5\pi}{4}\right) = P_2(3, 225°),$$

$$P_3\left(-5, \frac{\pi}{3}\right) = P_3(-5, 60°).$$

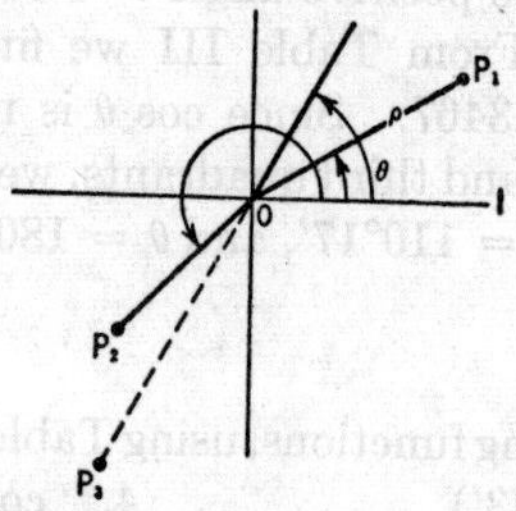

Fig. 19

A point may have more than one set of polar coordinates. For example, in the above illustration the point P_2 could also be represented by the coordinates $P_2(3, -135°) = P_2(-3, 45°)$.

Special graph paper, *polar coordinate paper*, has been devised to make this plotting easier. Fig. 20 illustrates the use of polar coordinate paper. The concentric circles about the origin measure the units of ρ. The angles are measured as shown on the paper.

25. Graphs in Polar Coordinates.

Suppose P is a variable point whose coordinates are given by (ρ, θ), and suppose its path is defined by an equation relating ρ

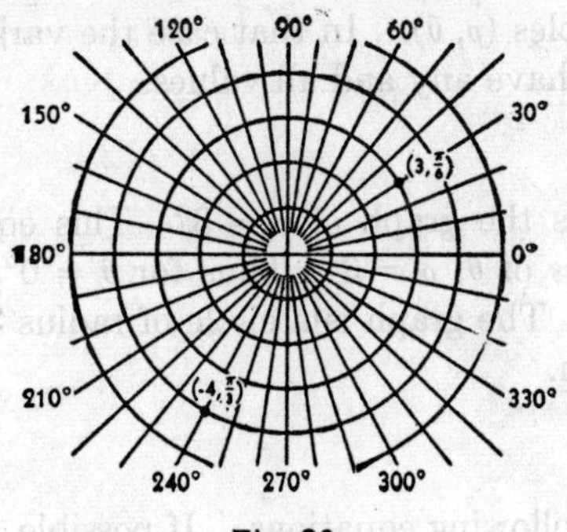

Fig. 20

and θ. The graph of the equation consists of those points in the plane whose coordinates (ρ, θ) satisfy the equation. To draw the graph, we plot a sequence of points and draw a smooth curve through these points.

Illustration.

Draw the graph of $\rho = 2 \cos \theta$.

Solution. We assign values to θ and find the corresponding values of ρ, giving the following table:

θ	$\cos \theta$	$\rho = 2 \cos \theta$
0°	1	2
30°	.87	1.74
60°	.5	1
90°	0	0
120°	−.5	−1
150°	−.87	−1.74
180°	−1	−2

Values from 180° to 360° give the same points. (Check this.)

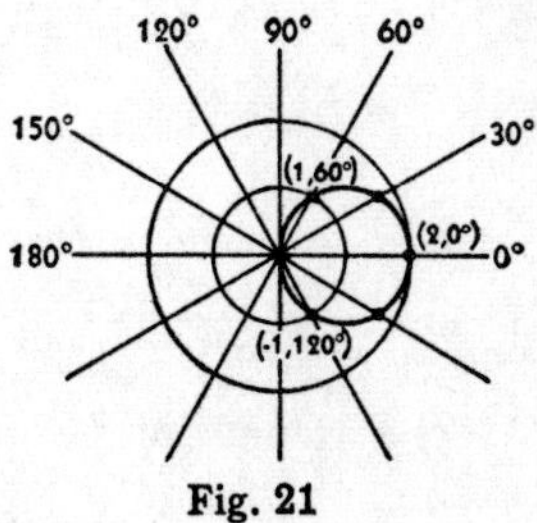

Fig. 21

We then plot the points (ρ, θ) and draw a smooth curve through them. We get the graph of Fig. 21.

The equation which defines the path of P may involve only

one of the variables (ρ, θ). In that case the variable which is not mentioned may have any and all values.

Illustration.

What is the graph of $\rho = 3$? This equation says that for *all values* of θ, $\rho = 3$. Thus for $\theta = 0°$, $\rho = 3$; $\theta = 30°$, $\rho = 3$; etc. The graph is a circle of radius 3 with the center at the origin.

26. Exercise 7.

Graph the following equations. If possible, use polar coordinate paper.

1. $\rho = 2 \sin \theta$.
2. $\rho = 5$.
3. $\rho = \tan \theta$.
4. $\rho = 3 \sec \theta$.
5. $\rho = 1 + \cos \theta$.
6. $\rho = \sin 2\theta$.

Chapter IV

VARIATIONS AND GRAPHS OF THE TRIGONOMETRIC FUNCTIONS

27. Definitions.

If there exists a relationship between two variables, y and x, such that whenever a value is given to x a corresponding value of y can be determined, we say that y *is a function of* x;

y is called the *dependent variable.*

x is called the *independent variable.*

This is the general concept of the word "function." We have been speaking of trigonometric functions and have defined these functions as certain ratios; the term "function" here also agrees with the above definition of a function.

Illustration.

We may, for example, write y = sin x. Then as we assign values to x we determine the values for y. We may compute the following table of values, arbitrarily choosing the values for x:

x	0°	30°	60°	90°	120°	150°	180°	210°	240°	270°	300°	330°	360°
y = sin x	0	.5	.87	1	.87	.5	0	−.5	−.87	−1	−.87	−.5	0

In the above illustration, the angle x is the independent variable; we may give it any value we please. If we let x assume the values from 0° to 360°, as we did in the illustration, then we say that x *varies from 0° to 360°*. From the above table we note that, as x varies from 0° to 90°, sin x varies from 0 to 1; 90° to 180°, 1 to 0; 180° to 270°, 0 to −1; 270° to 360°, −1 to 0. If the values of the function get *larger* as x gets *larger*, the function is said to *increase.* If the values of the function get *smaller* as x gets *larger*, the function is said to *decrease*. Thus the *variation of sin x* can be summarized:

If x increases from	0° to 90°	90° to 180°	180° to 270°	270° to 360°
then sin x	Inc. 0 to 1	Dec. 1 to 0	Dec. 0 to −1	Inc. −1 to 0

28. Periodicity of the Trigonometric Functions.

The angle x may assume any value; thus we need not stop at 360° as we did in the illustration of the last paragraph. We may let x be 400°, 550°, etc. However, from § 19, we have

any function of (n360° + α) = same function of α.

For example

$$\sin 30° = \sin (360° + 30°) = \sin 390° = \sin (720° + 30°) = \sin 750° = \text{etc.}$$

If a function thus repeats its values at a regular interval, it is said to be **periodic,** and its **period** is the interval. The sine of x is then a periodic function and has a period of 360° or 2π radians.

It is easily seen that all the trigonometric functions are periodic and that a period is 360°. We note, however, that tan θ and cot θ repeat themselves every 180°, so that they have a period of 180° as well as 360°. Because of this periodicity, all the information about the values of the trigonometric functions can be obtained by considering only angles from 0° to 360°.

29. Variation of Tan θ.

We recall that, by definition, tan θ = y/x. If we let x = 1, then tan θ = y. Let θ assume values from 0° to 90°. We notice

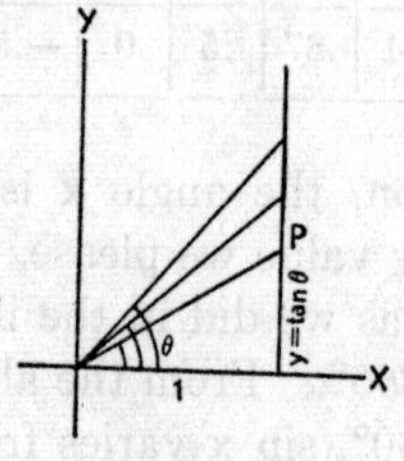

Fig. 22

that, as θ approaches 90°, the values of y get larger and larger and finally become greater than any specified number, however large, provided θ is sufficiently near to 90°. Such a condition is summarized in the statement

"tan θ becomes positively infinite as θ approaches 90°, if $\theta < 90°$; symbolically tan 90° = ∞."

We now see that the variation of the tangent of θ is given by

θ	0° to 90°	90° to 180°	180° to 270°	270° to 360°
tan θ	Inc. 0 to +∞	Inc. −∞ to 0	Inc. 0 to +∞	Inc. −∞ to 0

The variations of all the functions may be summarized in a table (I abbreviates "increases from" and D abbreviates "decreases from"):

θ	0° to 90°	90° to 180°	180° to 270°	270° to 360°
sin θ	I. 0 to 1	D. 1 to 0	D. 0 to −1	I. −1 to 0
cos θ	D. 1 to 0	D. 0 to −1	I. −1 to 0	I. 0 to 1
tan θ	I. 0 to +∞	I. −∞ to 0	I. 0 to +∞	I. −∞ to 0
cot θ	D. +∞ to 0	D. 0 to −∞	D. +∞ to 0	D. 0 to −∞
sec θ	I. 1 to +∞	I. −∞ to −1	D. −1 to −∞	D. +∞ to 1
csc θ	D. +∞ to 1	I. 1 to +∞	I. −∞ to −1	D. −1 to −∞

30. Graphs of the Functions.

If x is the variable angle, then y = sin x may be graphed on the x, y − plane; i.e., each pair of corresponding values of x and y can be considered as the coordinates of a point in a system of rectangular coordinates. The path of the point (x, y) is the graph of the function sin x. *To plot the points* (x, y) *first change* x *to radian measure and then use the same units of length on the* x *and* y *axes.* **Always plot x in radians.**

Illustration.

Draw the graph of the function y = sin x.

Solution: Obtain the values from the Table of Natural Functions correct to 2 decimal places. We construct the table:

x (degrees)	0°	30°	45°	60°	75°	90°	105°	120°	135°	150°	180°
x (radians)	0	.52	.79	1.05	1.31	1.57	1.83	2.09	2.36	2.62	3.14
y = sin x	0	.50	.71	.87	.97	1.00	.97	.87	.71	.50	0

This table can be duplicated for x from 180° to 360° if we remember that the sine of x is negative when x is in the third and fourth quadrants. Plotting x in radians against y, we get the graph of Fig. 23.

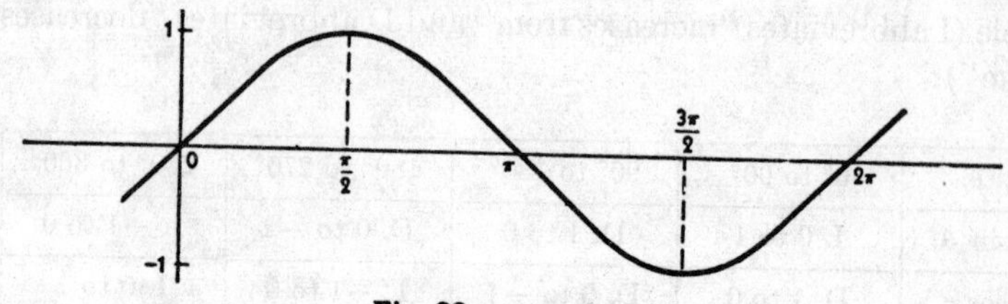

Fig. 23. y = sin x.

In a similar manner we can draw the graphs of all the trigonometric functions. The table of values is given below. The values of x are given at intervals of $\frac{\pi}{6}$ (or 30°) and are correct to two decimal places. Since the trigonometric functions are periodic, this table can be extended in either direction. The extreme right column gives x in radians correct to two decimal places. Using this table we draw the graphs of the trigonometric functions; they are given in Figs. 23–28.

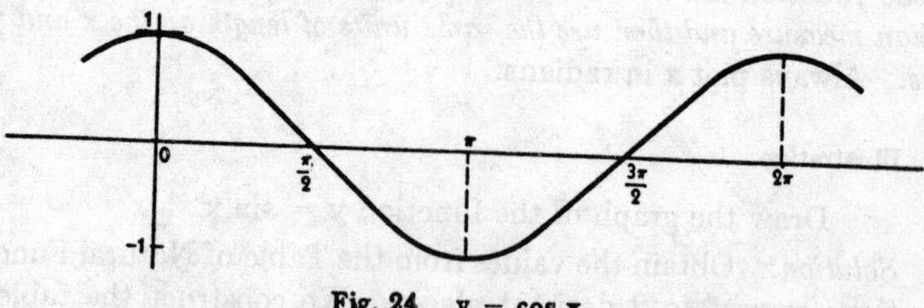

Fig. 24. y = cos x.

x	sin x	cos x	tan x	cot x	sec x	csc x	x
0	0	1.00	0	$\pm\infty$	1.00	$\pm\infty$	0
$\frac{\pi}{6}$	0.50	0.87	0.58	1.73	1.15	2.00	0.52
$\frac{\pi}{3}$	0.87	0.50	1.73	0.58	2.00	1.15	1.05
$\frac{\pi}{2}$	1.00	0	$\pm\infty$	0	$\pm\infty$	1.00	1.57
$\frac{2\pi}{3}$	0.87	-0.50	-1.73	-0.58	-2.00	1.15	2.09
$\frac{5\pi}{6}$	0.50	-0.87	-0.58	-1.73	-1.15	2.00	2.62
π	0	-1.00	0	$\pm\infty$	-1.00	$\pm\infty$	3.14
$\frac{7\pi}{6}$	-0.50	-0.87	0.58	1.73	-1.15	-2.00	3.67
$\frac{4\pi}{3}$	-0.87	-0.50	1.73	0.58	-2.00	-1.15	4.19
$\frac{3\pi}{2}$	-1.00	0	$\pm\infty$	0	$\pm\infty$	-1.00	4.71
$\frac{5\pi}{3}$	-0.87	0.50	-1.73	-0.58	2.00	-1.15	5.24
$\frac{11\pi}{6}$	-0.50	0.87	-0.58	-1.73	1.15	-2.00	5.76
2π	0	1.00	0	$\pm\infty$	1.00	$\pm\infty$	6.28

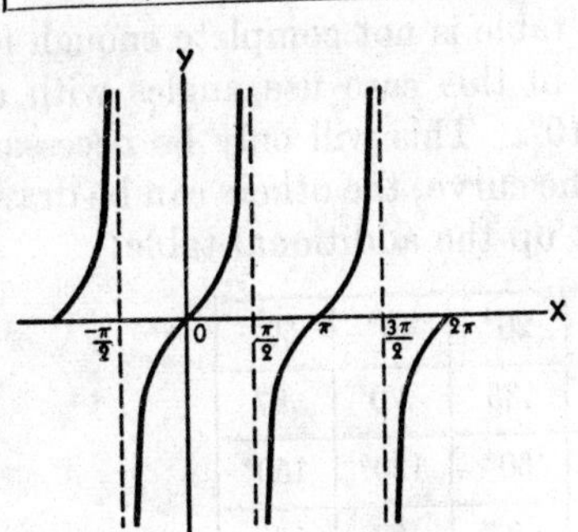

Fig. 25. y = tan x.

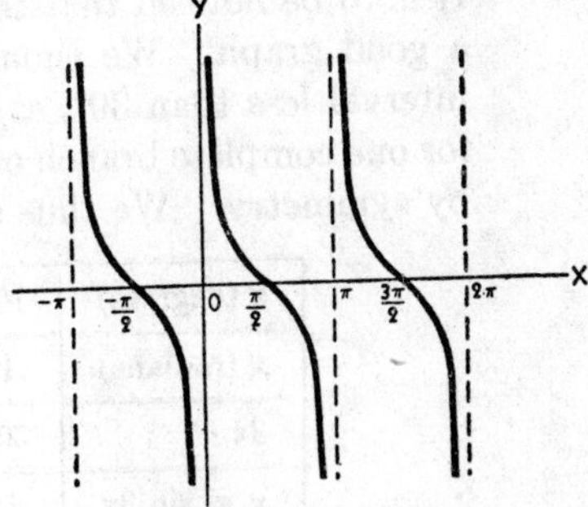

Fig. 26. y = cot x.

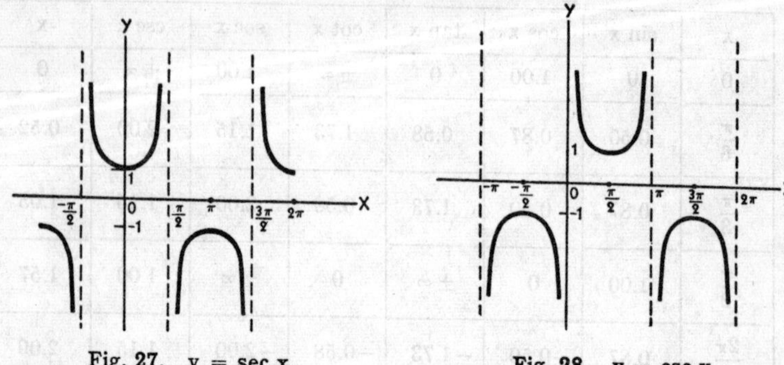

Fig. 27. y = sec x. Fig. 28. y = csc x.

31. Graphs of Other Functions.

Using the method of the last paragraph, i.e., making a table of values of y which correspond to values of the angle x and then plotting the points, we can draw the graph of any expression in x which involves trigonometric functions of x.

Illustration.

Graph the function $y = \sin 3x$ from $x = 0$ to $x = 2\pi$.

Solution: The table of values is

x (degrees)	0°	30°	60°	90°	120°	150°	180°	
x (radians)	0	.52	1.05	1.57	2.09	2.62	3.14	etc.
3x	0	90°	180°	270°	360°	450°	540°	
y = sin 3x	0	1	0	−1	0	1	0	

It is to be noticed that this table is not complete enough for a good graph. We should in this case use angles with an interval less than 30°, say 10°. This will only be necessary for one complete branch of the curve; the others can be drawn by symmetry. We thus set up the additional table:

x (degrees)	10°	20°	40°	50°
x (radians)	.17	.35	.70	.87
3x	30°	60°	120°	150°
y = sin 3x	.50	.87	.87	.50

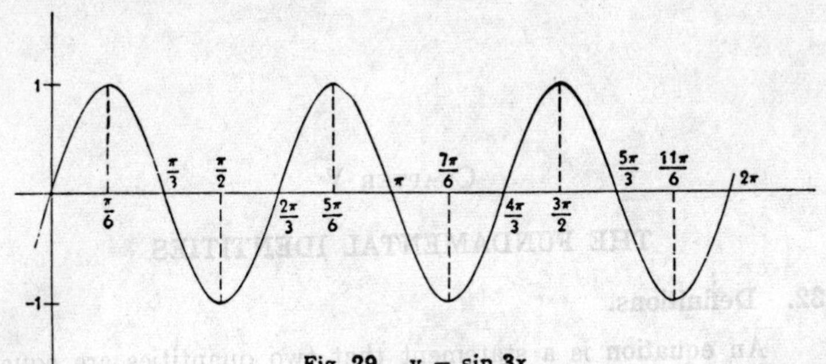

Fig. 29. y = sin 3x.

Chapter V

THE FUNDAMENTAL IDENTITIES

32. Definitions.

An **equation** is a statement that two quantities are equal. The quantities usually involve variables. An equation in which the two sides of the equality are equal for all defined values of the variables is called an **identical equation** or, briefly, an **identity.** An equation which is true for only certain values of the variables is called a **conditional equation** or, usually, just an **equation.**

Illustrations.

1. $a^2 - b^2 = (a - b)(a + b)$ is true for all values of a and b and is therefore an identity.

2. $x - 5 = 0$ is true only when $x = 5$ and is therefore a conditional equation.

33. The Fundamental Identities.

There are many relationships among the trigonometric functions which are true for all values of the angle. The most important of these are called the **fundamental identities.** There are eight of them:

The reciprocal relations:

(1) $\csc \theta = \dfrac{1}{\sin \theta}$;

(2) $\sec \theta = \dfrac{1}{\cos \theta}$;

(3) $\cot \theta = \dfrac{1}{\tan \theta}$;

The quotient relations:

(4) $\tan \theta = \dfrac{\sin \theta}{\cos \theta}$; (5) $\cot \theta = \dfrac{\cos \theta}{\sin \theta}$;

The Pythagorean relations:

(6) $\sin^2\theta + \cos^2\theta = 1$;
(7) $\tan^2\theta + 1 = \sec^2\theta$;
(8) $1 + \cot^2\theta = \csc^2\theta$.

These identities are proved directly from the definitions.

Proof of (1):

By the definitions we have $\csc\theta = \frac{r}{y}$ and $\sin\theta = \frac{y}{r}$.

Therefore $\csc\theta = \frac{r}{y} = \frac{1}{\frac{y}{r}} = \frac{1}{\sin\theta}$.

Proof of (5):

By the definitions we have $\sin\theta = \frac{y}{r}$, $\cos\theta = \frac{x}{r}$, $\cot\theta = \frac{x}{y}$.

Therefore $\frac{\cos\theta}{\sin\theta} = \frac{\frac{x}{r}}{\frac{y}{r}} = \frac{x}{r}\cdot\frac{r}{y} = \frac{x}{y} = \cot\theta$.

Proof of (6):

By the Pythagorean theorem we have $y^2 + x^2 = r^2$.

Dividing through by r^2 we get $\frac{y^2}{r^2} + \frac{x^2}{r^2} = 1$.

Then, from the definitions, $\sin^2\theta + \cos^2\theta = 1$.

The other identities may be proved in a similar manner.

The student should commit the fundamental identities to memory.

Note: The identities may take different forms. Thus, for example, since $\csc\theta = \frac{1}{\sin\theta}$, it is also true that $\sin\theta = \frac{1}{\csc\theta}$. Likewise, from $\sin^2x + \cos^2x = 1$ we obtain $\sin^2x = 1 - \cos^2x$ and $\cos^2x = 1 - \sin^2x$. The student should be able to recognize and use the identities in these slightly different forms.

34. Proofs of General Identities.

There are a large number of identities among the trigonometric functions. To prove these identities we use the fundamental identities of the last section.

Rule. *To prove an identity, alter* **only** *one member until it assumes the same form as the other member, which is* **not** *altered.*

Illustrations.

1. Prove the identity $\dfrac{1 - \tan^2 x}{1 + \tan^2 x} = 1 - 2 \sin^2 x.$

We shall alter only the left member:

Using $\tan x = \dfrac{\sin x}{\cos x}$: $\quad \dfrac{1 - \dfrac{\sin^2 x}{\cos^2 x}}{1 + \dfrac{\sin^2 x}{\cos^2 x}} =$

Simplifying: $\dfrac{\cos^2 x - \sin^2 x}{\cos^2 x} \cdot \dfrac{\cos^2 x}{\cos^2 x + \sin^2 x} =$

Using $\cos^2 x + \sin^2 x = 1$: $\quad \cos^2 x - \sin^2 x =$

Using $\cos^2 x = 1 - \sin^2 x$: $\quad 1 - 2 \sin^2 x = 1 - 2 \sin^2 x.$

2. Prove the identity $\quad \cos \alpha = \dfrac{\csc \alpha}{\cot \alpha + \tan \alpha}.$

We shall change the right member:

By (1), (4), and (5): $\quad = \dfrac{\dfrac{1}{\sin \alpha}}{\dfrac{\cos \alpha}{\sin \alpha} + \dfrac{\sin \alpha}{\cos \alpha}};$

Simplifying: $\quad = \dfrac{\dfrac{1}{\sin \alpha}}{\dfrac{\cos^2 \alpha + \sin^2 \alpha}{\sin \alpha \cos \alpha}}$

$$= \frac{1}{\sin \alpha} \cdot \frac{\sin \alpha \cos \alpha}{\cos^2 \alpha + \sin^2 \alpha};$$

By (6): $\quad = \cos \alpha.$

Suggestions for proving identities:

I. *If one member involves only one function it may be best to express everything on the other side in terms of this function.*

II. *Any function may be expressed in terms of any other function; in particular, all the functions can be easily expressed in terms of sines and cosines.*

III. *Avoid the use of radicals, whenever possible.*

35. Exercise 8.

I. Prove the fundamental identities (2), (3), (4), (7), (8)

II. Prove the following identities.

1. $\dfrac{\sin \alpha}{\csc \alpha} + \dfrac{\cos \alpha}{\sec \alpha} = 1.$

2. $\tan x + \cot x = \sec x \csc x.$

3. $\dfrac{1}{1 - \sin \theta} + \dfrac{1}{1 + \sin \theta} = 2 \sec^2\theta.$

4. $\sin^4 y - \cos^4 y = 2 \sin^2 y - 1.$ (HINT: factor the left member.)

5. $\dfrac{\sec x}{\cot x + \tan x} = \sin x.$

36. Trigonometric Equations in One Unknown.

Since the trigonometric functions are periodic, any equation involving trigonometric functions has infinitely many solutions. We shall, however, be concerned only with the finding of those solutions which are positive or zero and less than 360°.

Let us consider the following types of trigonometric equations:

1. Linear equations in one function of θ;
2. Quadratic equations in one function of θ;
3. Equations involving more than one function.

The solution of type 1 has been considered in §22.

Illustration.

Solve: $2 \sin \theta - 1 = 0.$

Solution: $2 \sin \theta = 1; \sin \theta = \frac{1}{2}.$

Recalling that $\sin 30° = \frac{1}{2}$ and that the sine is positive in quadrants I and II, we get $\theta = 30°, 150°.$

See also the illustration in §22.

Type 2 can be solved by the methods of quadratic equations, which the student has met in algebra.

Illustration.

Solve: $2\cos^2 x - \cos x - 1 = 0.$

Solution: Factor: $(2\cos x + 1)(\cos x - 1) = 0.$

Then $2\cos x + 1 = 0$; $\cos x = -\frac{1}{2}$; $x = 120°, 240°$. | $\cos x - 1 = 0$; $\cos x = 1$; $x = 0°$.

The solution of the type 3 can be divided into two classes:

I. It may be possible to factor the equation into factors each of which contains only one function and whose product is equal to *zero.*

Illustration.

Solve: $\sin x \sec^2 x - 2\sin x = 0.$

Solution: $\sin x(\sec^2 x - 2) = 0;$

$\sin x = 0$; $x = 0°, 180°$. | $\sec^2 x - 2 = 0$; $\sec^2 x = 2$; $\sec x = \pm\sqrt{2}$; $x = 45°, 135°, 225°, 315°$.

II. Solve the equation by the aid of identities.

Illustrations.

1. Solve $2\sin^2 x + \cos x - 1 = 0.$

Solution: Using $\sin^2 x = 1 - \cos^2 x$ we get

$$2 - 2\cos^2 x + \cos x - 1 = 0;$$
$$2\cos^2 x - \cos x - 1 = 0;$$
$$(2\cos x + 1)(\cos x - 1) = 0;$$

$\cos x = -\frac{1}{2}$; $x = 120°, 240°$. | $\cos x = 1$; $x = 0°$.

2. Solve $\sin x + 1 = \cos x.$

Solution: Using $\cos x = \sqrt{1 - \sin^2 x}$ we get

$$\sin x + 1 = \sqrt{1 - \sin^2 x};$$

Squaring both sides: $\sin^2 x + 2\sin x + 1 = 1 - \sin^2 x;$

$$2\sin^2 x + 2\sin x = 0;$$
$$2\sin x(\sin x + 1) = 0$$

$\sin x = 0$; $x = 0°, 180°$. | $\sin x = -1$; $x = 270°$.

check: $x = 0°: 0 + 1 = 1.$ | $x = 270°: -1 + 1 = 0.$

$x = 180°: 0 + 1 \neq -1$; $\therefore$ 180° is not a solution.

Warning.

To solve the equation the student may square both sides or multiply both sides by an expression involving the unknown. Both of these operations may lead to an *extraneous solution*, i.e., a solution which does not satisfy the original equation. It, therefore, becomes necessary to check all solutions if either of these two operations has been performed.

We caution against dividing through by the unknown; *this may cause the loss of a solution.*

Illustration.

Solve csc x tan x = tan x.

If we divide through by tan x we get: csc x = 1; x = 90°.

But there are two other solutions, x = 0° and x = 180°, obtained from tan x = 0.

37. Exercise 9.

Solve each equation using Table of Natural Functions if necessary:

1. $3 + 3 \cos x = 2 \sin^2 x$.
2. $\sqrt{3} \cot x + 1 = \csc^2 x$.
3. $\cot^2 x + \csc^2 x = 9$.
4. $3 \cot x - \cos^2 x \csc x - 2 \csc x + \sin x = 0$.
5. $3 \cos x = \sqrt{3} \sin x$.

CHAPTER VI

ADDITION FORMULAS

38. The Sine and Cosine of the Sum of Two Angles.

If α and β are any two angles, then

(1) $\sin(\alpha + \beta) = \sin\alpha\cos\beta + \cos\alpha\sin\beta$;

(2) $\cos(\alpha + \beta) = \cos\alpha\cos\beta - \sin\alpha\sin\beta$.

Proof. We shall prove the formulas for the case when α and β are positive acute angles. The proof is so worded that it will hold for any value of α and β if careful attention is paid to the sign of the directed line segments. Construct Fig. 30.

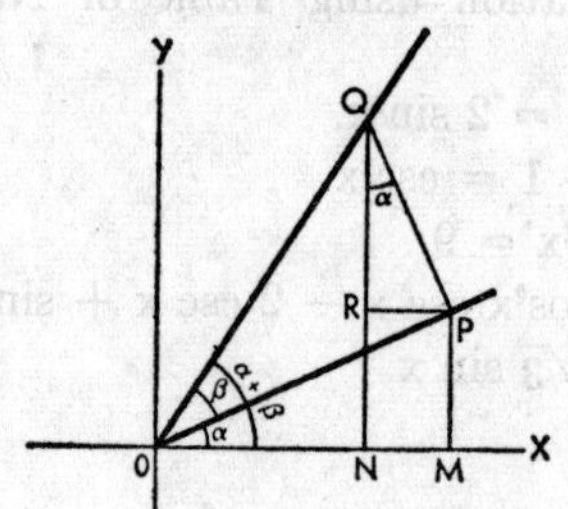

Fig. 30

The line $PQ \perp OP$ and $QN \perp OX$; therefore by elementary plane geometry $\angle RQP = \alpha$. The triangles ONQ, OMP, and PRQ are right triangles. Hence from

$$\triangle OMP: \quad \sin\alpha = \frac{PM}{OP} \quad \text{or} \quad PM = OP\sin\alpha; \qquad \text{(i)}$$

$$\cos\alpha = \frac{OM}{OP} \quad \text{or} \quad OM = OP\cos\alpha; \qquad \text{(ii)}$$

$$\triangle PRQ: \quad \sin\alpha = \frac{PR}{QP} \quad \text{or} \quad PR = QP\sin\alpha; \qquad \text{(iii)}$$

$$\cos \alpha = \frac{QR}{QP} \quad \text{or} \quad QR = QP \cos \alpha; \qquad \text{(iv)}$$

$$\triangle OPQ: \quad \sin \beta = \frac{QP}{OQ}; \cos \beta = \frac{OP}{OQ}. \qquad \text{(v)}$$

Then in $\triangle ONQ$, since $RN = PM$, we have

$$NQ = QR + RN = QR + PM.$$

Further

$$\sin (\alpha + \beta) = \frac{NQ}{OQ} = \frac{RN + QR}{OQ} = \frac{PM}{OQ} + \frac{QR}{OQ}.$$

Using (i) and (iv) above, we see that

$$\sin (\alpha + \beta) = \frac{OP \sin \alpha}{OQ} + \frac{QP \cos \alpha}{OQ} = \frac{OP}{OQ} \sin \alpha + \frac{QP}{OQ} \cos \alpha$$

and using (v), we find that

$$\sin (\alpha + \beta) = \sin \alpha \cos \beta + \sin \beta \cos \alpha.$$

From the same figure we see that

$$\cos (\alpha + \beta) = \frac{ON}{OQ} = \frac{OM}{OQ} - \frac{NM}{OQ};$$

and, since $NM = PR$, using (ii), (iii), and (v), we get

$$\cos (\alpha + \beta) = \frac{OP \cos \alpha}{OQ} - \frac{QP \sin \alpha}{OQ} = \cos \alpha \cos \beta - \sin \alpha \sin \beta.$$

We shall assume without further discussion that these formulas as well as those that follow in this chapter are true for all values of α and β, positive, negative, or zero, unless otherwise specified.

Illustration.

Find the exact value of cos 105° by expressing 105° as a sum.

Solution:
$$\begin{aligned}\cos 105° &= \cos (60° + 45°) \\ &= \cos 60° \cos 45° - \sin 60° \sin 45° \\ &= \frac{1}{2}\frac{\sqrt{2}}{2} - \frac{\sqrt{3}}{2}\frac{\sqrt{2}}{2} = \frac{\sqrt{2} - \sqrt{6}}{4}\end{aligned}$$

Illustration.

Prove $\sin(\pi + \theta) = -\sin\theta$.

Solution: By (1)

$$\begin{aligned}\sin(\pi + \theta) &= \sin\pi\cos\theta + \sin\theta\cos\pi \\ &= 0\cdot\cos\theta + \sin\theta\cdot(-1) \\ &= -\sin\theta.\end{aligned}$$

39. Addition Formulas for Tangent and Cotangent.

(3) $$\tan(\alpha + \beta) = \frac{\tan\alpha + \tan\beta}{1 - \tan\alpha\tan\beta}.$$

(4) $$\cot(\alpha + \beta) = \frac{\cot\alpha\cot\beta - 1}{\cot\alpha + \cot\beta}.$$

Proof of (3):

$$\tan(\alpha + \beta) = \frac{\sin(\alpha + \beta)}{\cos(\alpha + \beta)}$$

[by (1) and (2)]: $$= \frac{\sin\alpha\cos\beta + \cos\alpha\sin\beta}{\cos\alpha\cos\beta - \sin\alpha\sin\beta}$$

[Dividing both numerator and denominator by $\cos\alpha\cos\beta$]: $$= \frac{\dfrac{\sin\alpha\cos\beta}{\cos\alpha\cos\beta} + \dfrac{\cos\alpha\sin\beta}{\cos\alpha\cos\beta}}{\dfrac{\cos\alpha\cos\beta}{\cos\alpha\cos\beta} - \dfrac{\sin\alpha\sin\beta}{\cos\alpha\cos\beta}}$$

[Cancelling common factors]: $$= \frac{\dfrac{\sin\alpha}{\cos\alpha} + \dfrac{\sin\beta}{\cos\beta}}{1 - \dfrac{\sin\alpha\sin\beta}{\cos\alpha\cos\beta}}$$

[Using (4), § 33]: $$= \frac{\tan\alpha + \tan\beta}{1 - \tan\alpha\tan\beta}.$$

The proof of (4) is left to the student with the hint that in the proof he may divide both numerator and denominator by $\sin\alpha$ and $\sin\beta$.

Illustration.

Find the exact value of $\tan 75°$.

Solution:

$$\tan 75° = \tan (45° + 30°) = \frac{\tan 45° + \tan 30°}{1 - \tan 45° \tan 30°}$$

$$= \frac{1 + \frac{\sqrt{3}}{3}}{1 - \frac{\sqrt{3}}{3}} = \frac{\frac{3 + \sqrt{3}}{3}}{\frac{3 - \sqrt{3}}{3}} = \frac{3 + \sqrt{3}}{3 - \sqrt{3}} \cdot \frac{3 + \sqrt{3}}{3 + \sqrt{3}}$$

$$= \frac{9 + 6\sqrt{3} + 3}{9 - 3} = 2 + \sqrt{3}.$$

40. Functions of the Difference of Two Angles.

(5) $\sin (\alpha - \beta) = \sin \alpha \cos \beta - \cos \alpha \sin \beta.$

(6) $\cos (\alpha - \beta) = \cos \alpha \cos \beta + \sin \alpha \sin \beta.$

(7) $\tan (\alpha - \beta) = \dfrac{\tan \alpha - \tan \beta}{1 + \tan \alpha \tan \beta}.$

(8) $\cot (\alpha - \beta) = \dfrac{\cot \alpha \cot \beta + 1}{\cot \beta - \cot \alpha}.$

Proof of (5):

To obtain the formula for $\sin (\alpha - \beta)$ we substitute $(-\beta)$ for β in (1).

$\sin (\alpha - \beta) = \sin (\alpha + [-\beta]) = \sin \alpha \cos (-\beta) + \sin (-\beta) \cos \alpha.$

We recall from §21: $\cos (-\beta) = \cos \beta$; $\sin (-\beta) = -\sin \beta$;

Therefore $\sin (\alpha - \beta) = \sin \alpha \cos \beta - \sin \beta \cos \alpha.$

The other three formulas are proved in a similar manner.

41. Double Angle Formulas.

(9) $\sin 2\alpha = 2 \sin \alpha \cos \alpha.$

(10) $\cos 2\alpha = \cos^2 \alpha - \sin^2 \alpha;$

$= 2 \cos^2 \alpha - 1;$

$= 1 - 2 \sin^2 \alpha.$

(11) $\tan 2\alpha = \dfrac{2 \tan \alpha}{1 - \tan^2 \alpha}.$

(12) $\cot 2\alpha = \dfrac{\cot^2 \alpha - 1}{2 \cot \alpha}.$

Proof of (11):

To obtain formula (11) we let $\beta = \alpha$ in (3).

$$\tan 2\alpha = \tan(\alpha + \alpha) = \frac{\tan\alpha + \tan\alpha}{1 - \tan\alpha\tan\alpha} = \frac{2\tan\alpha}{1 - \tan^2\alpha}.$$

The other formulas are proved in a similar manner.

42. Half-Angle Formulas.

$$(13)\quad \sin\frac{\theta}{2} = \pm\sqrt{\frac{1 - \cos\theta}{2}}.$$

$$(14)\quad \cos\frac{\theta}{2} = \pm\sqrt{\frac{1 + \cos\theta}{2}}.$$

$$(15)\quad \tan\frac{\theta}{2} = \pm\sqrt{\frac{1 - \cos\theta}{1 + \cos\theta}} = \frac{1 - \cos\theta}{\sin\theta} = \frac{\sin\theta}{1 + \cos\theta}.$$

$$(16)\quad \cot\frac{\theta}{2} = \pm\sqrt{\frac{1 + \cos\theta}{1 - \cos\theta}} = \frac{\sin\theta}{1 - \cos\theta} = \frac{1 + \cos\theta}{\sin\theta}.$$

Proof of (13):

From (10) we have $\cos 2\alpha = 1 - 2\sin^2\alpha$.

Solving for $\sin^2\alpha$: $\sin^2\alpha = \dfrac{1 - \cos 2\alpha}{2}$

and $$\sin\alpha = \pm\sqrt{\frac{1 - \cos 2\alpha}{2}}.$$

Let $\theta = 2\alpha$; then $\dfrac{\theta}{2} = \alpha$, and we get (13).

To obtain (14) use the second part of (10).

Proof of (15):

The student should be able to follow the subsequent steps of this proof. We suggest that he determine at each step what formula has been used.

$$\tan\frac{\theta}{2} = \frac{\sin\frac{\theta}{2}}{\cos\frac{\theta}{2}} = \frac{\pm\sqrt{\frac{1 - \cos\theta}{2}}}{\pm\sqrt{\frac{1 + \cos\theta}{2}}} = \pm\sqrt{\frac{1 - \cos\theta}{1 + \cos\theta}}.$$

If we rationalize the denominator, we get

$$\pm\sqrt{\frac{1-\cos\theta}{1+\cos\theta}} = \frac{1-\cos\theta}{\sin\theta}.$$

If we rationalize the numerator, we get

$$\pm\sqrt{\frac{1-\cos\theta}{1+\cos\theta}} = \frac{\sin\theta}{1+\cos\theta}.$$

The $\pm$ sign is not necessary on the right-hand side of the last two formulas since $1 - \cos\theta$ and $1 + \cos\theta$ are always positive, while $\sin\theta$ will always have the same sign as $\tan\frac{\theta}{2}$.

The other formulas are proved in a similar manner.

43. Product Formulas.

(17) $2 \sin\alpha \cos\beta = \sin(\alpha+\beta) + \sin(\alpha-\beta)$.
(18) $2 \cos\alpha \cos\beta = \cos(\alpha+\beta) + \cos(\alpha-\beta)$.
(19) $2 \sin\alpha \sin\beta = \cos(\alpha-\beta) - \cos(\alpha+\beta)$.
(20) $2 \sin\beta \cos\alpha = \sin(\alpha+\beta) - \sin(\alpha-\beta)$.

Proofs:

The proofs of these formulas are obtained by adding and subtracting formulas (1), (2), (5), and (6). Thus for example

(1) $\sin(\alpha+\beta) = \sin\alpha\cos\beta + \cos\alpha\sin\beta$
(5) $\sin(\alpha-\beta) = \sin\alpha\cos\beta - \cos\alpha\sin\beta$

$\sin(\alpha+\beta) + \sin(\alpha-\beta) = 2\sin\alpha\cos\beta$,

which is formula (17) above.

Note: Formula (20) is equivalent to (17) since it merely interchanges α and β; remember that the formulas hold for any values of the angles.

44. Sums and Differences of Sines and Cosines.

(21) $\sin x + \sin y = 2 \sin\frac{x+y}{2} \cos\frac{x-y}{2}$.

(22) $\sin x - \sin y = 2 \cos\frac{x+y}{2} \sin\frac{x-y}{2}$.

(23) $\cos x + \cos y = 2 \cos\frac{x+y}{2} \cos\frac{x-y}{2}$.

(24) $\cos x - \cos y = -2 \sin \frac{x+y}{2} \sin \frac{x-y}{2}$.

Proofs:

The proofs of these formulas are obtained from formulas (17), (18), (19), and (20) by letting

$x = \alpha + \beta$ and $y = \alpha - \beta$.

We see that $x + y = 2\alpha$ and $x - y = 2\beta$, or

$$\alpha = \frac{x+y}{2} \text{ and } \beta = \frac{x-y}{2}.$$

Substituting directly into (17), (18), (19), (20) we get the above formulas.

Note: We do not give any addition formulas for the secant and cosecant. If they occur in any problem, first express them in terms of the sine and cosine respectively by using the reciprocal identities.

Illustration.

Prove the identity $\csc 2x = \frac{1}{2} \sec x \csc x$.

Solution: $\csc 2x = \frac{1}{\sin 2x} = \frac{1}{2 \sin x \cos x} = \frac{1}{2} \frac{1}{\sin x} \frac{1}{\cos x}$
$= \frac{1}{2} \sec x \csc x$.

45. Identities Involving Addition Formulas.

The addition formulas are used mainly to change the form of a given trigonometric expression. They are especially helpful in proving identities.

Suggestions for proving identities involving multiple angles.

I. *Express secant and cosecant in terms of sine and cosine.*
II. *Reduce the multiple angles by changing to functions of single angles.*
III. *Avoid introducing radicals, whenever possible.*

Illustrations.

1. Prove $\tan x = \frac{\sin 2x}{1 + \cos 2x}$.

Solution: We change the right member, using (9) and (10).

$$\tan x = \frac{2 \sin x \cos x}{1 + 2\cos^2 x - 1} = \frac{2 \sin x \cos x}{2 \cos^2 x} = \frac{\sin x}{\cos x} = \tan x.$$

2. Prove $\sin 3x = 3 \sin x - 4 \sin^3 x$.

Solution: $\sin (3x) = \sin (2x + x) = \sin 2x \cos x + \sin x \cos 2x$

$$= (2 \sin x \cos x) \cos x + \sin x (\cos^2 x - \sin^2 x)$$
$$= 2 \sin x \cos^2 x + \sin x \cos^2 x - \sin^3 x$$
$$= 3 \sin x \cos^2 x - \sin^3 x$$
$$= 3 \sin x (1 - \sin^2 x) - \sin^3 x$$
$$= 3 \sin x - 3 \sin^3 x - \sin^3 x$$
$$= 3 \sin x - 4 \sin^3 x.$$

3. Prove $\dfrac{\sin 5x + \sin 3x}{\cos 5x + \cos 3x} = \tan 4x.$

Solution: By (17) and (19):

$$\frac{\sin 5x + \sin 3x}{\cos 5x + \cos 3x} = \frac{2 \sin \frac{5x + 3x}{2} \cos \frac{5x - 3x}{2}}{2 \cos \frac{5x + 3x}{2} \cos \frac{5x - 3x}{2}}$$

$$= \frac{\sin 4x}{\cos 4x} = \tan 4x$$

4. Prove $\tan \left(\dfrac{x}{2} + 45^\circ\right) - \tan x = \sec x.$

Solution:

$$\tan \left(\frac{x}{2} + 45^\circ\right) - \tan x = \frac{\tan \frac{x}{2} + \tan 45^\circ}{1 - \tan \frac{x}{2} \tan 45^\circ} - \tan x$$

$$= \frac{\tan \frac{x}{2} + 1}{1 - \tan \frac{x}{2}} - \tan x = \frac{\frac{\sin x}{1 + \cos x} + 1}{1 - \frac{\sin x}{1 + \cos x}} - \tan x$$

$$= \frac{\sin x + 1 + \cos x}{1 + \cos x - \sin x} - \frac{\sin x}{\cos x}$$

$$= \frac{\cancel{\cos x \sin x} + \cos x + \cos^2 x - \sin x - \cancel{\sin x \cos x} + \sin^2 x}{\cos x\,(1 + \cos x - \sin x)}$$

$$= \frac{(\cos x - \sin x + 1)}{\cos x\,(1 + \cos x - \sin x)}$$

$$= \frac{1}{\cos x} = \sec x.$$

46. Equations Involving Multiple Angles.

To solve an equation which involves functions of a multiple angle we find one or more equations, each involving only one function of one angle, *whose solutions include all solutions of the given equation.*

Illustrations.

1. Find all the values of x, positive or zero and less than 360°, for which $\sin 2x + \cos x = 0$.

Solution: Using (9) we get

$$2 \sin x \cos x + \cos x = 0;$$
$$\cos x\,(2 \sin x + 1) = 0;$$

$$\begin{array}{l|l} \cos x = 0; & 2 \sin x + 1 = 0; \\ x = 90°, 270°. & \sin x = -\tfrac{1}{2}; \\ & x = 210°, 330°. \end{array}$$

2. Solve $\sin 3x = \dfrac{\sqrt{3}}{2}$.

Solution: The angle whose sine is $\dfrac{\sqrt{3}}{2}$ is $60° + n(360°)$ or $120° + n(360°)$.

Therefore $3x = 60°, 120°, 420°, 480°, 780°, 840°$;
and $x = 20°, 40°, 140°, 160°, 260°$, and $280°$.

3. Solve $\dfrac{\tan x + \tan 2x}{1 - \tan x \tan 2x} = 1$.

Solution: By (3) we have

$$\frac{\tan x + \tan 2x}{1 - \tan x \tan 2x} = \tan (x + 2x) = \tan 3x.$$

Therefore $\tan 3x = 1$;
$3x = 45°, 225°, 405°, 585°, 765°, 945°$;
and $x = 15°, 75°, 135°, 195°, 255°, 315°$.

4. Solve $\sin 3x + \sin 5x + \cos 2x - \cos 6x = 0$.

Solution: By (21) and (24) the left member becomes

$$2 \sin 4x \cos x + [-2 \sin 4x \sin(-2x)] = 0$$

and, since $\sin(-2x) = -\sin 2x$,

$$2 \sin 4x \cos x + 2 \sin 4x \sin 2x = 0;$$

$$2 \sin 4x (\cos x + \sin 2x) = 0;$$

$\sin 4x = 0;$

$4x = 0°, 180°, 360°, 540°, 720°, 900°, 1080°, 1260°;$

$x = 0°, 45°, 90°, 135°, 180°, 225°, 270°, 315°.$

$\cos x + \sin 2x = 0;$

and by Illustration 1 above

$x = 90°, 270°, 210°, 330°.$

47. Exercise 10.

Prove the following identities:

1. $\sin(30° + x) \cos(30° - x) + \cos(30° + x) \sin(30° - x) = \dfrac{\sqrt{3}}{2}.$
2. $\dfrac{\tan 150° - \tan 15°}{1 + \tan 150° \tan 15°} = -1.$ [Hint: Use formula (7)].
3. $\dfrac{2 \tan \alpha}{\tan 2\alpha} = 1 - \tan^2\alpha.$
4. $\cos 3x = 4 \cos^3 x - 3 \cos x.$
5. $\dfrac{\sin 7x - \sin 9x}{\cos 9x - \cos 7x} = \cot 8x.$
6. $\tan\left(45° - \dfrac{x}{2}\right) + \tan x = \sec x.$
7. $\tan \beta \sin 2\beta = 2 \sin^2\beta.$

Solve the following equations:

8. $\sin \dfrac{x}{2} = \dfrac{1}{2}.$
9. $\cos 5x + \cos x = \cos 3x.$
10. $\cos 2x - \sin x = 0.$
11. $5 \sin 2x = 3.$
12. $\cos 3x + \sin 2x - \sin 6x + \cos 5x = 0.$

CHAPTER VII

LOGARITHMS

48. Definitions.

The **logarithm** *of a number* **N** *to the base* **a** *is the exponent* **x** *of the power to which the base must be raised to equal the number* **N.** The base **a** must be positive and not equal to 1. We shall limit ourselves to the logarithms of positive numbers; thus **N** is always positive in our discussions. The statement "*The logarithm of* **N** *to the base* **a**" is abbreviated "$\log_a N$." We can then state the above definition in a slightly different form.

If $N = a^x$, *then* $x = \log_a N$.

Because of the above restrictions on **a** and **N** we limit ourselves to real numbers, and because of the definition stated above, the logarithm of **N** is unique. Thus every positive number has one and only one logarithm, and every logarithm represents one and only one number.

Illustrations.

The following statements, read across the page, are equivalent.

$3^2 = 9;\quad \log_3 9 = 2.$

$25^{1/2} = 5;\quad \log_{25} 5 = \frac{1}{2}.$

$2^{-3} = \frac{1}{8};\quad \log_2 \frac{1}{8} = -3.$

$(\frac{1}{3})^2 = \frac{1}{9};\quad \log_{1/3} \frac{1}{9} = 2.$

$27^{2/3} = 9;\quad \log_{27} 9 = \frac{2}{3}.$

$8^{-1/3} = \frac{1}{2};\quad \log_8 \frac{1}{2} = -\frac{1}{3}.$

49. Common Logarithms.

For computations in our number system it is convenient to take 10 as the base; the logarithms are then called common logarithms. Henceforth we shall use a = 10 and shall omit the base in the abbreviated symbol. Thus $\log_{10}N$ is simply written log N.

From the following table we can obtain the numbers whose logarithms are integers.

Exponential Form	Logarithmic Form
. . . .	. . . .
$10^3 = 1000$	$\log 1000 = 3$
$10^2 = 100$	$\log 100 = 2$
$10^1 = 10$	$\log 10 = 1$
$10^0 = 1$	$\log 1 = 0$
$10^{-1} = 0.1$	$\log 0.1 = -1$
$10^{-2} = 0.01$	$\log 0.01 = -2$
$10^{-3} = 0.001$	$\log 0.001 = -3$
. . . .	. . . .

This table gives the logarithms of the numbers which are integral powers of 10. The question now becomes: "What is the logarithm of a number which is not an integral power of 10?" For example, let us consider the problem of finding log 125. Since 125 is between 100 and 1000, it is natural to suppose, from the above table, that log 125 is between 2 and 3. The logarithm then is 2 + (a proper fraction). We shall express the proper fraction in decimal form. In general, we have

$$\textbf{log N} = \textbf{(an integer)} + \textbf{(decimal fraction} \geqq \textbf{0,} < \textbf{1).}$$

The integral part is called the **characteristic.** The decimal fraction is called the **mantissa.** Thus

$$\textbf{log N} = \textbf{characteristic} + \textbf{mantissa.}$$

In general, the mantissas are non-repeating infinite decimals which can be approximated correctly to as many places as desired. The approximations have been tabulated in four-place, five-place, or higher-place tables and these tables are called logarithmic tables.

Thus the mantissa, or decimal part, is found from tables. These values are **always positive.**

The characteristic is determined according to the following two rules:

Rule 1.

If the number **N** *is greater than* 1, *the characteristic of its logarithm is* one *less than the number of digits to the left of the decimal point.*

Rule 2.

If the number **N** *is less than* 1, *the characteristic of its logarithm is negative; if the first digit which is not zero occurs in the* k^{th} *decimal place, the characteristic is* $-\mathbf{k}$.

Since we shall combine the characteristic and mantissa to obtain the complete logarithm by

$$\log N = \text{characteristic} + \text{mantissa};$$

and, further, since the mantissa is always positive, we shall *always* write a negative characteristic, $-k$, as $(\mathbf{10} - \mathbf{k}) - \mathbf{10}$. Thus suppose the mantissa = .5732, then if the characteristics are 1, 0, -1, and -2, the logarithms are written 1.5732, 0.5732, 9.5732 − 10, and 8.5732 − 10, respectively.

Illustrations.

The characteristics of the logarithms of the numbers in the left column are given in the right column.

Number	Characteristic
135.2	2
57.35	1
2.693	0
0.3296	9–10
0.0735	8–10
0.000037	5–10

Antilogarithms.

If the logarithm is given, the problem becomes that of finding the number which corresponds to this logarithm. The number is called the **antilogarithm.** The characteristic of the given logarithm determines the position of the decimal point in the antilogarithm. In placing the decimal point we use in reverse order the two rules given previously for determining the characteristic of a number. Thus suppose the digits of the antilogarithm are 5932 and the characteristic is 1. Using Rule 1 we observe that there should be two digits before the decimal point in the antilogarithm and we place the decimal point after the "9," making the number 59.32. Or suppose that the digits of the antilogarithm are 3756 and the characteristic is $8 - 10$; that is, -2. Using Rule 2, we decide that the first non-zero digit after the decimal point in the antilogarithm should be the second digit after the decimal point; we get 0.03756.

50. Use of Tables.

To find the mantissa of a logarithm we use the Table of the Common Logarithms of Numbers in the back of this book (pp. 1–21). This table is also used to find the antilogarithm. The numbers .05903, .590300, 5903, 5903000 are said to have the **same sequence** of digits. (The initial and end zeros are disregarded.) The *mantissa* of the logarithm for each of these numbers is the *same;* the characteristics are, of course, different.

The first two pages give the mantissa of the logarithm of all numbers of one, two, and three digits. The remaining pages of this table give the mantissa of numbers containing four digits. Since the mantissa of the logarithm for numbers having the same sequence of digits is the same, the first two pages need not be used. They are *not to be used in finding the antilogarithms.*

I. *To find the logarithm of a given number.*

Illustration 1.

Find log 3.76.

Solution: 1. By Rule 1 the characteristic is 0.

2. On the second page of the table we find the first two digits (37) in the left column headed by "N." In the "37" row and in the column headed by "6" (the third digit of the number) we find 57519, which is the mantissa of the logarithm. Thus log 3.76 = 0.57519.

Illustration 2.

Find log 42.98.

Solution: 1. By Rule 1 the characteristic is 1.

2. On page 10 of the table we find 429 in the left column headed by "N." In the "429" row and in the column headed by "8" (the fourth digit of the number) we find 327. These digits are the **last three** digits of the mantissa. To find the **first two** digits of the mantissa look in the second column under "L." The first two digits can be found either in the same row or in the first row *above* this row which has the leading digits given. In this example we find 63 as the first two digits in the "427" row. Thus the complete mantissa is 63327, and we get log 42.98 = 1.63327.

The meaning of the ($_*$) *in the table.*

Some numbers in the table are preceded by an asterisk. This means that the *first two digits* of the mantissa are found in the "L" column directly below the row in which the number appears. Thus log 407.7 = 2.61034. The first two digits of the mantissa, 61, being found in the "408" row.

Note: Many tables are printed in this manner to save space, especially seven- or eight-place tables. Thus the mastery of this table will enable the student to read many other kinds of similarly printed tables.

II. *To find the antilogarithm of a given logarithm.*

Illustration 1.

Find N, if log N = 8.57461 − 10.

Solution: 1. Find the mantissa 57461 in the table; it appears in the column headed by "5" and in the row which has 375 at the left under "N." Thus the sequence of digits is 3755.

2. The characteristic is 8 − 10 or −2. Using Rule 2 for the characteristic the first significant digit after the decimal point should occur in the second place. Therefore, if Log N = 8.57461 − 10, N = 0.03755.

Illustration 2.

If log N = 1.27114, N = 18.67.

51. Interpolation.

The logarithm to a five digit (or higher digit) number can be found by linear interpolation. For example, log 65362 is between log 65360 and log 65370 and the mantissa is therefore between 81531 and 81538 as found in the table; to get the most nearly exact value we interpolate. This interpolation can be carried out in the same manner as was done in the case of the trigonometric functions, see §6, C. However, for greater speed in computation, tables of proportional parts are furnished with most tables of logarithms.

How To Use the Table of Proportional Parts.

The extreme right-hand column, headed by "P P," is the table of proportional parts. In this column are found "groups" of numbers each having a heading such as, for example, "22," "21," "20," etc., on page 6 of the table. These numbers represent the *tabular difference.* The tabular difference is found by subtracting two consecutive mantissas. In each group of numbers the integers 1 to 9 in a column on the *left* represent the *fifth* digit in the number. The numbers on the *right* represent the corresponding proportional part for that tabular difference.

Illustration 1.

Find log 224.58.

Solution: The characteristic is 2.

The mantissa is between those for 22450 and 22460, which are 35122 and 35141, respectively. The tabular difference = 35141 − 35122 = 19; (this subtraction should be done mentally). In the "P P" column find the group headed by "19." The fifth digit of the number (224.58) is 8; therefore look opposite 8 and read 15.2. *This is the amount to be added to the* **smaller** *mantissa* given above (35122). The decimal point in 15.2 indicates the position of the last digit in the given table; thus we add

	19
1	1.9
2	3.8
3	5.7
4	7.6
5	9.5
6	11.4
7	13.3
8	15.2
9	17.1

$$\begin{array}{r} 35122 \\ 152 \\ \hline 351372 \end{array}$$

and round off to five places giving 35137,

which is the correct mantissa. Therefore log 224.58 = 2.35137.

The student should practice the use of the Prop. Parts Table until he can do the addition and subtraction mentally. However, he should always strive for accuracy, and if necessary, should write out each step.

Illustration 2.

Given log N = 7.55469 − 10. Find N.

Solution: The mantissa 55469 is between 55461 and 55473 as found in the table corresponding to N = 3586 and N = 3587, respectively. The tabular difference = 55473 − 55461 = 12. The Prop. Parts difference = 55469 − 55461 = 8; (**always** subtract the **smaller found** mantissa from the **given** mantissa to find this difference). In the Prop. Parts Table in the group headed by 12 we find 8.4 on the *right;* this is the closest number to 8. To the left of this number is found the number 7, which is therefore the fifth digit of N. The first four are those which correspond to the *smaller* mantissa found in the table, i.e., 3586. Therefore the digits of the number are 35867. The characteristic is 7 − 10 = −3. Thus we get N = 0.0035867.

52. Exercise 11.

Find the logarithms of the following numbers:

1. 83.69.	4. 527.2.	7. 0.028753.
2. 156000.	5. 0.003963.	8. 0.0000057.
3. 1.694.	6. 0.9467.	9. 76588.

Find the antilogarithms of the following logarithms:

10. 2.88110.	13. 8.70375 − 10.	16. 1.99326.
11. 0.20276.	14. 6.52306 − 10.	17. 4.64217.
12. 5.77583.	15. 9.44252 − 10.	18. 3.93502.

53. Properties of Logarithms.

We recall the following laws of exponents:

I. $a^x a^y = a^{x+y}$.

II. $\dfrac{a^x}{a^y} = a^{x-y}$.

III. $(a^x)^k = a^{kx}$.

IV. $\sqrt[q]{a^x} = a^{x/q}$.

Corresponding to these laws of exponents we have the following very useful properties of logarithms.

Property I. *The logarithm of a product is equal to the sum of the logarithms of its factors;*

$$\log_a MN = \log_a M + \log_a N.$$

Proof:

Let $x = \log_a M$ and $y = \log_a N$.
By definition: $M = a^x$ and $N = a^y$.
Then $MN = a^x a^y = a^{x+y}$ by (I).
By definition: $\log_a MN = x + y = \log_a M + \log_a N$.

Note: $\log_a MNP = \log_a M + \log_a N + \log_a P$ and similarly for more factors.

Property II. *The logarithm of a quotient is equal to the logarithm of the numerator minus the logarithm of the denominator;*

$$\log_a \frac{M}{N} = \log_a M - \log_a N.$$

This is proved by using (II); we leave the proof to the student.

Property III. *The logarithm of the* k^{th} *power of a number equals* k *times the logarithm of the number;*

$$\log_a N^k = k \log_a N.$$

Proof:

Let $x = \log_a N$, then $N = a^x$.
By (III): $N^k = (a^x)^k = a^{kx}$.
Then by definition: $\log_a N^k = kx = k \log_a N$.

Property IV. *The logarithm of the* q^{th} *root of a number is equal to the logarithm of the number divided by* q;

$$\log_a \sqrt[q]{N} = \frac{1}{q} \log_a N.$$

This property can be proved by using (IV).

Illustration.

Express $\log \dfrac{(\sqrt{59.32})\,(0.436)}{(3.294)^3(3.1416)}$ as the algebraic sum of logarithms.

Solution:

$$\log \frac{(\sqrt{59.32})\,(0.436)}{(3.294)^3(3.1416)}$$

[By Property II]:

$$= \log (\sqrt{59.32})(0.436) - \log (3.294)^3(3.1416)$$

[By Property I]:

$$= \log \sqrt{59.32} + \log 0.436 - [\log(3.294)^3 + \log 3.1416]$$

[By Props. III, IV]:

$$= \tfrac{1}{2} \log 59.32 + \log 0.436 - 3 \log 3.294 - \log 3.1416.$$

54. Computations Using Logarithms.

In carrying out computations using logarithms it is desirable to have a systematic form in which to display the work. Only a systematic form can be read by a person who has not carried out the actual computations. We recommend the form given in the illustrations. The student should check each step carefully.

Illustration 1.

Find N if $N = \dfrac{5.367}{(12.93)(0.06321)}$

Solution: We employ logarithms and, using the four properties in the last section, we get

$$\log N = \log 5.367 - [\log 12.93 + \log 0.06321].$$

(1)	log 5.367 = 0.72973	
(2)	log 12.93 = 1.11160	
(3)	log 0.06321 = 8.80079 − 10	
(4)	(2) + (3) = 9.91239 − 10	
(5)	(1) − (4) = 0.81734	
		N = 6.5666

Note: To subtract (4) from (1) we first change 0.72973 to 10.72973 − 10.

Illustration 2.

Find N if $N = \frac{(\sqrt[3]{0.9573})(3.21)^2}{98.32}$.

Solution:

$\log N = \frac{1}{3}\log 0.9573 + 2\log 3.21 - \log 98.32.$

(1)	$\frac{1}{3}\log 0.9573 = \frac{1}{3}[9.98105 - 10]$	$= 9.99368 - 10$
(2)	$2\log 3.21 = 2[0.50651]$	$= 1.01302$
(3)	(1) + (2)	$= 11.00670 - 10$
(4)	log 98.32	$= 1.99264$
(5)	(3) − (4)	$= 9.01406 - 10$
		N = 0.10329

Note: To find

$\frac{1}{3}[9.98105 - 10] = \frac{1}{3}[29.98105 - 30] = 9.99368 - 10$

always rearrange a negative characteristic so that *after* dividing the result will be x.xxxxx − **10.**

Illustration 3.

Find N if $N = \sqrt{\frac{(0.35926)^3}{673.52}}$.

Solution:

$\log N = \frac{1}{2}[3\log 0.35926 - \log 673.52].$

(1)	$3\log 0.35926 = 3[9.55541 - 10]$	$= 28.66623 - 30$
(2)	log 673.52	$= 2.82835$
(3)	(1) − (2)	$= 25.83788 - 30$
(4)	$\frac{1}{2}$ (3) $= \frac{1}{2}[15.83788 - 20]$	$= 7.91894 - 10$
		N = 0.0082974

Note: In going from (3) to (4) we made the change 25.83788 − 30 = 15.83788 − 20. This was done so that after dividing by 2 we would have x.xxxxx — 10.

55. Cologarithms.

The cologarithm of a number is defined as the logarithm of the reciprocal of the number.

$$\textbf{colog N} = \log \frac{1}{\text{N}} = \log 1 - \log \text{N} = 0 - \log \text{N}$$

$$= [\mathbf{10.0000} - \mathbf{10}] - \textbf{log N}.$$

Illustration.

Find colog 35.7.

Solution:

$$\begin{aligned} \text{colog } 35.7 &= \log 1 - \log 35.7 \\ &= 10.00000 - 10 \\ \text{minus} \quad & \quad\ \underline{1.55267 \qquad\quad} \\ &= 8.44733 - 10. \end{aligned}$$

To find a cologarithm mentally:

Subtract each digit of the logarithm of the number, except the last one, from 9 and the last one from 10 and subtract 10 from the result. Thus in the above illustration: $9 - 1 = 8; 9 - 5 = 4; 9 - 5 = 4; 9 - 2 = 7; 9 - 6 = 3; 10 - 7 = 3$ giving $8.44733 - 10$.

Cologarithms are used in computation when it appears to be desirable to add all the logarithms instead of subtracting some of them.

Illustration.

$$\log \frac{306}{(98.1)(1.52)} = \log 306 - \log 98.1 - \log 1.52$$

$$= \log 306 + \text{colog } 98.1 + \text{colog } 1.52.$$

(1)	log 306 = 2.48572
(2)	colog 98.1 = 8.00833 − 10
(3)	colog 1.52 = 9.81816 − 10
(4)	sum = 20.31221 − 20

56. Exercise 12.

Evaluate to four significant figures by means of logarithms.

1. $(73.29)^2(0.009736)(\sqrt{0.3527})$.

2. $\sqrt[3]{\dfrac{(643.3)(\sqrt{3.288})}{(0.00325)(4325)}}$.

3. $\dfrac{(7.0894)^3(9.2327)}{(8741.3)(0.21015)}$.

4. $\dfrac{(11.238)(21.332)}{(1.0096)(10.761)}$.

57. Logarithms of Trigonometric Functions.

The trigonometric functions are constants which we can find accurately to four places by the table. For example, sin 26°10′ = .4410. We can, therefore, find the logarithm of the trigonometric function:

$$\log \sin 26°10' = \log 0.4410 = 9.64444 - 10.$$

This procedure, however, would necessitate first using the Table of Natural Trigonometric Functions to find sin 26°10′ and then using the logarithm table to find log 0.4410. To eliminate such a two-step procedure, tables which give the logarithms of the trigonometric functions directly have been computed. They are given in the Table of the Logarithms of the Trigonometric Functions.

The Table of the Logarithms of the Trigonometric Functions has been divided into two tables. Pages 23–46 contain the logarithms of the trigonometric functions from 0° to 1° and 89° to 90° for every second, and from 1° to 6° and 84° to 89° for every ten seconds. Pages 47–92 contain the logarithms of the trigonometric functions from minute to minute. The second part of the table, which is the most frequently used table, will be explained first. The angles from 0° to 45° are labelled at the *top* of the pages with the minutes for these angles given in the *left*-hand column and the heading of the columns is to be taken from the *top* of the pages. The angles from 45° to 90° are labelled at the *bottom* of the pages with the minutes in the *right*-hand column and the heading of the columns is to be taken from the *bottom* of the pages. The interpolation is done by the proportional parts table, given under the heading "P P," as was done in the table of common logarithms. The tabular differences are given in the table in the columns headed by "d" to the right of the column of "L Sin" and "L Cos";

"L Tan" and "L Cot" have a common difference, and the "c d" column between them refers to both. The "P P" column gives the proportional parts differences for each second.

The sines and cosines are always less than 1, and therefore the logarithms of these trigonometric functions always have a negative characteristic. Similarly, $\tan \theta$ for $\theta < 45°$ and $\cot \theta$ for θ between 45° and 90° are less than 1, and the logarithms of these functions have a negative characteristic. The first part of the characteristic for the logarithm of these functions is explicitly given in the table; it is, therefore, necessary to subtract 10 from the logarithm of the above mentioned functions as given in the table. The characteristic for the logarithm of $\tan \theta$ for θ between 45° and 90° and $\cot \theta$ for $\theta < 45°$ as given in the table is the exact characteristic, and no subtraction is to be made in this case. No logarithms are given for the secant and cosecant; if these functions occur in any computation, express them by means of the reciprocal identities in terms of sine and cosine and use cologarithms. Thus $\log \csc \theta = \text{colog} \sin \theta$.

Illustrations.

1. Find log sin 23°43′37″.

Solution: Page 71 of the tables gives 23° at the *top;* therefore the minutes (43) are found in the *left* column. In the "L Sin" column opposite 43′ read 9.60446. In the "d" column between 43′ and 44′ read 28; this is the tabular difference. In the "P P" column and in the column headed by 28,

opposite 30 read 14.0,
opposite 7 read 3.3;

therefore the proportional part for 37″ is 17.3, or 17 to the nearest integer. Adding this to 9.60446, the result gives log sin 23°43′37″ = 9.60463 − 10.

2. Find log cos 60°36′43″.

Solution: 60° is found at the *bottom* of p. 77; therefore the *right*-hand minute column is used. Opposite 36 in the minute column and in the column headed by "L Cos" at the *bottom* of the page read 9.69100. The "d" column gives the tabular difference as 23. Under "P P" in the "23" column

opposite 40 read 15.3,
opposite 3 read 1.2;

therefore the proportional part difference is 16.5, (17). This amount is to be subtracted from 9.69100 because the cosine is a decreasing function (the cotangent is also a decreasing function, whereas the sine and tangent are increasing functions). Thus

$\log \cos 60°36'43'' = 9.69083 - 10.$

3. Given $\log \tan \theta = 9.74627 - 10$. Find θ.

Solution: In the "L Tan" columns the number just smaller than 9.74627 is 9.74613 as found on p. 77 and corresponding to 29°8′. The tabular difference as found in the "c d" column is 30; the proportional parts difference is 9.74627 − 9.74613 = 0.00014, or simply 14. (Only the last two digits are affected in this difference and the subtraction should be done mentally.) Under "P P" in the "30" column

opposite 10.0 read at the left 20,

opposite 4.0 read at the left 8;

therefore $\theta = 29°8'28''$.

4. Given $\log \cot \theta = 9.81672 - 10$. Find θ.

Solution: In the "L Cot" column search for the numbers closest to 9.81672. On p. 81 in the "L Cot" column headed at the bottom of the page find 9.81693 and 9.81666 corresponding to 56°44′ and 56°45′, respectively. The tabular difference, as found in the "c d" column, is 27. The difference between 9.81693 (always use the logarithm corresponding to the *smaller* angle) and 9.81672 (the *given* logarithm) is 21. In the "P P" column under 27

opposite 18.0 (on the right) read 40 (on the left),

opposite 3.2 (on the right) read 7 (on the left);

therefore the proportional part is 47″ and $\theta = 56°44'47''$.

Note: If it is desirable to find the angle to the nearest tenth of a second it could be accomplished by the proportional parts as follows:

	18.0	is the prop. part for 40
	2.7	is the prop. part for 6
	.32	is the prop. part for .7
thus	21.02	is the prop. part for 46.7

and $\theta = 56°44'46.7''$.

58. The Small Angles.

Pages 23–46 contain the logarithms of the trigonometric functions for the small angles to the seconds. The general arrangement of this table is similar to that just described. Again the degrees are given at the tops and bottoms of the pages. The trigonometric functions are also labelled at the tops and the bottoms of the pages. On pp. 23–35, the minutes and each ten seconds are given in columns at the left and right, headed ′ and ″, and the odd seconds are given in a horizontal row at the top and bottom of each page. On pp. 36–46, the minutes are given in columns at the left and right, headed ′, and each ten seconds is given in a horizontal row at the top and bottom of each page. The columns of minutes on the left read downward; the horizontal rows at the top, from left to right; these go with the degrees at the top of the pages. The columns of minutes at the right and the horizontal rows at the bottom, read in the opposite directions, go with the degrees at the bottoms of the pages. On pp. 36–46 the odd seconds are found by use of the "P P" columns. On pp. 24–25, the proportional parts are given at the top of the pages; on all other pages they occur in the usual right hand column.

Illustrations.

1. Find log sin 0°37′24.37″.

Solution: Page 30 gives log sin 0°37′24″ = 8.03659 − 10. Tabular difference is 19. "P P" column reads

proportional part for 3 = 5.7,

$\frac{1}{10}$ proportional part for 7 = 1.33;

therefore proportional part for .37 = 7.03 or 7.

Thus log sin 0°37′24.37″ = 8.03666 − 10.

2. Find log tan 0°42′17.48″.

Solution: From p. 33: log tan 0°42′17″ = 8.08992 − 10. Tabular difference is 17 and from the "P P" column the proportional part for .48 is 8.16 or 8. Thus

log tan 0°42′17.48″ = 8.09000 − 10.

3. Find log cos 89°43′26.4″.

Solution: From p. 26: log cos 89°43′26″ = 7.68296 − 10. The tabular difference is 44 and the "P P" column gives the proportional part for 4 as 17.6 or 18. Since the cosine is a decreasing function, this is to be subtracted from the above logarithm. Therefore, the result is log cos 89°43′26.4″ = 7.68278 − 10.

4. Find log cot 84°34′32″.

Solution: From p. 45: log cot 84°34′30″ = 8.97758 − 10. The tabular difference is 23 and the "P P" column gives the proportional part for 2 as 4.6 or 5. Subtract this from the above logarithm and obtain the result, log cot 84°34′32″ = 8.97753 − 10.

5. Given log tan θ = 8.74875 − 10. Find θ.

Solution: From p. 41: log tan 3°12′30″ = 8.74861 − 10. The tabular difference is 38 and the prop. parts difference (8.74875 − 8.74861) is 14. From the "P P" column under 38: the proportional part for 3.7 is 14.6. Thus θ = 3°12′33.7″.

S and T Tables.

When a very small angle is to be found by means of its logarithmic sine or tangent (angle near 90° by means of its logarithmic cosine or cotangent), and accuracy is desired, the cologarithms of S or T should be used. These are given in the columns headed "C S" and "C T," pp. 48–50. When the logarithm of the sine or tangent of very small angles (cosine or cotangent of angles near 90°) expressed to many decimal fractions of a second, the S and T table, pp. 2–21, should be used. The formulas for their use are as follows.

To find the logarithm:

$$\log \sin \theta = \log \theta'' + S;$$
$$\log \tan \theta = \log \theta'' + T;$$
$$\log \cos \theta = \log (90^\circ - \theta)'' + S;$$
$$\log \cot \theta = \log (90^\circ - \theta)'' + T;$$

where θ'' = number of seconds in θ and $(90^\circ - \theta)''$ = number of seconds in $90^\circ - \theta$.

To find the angle:

$$\log \theta'' = \log \sin \theta + CS;$$
$$\log \theta'' = \log \tan \theta + CT;$$
$$\log (90^\circ - \theta)'' = \log \cos \theta + CS;$$
$$\log (90^\circ - \theta)'' = \log \cot \theta + CT;$$

where the notation is the same as above.

Illustrations.

1. Find log sin 3.4785″.

Solution:

From p. 8: log 3.4785 = 0.54139
From p. 2: S = 4.68557 − 10
log sin 3.4785″ = 5.22696 − 10.

2. Given log sin θ = 6.82973 − 10. Find θ.

Solution: The value of θ (see p. 48) lies between 0°2′ and 0°3′, or between 120″ and 180″, and, corresponding to this,

CS = 5.31443
log sin θ = 6.82973 − 10
log θ'' = 2.14416.

The number corresponding to the logarithm 2.14416 is (p. 4) 139368. Therefore, θ = 139.368″ = 0°2′19.368″.

Change of Trigonometric Functions.

It is sometimes required to find the logarithm of one trigonometric function from that of another, without requiring the angle. To facilitate this, special proportional tables, headed with the tabular differences of both functions, are given wherever the space admits.

Illustration.

Given, log tan x = 9.67644. Find log cos x.

Solution: The difference between the given logarithm and that given in the table, 9.67622 − 10 (see p. 73, opposite 25°23′) is 22. The tabular differences of the two logarithmic functions at this place are 32 and 6. In the proportional parts table for $\frac{6}{32}$, 22 corresponds to 4; this, subtracted from the tabular logarithmic cosine 9.95591 − 10, gives the required

log cos x = 9.95587 − 10.

59. Exercise 13.

Find the logarithm of the function from the table.

1. sin 21°17′.	5. sin 2°23′43.7″.	9. sin 63°22′37″.
2. cos 37°43′.	6. cos 88°52′12.6″.	10. cos 74°44′28″.
3. tan 19°36′.	7. tan 1°39′22.3″.	11. tan 81°38′54″.
4. cot 47°12′.	8. cot 89°01′24.8″.	12. cot 57°47′12″.

Find the angle θ from the table.

13. $\log \sin \theta = 9.80153 - 10.$
14. $\log \cos \theta = 9.79342 - 10.$
15. $\log \tan \theta = 0.08476.$
16. $\log \cot \theta = 9.90897 - 10.$
17. $\log \sin \theta = 8.37752 - 10.$
18. $\log \cos \theta = 7.98223 - 10.$
19. $\log \tan \theta = 8.67627 - 10.$
20. $\log \cot \theta = 8.26378 - 10.$

CHAPTER VIII

THE RIGHT TRIANGLE AND APPLICATIONS

In §8 we introduced the solution of a right triangle and discussed this solution for rather simple numbers. We shall now turn to the solution of a right triangle where the use of logarithms becomes desirable. It is suggested that the student review §8 before taking up this chapter.

60. Logarithmic Solution of a Right Triangle.

We recall the following formulas for a right triangle. For the lettering see Fig. 31.

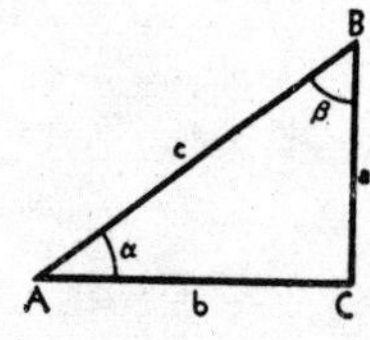

Fig. 31

(1) $\sin \alpha = \frac{a}{c} = \cos \beta.$ (2) $\cos \alpha = \frac{b}{c} = \sin \beta.$

(3) $\tan \alpha = \frac{a}{b} = \cot \beta.$ (4) $\cot \alpha = \frac{b}{a} = \tan \beta.$

(5) $a^2 + b^2 = c^2.$ (6) $\alpha + \beta = 90°.$

For logarithmic computation we write (5) in the forms

(7) $a = \sqrt{(c - b)(c + b)}$; (8) $b = \sqrt{(c - a)(c + a)}.$

These formulas are obtained from (5) by transposing first b^2 and then a^2 and factoring the right hand side; i.e., $a^2 = c^2 - b^2 = (c - b)(c + b)$. We avoid using (7) or (8) in solving the triangle; however, the formulas are used in checking the solution. The computing form is given in the illustrations. *The student should write down the complete form before looking up any logarithms.*

Illustration 1.

Solve the right triangle ABC if $\beta = 37°23'43''$ and $a = 1.3572$.

Solution:

Data: $a = 1.3572$; $\beta = 37°23'43''$		
Formulas	**Computation**	
$\alpha = 90° - \beta$	$\alpha = 90° - 37°23'43'' =$	$52°36'17''$
$\frac{b}{a} = \tan\beta$, or $b = a\tan\beta$.	$\log a = 0.13264$ $\log\tan\beta = 9.88334 - 10$ $\log b = 0.01598$	$b = 1.0375$
$\frac{a}{c} = \cos\beta$, or $c = \frac{a}{\cos\beta}$	$\log a = 10.13264 - 10$ $\log\cos\beta = 9.90008 - 10$ $\log c = 0.23256$	$c = 1.7083$
check: $a = \sqrt{(c-b)(c+b)}$ $c - b = 0.6708$ $c + b = 2.7458$ $\log a = 0.13264$ ← check →	$\log(c-b) = 9.82659 - 10$ $\log(c+b) = 0.43867$ sum $= 0.26526$ $\frac{1}{2}$ sum $= 0.13263$	

Illustration 2.

Solve the right triangle ABC if $b = 21.634$ and $c = 47.176$.

Solution:

Data: $b = 21.634$; $c = 47.176$.		
Formulas	**Computation**	
$\cos\alpha = \frac{b}{c}$	$\log b = 11.33514 - 10$ $\log c = 1.67372$ $\log\cos\alpha = 9.66142 - 10$	$\alpha = 62°42'15''$
$\sin\alpha = \frac{a}{c}$, or $a = c\sin\alpha$	$\log c = 1.67372$ $\log\sin\alpha = 9.94873 - 10$ $\log a = 11.62245 - 10$	$a = 41.923$
$\beta = 90° - \alpha$	$\beta = 90° - 62°42'15'' =$	$27°17'45''$.
check: $b = \sqrt{(c-a)(c+a)}$ $c - a = 5.253$ $c + a = 89.099$ $\log b = 1.33514$ ← check →	$\log(c-a) = 0.72041$ $\log(c+a) = 1.94988$ sum $= 2.67029$ $\frac{1}{2}$ sum $= 1.33514$	

61. Exercise 14.

Solve the right triangle ABC and check:

1. a = 13.74; $\alpha = 21°53'$.
2. b = 6.397; $\alpha = 57°17'$.
3. c = 0.37116; $\alpha = 40°03'41''$.
4. a = 113.23; b = 100.75.
5. c = 93.588; a = 45.031.

62. Angle of Elevation and Depression.

In Fig. 32, let O be the point of the observer and C, the object which is sighted. If the observer is **below** the object he sights, the angle made by the line of sight and the horizontal is called the **angle of elevation.** If the observer is **above** the object he sights, the angle made by the line of sight and the horizontal is called the **angle of depression.**

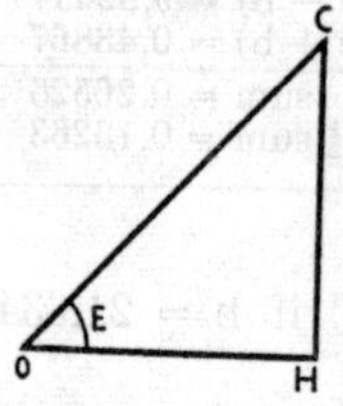

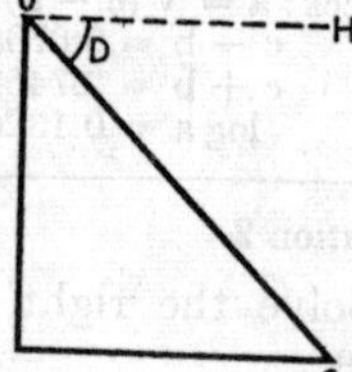

Fig. 32

Many problems involving right triangles arise from these angles. In solving any problem from descriptive data we suggest the following steps:

I. *Construct a figure.*
II. *Label the figure, introducing single letters to represent unknown angles or lengths.*
III. *Outline the solution. If more than one triangle is involved, clearly indicate which triangle and formulas are used.*
IV. *Arrange the computation form and solve for the desired unknown.*

Illustration.

If the angle of elevation of the sun is 49°27′ find the height of a flagpole whose horizontal shadow is 83.59 feet.

Solution:

Construct Fig. 33. BC represents the flagpole; AC, the shadow. The angle of elevation of the sun is at A; we represent it by α. We let AC = b and solve for BC = a.

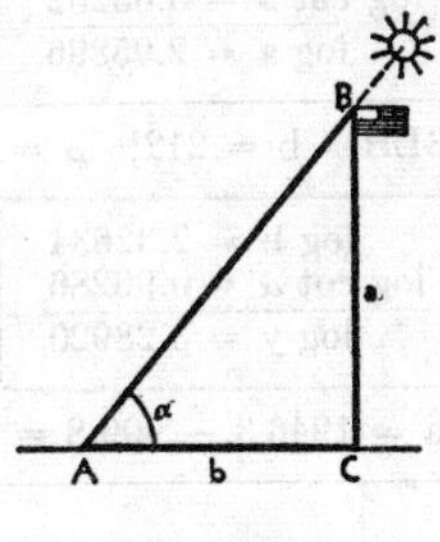

Fig. 33

In $\triangle ABC$: b = 83.59; α = 49°27′.		
$\frac{a}{b} = \tan \alpha$, or $a = b \tan \alpha$	log b = 1.92215 log tan α = 0.06773 log a = 1.98988	a = 97.70 ft.

Illustration.

From the top of a lighthouse 212 feet above a lake, the keeper spots a boat sailing directly towards him. He observes the angle of depression of the boat to be 6°13′ and then later to be 13°7′. Find the distance the boat has sailed between the observations.

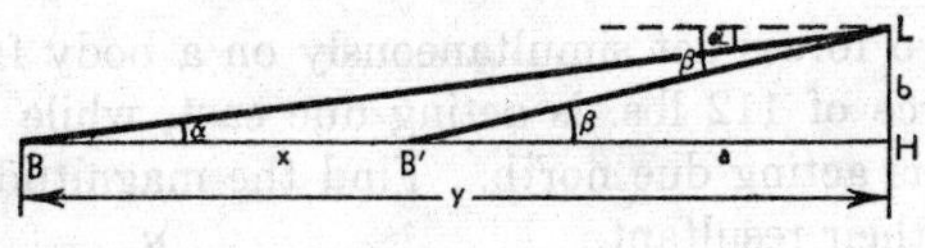

Fig. 34

Solution.

Construct Fig. 34. B and B′ represent the two positions of the boat. LH represents the lighthouse; angle α, the first angle of depression and β, the second. The line BH is parallel to the horizontal and therefore the angle at B equals α and the angle at B′ equals β. We letter the figure as shown.

In $\triangle B'LH$: $b = 212$; $\beta = 13°7'$.		
$\frac{a}{b} = \cot\beta$, or $a = b\cot\beta$.	$\log b = 2.32634$ $\log\cot\beta = 0.63262$ $\log a = 2.95896$	$a = 909.8$
In $\triangle BLH$: $b = 212$; $\alpha = 6°13'$.		
$\frac{y}{b} = \cot\alpha$, or $y = b\cot\alpha$.	$\log b = 2.32634$ $\log\cot\alpha = 0.96286$ $\log y = 3.28920$	$y = 1946.3$
$x = y - a = 1946.3 - 909.8 = 1036.5$ ft.		

63. Right Angle Vectors.

A **vector** is a quantity which has magnitude and direction. Such a quantity can be represented by an arrow; the length of the arrow represents its magnitude and the head indicates its direction. The **sum** of two vectors is called the **resultant.** In general, the resultant is the diagonal of the parallelogram of which the two vectors are the adjacent sides. The two vectors are called the **components** of the resultant along their lines. Vectors are used to represent forces, velocities, acceleration, and other physical quantities.

A special case occurs when the two vectors are at right angles to each other. Then the resultant becomes the diagonal of a rectangle, and we can employ the right triangle to solve problems involving such vectors.

Illustration.

Two forces act simultaneously on a body free to move. One force of 112 lbs. is acting due east, while the other of 88 lbs. is acting due north. Find the magnitude and direction of their resultant.

Solution.

Construct Fig. 35.

$OA = b = 112$ lbs.

$OB = 88$ lbs. $= RA = a$.

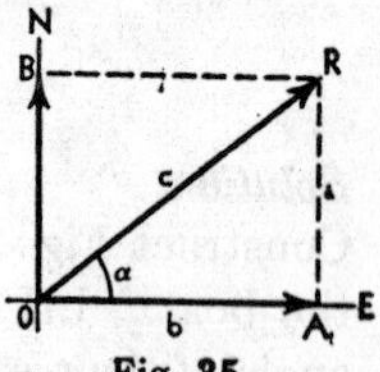

N B R c a α O b A E

Fig. 35

In $\triangle OAR$: $a = 88$; $b = 112$.		
$\frac{a}{b} = \tan \alpha$.	$\log a = 1.94448$ $\log b = 2.04922$ $\log \tan \alpha = 9.89526 - 10$	$\alpha = 38°9'25''$
$\frac{a}{c} = \sin \alpha$, or $c = \frac{a}{\sin \alpha}$	$\log a = 11.94448 - 10$ $\log \sin \alpha = 9.79086 - 10$ $\log c = 2.15362$	$c = 142.44$
Therefore the resultant is 142.44 lbs. and its direction is 38°9′25″ north of east.		

Illustration.

A force of 315 lbs. is acting at an angle of 67° with the horizontal. What are its horizontal and vertical components?

Solution:

Construct Fig. 36.
OR = vector force = e.
b = OA = horizontal component.
a = OB = vertical component.

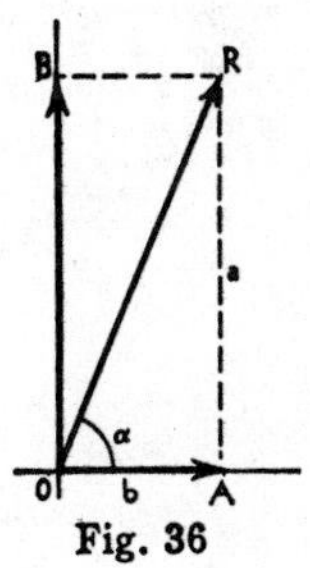

Fig. 36

In $\triangle OAR$: $c = 315$; $\alpha = 67°$.		
$\frac{a}{c} = \sin \alpha$, or $a = c \sin \alpha$.	$\log c = 2.49831$ $\log \sin \alpha = 9.96403 - 10$ $\log a = 2.46234$	$a = 289.96$ lbs.
$\frac{b}{c} = \cos \alpha$, or $b = c \cos \alpha$.	$\log c = 2.49831$ $\log \cos \alpha = 9.59188 - 10$ $\log b = 2.09019$	$b = 123.08$ lbs.

64. Exercise 15.

1. What is the angle of elevation of the sun when a tree 136 ft. high casts a shadow 159.6 ft. long?
2. From the top of a lighthouse 205 ft. above sea level, the keeper finds the angle of depression of a ship to be 8°44′.

Find the horizontal distance of the ship from the light house.

3. A shell is fired from a gun raised to an angle of 37°43′ with the horizontal. The muzzle velocity is 1125 ft. per second. Find the horizontal and vertical components of this velocity.

4. A ship is sailing due west at a rate of 23.46 miles per hour. A man walks due south across the deck at the rate of 3.15 miles per hour. Find the direction of his motion and his speed with respect to the surface of the earth.

Chapter IX

OBLIQUE TRIANGLES

65. Cases.

An **oblique** triangle is a triangle in which no one of the angles is a right angle. The triangle has **six parts**, the three sides a, b, and c, and the three angles α, β, and γ. If we have given any three parts of which at least one is a side, we can solve the triangle, i.e., find the remaining parts and construct the triangle.

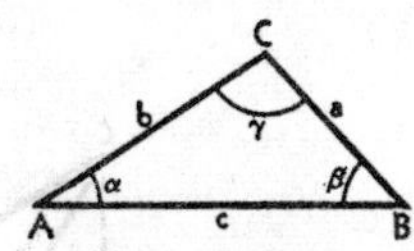

Fig. 37

There are four distinct cases.

Case I. *Given two angles and a side.*

Case II. *Given two sides and an angle opposite one of them.*

Case III. *Given two sides and the included angle.*

Case IV. *Given three sides.*

To solve the triangle we employ certain laws of the trigonometric functions. Before we return to these laws, we recall from plane geometry the elementary fact

A. $\alpha + \beta + \gamma = 180°$.

66. The Law of Cosines.

In any triangle, the square of any side is equal to the sum of the squares of the other sides minus twice their product times the cosine of their included angle.

$$a^2 = b^2 + c^2 - 2bc \cos \alpha;$$

B. $$b^2 = a^2 + c^2 - 2ac \cos \beta;$$

$$c^2 = a^2 + b^2 - 2ab \cos \gamma.$$

We can solve these formulas for the cosines of the angles; we obtain

$$\cos \alpha = \frac{b^2 + c^2 - a^2}{2bc};$$

C. $$\cos \beta = \frac{a^2 + c^2 - b^2}{2ac};$$

$$\cos \gamma = \frac{a^2 + b^2 - c^2}{2ab}.$$

Proof: Let a be any side and let β be an acute angle of the triangle. Drop a perpendicular from C to $\overline{AB}$, or $\overline{AB}$ extended. If α is acute we obtain Fig. 38; if α is obtuse we obtain Fig. 39. Let $\overline{AD} = m$ and $\overline{DC} = h$.

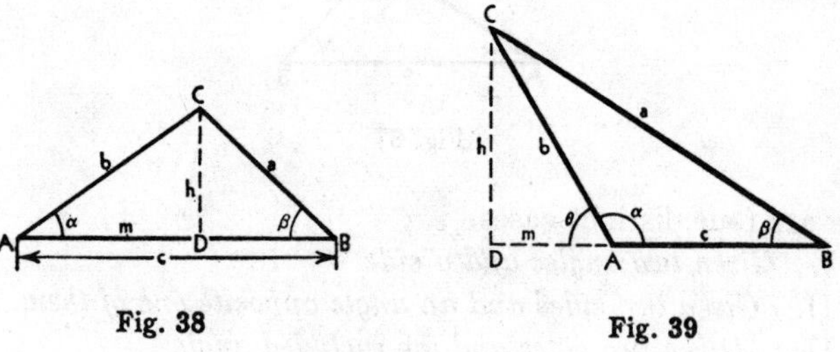

Fig. 38 Fig. 39

In either figure we have the following properties by the Pythagorean theorem.

In right $\triangle ADC$: $b^2 = h^2 + m^2$; $h^2 = b^2 - m^2$. (1)

In right $\triangle BDC$: $a^2 = h^2 + (\overline{BD})^2$. (2)

Substituting (1) into (2): $a^2 = b^2 - m^2 + (\overline{BD})^2$. (3)

I. If α is acute (see Fig. 38.),

$\overline{BD} = c - m$, ($\overline{BD}$ is not a directed line segment)

and in right triangle ADC: $\cos \alpha = \frac{m}{b}$; $m = b \cos \alpha$.

Substituting in (3) we obtain

$$\begin{aligned} a^2 &= b^2 - m^2 + (c - m)^2 \\ &= b^2 - m^2 + c^2 - 2cm + m^2 \\ &= b^2 + c^2 - 2bc \cos \alpha. \end{aligned}$$

II. If α is obtuse (see Fig. 39),

$$\overline{BD} = c + m.$$

In right $\triangle ADC$: $\cos \theta = \frac{m}{b}$ and, since $\alpha = 180° - \theta$,

$$\cos \alpha = -\cos \theta = -\frac{m}{b}; \; m = -b \cos \alpha.$$

Substituting in (3):

$$\begin{aligned} a^2 &= b^2 - m^2 + (c + m)^2 \\ &= b^2 - m^2 + c^2 + 2cm + m^2 \\ &= b^2 + c^2 - 2bc \cos \alpha. \end{aligned}$$

67. The Use of Law of Cosines in Solving Triangles.

The law of cosines is not suited for logarithmic computation as the formulas involve the sum and difference of terms. However, the law of cosines may be used when the numbers are convenient or if a computing machine is available.

A. Case III. *Given two sides and the included angle.*

Illustration.

Solve the triangle ABC if $a = 5$, $b = 7$, and $\gamma = 60°$.

Solution: By the law of cosines:

$$\begin{aligned} c^2 &= a^2 + b^2 - 2ab \cos \gamma = 25 + 49 - (2)(5)(7)(\tfrac{1}{2}) \\ &= 25 + 49 - 35 = 39. \end{aligned}$$

$$\boxed{c = \sqrt{39} = 6.245.}$$

$$\cos \alpha = \frac{b^2 + c^2 - a^2}{2bc} = \frac{49 + 39 - 25}{(2)(7)(\sqrt{39})} = \frac{3}{26}\sqrt{39} = \frac{18.735}{26}$$

$$= .7206; \quad \boxed{\alpha = 43°54'.}$$

$$\cos \beta = \frac{a^2 + c^2 - b^2}{2ac} = \frac{25 + 39 - 49}{(2)(5)(\sqrt{39})} = \frac{6.245}{26} = .2402.$$

$$\boxed{\beta = 76°06'.}$$

check: $\alpha + \beta + \gamma = 60° + 43°54' + 76°06' = 180°00'.$

B. Case IV. *Given three sides.*

Illustration.

Solve the triangle ABC if a = 3, b = 4, c = 6.

Solution: By the law of cosines:

$$\cos \alpha = \frac{b^2 + c^2 - a^2}{2bc} = \frac{16 + 36 - 9}{(2)(4)(6)} = \frac{43}{48} = .8958.$$

$\boxed{\alpha = 26°23'.}$

$$\cos \beta = \frac{a^2 + c^2 - b^2}{2ac} = \frac{9 + 36 - 16}{(2)(3)(6)} = \frac{29}{36} = .8056.$$

$\boxed{\beta = 36°20'.}$

$$\cos \gamma = \frac{a^2 + b^2 - c^2}{2ab} = \frac{9 + 16 - 36}{(2)(3)(4)} = -\frac{11}{24} = -.4583.$$

$\boxed{\gamma = 117°17'.}$

check: $\alpha + \beta + \gamma = 26°23' + 36°20' + 117°17' = 180°00'.$

68. The Law of Sines.

In any triangle, any two sides are proportional to the sines of the opposite angles.

$$\text{D.} \quad \frac{a}{\sin \alpha} = \frac{b}{\sin \beta} = \frac{c}{\sin \gamma}.$$

There are three equations in this formula.

$$\frac{a}{\sin \alpha} = \frac{b}{\sin \beta}; \quad \frac{a}{\sin \alpha} = \frac{c}{\sin \gamma}; \quad \frac{b}{\sin \beta} = \frac{c}{\sin \gamma}.$$

Proof: Construct Fig. 40 and Fig. 41, in the same manner as Figs. 38 and 39.

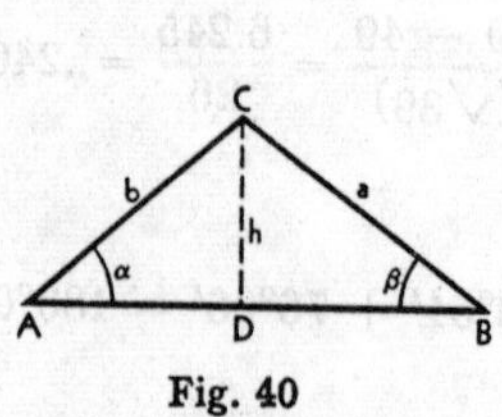

Fig. 40

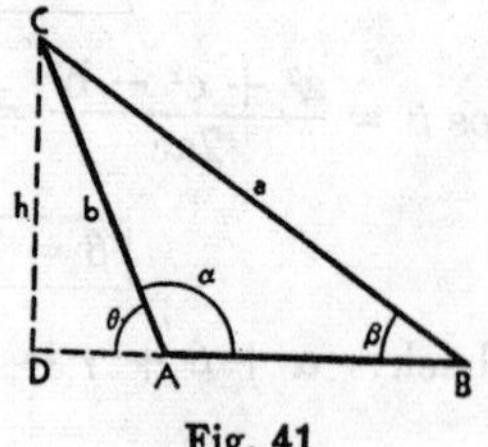

Fig. 41

I. If α is acute (see Fig. 40):

In right $\triangle ADC$: $\sin \alpha = \frac{h}{b}$; $h = b \sin \alpha$. (1)

In right $\triangle BDC$: $\sin \beta = \frac{h}{a}$; $h = a \sin \beta$. (2)

From (1) and (2) we get

$$a \sin \beta = b \sin \alpha.$$

Dividing both sides by $\sin \alpha \sin \beta$:

$$\frac{a}{\sin \alpha} = \frac{b}{\sin \beta}.$$

II. If α is obtuse (see Fig. 41):

In right $\triangle ADC$: $\sin \theta = \frac{h}{b}$; and since $\alpha = 180° - \theta$ we hav

$\sin \alpha = \sin \theta = \frac{h}{b}$; $h = b \sin \alpha$. (3)

In right $\triangle BDC$: $\sin \beta = \frac{h}{a}$; $h = a \sin \beta$. (4)

From (3) and (4) we again obtain

$$a \sin \beta = b \sin \alpha$$

or

$$\frac{a}{\sin \alpha} = \frac{b}{\sin \beta}.$$

69. The Solution of Case I. *Given two angles and a side.*

We use the law of sines to solve this case.

1. Find the third angle by A, $\alpha + \beta + \gamma = 180°$.
2. Find the unknown sides by the law of sines.

Illustration.

Solve the triangle ABC, if $\alpha = 39°28'31''$, $\gamma = 110°43'12''$, $a = 36.482$.

Solution:

Data: $\alpha = 39°28'31''$; $\gamma = 110°43'12''$; $a = 36.482$.		
Formulas	**Computation**	
$\beta = 180° - (\alpha + \gamma)$	$\beta = 180° - 150°11'43''$ =	$29°48'17''$
$\frac{b}{\sin \beta} = \frac{a}{\sin \alpha}$ $b = \frac{a \sin \beta}{\sin \gamma}$	$\log a = 1.56207$ $\log \sin \beta = 9.69639 - 10$ sum $= 11.25846 - 10$ $\log \sin \alpha = 9.80328 - 10$ $\log b = 1.45518$	$b = 28.522$
$\frac{c}{\sin \gamma} = \frac{a}{\sin \alpha}$ $c = \frac{a \sin \gamma}{\sin \alpha}$	$\log a = 1.56207$ $\log \sin \gamma = 9.97096 - 10$ sum $= 11.53303 - 10$ $\log \sin \alpha = 9.80328 - 10$ $\log c = 1.72975$	$c = 53.672$

Note: Cologarithms may be used if desired.

70. Solution of Case II, the Ambiguous Case. *Given two sides and an angle opposite one of them.*

Case II is called the ambiguous case because with the given data there may exist two triangles, one triangle, or no triangle. Let the given parts be a, b, and α; then the possibilities are shown in Figs. 42 to 47.

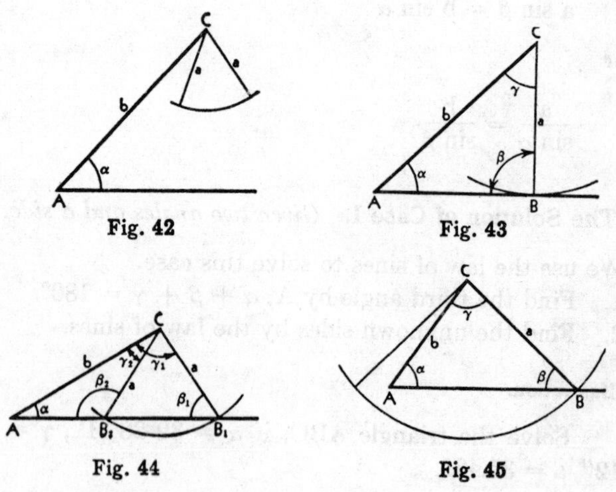

Fig. 42 Fig. 43

Fig. 44 Fig. 45

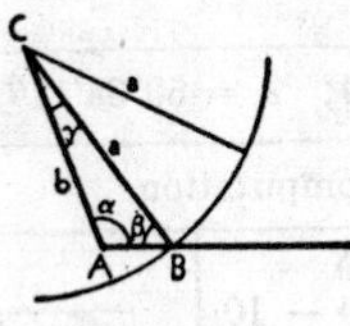

Fig. 46

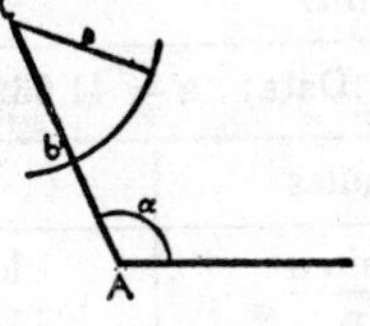

Fig. 47

We can summarize the conditions of the figures as follows:

If $\alpha < 90°$ $\begin{cases} a < b \sin \alpha; \text{ no solution.} \\ a = b \sin \alpha; \text{ one solution, } \beta = 90°. \\ a > b \sin \alpha;\ a < b; \text{ two solutions.} \\ a \geqq b; \text{ one solution.} \end{cases}$

If $\alpha \geqq 90°$ $\begin{cases} a > b; \text{ one solution.} \\ a \leqq b; \text{ no solution.} \end{cases}$

This case may be solved by the use of the law of sines. Let a, b, and α be the given parts.

1. Find β by use of $\sin \beta = \dfrac{b \sin \alpha}{a}$. If

   (i) $\log \sin \beta > 0$, then there is no solution;
   (ii) $\log \sin \beta = 0$, then $\beta = 90°$, and there is only one solution;
   (iii) $\log \sin \beta < 0$, there are two solutions for β, one obtuse and one acute.

2. Compute γ by using $\gamma = 180° - (\alpha + \beta)$. If there are two values for β, use both values unless you find $\alpha + \beta \geqq 180°$; for such values there is no triangle.

3. Compute $c = \dfrac{a \sin \gamma}{\sin \alpha}$ for each possible γ.

Illustration.

Solve the triangle ABC if a = 41.632. b = 34.794, α = 50°34′11″.

Solution:

Data: a = 41.632; b = 34.794; α = 50°34′11″		
Formulas	**Computation**	
$\sin\beta = \dfrac{b \sin\alpha}{a}$	log b = 1.54150 log sin α = 9.88784 − 10 colog a = 8.38057 − 10 log sin β = 9.80991 − 10 β_2 = 180° − β_1	β_1 = 40°12′16″ β_2 = 139°47′44″
$\gamma = 180° - (\alpha + \beta)$	γ_1 = 180° − (50°34′11″ + 40°12′16″) = 89°13′33″ γ_2 = 180° − (50°34′11″ + 139°47′44″) = impossible	
$c = \dfrac{a \sin\gamma}{\sin\alpha}$	log a = 1.61943 log sin γ = 9.99996 − 10 colog sin α = 0.11216 log c = 1.73155	c = 53.895

Illustration.

Solve the triangle ABC if c = .5126, b = .7531, γ = 36°53′.

Solution:

Data: c = .5126; b = .7531; γ = 36°53′.		
Formulas	**Computation**	
$\sin\beta = \dfrac{b \sin\gamma}{c}$	log b = 9.87685 − 10 log sin γ = 9.77829 − 10 sum = 19.65514 − 20 log c = 9.70978 − 10 log sin β = 9.94536 − 10 β_2 = 180° − β_1	β_1 = 61°51′26″ β_2 = 118°08′34″
$\alpha = 180° - (\gamma + \beta)$	α_1 = 180° − 98°44′26″ = 81°15′34″ α_2 = 180° − 155°01′34″ = 24°58′26″	
$a_1 = \dfrac{c \sin\alpha_1}{\sin\gamma}$	log c = 9.70978 − 10 log sin α_1 = 9.99493 − 10 sum = 19.70471 − 20 log sin γ = 9.77829 − 10 log a_1 = 9.92642 − 10	a_1 = .84415
$a_2 = \dfrac{c \sin\alpha_2}{\sin\gamma}$	log c = 9.70978 − 10 log sin α_2 = 9.62553 − 10 sum = 19.33531 − 20 log sin γ = 9.77829 − 10 log a_2 = 9.55702 − 10	a_2 = .36059

71. The Law of Tangents.

In any triangle the difference of any two sides, divided by their sum equals the tangent of one half of the difference of the opposite angles divided by the tangent of one half of their sum:

$$\text{E.}\quad \frac{a-b}{a+b} = \frac{\tan\frac{1}{2}(\alpha-\beta)}{\tan\frac{1}{2}(\alpha+\beta)};\ \frac{a-c}{a+c} = \frac{\tan\frac{1}{2}(\alpha-\gamma)}{\tan\frac{1}{2}(\alpha+\gamma)};$$

$$\frac{b-c}{b+c} = \frac{\tan\frac{1}{2}(\beta-\gamma)}{\tan\frac{1}{2}(\beta+\gamma)}.$$

Note: The difference of any two sides may be taken in any order, but the difference of the corresponding angles must be in the same order. Always take the order in which the difference of the sides will be positive.

Proof: This law is proved by using the law of sines and the addition formulas.

From the law of sines, $\frac{a}{c} = \frac{\sin\alpha}{\sin\gamma}$.

Subtracting 1 from both sides, we get $\frac{a}{c} - 1 = \frac{\sin\alpha}{\sin\gamma} - 1$

$$\text{or}\quad \frac{a-c}{c} = \frac{\sin\alpha - \sin\gamma}{\sin\gamma}. \qquad (1)$$

Adding 1 to both sides, we get $\frac{a}{c} + 1 = \frac{\sin\alpha}{\sin\gamma} + 1$

$$\text{or}\quad \frac{a+c}{c} = \frac{\sin\alpha + \sin\gamma}{\sin\gamma}. \qquad (2)$$

Dividing each side of (1) by the corresponding side of (2) we obtain

$$\frac{a-c}{a+c} = \frac{\sin\alpha - \sin\gamma}{\sin\alpha + \sin\gamma}$$

Using (22) and (21) of §44,

$$\frac{a-c}{a+c} = \frac{2\cos\frac{1}{2}(\alpha+\gamma)\sin\frac{1}{2}(\alpha-\gamma)}{2\sin\frac{1}{2}(\alpha+\gamma)\cos\frac{1}{2}(\alpha-\gamma)} = \frac{\tan\frac{1}{2}(\alpha-\gamma)}{\tan\frac{1}{2}(\alpha+\gamma)}.$$

72. The Solution of Case III by Using the Law of Tangents.

Let b, c, and α be the given parts.

1. Compute $\frac{1}{2}(\beta+\gamma) = \frac{1}{2}(180° - \alpha)$.
2. Find $\frac{1}{2}(\beta-\gamma)$ by using the law of tangents.
3. Compute β and γ:

$$\beta = \tfrac{1}{2}(\beta + \gamma) + \tfrac{1}{2}(\beta - \gamma);\ \gamma = \tfrac{1}{2}(\beta + \gamma) - \tfrac{1}{2}(\beta - \gamma).$$

4. Find a by using the law of sines.
5. Check by the law of sines.

Illustration.

Solve the triangle ABC if b = 0.3562; c = 0.9378; α = 62°37′10″.

Solution:

Data: b = 0.3562; c = 0.9378; α = 62°37′10″.		
Formulas	**Computation**	
$\frac{1}{2}(\gamma + \beta) = \frac{1}{2}(180° - \alpha)$	$\frac{1}{2}(\gamma + \beta) = \frac{1}{2}(180° - 62°37'10'') = 58°41'25''$	
$\tan \frac{1}{2}(\gamma - \beta) = \dfrac{(c - b)\tan \frac{1}{2}(\gamma + \beta)}{c + b}$ c = 0.9378 b = 0.3562 c − b = 0.5816 c + b = 1.2940	log (c − b) = 9.76462 − 10 log tan $\frac{1}{2}(\gamma + \beta)$ = 0.21593 colog (c + b) = 9.88807 − 10 log tan $\frac{1}{2}(\gamma - \beta)$ = 9.86862 − 10 $\frac{1}{2}(\gamma - \beta)$ = 36°27′46″	γ = 95°9′11″ β = 22°13′39″
$a = \dfrac{c \sin \alpha}{\sin \gamma}$	log c = 9.97211 − 10 log sin α = 9.94840 − 10 colog sin γ = 0.00176 log a = 9.92227 − 10	a = 0.83612
check: log a = 9.92227 − 10 log sin α = 9.94840 − 10 Diff. = 9.97387 − 10	log b = 9.55169 − 10 log sin β = 9.57782 − 10 Diff. = 9.97387 − 10	log c = 9.97211 − 10 log sin γ = 9.99824 − 10 Diff. = 9.97387 − 10

73. Half-Angle Formulas.

We define the following two quantities:

s denotes one half the perimeter of the triangle;
r denotes the radius of the inscribed circle.

Then in any triangle ABC we have

$$s = \tfrac{1}{2}(a + b + c) \quad \text{and} \quad r = \sqrt{\frac{(s - a)(s - b)(s - c)}{s}}.$$

Further

$$\text{F.} \quad \tan\frac{\alpha}{2} = \frac{r}{s - a};\ \tan\frac{\beta}{2} = \frac{r}{s - b};\ \tan\frac{\gamma}{2} = \frac{r}{s - c}.$$

We shall assume the truth of these formulas without further proof. The proof may be found in most standard textbooks.

74. The Solution of Case IV by Half-Angle Formulas.

1. Compute s, (s − a), (s − b), (s − c), and log r.
2. Find α, β, and γ by the half-angle formulas.
3. Check by using $\alpha + \beta + \gamma = 180°$.

Illustration.

Solve the triangle ABC if

$$a = 163.6,\ b = 397.5,\ c = 253.7.$$

Solution:

Data: a = 163.6; b = 397.5; c = 253.7.		
Formulas	**Computation**	
$s = \frac{1}{2}(a + b + c)$.	a = 163.6 b = 397.5 c = 253.7 2s = 814.8 s = 407.4 ← check → s = 407.4	s − a = 243.8 s − b = 9.9 s − c = 153.7
$r^2 = \frac{(s-a)(s-b)(s-c)}{s}$	log (s − a) = 2.38703 log (s − b) = 0.99564 log (s − c) = 2.18667 log N = 5.56934 log s = 2.61002 2log r = 2.95932 log r = 1.47966	
$\tan \frac{1}{2}\alpha = \frac{r}{s-a}$.	log r = 11.47966 − 10 log (s − a) = 2.38703 log tan ½ α = 9.09263 − 10 ½ α = 7°03′21″	α = 14°06′42″
$\tan \frac{1}{2}\beta = \frac{r}{s-b}$.	log r = 1.47966 log (s − b) = 0.99564 log tan ½ β = 0.48402 ½ β = 71°50′11″	β = 143°40′22″
$\tan \frac{1}{2}\gamma = \frac{r}{s-c}$.	log r = 11.47966 − 10 log (s − c) = 2.18667 log tan ½ γ = 9.29299 − 10 ½ γ = 11°06′28″	γ = 22°12′56″
check: $\alpha + \beta + \gamma = 180°$		sum = 180°00′00″

75. The Area of the Oblique Triangle.

We can solve for the area of the oblique triangle in the following cases:

Case I. *Given two angles and a side.*
Case III. *Given two sides and the included angle.*
Case IV. *Given three sides.*

The area in terms of two sides and the included angle is given by

(1) $\mathbf{K = \frac{1}{2}\,bc \sin \alpha;\ K = \frac{1}{2}\,ac \sin \beta;\ K = \frac{1}{2}\,ab \sin \gamma.}$

Proof.

Construct Fig. 48.

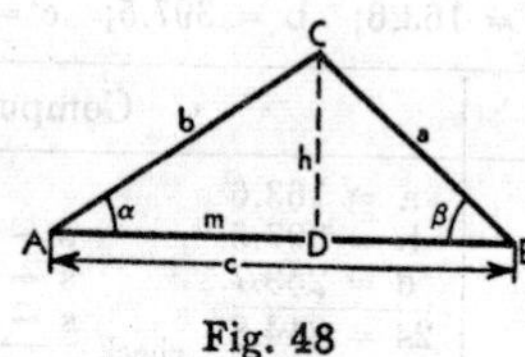

Fig. 48

Let b and c be the given sides; the included angle is α; h is the altitude.

In ACD: $\sin \alpha = \dfrac{h}{b}$; $h = b \sin \alpha$.

From plane geometry: $K = \frac{1}{2}\,hc = \frac{1}{2}\,bc \sin \alpha$.

Similarly for the other formulas

The area in terms of one side and the angles is given by

(2) $\mathbf{K = \dfrac{a^2 \sin \beta \sin \gamma}{2 \sin \alpha};\ K = \dfrac{b^2 \sin \alpha \sin \gamma}{2 \sin \beta};\ K = \dfrac{c^2 \sin \alpha \sin \beta}{2 \sin \gamma}.}$

Proof.

From the law of sines, we have

$$\frac{c}{\sin \gamma} = \frac{b}{\sin \beta} \quad \text{or} \quad c = \frac{b \sin \gamma}{\sin \beta}.$$

Substituting into (1) we get

$$K = \tfrac{1}{2}\,bc \sin \alpha = \tfrac{1}{2}\,b\,\frac{b \sin \gamma}{\sin \beta} \cdot \sin \alpha = \frac{b^2 \sin \alpha \sin \gamma}{2 \sin \beta}.$$

The area in terms of the sides is given by

(3) $\mathbf{K = \sqrt{s(s-a)(s-b)(s-c)}}$,*

where $s = \frac{1}{2}(a + b + c)$.

Note: To find the area in Case II, one must first find the other angles of the triangle, using the law of sines. Then either (1) or (2) may be used to find the area.

Illustration.

Find the area of the triangle ABC if

$$a = 163.6,\ b = 397.5,\ c = 253.7.$$

Solution:

Data: $a = 163.6,\quad b = 397.5,\quad c = 253.7.$

Formula: $K = \sqrt{s(s-a)(s-b)(s-c)}.$

$a = 163.6$		$\log(s-a) = 2.38703$
$b = 397.5$	$s - a = 243.8$	$\log(s-b) = 0.99564$
$c = 253.7$	$s - b = 9.9$	$\log(s-c) = 2.18667$
$2s = 814.8$	$s - c = 153.7$	$\log s = 2.61002$
$s = 407.4$ ← check →	$s = 407.4$	$2 \log K = 8.17936$
		$\log K = 4.08968$ **K = 12294.**

76. Summary.

Cases	Solution
I. Given two angles and a side.	Solve by formula (A) and the law of sines. Find the area by (2).
II. Given two sides and an opposite angle.	Ambiguous case. Solve by the law of sines.
III. Given two sides and the included angle.	Find the angles by law of tangents; then find the third side by the law of sines. If only third side is required, use the law of cosines. Find the area by (1).
IV. Given three sides.	Solve by half-angle formulas. Law of cosines may be used. Find area by (3).

* For the proof of this formula see, for example, Hart, *Plane and Spherical Trigonometry*, p. 140.

77. Exercise 16.

Solve the following triangles and find the area.

1. a = 598.36; $\alpha = 66°39'17''$; $\beta = 69°30'53''$.
2. c = 273.61; a = 392.36; $\gamma = 37°14'26''$.
3. a = 0.38954; c = 0.52927; $\beta = 67°12'16''$.
4. a = 9.782; b = 10.600; c = 13.878.
5. b = 4; $c = \sqrt{3}$; $\alpha = 30°$.

78. Some Applications.

1. **Vectors.** The resultant of two vectors is in general the diagonal of a parallelogram of which the two vectors form adjoining sides. If we have the magnitudes of two vectors and their directions relative to each other, we may use the oblique triangle to compute the resultant and its direction.

Illustration.

Two forces of 50 lbs. and 30 lbs. have an included angle of 60°. Find the magnitude and direction of their resultant.

Solution:

Construct the parallelogram and label it as in Fig. 49.

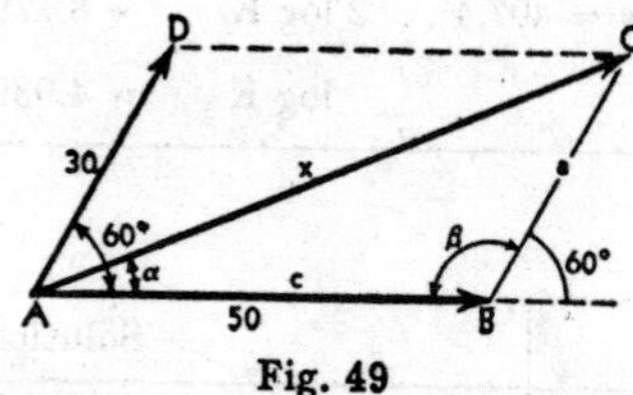

Fig. 49

Since $\overline{AD}$ is parallel to $\overline{BC}$ we have $\angle ABC = \beta = 180° - 60° = 120°$.

By the law of cosines:

$$x^2 = c^2 + a^2 - 2\,ac\cos\beta$$
$$= 2500 + 900 - 2(50)(30)(-\tfrac{1}{2})$$
$$= 2500 + 900 + 1500 = 4900.$$

$$\boxed{x = 70 \text{ lbs.}}$$

$$\cos\alpha = \frac{x^2 + c^2 - a^2}{2bc} = \frac{4900 + 2500 - 900}{2(70)(50)} = \frac{13}{14} = .9286.$$

$$\boxed{\alpha = 21°47'.}$$

2. Surveying. Oblique triangles are used in many problems of surveying; we shall illustrate with some examples.

Illustration.

A surveyor desires to run a straight line due east past an obstruction. See Fig. 50. He measures AB = 780.6 ft., S25°19′E (25°19′ east of south) and then runs BC in the direction N45°47′E. Find the length AC if C is due east of A.

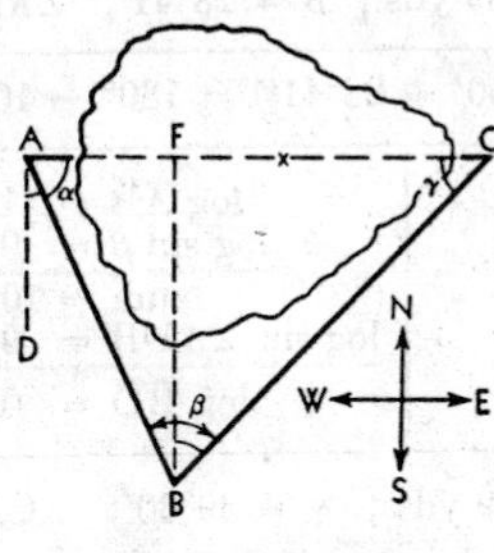

Fig. 50

Solution:

Given: ∠DAB = 25°19′; AB = 780.6 ft.; ∠FBC = 45°47′.

$\alpha = 90° - \angle DAB = 90° - 25°19' = 64°41'.$

$\gamma = 90° - \angle FBC = 90° - 45°47' = 44°13'.$

$\beta = 180° - (\alpha + \gamma) = 180° - 108°54' = 71°06'.$

$x = \dfrac{\overline{AB}\sin\beta}{\sin\gamma}$	$\log \overline{AB} = 2.89243$ $\log \sin\beta = 9.97593 - 10$ sum $= 12.86836 - 10$ $\log \sin\gamma = 9.84347 - 10$ $\log x = 3.02489$ **x = 1059. ft.**

Illustration.

To find the distance between inaccessible points C and D on one side of a river, a surveyor on the other side takes the following readings (see Fig. 51): $\overline{AB}$ = 15.69 yards; $\alpha = 32°19'$; $\beta = 28°41'$; $\gamma = 39°30'$; $\delta = 36°20'$. Find $\overline{CD}$.

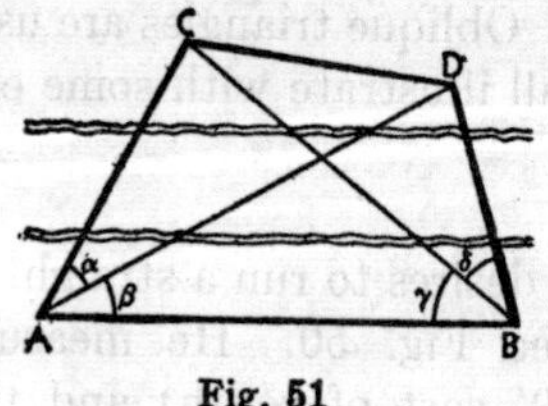

Fig. 51

Solution:

In △ABD: $\overline{AB} = 15.69$ yds.; $\beta = 28°41'$; $\angle ABD = \delta + \gamma = 75°50'$.	
$\angle ADB = 180° - (75°50' + 28°41') = 180° - 104°31' = 75°29'$.	
$\overline{BD} = \dfrac{\overline{AB}\sin\beta}{\sin\angle ADB}$	$\log\overline{AB} = 1.19562$ $\log\sin\beta = 9.68121 - 10$ sum $= 10.87683 - 10$ $\log\sin\angle ADB = 9.98591 - 10$ $\log\overline{BD} = 0.89092$ $\qquad\overline{BD} = 7.779$
In △ABC: $\overline{AB} = 15.69$ yds.; $\gamma = 39°30'$; $\angle CAB = \alpha + \beta = 61°00'$.	
$\angle ACB = 180° - 100°30' = 79°30'$.	
$\overline{BC} = \dfrac{\overline{AB}\sin\angle CAB}{\sin\angle ACB}$	$\log\overline{AB} = 1.19562$ $\log\sin\angle CAB = 9.94182 - 10$ sum $= 11.13744 - 10$ $\log\sin\angle ACB = 9.99267 - 10$ $\log\overline{BC} = 1.14477$ $\qquad\overline{BC} = 13.956$
In △BCD: $\overline{BC} = 13.95$ yds.; $\overline{BD} = 7.779$ yds.; $\delta = 36°20'$.	
$\overline{CD}^2 = \overline{BC}^2 + \overline{BD}^2$ $- 2(BD)(BC)\cos\delta$ $= \overline{BC}^2 + \overline{BD}^2 - R$	$2\log\overline{BC} = 2.28954$ $\qquad\overline{BC}^2 = 194.78$ $2\log\overline{BD} = 1.78184$ $\qquad\overline{BD}^2 = 60.51$ $\qquad$ sum $= 255.29$ $\log 2 = 0.30103$ $\log BD = 0.89092$ $\log BC = 1.14477$ $\log\cos\delta = 9.90611 - 10$ $\log R = 2.24283$ $\qquad R = 174.92$ $\qquad\overline{CD}^2 = 80.37$ $CD = 8.9648$ yds.

CHAPTER X

INVERSE TRIGONOMETRIC FUNCTIONS

79. Definitions.

The expression **arcsin x** *means an* **angle whose sine is x.** We express this by the equivalent equations

$$\mathbf{x = \sin y} \quad \text{and} \quad \mathbf{y = \arcsin x.}$$

The first of these equations expresses x as a function of y; the second equation expresses y as a function of x. Thus we say that sin y and arcsin x are **inverse functions.**

We stress that *arcsin x* stands for an *angle*.

Note: The angle arcsin x is also denoted by $\sin^{-1}x$. In this notation we caution the student against using -1 as an exponent.

In arcsin x, the variable x necessarily falls between -1 and $+1$; i.e., $-1 \leqq x \leqq 1$. However, to each of these values of x there corresponds infinitely many values of arcsin x.

Illustrations.

1. Find arcsin $\frac{1}{2}$.

   Solution: Let y = arcsin $\frac{1}{2}$; then sin y = $\frac{1}{2}$. Since sin 30° = $\frac{1}{2}$, we have y = 30°, 150°. Because the sine has a period of 360°, any angle which we can obtain by adding any integral multiple of 360° to 30° or 150° would also satisfy this equation. We thus get

$$\arcsin \tfrac{1}{2} = \begin{cases} 2n\pi + \frac{1}{6}\pi \\ 2n\pi + \frac{5}{6}\pi \end{cases} \text{or} \begin{cases} n(360°) + 30° \\ n(360°) + 150° \end{cases},$$

   where n may be any integer, positive, negative, or zero.

2. Find arctan $(-\sqrt{3})$.

Solution: $y = \arctan(-\sqrt{3})$; $\tan y = -\sqrt{3}$; $y = 120°$, $300°$. Since the tangent has period π, we have

$$\arctan(-\sqrt{3}) = n\pi + \tfrac{2}{3}\pi.$$

3. Find $\sin \arccos \frac{4}{5}$, if $\arccos \frac{4}{5}$ is in quadrant I.

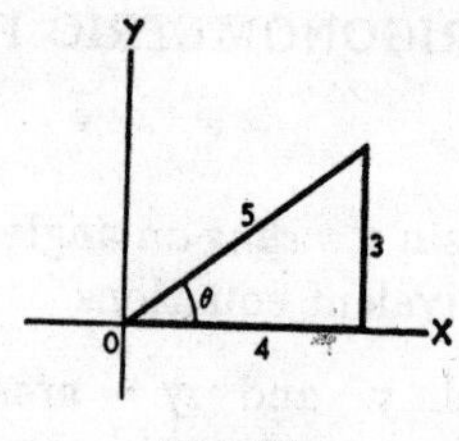

Fig. 52

Solution: Let $\theta = \arccos \frac{4}{5}$; then $\cos \theta = \frac{4}{5}$. We can then construct the triangle of Fig. 52. From the triangle we get $\sin \theta = \frac{3}{5}$; therefore $\sin \arccos \frac{4}{5} = \frac{3}{5}$.

80. Exercise 17.

Find the values for each expression.

1. $\arcsin 1$.
2. $\operatorname{arccot} \sqrt{3}$.
3. $\arccos(-\frac{1}{2})$.
4. $\operatorname{arcsec} 2$.
5. $\arcsin(\frac{1}{2}\sqrt{2})$.
6. $\operatorname{arcsec} \frac{1}{2}$.
7. $\arcsin .7009$.
8. $\arccos .3090$.

Assume the angle involved is between 0° and 180° and find the function.

9. $\sin \arcsin \frac{2}{3}$.
10. $\cos \arcsin \frac{4}{5}$.
11. $\cot \arctan(-\frac{1}{3})$.
12. $\sec \arcsin x$.
13. $\csc \arctan y$.
14. $\tan \arccos z$.

81. Principal Values.

Definition.

The principal value of $\arcsin x$, $\operatorname{arccsc} x$, $\arctan x$, or $\operatorname{arccot} x$ is that value which is **smallest numerically.**

Thus if $x \geqq 0$, the principal value is that single value of the angle lying between 0 and $\frac{\pi}{2}$, inclusive; if $x < 0$, the principal value is negative and not less than $-\frac{\pi}{2}$.

The principal value of arccos x or arcsec x is the **smallest positive** value of the angle. This value lies between 0 and $\frac{\pi}{2}$, inclusive, if $x \geqq 0$; between $\frac{\pi}{2}$ and π, if $x < 0$.

Note: The principal value of arccos x and arcsec x cannot be defined as are the principal values of the other inverse functions because, for example, arccos $\frac{1}{2}$ is 60° or −60°, so that there are two *numerically* smallest values of arc cos $\frac{1}{2}$.

Most textbooks make the distinction between any value of the inverse function and the principal value by beginning the name of the inverse function with a *small* letter to indicate *any* value and with a *capital letter to indicate the principal value.* We may summarize in the following table

$$-\frac{\pi}{2} \leqq \text{Arcsin } x \leqq \frac{\pi}{2}$$

$$0 \leqq \text{Arccos } x \leqq \pi$$

$$-\frac{\pi}{2} \leqq \text{Arctan } x \leqq \frac{\pi}{2}$$

$$-\frac{\pi}{2} \leqq \text{Arccot } x \leqq \frac{\pi}{2}$$

$$0 \leqq \text{Arcsec } x \leqq \pi$$

$$-\frac{\pi}{2} \leqq \text{Arccsc } x \leqq \frac{\pi}{2}.$$

Illustrations.

1. Arccot $\sqrt{3}$ = 30°.
2. Arccot $(-\sqrt{3})$ = −30°.
3. Arcsin .0378 = 2°10′.
4. Arcsec (−1.675) = 126°40′.
5. Arctan 1 = 45°.
6. $\text{Tan}^{-1}(-1) = -45°$.
7. $\text{Cos}^{-1}(-\sqrt{\tfrac{3}{2}}) = 150°$.

8. Find cos (Arccos x − Arccos 3y).

Solution: Let Arccos x = α and Arccos 3y = β. We can then construct the triangles of Figs. 53 and 54.

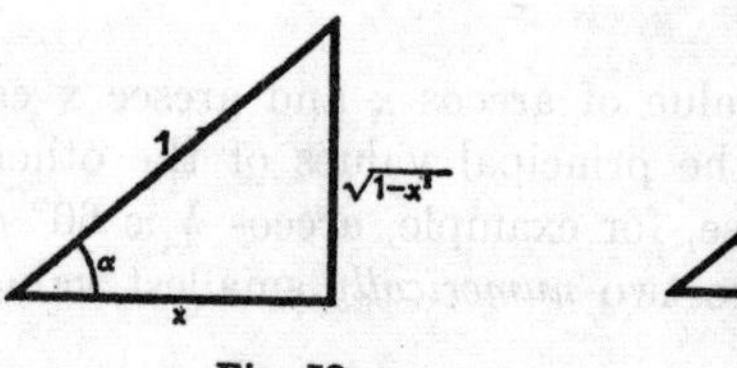

Fig. 53

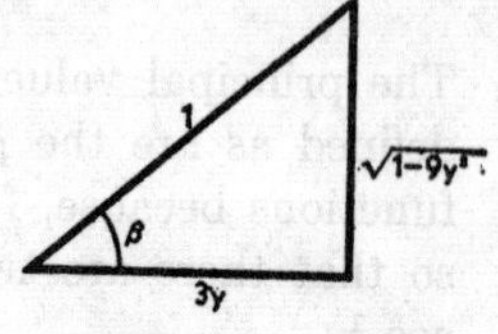

Fig. 54

$$\begin{aligned}\cos(\alpha-\beta) &= \cos\alpha\cos\beta+\sin\alpha\sin\beta\\ &= (x)(3y)+(\sqrt{1-x^2})(\sqrt{1-9y^2})\\ &= 3xy+\sqrt{(1-x^2)(1-9y^2)}.\end{aligned}$$

Note: sin α and sin β are both positive since α and β must be between 0 and π; we thus take only the positive radical.

9. Find sin ($\frac{1}{2}$ Arcos $\frac{5}{13}$).

Solution: Let Arccos $\frac{5}{13} = \alpha$; then we have

$$\sin\left(\tfrac{1}{2}\alpha\right)=\sqrt{\frac{1-\cos\alpha}{2}}=\sqrt{\frac{1-\frac{5}{13}}{2}}=\sqrt{\frac{8}{26}}=\frac{2}{13}\sqrt{13}.$$

82. Exercise 18.

Express each principal value in degrees and radians.

1. Arcsin ($-\frac{1}{2}$).
2. Arccos ($\frac{1}{2}\sqrt{2}$).
3. Arcsec 2.
4. Arctan .3281.

Find the value of each quantity.

5. sin (2 Arccos $\frac{5}{13}$).
6. tan ($\frac{1}{2}$ Arcsin $\frac{3}{5}$).
7. sin (Arccos y + Arcsin x).
8. cos (Arctan x + Arccot y).

Part Two

SPHERICAL TRIGONOMETRY

Chapter XI

SPHERICAL GEOMETRY

83. Introduction.

In the following chapters we shall discuss schematic methods for solving all the essential problems of spherical trigonometry. These methods may be grasped by the reader who has not previously had a course in solid geometry. However, in order to give a clear understanding of the rigid proofs which underlie these methods, and better to acquaint the student with the various fields of application, this chapter presents the basic theorems of spherical geometry. For a detailed discussion of these theorems the reader should consult a textbook on solid geometry.

84. Theorems of Solid Geometry.

Because the reader lives continually in a three dimensional world, many of the properties of space and the relationships of lines and planes in space are intuitively familiar to him and need no further proof to be acceptable. For the systematic mathematician, however, their rigid proof is necessary.

Where the trigonometry studied thus far dealt with the properties of triangles within a plane, no comment was made as to just what was understood by the word plane. In passing over to the solid geometry it becomes necessary to be more specific, since we shall be dealing with more than one plane. The definition most often encountered in elementary textbooks is:

Definition. *A* **plane** *is a surface such that a straight line joining any two points of the surface lies entirely in the surface.*

Consequently, any line which has two points in common with a plane lies in the plane. If it has only one point in common with the plane, it is said to intersect the plane.

Any number of planes can be passed through a single line.

If, however, we specify that a plane which passes through the given line must also pass through a specific point not on the line, we see immediately that there can be only one such plane. We say therefore that the plane is completely determined by a line and a point not on the line. More completely, *a plane is determined by*

a) *a line and a point not on the line,*
b) *two intersecting lines,*
c) *two parallel lines,*
d) *three points, not in a straight line.*

Planes have many of the relationships with respect to each other that lines have. For example, planes intersect each other, and one of the very basic theorems of solid geometry is:

Theorem 1. *If two planes intersect, their intersection is a straight line.*

Two intersecting planes form four *space angles* even as two intersecting straight lines in a plane form four plane angles; and in both instances the two angles opposite each other are equal.

If a line intersects a plane, it may be perpendicular to the plane; that is the case if it is perpendicular to every line in the plane which passes through its point of intersection with the plane. If, however, we wish to construct a line which is perpendicular to a given plane, or conversely, a plane which is perpendicular to a given line, it is only necessary to insure that the line is perpendicular to *two* lines in the plane which pass through the point of intersection of the given line and plane. This is expressed in the following theorem.

Theorem 2. *A straight line perpendicular to each of two straight lines in a plane at their intersection is perpendicular to the plane.*

Proof. Let the line AA′ be perpendicular to OB and OC (Fig. 55). Let OD represent any other line that lies in the plane of O, B and C and which passes through the intersection O. Suppose that D is the point of intersection of this line with the line BC. The points A and A′ are determined so that OA = OA′. Then $\triangle ABC \cong \triangle A'BC$ (three sides equal). Consequently, $\triangle ABD \cong \triangle A'BD$ (two sides and the included angle equal; AB

= A′B, BD = BD, ∠ABD = ∠A′BD). Therefore, AD = A′D and △AOD ≅ △A′OD (three sides equal). Then ∠AOD = ∠A′OD. But, since ∠AOD + ∠A′OD = 180°, ∠AOD = 90°.

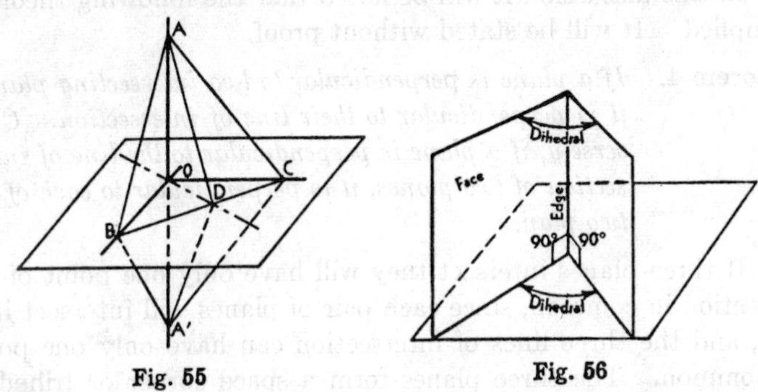

Fig. 55 Fig. 56

The space angles which are formed when two planes intersect are called **dihedral angles.** The line along which the planes intersect shall be called the **edge** of the dihedral angle and each of the planes a **face.** To determine the magnitude of a dihedral angle we first pass a plane perpendicular to the edge; the intersections of this plane with the faces of the dihedral angle will be straight lines (by Theorem 1) and thus will form a plane angle in this plane of intersection. See Fig. 56. We now formulate the following definition:

The **magnitude** *of a dihedral angle is the magnitude of the plane angle formed by the intersections of the faces of the dihedral angle with a plane perpendicular to the edge of the dihedral angle.*

If we intersect the edge of the dihedral angle at two distinct points by planes which are perpendicular to the edge, then these planes are parallel to each other. The lines in which these parallel planes intersect a face of the dihedral angle will be parallel straight lines. Whatever the point at which these planes are passed through the edge, the angles which are formed by their lines of intersection with the faces of the dihedral angle will be equal. This fact is due to

Theorem 3. *If two angles which are not in the same plane have their sides parallel each to each and extend in the same direction from their vertices, the angles are equal.*

We see, therefore, that a dihedral angle may be measured by passing a plane through any point of the edge and perpendicular to the edge.

In this discussion it will be noted that the following theorem is implied. It will be stated without proof.

Theorem 4. *If a plane is perpendicular to two intersecting planes, it is perpendicular to their line of intersection. Conversely, if a plane is perpendicular to the line of intersection of two planes, it is perpendicular to each of the two planes.*

If three planes intersect they will have only one point of intersection in common, since each pair of planes will intersect in a line, and the three lines of intersection can have only one point in common. The three planes form a space corner or **trihedral angle,** and the point of intersection is called the **vertex.** A trihedral angle has three **face angles** formed by the possible pairs of lines of intersection of the planes. The faces of a trihedral angle form dihedral angles. See Fig. 57. The inside corner of an ordinary box forms a trihedral angle each of whose dihedral angles is equal to 90°, and each of whose face angles is 90°.

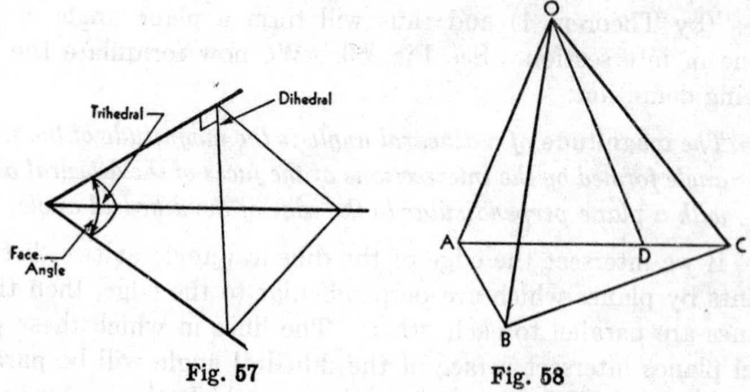

Fig. 57 Fig. 58

We shall now state a basic theorem regarding trihedrals which we shall subsequently find important.

Theorem 5. *The sum of two face angles of a trihedral angle is greater than the third face angle.*

Proof. In Fig. 58 we have the trihedral angle O–ABC with vertex

at O. Construct OD in the plane of AOC so that $\angle AOD = \angle AOB$, and $OD = OB$. Now pass a plane through A, B, and D; we can assume that C lies in this plane. We find that
$AD = AB$ since $\triangle AOD \cong \triangle AOB$.
From plane geometry $AB + BC > AC$. (1)
However, $AC = AD + DC = AB + DC$. (2)
From (1) and (2) we get $BC > DC$;
then $\angle BOC > \angle DOC$ and consequently,
$\angle AOB + \angle BOC > \angle AOB + \angle DOC$,
or $\angle AOB + \angle BOC > \angle AOD + \angle DOC = \angle AOC$.

85. The Sphere.

Definition. *A* **spherical surface** *is a curved surface every point of which is at an equal distance from one single point inside the surface.*

This point inside the surface is the **center,** and a line joining the center with any point on the surface is a **radius.** In the future, when we speak of a sphere, we shall mean the spherical surface, since spherical trigonometry deals only with the properties of lines and angles on the surface.

A very characteristic property of the sphere is stated in the following theorem.

Theorem 6. *If a plane intersects a sphere, the intersection is always a circle.*

Proof. In Fig. 59 let us suppose that π represents an arbitrary plane intersecting the sphere with center at O. Let M and N be arbitrary points on the intersection of the plane and the sphere, and let P be the foot of the perpendicular to π from O. The two right triangles OPM and OPN are congruent since OM and ON are radii. Consequently, $PM = PN$. This result will hold for all points on the intersection. It is therefore a circle with center at P.

If the plane which intersects the sphere passes through the center of the sphere, the intersection is called a **great circle**; if not, it is a **small circle.** For example, the equator and the meridians on the earth are great circles; the parallels of latitude are small circles.

Through any two points on a sphere, not extremities of a

diameter, one and only one circle of the sphere can be drawn, because those two points together with the center are sufficient to determine a single plane of intersection. (See §84.)

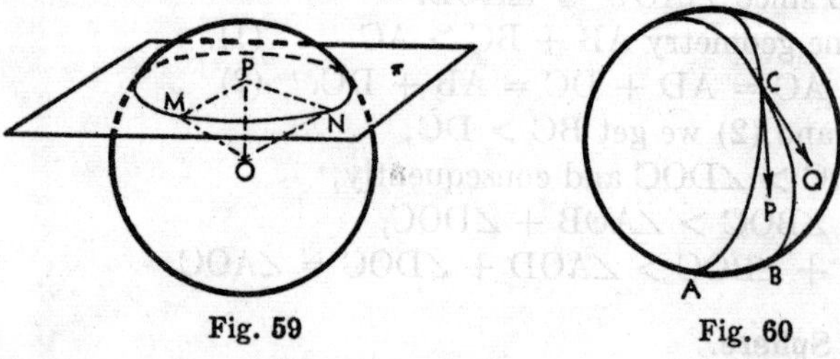

Fig. 59

Fig. 60

The arc of a great circle is measured as in plane geometry in degrees, minutes, and seconds by the angle subtended at its center.

The **poles** of a great circle are the two ends of the diameter of the sphere which is perpendicular to the plane of the great circle. The **polar distance** is the great circle arc between the given great circle and its pole and is therefore 90°. The poles of small circles are similarly defined; the polar distance here, however, is the shorter of the two great circle arcs between the small circle and its poles. If a straight line has only one point in common with a sphere, it is defined as a **tangent** to the sphere. The tangent to a sphere will also be a tangent to the great circle which is the intersection of the sphere with a plane containing this tangent line and the center of the sphere.

When two circles on a sphere intersect, the **angle** between them is defined to be the **angle between their tangents.** In Fig. 60 the angle between the arcs CA and CB is defined to be the plane angle between the tangents CP and CQ.

In plane geometry we learned that the tangent to a circle is perpendicular to the radius which connects the center of the circle with the point of contact of the tangent. We have an analogous theorem in spherical geometry.

Theorem 7. *If a line is perpendicular to a radius of a sphere at its its outer extremity, it is a tangent to the sphere.*

Proof. Consider the sphere with center O (Fig. 61) and the line l which is perpendicular to the radius OA at the extremity A. Let B be any other point on the line l. From plane geometry we know that OB as the hypotenuse of a right triangle must be greater than OA and therefore greater than the radius of the sphere. Consequently, no other point on l can be at a distance OA from the center; that is, no other point of l can lie on the sphere.

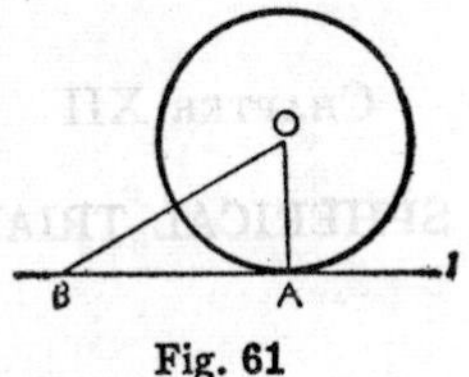

Fig. 61

At any point on a sphere there are innumerable tangent lines in many directions. All these lines lie in one plane which is defined to be the **tangent plane** to the sphere at the point.

We have already seen that the angle between two circles on a sphere is measured by the plane angle between the tangents to the circles at their intersection. If the circles are great circles, the angle between them is referred to as a **spherical angle** and their intersection as the **vertex**. An important relation between spherical angle and spherical arc is given by the following theorem.

Theorem 8. *A spherical angle has the same measure as the arc of the great circle drawn so that its pole is the vertex of the angle and included between the sides of the angle.*

Proof. Consider Fig. 62.

The line OA is perpendicular to OB by the definition of a pole.

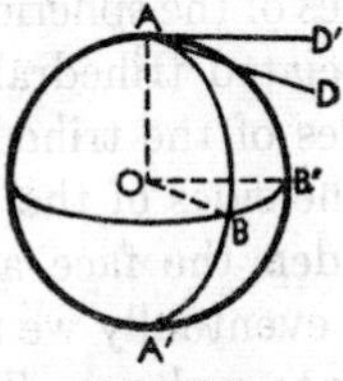

Fig. 62

OA is perpendicular to AD by Theorem 7. Therefore, AD is parallel to OB. Similarly, AD′ is parallel to OB′. Then arc BB′ = ∠BOB′ = ∠DAD′. This last equality is based on Theorem 3.

The part of the spherical surface bounded by the semicircumferences ABA′ and AB′A′ is called a **lune.** It has two equal spherical angles, one at A, the other at A′.

CHAPTER XII

THE SPHERICAL TRIANGLE

86. Introduction.

In the preceding chapter we discussed the trihedral angle. If we consider a trihedral angle which has its vertex at the center of a sphere, then its faces will intersect the sphere in arcs which we know to be great circle arcs, since they lie in planes which pass through the center of the sphere and since all such planes intersect the sphere in great circles. *The three arcs thus formed are the sides of a* **spherical triangle.** The sides of a spherical triangle then subtend the face angles of a trihedral angle (called its **associated** trihedral angle) which are central angles in great circles. It follows from plane geometry that the sides and the face angles are equal when they are measured in circular measure; that is, in degrees, minutes, and seconds, or in radians. We shall consequently *measure both the sides and the angles of a spherical triangle in circular measure.* Spherical triangles may be right or oblique, equilateral, isosceles, or scalene, even as plane triangles are.

Many of the properties of the spherical triangle may be proved or deduced from its associated trihedral. For example, suppose we increase the face angles of the trihedral associated with some arbitrary triangle; then the sides of the triangle will increase correspondingly. As we widen the face angles we flatten out the trihedral more and more; eventually we must reach a stage where it has been flattened out into a plane. The sum of the face angles of the trihedral, and consequently of the sides of the spherical triangle, which now has degenerated into a circle, will then be 360°. In order to eliminate such degenerate triangles from our discussion we shall restrict ourselves by the following agreement.

Proposition 1. *The sum of the sides of a spherical triangle is less than 360°; that is,* $a + b + c < 360°$.

Again consider an arbitrary spherical triangle and its associated trihedral angle. The triangle may be such that we can, without disobeying Proposition 1, increase one side until it is greater than 180°. In this case the resulting figure will look very much like the spherical triangle AB′BC in Fig. 63 and will again be a degenerate sort of spherical triangle, all the properties of which can be found by discussing the spherical triangle ABC in which the side AB is the shorter arc of the great circle AB′B.

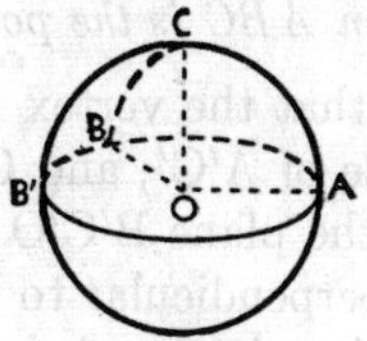

Fig. 63

We shall therefore use the following proposition.

Proposition 2. *Any side of a spherical triangle is less than 180°.*

In Theorem 5 of the preceding chapter we proved that the sum of two face angles of trihedral angle is greater than the third. Since the side of a spherical triangle is equal to the corresponding face angle of its associated trihedral, we find the following theorem obvious.

Theorem 9. *Any side of a spherical triangle is less than the sum of the other two sides.*

If we join the vertices of a spherical triangle by straight lines, a plane triangle is formed, in which we know that the order of magnitude of the angles is the same as that of the respective opposite sides. Thus if $a < b < c$, then $A < B < C$. In passing from the plane triangle ABC to the spherical triangle ABC, this property is preserved.

Theorem 10. *In any spherical triangle the order of magnitude of the angles is the same as that of the respective opposite sides.*

87. Polar Triangles.

Definition. If arcs of great circles are drawn with the vertices of a spherical triangle as poles, these arcs form a second triangle called the **polar triangle** of the first.

As a matter of convention we shall designate the angles of the polar triangle of a spherical triangle ABC by A′, B′, C′ and the sides by a′, b′, c′. (Fig. 64.)

Theorem 11. *If A′B′C′ is the polar triangle of a spherical triangle ABC, then ABC is the polar triangle of A′B′C′.*

Proof. Let us suppose that the vertex A (Fig. 64) is the pole of the arc B′C′; B, the pole of A′C′; and C the pole of A′B′. Then AO is perpendicular to the plane B′C′O and therefore to the line C′O. Likewise, BO is perpendicular to the line C′O. Since C′O is perpendicular to two lines in the plane ABO, it is perpendicular to the entire plane. Then C′ is the pole of arc AB. Similarly, B′ and A′ are poles of arc AC and arc BC, respectively. Therefore ABC is the polar triangle of A′B′C′.

We shall say that the spherical triangles ABC and A′B′C′ are **mutually** polar triangles.

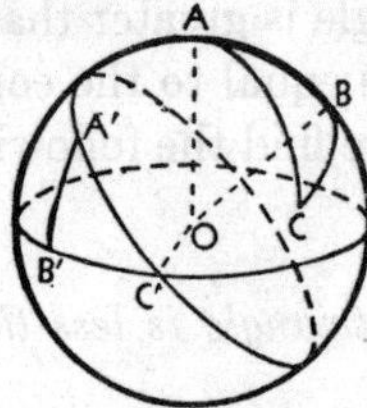

Fig. 64

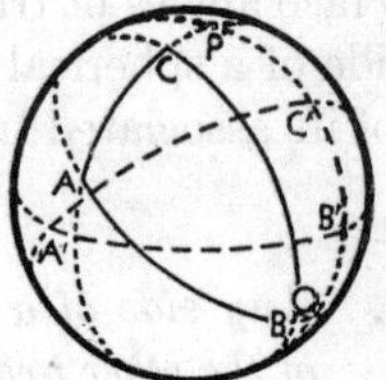

Fig. 65

Theorem 12. *In two mutually polar triangles, an angle of one is the supplement of the side opposite the corresponding angle of the other.*

Proof. In Fig. 65, ABC is a spherical triangle and A′B′C′ its polar triangle. Since A is the pole of arc PC′B′Q, it follows that arc PQ equals angle A. By Theorem 11, B′ will be the pole of arc ACP; therefore arc B′P = 90°, since the plane of any great circle through B′ is perpendicular to the plane of ACP. Likewise,

C′ being the pole of arc ABQ, $\widehat{C'Q} = 90°$. Combining these two results we have

$$\widehat{PQ} = \widehat{PB'} + \widehat{B'Q} = 90° + (90° - \widehat{C'B'}) = 180° - a'.$$

Since $\widehat{PQ} = \angle A$, we find that $A = 180° - a'$.
Similarly, we may prove that
$B = 180° - b'$ and $C = 180° - c'$.

This proves the theorem.

As a result of Theorem 11, we can apply the preceding proof to A′B′C′ to show that $A' = 180° - a$, $B' = 180° - b$, $C' = 180° - c$. The reader should note that the above equations may be solved to express the sides of a spherical triangle as supplements of the opposite angles of the polar triangle; i.e., $a = 180° - A'$, etc. We shall make frequent use of these relationships and thus summarize them in formula form.

$$\mathbf{A = 180° - a',\ A' = 180° - a;\ a = 180° - A',\ a' = 180° - A;}$$
$$\mathbf{B = 180° - b',\ B' = 180° - b;\ b = 180° - B',\ b' = 180° - B;}$$
$$\mathbf{C = 180° - c',\ C' = 180° - c;\ c = 180° - C',\ c' = 180° - C.}$$

Theorem 13. *Any angle in a spherical triangle is less than 180°.*

Proof. Proposition 2 of §86 states that the side of a spherical triangle is less than 180°. Applying this to the polar triangle A′B′C′ of a triangle ABC, we see that $a' < 180°$. By Theorem 12, we know that $A = 180° - a'$. Since $a' < 180°$, $A < 180°$. A similar argument for the other angles completely proves the theorem.

Theorem 14. *The sum of the angles of a spherical triangle is greater than 180° and less than 540°.*

Proof. Let triangle A′B′C′ be the polar of triangle ABC. By Theorem 12, $A + a' = 180°$, $B + b' = 180°$, $C + c' = 180°$. Adding, we get

$$A + B + C + (a' + b' + c') = 540° \quad \text{or}$$

(1) $A + B + C = 540° - (a' + b' + c')$.

Therefore $A + B + C < 540°$. This proves the latter half of the theorem. To prove the remainder of the statement we use Proposition 1:

$$a' + b' + c' < 360°.$$

Combining this statement with (1) we see that

$$A + B + C > 540° - 360° = 180°.$$

Note: This theorem may be rather surprising to the reader since it is a radical departure from the familiar law, $A + B + C = 180°$, of plane triangles.

88. The Sine Law.

In a spherical triangle the sines of the angles are proportional to the sines of their opposite sides;

$$\frac{\sin A}{\sin a} = \frac{\sin B}{\sin b} = \frac{\sin C}{\sin c}.$$

Proof. In Fig. 66 ABC is a spherical triangle with its associated trihedral. Planes are passed through C perpendicular to OB and OA. The angles so formed are $\angle CPR = \angle B$ and $\angle CQR = \angle A$. By Theorem 4, the line CR is perpendicular to the plane OAB.

In $\triangle OCP$: $\sin a = \frac{CP}{CO}$.

In $\triangle PCR$: $\sin B = \frac{CR}{CP}$.

Therefore

(1) $$\sin a \sin B = \frac{CP}{CO}\frac{CR}{CP} = \frac{CR}{CO}.$$

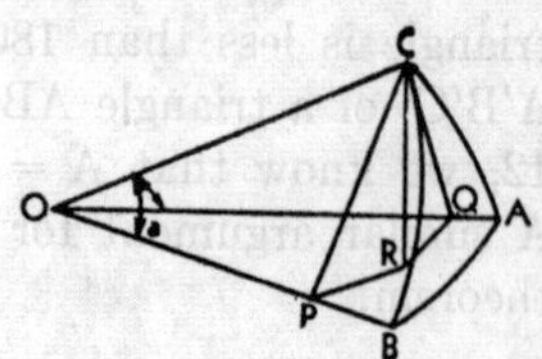

Fig. 66

In $\triangle OCQ$: $\sin b = \frac{CQ}{CO}$.

In $\triangle QCR$: $\sin A = \frac{CR}{CQ}$.

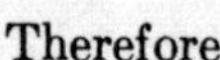
Therefore

(2) $$\sin b \sin A = \frac{CQ}{CO}\frac{CR}{CQ} = \frac{CR}{CO}.$$

Equating (1) and (2), since their right hand sides are equal, we find that

$$\sin a \sin B = \sin b \sin A$$

or

$$\frac{\sin B}{\sin b} = \frac{\sin A}{\sin a}.$$

By constructing planes through A perpendicular to OC and OB, the reader can prove similarly that

$$\frac{\sin B}{\sin b} = \frac{\sin C}{\sin c}.$$

Combining this with the preceding equation we obtain the complete law of sines.

89. The Cosine Law for Sides.

The cosine of a side of a spherical triangle is equal to the product of the cosines of the other two sides plus the product of the sines of those two sides multiplied by the cosine of their included angle;

$$\cos a = \cos b \cdot \cos c + \sin b \cdot \sin c \cdot \cos A,$$

$$\cos b = \cos a \cdot \cos c + \sin a \cdot \sin c \cdot \cos B,$$

$$\cos c = \cos a \cdot \cos b + \sin a \cdot \sin b \cdot \cos C.$$

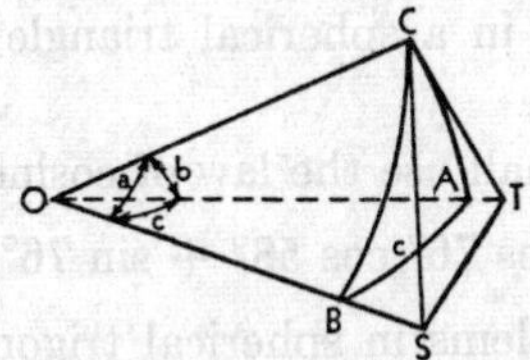

Fig. 67

Proof. In Fig. 67 O–ABC is a spherical triangle with its associated trihedral angle. A tangent plane to the sphere is constructed at C so that triangles SOC and TOC are right triangles and $\angle SCT = \angle C$. Applying the cosine law for plane triangles to triangle SOT, we find that

$$\overline{ST}^2 = \overline{OS}^2 + \overline{OT}^2 - 2\,\overline{OS} \cdot \overline{OT} \cdot \cos c,$$

and to triangle SCT,

$$\overline{ST}^2 = \overline{SC}^2 + \overline{TC}^2 - 2\,\overline{SC} \cdot \overline{TC} \cdot \cos C$$

Subtracting the second equation from the first, we get

$$0 = (\overline{OS}^2 - \overline{SC}^2) + (\overline{OT}^2 - \overline{TC}^2) - 2\,\overline{OS}\cdot\overline{OT}\cdot\cos c + 2\,\overline{SC}\cdot\overline{TC}\cdot\cos C,$$

and, using the Pythagorean theorem, we find that

$$0 = 2\,\overline{OC}^2 - 2\,\overline{OS}\cdot\overline{OT}\cdot\cos c + 2\,\overline{SC}\cdot\overline{TC}\cdot\cos C.$$

Division by $2\,\overline{OS}\cdot\overline{OT}$ gives

$$0 = \frac{OC}{OS}\cdot\frac{OC}{OT} - \cos\ c + \frac{SC}{OS}\cdot\frac{TC}{OT}\cos C,$$

which can be written

$$0 = \cos a\cdot\cos b - \cos c + \sin a\cdot\sin b\cdot\cos C,$$

or

$$\cos c = \cos a\cdot\cos b + \sin a\cdot\sin b\cdot\cos C.$$

This is one form of the cosine law.

The reader can, by constructing tangent planes to the sphere at A and B, derive in a similar manner to that above the other formulas of the law.

Illustration.

Find the side b in a spherical triangle if $a = 76°$, $c = 58°$, and $B = 117°$.

Solution. We shall use the law of cosines in its second form:

$$\cos b = \cos 76° \cos 58° + \sin 76° \sin 58° \cos 117°.$$

The solution of problems in spherical trigonometry by means of the cosine law becomes rather laborious, for we have no means of logarithmically handling a sum of two numbers. It will be necessary, therefore, to compute each of the two products in the right-hand member separately by logarithms, to add the result, and to determine b by a table of natural trigonometric functions. The complete solution of the problem will then have the following form.

log cos 76°	= 9.38368 − 10	log sin 76°	= 9.98690 − 10
log cos 58°	= 9.72421 − 10	log sin 58°	= 9.92842 − 10
log product	= 9.10789 − 10	log cos 117°	= 9.65705 − 10 (n)
1st product	= .12820	log product	= 9.57237 − 10 (n)
2nd product	= −.37357 ⟵	2nd product	= −.37357
cos c	= −.24537		

b = 104°12′13″ (from a table of natural cosines).*

90. The Cosine Law for Angles.

In any spherical triangle.

$$\cos A = -\cos B \cos C + \sin B \sin C \cos a,$$
$$\cos B = -\cos A \cos C + \sin A \sin C \cos b,$$
$$\cos C = -\cos A \cos B + \sin A \sin B \cos c.$$

The cosine law of §89 may be applied to the polar triangle A′B′C′ of a triangle ABC, giving (in the third form)

$$\cos c' = \cos a' \cos b' + \sin a' \sin b' \cos C'.$$

By Theorem 12 we may replace the sides and angle of this equation by the supplements of the corresponding angles and sides of the triangle ABC, thereby obtaining

$$\cos (180° - C) = \cos (180° - A) \cos (180° - B) + \sin (180° - A) \sin (180° - B) \cos (180° - c)$$

or

$$-\cos C = \cos A \cos B - \sin A \sin B \cos c.$$

Multiplication of both sides of the equation by -1 gives us the third form of the law of cosines for angles

$$\cos C = -\cos A \cos B + \sin A \sin B \cos c.$$

Similarly the reader may derive the other forms of the law.

The above derivation illustrates a method we shall encounter frequently in the study of spherical trigonometry, that of solving problems or deriving equations regarding a triangle by first applying known formulas to the polar triangle of the given triangle. This correspondence, where every relation between the angles of one triangle implies a corresponding relation between the sides of the second triangle and every relation between the sides of the first implies a corresponding relation between the angles of the second, illustrates a principle which is encountered in many fields of mathematics. This principle is known as the **principle of duality.**

* The symbol (n) signifies that the trigonometric function of which the logarithm appears here is negative. Since logarithms deal only with the numerical value of numbers, this symbol is used to indicate the sign which the resulting antilogarithm should be.

91. The Haversine Law.

The definition of the haversine is

$$\text{hav } \theta = \frac{1 - \cos \theta}{2}.$$

This definition holds for any angle θ. Let us consider the angle C of a spherical triangle and solve for cos C; we get

$$\cos C = 1 - 2 \text{ hav } C.$$

Substituting this in the cosine law for sides, we obtain

$$\begin{aligned} \cos c &= \cos a \cos b + \sin a \sin b\ (1 - 2 \text{ hav } C) \\ &= (\cos a \cos b + \sin a \sin b) - 2 \sin a \cdot \sin b \cdot \text{hav } C \\ &= \cos (a - b) - 2 \sin a \cdot \sin b \cdot \text{hav } C. \end{aligned}$$

Consider the side c in the definition of the haversine. If we solve for cos c, we get

$$\cos c = 1 - 2 \text{ hav } c.$$

Similarly for (a − b),

$$\cos (a - b) = 1 - 2 \text{ hav } (a - b).$$

When we substitute these into the above equation, we get

$$1 - 2 \text{ hav } c = 1 - 2 \text{ hav } (a - b) - 2 \sin a \cdot \sin b \cdot \text{hav } C,$$

or, after simplifying,

$$\text{hav } c = \text{hav } (a - b) + \sin a \cdot \sin b \cdot \text{hav } C.$$

This equation is known as the **haversine law.** We shall give a complete statement of this law.

In any spherical triangle

$$\begin{aligned} \textbf{hav a} &= \textbf{hav(b − c)} + \textbf{sin b} \cdot \textbf{sin c} \cdot \textbf{hav A}, \\ \textbf{hav b} &= \textbf{hav(a − c)} + \textbf{sin a} \cdot \textbf{sin c} \cdot \textbf{hav B}, \\ \textbf{hav c} &= \textbf{hav(a − b)} + \textbf{sin a} \cdot \textbf{sin b} \cdot \textbf{hav C}. \end{aligned}$$

Note: For a table of haversines see pp. 117–120 in the back section of this book. In this table two sets of numbers are given in each column. The left number represents the natural haversine values; the right number, the logarithm of the haversine. In the logarithm the characteristic is omitted; it is to be determined from the natural value (i.e., the left number). For example,

$$\text{hav } 52°20' = .1945 \text{ and } \log \text{hav } 52°20' = 9.2888 - 10.$$

92. The Haversine Nomogram.

A **nomogram** (or *alignment chart*) is a combination of graded scales usually arranged either parallel or perpendicular to each other which can be combined in such a way as to solve one or more algebraic equations. The most widely used and most applicable of these types of scales is the slide rule. Most nomograms are stationary charts and are constructed for the purpose of solving only one specific equation. Since the accuracy of nomograms depends entirely on the accuracy with which the scales are constructed and the ability of the individual to read charts, their use is not recommended in problems where a great deal of accuracy is required.

In this section a nomogram will be explained which the reader may easily construct and use to check the results of exercises.

From plane trigonometry (see Chapter VII), we have

$$\cos (a + b) = \cos a \cos b - \sin a \sin b,$$

$$\cos (a - b) = \cos a \cos b + \sin a \sin b.$$

From the first of these equations

$$\cos a \cos b = \cos (a + b) + \sin a \sin b.$$

Substituting this into the second equation above, we get

$$\cos (a - b) = \cos (a + b) + 2 \sin a \sin b,$$

or

$$\sin a \sin b = \frac{\cos (a - b) - \cos (a + b)}{2}$$

$$= \frac{1 - 2 \text{ hav } (a - b) - 1 + 2 \text{ hav } (a + b)}{2}$$

$$= \text{hav } (a + b) - \text{hav } (a - b).$$

If we substitute this result into the haversine law of the preceding section, we obtain

$$\text{hav } c = \text{hav } (a - b) + [\text{hav } (a + b) - \text{hav } (a - b)] \text{ hav } C,$$

which can be solved for hav C; thus

$$\text{hav } C = \frac{\text{hav } c - \text{hav } (a - b)}{\text{hav } (a + b) - \text{hav } (a - b)}.$$

On the basis of this formula we construct the haversine nomogram: By means of a table of haversines, draw a haversine scale as illustrated in Fig. 68.

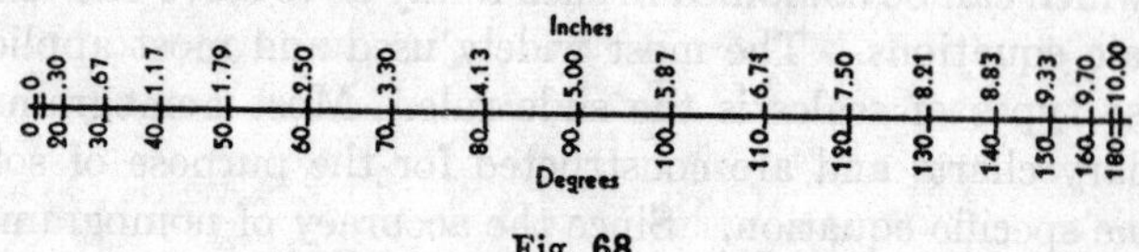

Fig. 68

Three such scales are placed, two vertical and one horizontal, and labeled as indicated in Fig. 69.

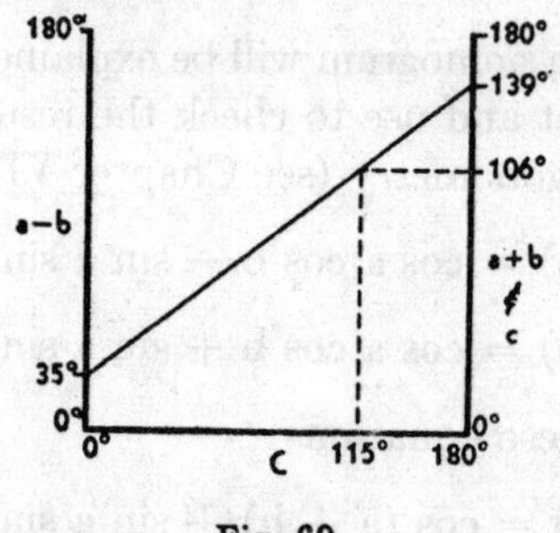

Fig. 69

In the following illustration it is demonstrated how an angle of the triangle may be found if the three sides are given. The nomogram may also be used to find the third side if two sides and the included angle are given.

Illustration.

Let the three sides of a spherical triangle be

$$a = 87°, \ b = 52°, \text{ and } c = 106°.$$

Then $a - b = 35°$ and $a + b = 139°$.

Draw a line connecting 35° on the left hand scale, $(a - b)$, to 139° on the right hand scale, $(a + b)$. Draw a line at 106° on the vertical scales parallel to the horizontal scale. Now read on the horizontal scale, (C), the coordinate of the point of intersection of the two lines. This coordinate is approximately 115°. Then C is approximately 115°.

Exercise 19.

1. Under what conditions will a spherical triangle be identical with its polar?
2. If each of the face angles of the associated trihedral of a spherical triangle is 90°, what is the value of each angle of the triangle?
3. Show that the sum of the angles of a right spherical triangle is less than 360°.
4. Given B = 65°15′, b = 47°42′, C = 79°; find c by the sine law.
5. Given a = 118°42′, c = 68°34′25″, B = 65°29′; solve for b by the cosine law for sides.
6. Solve Exercise 5 by the haversine law.

Chapter XIII

RIGHT SPHERICAL TRIANGLES

93. Spherical Triangles with More than One Right Angle.

Contrary to the fact that a plane triangle contains only one right angle or obtuse angle, a spherical triangle may very well have two or even three right or obtuse angles and still satisfy Theorem 14. The triangle with two or three right angles presents no problem if we remember that the angles of a spherical triangle

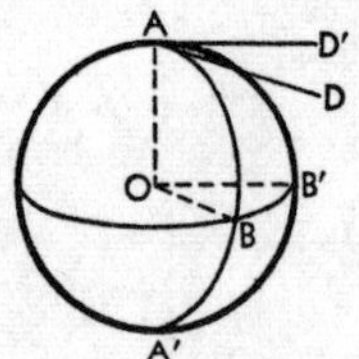

Fig. 70

are in measure the same as the dihedral angles of the associated trihedral. In Fig. 70 let △ABB′ have a right angle at B and at B′. Then the face AOB of the dihedral angle at B will be perpendicular to the plane of OB′B, and the face AOB′ of the dihedral angle at B′ will also be perpendicular to the plane OB′B. These two faces have the line OA in common. Since OA is the intersection of two planes perpendicular to the plane OB′B, then, by Theorem 4, OA must be perpendicular to the plane of OB′B. The angles AOB′ and AOB then are right angles, and consequently the arcs AB′ and AB are quadrants of great circles. (By "quadrant," we mean one-fourth the circumference of a circle; that is, a 90°-arc.)

We see also that the angle A of the spherical triangle and the face angle BOB′ both measure the same dihedral angle. Since arc BB′ subtends this angle BOB′ and is also the side opposite angle A in the spherical triangle ABB′, we have the result:

If a spherical triangle has two right angles, the sides opposite these angles are quadrants and the third angle has the same measure as its opposite side.

It follows from this statement that, if all three angles in a spherical triangle are right angles, the measure of each side is 90°. When, however, a spherical triangle has only one right angle, the side opposite it will in general not be equal to 90°.

94. Formulas of the Right Spherical Triangle.

There being no further problems to be solved concerning the angles and sides of birectangular and trirectangular triangles, we shall concern ourselves with right spherical triangles having only one right angle and, in particular, with right spherical triangles in which both of the oblique angles are less than 180°. (See Theorem 13.)

The magnitudes of the angles of a spherical triangle and also of the sides, provided they are determined in circular measure (i.e., in degrees or radians) are independent of the radius of the sphere on which the triangle is formed. For example, a great circle arc between the north pole and the equator is 90° whether measured on a small globe or on the actual earth. For simplicity, we shall assume in this discussion that the spheres are of unit radius. We shall also, as a matter of convention, use the letter C to designate the right angle of a spherical triangle. (See Fig. 71.) It follows from this fact, in the trihedral angle O − ABC associated with a right spherical triangle ABC, the plane AOC will be perpendicular to the plane BOC.

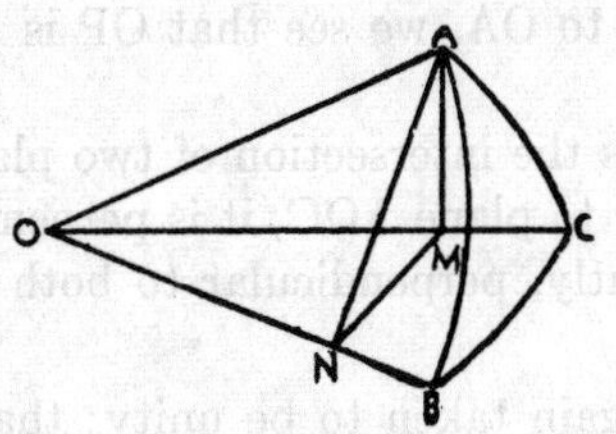

Fig. 71

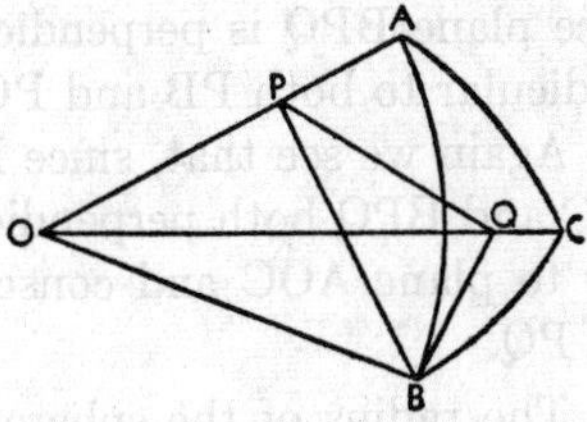

Fig. 72

The plane in which angle ANM lies (Fig. 71) is constructed perpendicular to OB and, consequently, to the entire plane BOC. Then angle ANM is the plane angle which measures the dihedral angle having OB as edge. The angle B of the spherical triangle ABC is by definition equal to the dihedral angle with edge OB, so that

$$\angle B = \angle ANM.$$

Furthermore, since AM is the intersection of two planes ANM and AOC both perpendicular to plane BOC, then, by Theorem 4, AM is perpendicular to OC. From the figure we can now read the following relationships, remembering that the radius of the sphere is unity.

In $\triangle AOM$:
(right angle at M) $\sin b = \dfrac{AM}{OA} = \dfrac{AM}{1} = AM;$

$\cos b = \dfrac{OM}{OA} = \dfrac{OM}{1} = OM.$

In $\triangle AON$:
(right angle at N) $\sin c = \dfrac{AN}{OA} = \dfrac{AN}{1} = AN;$

$\cos c = \dfrac{ON}{OA} = \dfrac{ON}{1} = ON.$

In $\triangle MON$:
(right angle at M) $\sin a = \dfrac{MN}{OM} = \dfrac{MN}{\cos b};$

therefore $MN = \sin a \cos b.$

In Fig. 72 ABC represents the same right spherical triangle as in Fig. 71. Here, however, the plane BPQ is constructed perpendicular to OA so that angle BPQ is the plane angle which measures the dihedral angle having OA as edge. Consequently, angle BPQ measures the spherical angle A of the triangle ABC. Since plane BPQ is perpendicular to OA, we see that OP is perpendicular to both PB and PQ.

Again we see that, since BQ is the intersection of two planes BOC and BPQ both perpendicular to plane AOC, it is perpendicular to plane AOC and consequently, perpendicular to both OQ and PQ.

The radius of the sphere is again taken to be unity; that is OB = 1. From the figure we read:

In $\triangle$BOQ: (right angle at Q) $\sin a = \dfrac{BQ}{OB} = BQ,$

$$\cos a = \frac{OQ}{OB} = OQ.$$

In $\triangle$BOP: (right angle at P) $\sin c = \dfrac{BP}{OB} = BP,$

$$\cos c = \frac{OP}{OB} = OP.$$

In $\triangle$POQ: (right angle at P) $\sin b = \dfrac{PQ}{OQ} = \dfrac{PQ}{\cos a};$

therefore $\quad PQ = \sin b \cos a.$

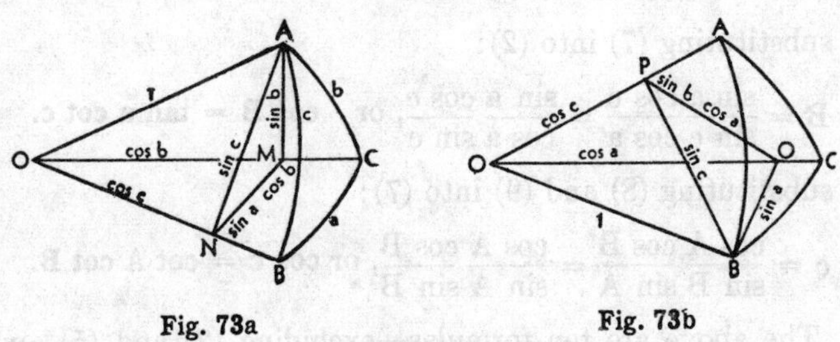

Fig. 73a Fig. 73b

By writing these results into the figures, we arrive at Fig. 73.

In triangle AMN of Fig. 73a, which has a right angle at M, we read the following relationships:

$$\sin B = \frac{\sin b}{\sin c}, \text{ or } \qquad \sin b = \sin c \sin B. \qquad (1)$$

$$\cos B = \frac{\sin a \cos b}{\sin c}, \qquad (2)$$

$$\tan B = \frac{\sin b}{\sin a \cos b} = \frac{\tan b}{\sin a}, \text{ or } \qquad \sin a = \tan b \cot B. \qquad (3)$$

In triangle BPQ of Fig. 73b, which has a right angle at Q, we read the following relationships:

$$\sin A = \frac{\sin a}{\sin c}, \text{ or } \qquad \sin a = \sin c \sin A. \qquad (4)$$

$$\cos A = \frac{\sin b \cos a}{\sin c}, \tag{5}$$

$$\tan A = \frac{\sin a}{\sin b \cos a} = \frac{\tan a}{\sin b} \qquad \mathbf{\sin b = \tan a \cot A.} \tag{6}$$

In triangle POQ of Fig. 73b, which has a right angle at P,

$$\cos b = \frac{\cos c}{\cos a}, \text{ or} \qquad \mathbf{\cos c = \cos a \cos b.} \tag{7}$$

By substituting (1) into (5): $\mathbf{\cos A = \sin B \cos a.}$ (8)

By substituting (4) into (2): $\mathbf{\cos B = \sin A \cos b.}$ (9)

By substituting (7) into (5):

$$\cos A = \frac{\sin b}{\sin c}\frac{\cos c}{\cos b} = \frac{\sin b}{\cos b}\frac{\cos c}{\sin c}, \text{ or} \qquad \mathbf{\cos A = \tan b \cot c.} \tag{10}$$

By substituting (7) into (2):

$$\cos B = \frac{\sin a}{\sin c}\frac{\cos c}{\cos a} = \frac{\sin a}{\cos a}\frac{\cos c}{\sin c}, \text{ or} \qquad \mathbf{\cos B = \tan a \cot c.} \tag{11}$$

By substituting (8) and (9) into (7):

$$\cos c = \frac{\cos A}{\sin B}\frac{\cos B}{\sin A} = \frac{\cos A}{\sin A}\frac{\cos B}{\sin B}, \text{ or } \mathbf{\cos c = \cot A \cot B.} \tag{12}$$

The above are ten formulas—excluding (2) and (5)—which are necessary for the complete solution of the right spherical triangle. These formulas can also be obtained by application of the sine and cosine law to the right triangle.

Illustration 1.

Cos C = $-\cos A \cos B + \sin A \sin B \cos c$.

Since C = 90°, cos C = 0. Therefore

$$\cos c = \frac{\cos A \cos B}{\sin A \sin B} = \cot A \cot B.$$

This is formula (12).

Illustration 2.

From the law of sines: $\dfrac{\sin A}{\sin a} = \dfrac{\sin C}{\sin c}$.

Since C = 90°, sin C = 1. Therefore

sin c sin A = sin a.

This is formula (4).

95. Napier's Rules.

With the formulas of the preceding section we can completely solve the problem of the right spherical triangle. To memorize these formulas, however, is rather confusing, and their derivation is not particularly obvious. Therefore, to facilitate their use, two rules are presented here. These were first formulated by John Napier, the inventor of logarithms.

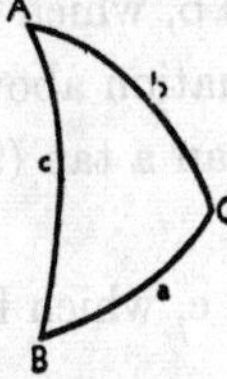

Fig. 74

To understand Napier's rules, note first that in the spherical triangle of Fig. 74, the sides and angles, excluding the right angle C, form a cycle a, b, A, c, B. While retaining this same circular order, we replace the hypotenuse and the two angles by their complements so that the cycle we shall use is made up of a, b, 90° − A, 90° − c, 90° − B. We use $\bar{A}$, $\bar{c}$, and $\bar{B}$ to represent the complements of A, c, and B. The five quantities $\bar{c}$, $\bar{B}$, a, b, and $\bar{A}$ are called "parts." If we consider any three of these parts, we see that either the three are together in the cycle or else one of them is separated from the other two. In the first case, the middle one of the three is called the "middle" part and the two on each side are said to be "adjacent." In the second case, the one which is separated from the others is called the "middle" part and the other two are said to be "opposite." We now state the rules.

Rule 1. *The sine of a middle part is equal to the product of the cosines of the opposite parts.*

Rule 2. *The sine of a middle part is equal to the product of the tangents of the adjacent parts.*

Illustration 1.

In the three parts $\bar{B}$, a, and $\bar{c}$, the angle $\bar{B}$ is the middle part and a and $\bar{c}$ are adjacent, since the three have the order a, $\bar{B}$, $\bar{c}$ in the cycle. In the three parts $\bar{B}$, b, and $\bar{A}$, the angle $\bar{B}$ is again the middle part and b and $\bar{A}$ are opposite. Applying Napier's rules we obtain

$$\sin \bar{B} = \cos \bar{A} \cos b \quad \text{and} \quad \sin \bar{B} = \tan a \tan \bar{c}.$$

Rewriting these equations,

$$\sin (90° - B) = \cos (90° - A) \cos b$$

or

$$\cos B = \sin A \cos b, \text{ which is formula (9);}$$

and for the second equation above

$$\sin (90° - B) = \tan a \tan (90° - c)$$

or

$$\cos B = \tan a \cot c, \text{ which is formula (11)}$$

Illustration 2.

If we choose the side a to be in the middle then the angle B and the side b are adjacent; the angle A and the side c are opposite. The parts to be used in Napier's rules are then: a as the middle part, $\bar{B}$ and b as the adjacent parts, and $\bar{A}$ and $\bar{c}$ as the opposite parts. By Rule 1: $\sin a = \cos \bar{A} \cos \bar{c} = \sin A \sin c$, which is formula (4). By Rule 2: $\sin a = \tan \bar{B} \tan b = \cot B \tan b$, which is formula (3).

The reader has no doubt noted that the formula as determined by Napier's rules must be changed into a more convenient form for computation. The parts that appear with bars are replaced by the angles of the given triangle by means of the laws concerning cofunctions of complementary angles. (See III of §3.)

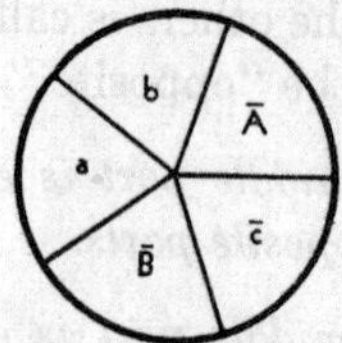

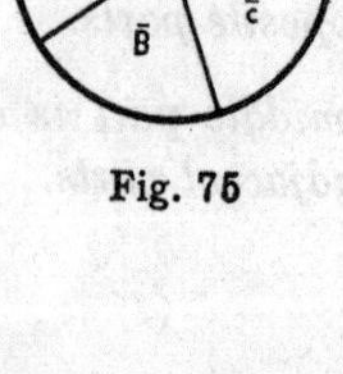

Fig. 75

Fig. 75 will be found very helpful in the use of Napier's rules. It shows the five essential parts used in the rules, and the circular arrangement makes it very simple to determine the parts adjacent and opposite to each of the five parts.

96. Suggestions for the Solution of the Right Spherical Triangle.

When any two parts (other than the right angle) of a right spherical triangle are known, the remaining three parts can be computed. In this computation, it will be found well worth while, for reasons of neatness and of accuracy, to observe carefully the following suggestions.

a) Set down a complete skeleton form of the solutions before obtaining any logarithms from the tables. The two forms most commonly employed are illustrated in §98.

b) Exhibit clearly the formula used to compute each unknown part, both as it is determined by Napier's rules and as it appears after the trigonometric functions of the complementary angles or sides have been changed to the corresponding cofunctions of the actual angles or sides.

c) Compute all unknown parts directly from the given data and not from previously computed parts.

d) Check the resulting solutions by substituting them in a formula that requires the use of all three solutions at once. Such a formula can be found by the use of Napier's rules.

97. Rules for Quadrants.

We have seen in Chapter XII that an angle or side of a spherical triangle may have any magnitude between 0° and 180°; that is, an angle or side may represent either a first or a second quadrant plane angle. When we use the cosine, tangent, or cotangent functions to compute an unknown part, the sign of the function determines for us in which quadrant that part lies. If the cosine, tangent, or cotangent of a side or angle is negative, the side or angle is in the second quadrant; if positive, it belongs to the first quadrant. When, however, the sine function is used, there will sometimes be an ambiguity, since the sine of an angle is positive in both quadrants; a given sine may represent two angles or sides one in each quadrant.

To determine whether one or both of the two angles found by means of the sine function is a solution of a triangle, we use the following rules.

Rule 1. *In a right spherical triangle an oblique angle and the side opposite are in the same quadrant.*

Consider, as a proof of the rule, equation (9): $\cos B = \sin A \cos b$, where B can represent either of the angles which is not the right angle. Since $A < 180°$, sin A is always positive; consequently, cos B has the same sign as cos b. That is, if cos b is negative, then cos B is also negative, and both the side b and the angle B are in the second quadrant; or, if one of the two functions is positive, then both are positive, and the angle and its opposite side are in the first quadrant.

Illustration.

In a right spherical triangle, let $A = 35°$ and $c = 114°$. To determine the side a we use equation (4), which may be obtained by applying Napier's first rule to the side a as the middle part;

$$\sin a = \cos \overline{A} \cos \bar{c} = \cos (90° - A) \cos (90° - c)$$
$$= \sin A \sin c.$$

Then $\sin a = \sin 35° \sin 114°$.

$$\log \sin 35° = 9.75859 - 10$$
$$\log \sin 114° = 9.96073 - 10$$
$$\log \sin a = 9.71932 - 10.$$
$$a = 31°36' \quad \text{or} \quad a = 148°24'.$$

By applying the rule we see that the first value is the correct solution.

Rule 2. *When the hypotenuse of a right spherical triangle is less than 90°, the legs are in the same quadrant; when the hypotenuse is greater than 90°, the legs are in different quadrants.*

As proof of this, consider equation (7): $\cos c = \cos a \cos b$. When $c < 90°$, cos c is positive; then, either cos a and cos b are both positive, and consequently a and b are both in the first quadrant, or cos a and cos b are both negative, and a and b are both in the second quadrant. On the other hand, when $c > 90°$, cos c is negative; then, either cos a is positive and cos b is negative, or the converse. In either case, we see that a and b are in different quadrants.

Illustration.

Returning to the preceding illustration, let us now determine the side b by using equation (6). This equation may be obtained by applying Napier's second rule to the side b as the middle part;

$$\sin b = \tan a \tan \overline{A} = \tan a \cot A.$$

Then $\sin b = \tan 31°36' \cot 35°$;

$$\begin{aligned} \log \tan 31°36' &= 9.78902 - 10 \\ \log \cot 35° &= 0.15477 \\ \hline \log \sin b &= 9.94379 - 10. \\ b = 61°28'26'' &\quad \text{or} \quad 118°31'34''. \end{aligned}$$

Now we apply Rule 2. Since the hypotenuse is greater than 90° (c = 114°) and the leg a is in the first quadrant, the other leg, b, must be in the second quadrant. Then the second solution above is the correct one.

Rule 3. *When the two given parts are a leg and its opposite angle, there are always two solutions.*

Consider a lune AA′ (Fig. 76) which is intersected by a great circle arc BC perpendicular to one side of the lune. In the resulting triangles, ABC and A′BC, the angles A and A′, being vertex angles of the lune, are equal. The arc a is the opposite side in each triangle to A and A′. We observe that, unless BC divides AA′ exactly in half, it divides the lune into two different triangles each of which contains the angle A and the side a. Thus,

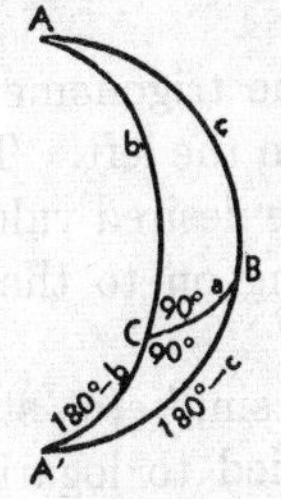

Fig. 76

when an angle and the side opposite it are given, there are always two triangles, which together form a lune, that satisfy these conditions.

From the figure it is also evident that if b, c, and B are the remaining parts of $\triangle ABC$, then the remaining parts of the triangle A′BC are $180° - b$, $180° - c$, and $180° - B$.

98. Model Solutions.

In order that the reader may get a clear concept of the methods of solution of right spherical triangles, we present here two model forms. Either of these, if rigidly adhered to, will be helpful in organizing the problem and in performing the computation with a minimum amount of effort and of possibility of error.

The first form is favored by the U. S. Naval Academy and other institutions at which Naval Training Programs are given and at which the use of Bowditch's *The American Practical Navigator* is recommended for computational work. The second form has been found more convenient for use in college trigonometry courses, where extensive secant, cosecant, and haversine tables are not ordinarily recommended for the student's use and where conversion to the conventional sine, cosine, and tangent tables becomes laborious.

Note: If tables for $\log \sec \theta$ and $\log \csc \theta$ (used in First Model Solution) are not available, the reader may use colog $\cos \theta$ and colog $\sin \theta$, respectively. That is, $\log \sec \theta = \log \dfrac{1}{\cos \theta} = \text{colog} \cos \theta$, and similarly for $\log \csc \theta$.

The reader should notice the following characteristics of the first model solution.

a) In each line are the trigonometric functions of only one angle, that given on the left. The reader will avoid error if he obtains all the desired values for this angle from the table before passing on to the next line in the computation.

b) The expressions "l sin," etc., stand for "log sin," etc.

c) The check is applied to logarithms of functions of the parts. The student should take care to be accurate in

First Model Solution.

Given: a = 46°45′18″, A = 59°12′36″.

By Rule 3 this problem will have two solutions.

Formulas:

for c: sin c = sin a csc A .... from Formula (4),
for B: sin B = sec a cos A .... from Formula (8),
for b: sin b = tan a cot A .... from Formula (6),
check: sin b = sin c sin B .... from Formula (1).

Solution:

a = 46°45′18″	l sin 9.86238 − 10	l sec 0.16425	l tan 0.02663
A = 59°12′36″	l csc 0.06599	l cos 9.70918 − 10	l cot 9.77516 − 10
c_1 = 57°59′23″	l sin 9.92837 − 10	....	....
c_2 = 122°0′37″		....	....
B_1 = 48°20′49″	l sin 9.87343 − 10 ←	l sin 9.87343 − 10	....
B_2 = 131°39′11″	....	....	....
b_1 = 39°18′49″	l sin 9.80180 − 10 ⟵	check ⟶	l sin 9.80179 − 10
b_2 = 140°41′11″			

Second Model Solution.

Given: c = 109°41′18″ and B = 27°26′54″.

To find b: sin b = cos $\bar{c}$ cos $\bar{B}$ or sin b = sin c sin B	l sin c = 9.97384 − 10 l sin B = 9.66366 − 10 l sin b = 9.63750 − 10	~~b = 154°16′42″~~ [by Rule 2.] b = 25°43′18″
To find a: sin $\bar{B}$ = tan $\bar{c}$ tan a or tan a = tan c cos B	l tan c = 0.44633 (n) l cos B = 9.94814 − 10 l tan a = 0.39447 (n)	a = 111°57′36″
To find A: sin $\bar{c}$ = tan $\bar{A}$ tan $\bar{B}$ or cot A = cos c tan B	l cos c = 9.52750 − 10 (n) l tan B = 9.71552 − 10 l cot A = 9.24302 − 10 (n)	A = 99°55′36″
Check: sin b = tan a tan $\bar{A}$ or sin b = tan a cot A	l tan a = 0.39447 (n) l cot A = 9.24302 − 10 (n) l sin b = 9.63749 − 10 l sin b = 9.63750 − 10	Checks!

finding the angles and sides from the derived logarithms, as this part of the computation is not included in the check.

Note: In the second column of the first model form we have written "l sin 9.86238 − 10." This is a breach of rigorous mathematical expression and is done solely for rapid computation. Of course, we mean l sin a = 9.86238 − 10, a being given in the first column. Similarly for the other parts of the form.

The reader should notice the following characteristics of the second model form.

a) The solution is set up in three columns. In the first column are the formulas used in the solution of each part, both as they are read from Napier's rules and after they have been transformed. In the second column are the logarithmic computations. In the third are the final results.

b) As before, "l sin," etc., are used for "log sin," etc.

c) The check is applied to the logarithms as in the first model form.

Note: Rule 2 was applied to b after a was found.

99. The Isosceles and Quadrantal Triangles.

Definition. An **isosceles triangle** is one with two equal sides.

Let a = b in the triangle ABC. If a great circle is passed through C and the midpoint M of AB, the triangle is divided into two symmetrical right triangles with right angles at M. If any two of the distinct parts (a, c, α, γ) are given we can then solve the triangle ABC by solving first one of the resulting right triangles.

Definition. A **quadrantal triangle** is a spherical triangle with one *side* equal to 90°.

By Theorem 12, we see that in the polar triangle of a quadrantal triangle the angle C′ opposite the side c′ which corresponds to the quadrantal side c must be 90°. Therefore, the polar triangle of a quadrantal triangle is a right spherical triangle and can be solved by the methods just discussed. We first solve the polar triangle and use these solutions and Theorem 12 to solve the quadrantal triangle.

Illustration.

Given: c = 90°, a = 100°, C = 65°		
In the polar triangle: C′ = 180° − c = 90°, A′ = 180° − a = 80°, c′ = 180° − C = 115°.		
To find B′: $\sin \overline{c}' = \tan \overline{A}' \tan \overline{B}'$, $\cot B' = \cos c' \tan A'$	l cos c′ = 9.62595 − 10 (n) l tan A′ = 0.75368 l cot L′ = 0.37963 (n)	B′ = 157°21′10″
To find b′: $\sin \overline{A}' = \tan b' \tan \overline{c}'$, $\tan b' = \tan c' \cos A'$	l tan c′ = 0.33133 (n) l cos A′ = 9.23967 − 10 l tan b′ = 9.57100 − 10 (n)	b′ = 20°25′29″
To find a′: $\sin a' = \cos \overline{c}' \cos \overline{A}'$, $\sin a' = \sin c' \sin A'$	l sin c′ = 9.95728 − 10 l sin A′ = 9.99335 − 10 l sin a′ = 9.95063 − 10	~~a′ = 63°11′40″~~ [by Rule 2.] a′ = 116°48′20″
Check: $\sin a' = \tan b' \tan \overline{B}'$, $\sin a' = \tan b' \cot B'$	l tan b′ = 9.57100 − 10 (n) l cot B′ = 0.37963 (n) l sin a′ = 9.95063 − 10 l sin a′ = 9.95063 − 10	checks!
In the original triangle: b = 180° − B′ B = 180° − b′ A = 180° − a′	b = 180° − 157°21′10″ B = 180° − 20°25′29″ A = 180° − 116°48′20″	b = 22°38′50″ B = 159°34′31″ A = 63°11′40″

100. Exercise 20.

1. Use Napier's rules to write three formulas each involving a and b.
2. Use Napier's rules to write three formulas each involving A and B.
3. Given A = 62°35′ and c = 71°17′; solve for the remaining parts of the right triangle.
4. Given a = 49°45′, b = 61°18′; solve for the remaining parts of the right triangle.
5. Given a = 90°, b = 55°30′, B = 40°50′; solve for A, C, and c.

Chapter XIV

THE OBLIQUE SPHERICAL TRIANGLE

101. Introduction.

A spherical triangle is completely determined when three of its six parts are given. This has already been found to be true in the case of the right spherical triangle in which case one of the known parts is the right angle, so that it is necessary to give only two other parts to determine the triangle. Although the triangle is completely determined by these given parts, it is not always uniquely determined; for, as we have already seen, if the two given parts of a right triangle are a leg and its opposite angle, two solutions are possible. Such an ambiguous case will also be encountered in the oblique triangle.

According to the parts which are given, we may consider under the following six cases, the problems of solving spherical triangles:

I. *Given two sides and the included angle.*
II. *Given two angles and the included side.*
III. *Given two sides and an angle opposite one of these sides.*
IV. *Given two angles and a side opposite one of these angles.*
V. *Given three sides.*
VI. *Given three angles.*

Each of these six cases can be solved by more than one method. To avoid confusion, however, only one method will be given in detail, with the exception of Case I, for which two methods are given since it has the widest practical application.

102. The Half-Angle Formulas.*

In this and the following two sections we shall derive the formulas necessary to solve all problems of oblique spherical tri-

* If the reader merely wishes to learn how to solve oblique spherical triangles, he may omit §§ 102, 103, and 104.

angles. In this section we develop the half-angle formulas for spherical triangles.

By solving for cos A in the law of cosines for sides (see §89) we obtain

$$\cos A = \frac{\cos a - \cos b \cos c}{\sin b \sin c}.$$

If we subtract both the left and right hand sides from 1, we get

$$1 - \cos A = 1 - \frac{\cos a - \cos b \cos c}{\sin b \sin c}$$

$$= \frac{\sin b \sin c + \cos b \cos c - \cos a}{\sin b \sin c}.$$

If we use equation (6), §40, to replace $\sin b \sin c + \cos b \cos c$ by $\cos (b - c)$, the above equation becomes

$$1 - \cos A = \frac{\cos (b - c) - \cos a}{\sin b \sin c}.$$

Since $\frac{1}{2}(a + b - c) + \frac{1}{2}(a - b + c) = a$ and $\frac{1}{2}(a + b - c) - \frac{1}{2}(a - b + c) = b - c$, we can use equation (24), §44, to replace the numerator by $2 \sin \frac{1}{2}(a + b - c) \sin \frac{1}{2}(a - b + c)$. Now, by using equation (10), §41, to change the left hand side, we may write our equation in the form

$$2 \sin^2 \tfrac{1}{2} A = \frac{2 \sin \frac{1}{2}(a + b - c) \sin \frac{1}{2}(a - b + c)}{\sin b \sin c}.$$

Half the sum of the sides of a triangle is commonly denoted by s:

$$s = \tfrac{1}{2}(a + b + c).$$

Subtracting a, b, and c from both members we obtain, respectively,

$$s - a = \tfrac{1}{2}(b + c - a),$$

$$s - b = \tfrac{1}{2}(a + c - b),$$

$$s - c = \tfrac{1}{2}(a + b - c).$$

Substituting this into the above equation and taking the square root, we obtain

$$\sin \tfrac{1}{2} A = \sqrt{\frac{\sin (s - b) \sin (s - c)}{\sin b \sin c}}. \qquad \text{(1a)}$$

By a similar argument we may arrive at

$$\sin \tfrac{1}{2} B = \sqrt{\frac{\sin (s-a) \sin (s-c)}{\sin a \sin c}}, \tag{1b}$$

$$\sin \tfrac{1}{2} C = \sqrt{\frac{\sin (s-a) \sin (s-b)}{\sin a \sin b}}. \tag{1c}$$

Beginning in the same manner as above, but adding each side of the original equation to 1, we obtain

$$1 + \cos A = \frac{1 + \cos a - \cos b \cos c}{\sin b \sin c}$$

$$= \frac{\cos a - (\cos b \cos c - \sin b \sin c)}{\sin b \sin c}$$

$$= \frac{\cos a - \cos (b + c)}{\sin b \sin c}.$$

Now, using equation (24), §44, and the same procedure as in the derivation just completed, we arrive at the formula:

$$\cos \tfrac{1}{2} A = \sqrt{\frac{\sin s \sin (s-a)}{\sin b \sin c}}. \tag{2a}$$

By symmetry

$$\cos \tfrac{1}{2} B = \sqrt{\frac{\sin s \sin (s-b)}{\sin a \sin c}}, \tag{2b}$$

$$\cos \tfrac{1}{2} C = \sqrt{\frac{\sin s \sin (s-c)}{\sin a \sin b}}. \tag{2c}$$

Dividing the left-hand side of (1a) by the left-hand side of (2a), and the right-hand side of (1a) by the right-hand side of (2a), we get

$$\tan \tfrac{1}{2} A = \sqrt{\frac{\sin (s-b) \sin (s-c)}{\sin s \sin (s-a)}}.$$

Multiplying numerator and denominator under the radical sign by sin (s – a) and removing sin (s – a) from the denominator of the radical, we have

$$\tan \tfrac{1}{2} A = \frac{p}{\sin (s-a)} \tag{3a}$$

where

$$p = \sqrt{\frac{\sin (s - a) \sin (s - b) \sin (s - c)}{\sin s}}.$$

Similarly we may derive the formulas:

$$\tan \tfrac{1}{2} B = \frac{p}{\sin (s - b)} \tag{3b}$$

and

$$\tan \tfrac{1}{2} C = \frac{p}{\sin (s - c)}. \tag{3c}$$

Equations (3) are the formulas which are usually used in solving the spherical triangle.

103. Napier's Analogies.

We shall now derive four formulas called Napier's analogies, which are essential in solving spherical triangles. By substituting $\alpha = \frac{1}{2} A$ and $\beta = \frac{1}{2} B$ in equations (1), §38; (5), §40, we obtain

$$\sin \tfrac{1}{2} (A + B) = \sin \tfrac{1}{2} A \cos \tfrac{1}{2} B + \cos \tfrac{1}{2} A \sin \tfrac{1}{2} B,$$

$$\sin \tfrac{1}{2} (A - B) = \sin \tfrac{1}{2} A \cos \tfrac{1}{2} B - \cos \tfrac{1}{2} A \sin \tfrac{1}{2} B.$$

If we divide the second equation by the first, the resulting equation is

$$\frac{\sin \frac{1}{2} (A - B)}{\sin \frac{1}{2} (A + B)} = \frac{\sin \frac{1}{2} A \cos \frac{1}{2} B - \cos \frac{1}{2} A \sin \frac{1}{2} B}{\sin \frac{1}{2} A \cos \frac{1}{2} B + \cos \frac{1}{2} A \sin \frac{1}{2} B}.$$

If the numerator and denominator of the right-hand side are divided by $\sin \frac{1}{2} A \sin \frac{1}{2} B$, the equation then becomes

$$\frac{\sin \frac{1}{2} (A - B)}{\sin \frac{1}{2} (A + B)} = \frac{\cot \frac{1}{2} B - \cot \frac{1}{2} A}{\cot \frac{1}{2} B + \cot \frac{1}{2} A}.$$

From formulas (3a) and (3b)

$$\cot \tfrac{1}{2} A = \frac{\sin (s - a)}{p}, \cot \tfrac{1}{2} B = \frac{\sin (s - b)}{p};$$

therefore

$$\frac{\sin \frac{1}{2} (A - B)}{\sin \frac{1}{2} (A + B)} = \frac{\sin (s - b) - \sin (s - a)}{\sin (s - b) + \sin (s - a)}.$$

Since $(s-b)+(s-a)=c$ and $(s-b)-(s-a)=a-b$, we may use equations (22), (21), §44, to replace

$$\sin(s-b)-\sin(s-a) \text{ by } 2\cos\tfrac{1}{2}c\sin\tfrac{1}{2}(a-b)$$

and $$\sin(s-b)+\sin(s-a) \text{ by } 2\sin\tfrac{1}{2}c\cos\tfrac{1}{2}(a-b).$$

The above equation now becomes

$$\frac{\sin\frac{1}{2}(A-B)}{\sin\frac{1}{2}(A+B)}=\frac{2\cos\frac{1}{2}c\sin\frac{1}{2}(a-b)}{2\sin\frac{1}{2}c\cos\frac{1}{2}(a-b)};$$

and after simplifying, this becomes

$$\frac{\sin\frac{1}{2}(A-B)}{\sin\frac{1}{2}(A+B)}=\frac{\tan\frac{1}{2}(a-b)}{\tan\frac{1}{2}c}. \tag{4}$$

Now suppose we begin with the equations

$$\cos\tfrac{1}{2}(A-B)=\cos\tfrac{1}{2}A\cos\tfrac{1}{2}B+\sin\tfrac{1}{2}A\sin\tfrac{1}{2}B$$

and $$\cos\tfrac{1}{2}(A+B)=\cos\tfrac{1}{2}A\cos\tfrac{1}{2}B-\sin\tfrac{1}{2}A\sin\tfrac{1}{2}B.$$

Let us follow the same general procedure as before. First we divide the first equation by the second. Then we divide both numerator and denominator of the resulting right-hand member by $\sin\frac{1}{2}A\sin\frac{1}{2}B$, obtaining

$$\frac{\cos\frac{1}{2}(A-B)}{\cos\frac{1}{2}(A+B)}=\frac{\cot\frac{1}{2}A\cot\frac{1}{2}B+1}{\cot\frac{1}{2}A\cot\frac{1}{2}B-1}.$$

Using equations (3a) and (3b), we find that the equation becomes

$$\frac{\cos\frac{1}{2}(A-B)}{\cos\frac{1}{2}(A+B)}=\frac{\dfrac{\sin(s-a)\sin(s-b)}{p^2}+1}{\dfrac{\sin(s-a)\sin(s-b)}{p^2}-1}.$$

Recalling the value for p (see equation [3a], §102), we may simplify the right-hand member and obtain

$$\frac{\cos\frac{1}{2}(A-B)}{\cos\frac{1}{2}(A+B)}=\frac{\sin s+\sin(s-c)}{\sin s-\sin(s-c)}.$$

Noticing that $s+(s-c)=(a+b)$ and $s-(s-c)=c$, we may proceed, analogously to the method employed in deriving equation (4), to obtain

$$\frac{\cos\frac{1}{2}(A-B)}{\cos\frac{1}{2}(A+B)}=\frac{\tan\frac{1}{2}(a+b)}{\tan\frac{1}{2}c}. \tag{5}$$

With the use of polar triangles we may now derive two more formulas analogous to the two which have just been derived. If A′B′C′ is the polar triangle of ABC, we may write

$$\frac{\sin \frac{1}{2}(A' - B')}{\sin \frac{1}{2}(A' + B')} = \frac{\tan \frac{1}{2}(a' - b')}{\tan \frac{1}{2} c'}$$

and

$$\frac{\cos \frac{1}{2}(A' - B')}{\cos \frac{1}{2}(A' + B')} = \frac{\tan \frac{1}{2}(a' + b')}{\tan \frac{1}{2} c'}.$$

From Theorem 12, §87, we recall that

$$\tfrac{1}{2}(A' - B') = \tfrac{1}{2}(180° - a - 180° + b) = -\tfrac{1}{2}(a - b).$$

$$\tfrac{1}{2}(A' + B') = 180° - \tfrac{1}{2}(a + b),$$

$$\tfrac{1}{2} C' = 90° - \tfrac{1}{2} c,$$

$$\tfrac{1}{2}(a' - b') = -\tfrac{1}{2}(A - B),$$

$$\tfrac{1}{2}(a' + b') = 180° - \tfrac{1}{2}(A + B),$$

$$\tfrac{1}{2} c' = 90° - \tfrac{1}{2} C.$$

Substituting these relations in the preceding formulas, we obtain

$$\frac{\sin \frac{1}{2}(a - b)}{\sin \frac{1}{2}(a + b)} = \frac{\tan \frac{1}{2}(A - B)}{\cot \frac{1}{2} C}, \tag{6}$$

$$\frac{\cos \frac{1}{2}(a - b)}{\cos \frac{1}{2}(a + b)} = \frac{\tan \frac{1}{2}(A + B)}{\cot \frac{1}{2} C}. \tag{7}$$

104. Gauss' Formulas. Rule of Quadrants.

It is desirable here to derive Gauss' formulas, which are particularly useful in checking the solution of the triangle. In §103 we made use of the equation

$$\sin \tfrac{1}{2}(A - B) = \sin \tfrac{1}{2} A \cos \tfrac{1}{2} B - \cos \tfrac{1}{7} A \sin \tfrac{1}{2} B.$$

By means of the half-angle formulas (1a), (1b), (2a), and (2b) of this chapter and the addition formulas of Chapter VI we may transform the right hand side of the above equation as follows:

$$\sin \tfrac{1}{2}(A - B) = \sqrt{\frac{\sin^2 (s - b) \sin (s - c) \sin s}{\sin a \sin b \sin^2 c}} - \sqrt{\frac{\sin s \sin^2 (s - a) \sin (s - c)}{\sin a \sin b \sin^2 c}}$$

$$= \frac{\sin(s-b)}{\sin c}\sqrt{\frac{\sin s \sin(s-c)}{\sin a \sin b}} - \frac{\sin(s-a)}{\sin c}\sqrt{\frac{\sin s \sin(s-c)}{\sin a \sin b}}$$

$$= \frac{\sin(s-b) - \sin(s-a)}{\sin c} \cos \tfrac{1}{2} C$$

$$= \frac{2 \sin \frac{1}{2}(a-b) \cos \frac{1}{2}(2s-a-b)}{2 \sin \frac{1}{2} c \cos \frac{1}{2} c} \cos \tfrac{1}{2} C$$

$$= \frac{\sin \frac{1}{2}(a-b) \cos \frac{1}{2} c}{\sin \frac{1}{2} c \cos \frac{1}{2} c} \cos \tfrac{1}{2} C;$$

$$\sin \tfrac{1}{2}(A-B) = \frac{\sin \frac{1}{2}(a-b)}{\sin \frac{1}{2} c} \cos \tfrac{1}{2} C. \tag{8a}$$

The reader should in a similar manner derive

$$\sin \tfrac{1}{2}(A+B) = \frac{\cos \frac{1}{2}(a-b)}{\cos \frac{1}{2} c} \cos \tfrac{1}{2} C, \tag{8b}$$

$$\cos \tfrac{1}{2}(A-B) = \frac{\sin \frac{1}{2}(a+b)}{\sin \frac{1}{2} c} \sin \tfrac{1}{2} C, \tag{8c}$$

$$\cos \tfrac{1}{2}(A+B) = \frac{\cos \frac{1}{2}(a+b)}{\cos \frac{1}{2} c} \sin \tfrac{1}{2} C. \tag{8d}$$

Any one of these formulas will be found useful in checking the solution of a spherical triangle, since each of the four involves all the sides and all the angles of the spherical triangle.

There is also a rule to which we must sometimes refer when solving a spherical triangle. To obtain this rule, consider equation (5):

$$\frac{\cos \frac{1}{2}(A-B)}{\cos \frac{1}{2}(A+B)} = \frac{\tan \frac{1}{2}(a+b)}{\tan \frac{1}{2} c}.$$

From Theorem 13 of Chapter XII, we find that $A - B$ must be less than 180° and, from Proposition 2, that $\frac{1}{2} c < 90°$. Consequently, $\cos \frac{1}{2}(A-B)$ and $\tan \frac{1}{2} c$ will always be positive. In the above equation we find then that $\cos \frac{1}{2}(A+B)$ and $\tan \frac{1}{2}(a+b)$ must have the same sign. Since the cosine and tangent are both positive in the first quadrant and negative in the second, it follows that $\frac{1}{2}(A+B)$ and $\frac{1}{2}(a+b)$ must lie either both in the first quadrant or both in the second quadrant. This gives us the following rule:

Rule of Quadrants. *In any spherical triangle, one-half the sum of two angles is in the same quadrant as one-half the sum of the sides opposite.*

105. The Cases.

We shall now discuss the methods of solution for the six cases of the oblique spherical triangle. Before we consider the individual cases we summarize the formulas that we shall use.

The Law of Sines: $$\frac{\sin A}{\sin a} = \frac{\sin B}{\sin b} = \frac{\sin C}{\sin c}. \qquad (1)$$

The Cosine Law:

$$\begin{aligned} \cos a &= \cos b \cos c + \sin b \sin c \cos A, \\ \cos b &= \cos a \cos c + \sin a \sin c \cos B, \qquad (2) \\ \cos c &= \cos a \cos b + \sin a \sin b \cos C. \end{aligned}$$

The Half-Angle Formulas:

$$\tan \tfrac{1}{2} A = \frac{p}{\sin (s - a)}, \ \tan \tfrac{1}{2} B = \frac{p}{\sin (s - b)}, \qquad (3)$$

$$\tan \tfrac{1}{2} C = \frac{p}{\sin (s - c)},$$

where

$$p = \sqrt{\frac{\sin (s - a) \sin (s - b) \sin (s - c)}{\sin s}}; \ s = \tfrac{1}{2} (a + b + c).$$

The Four Analogies of Napier:

$$\frac{\sin \frac{1}{2} (A - B)}{\sin \frac{1}{2} (A + B)} = \frac{\tan \frac{1}{2} (a - b)}{\tan \frac{1}{2} c}, \qquad (4)$$

$$\frac{\cos \frac{1}{2} (A - B)}{\cos \frac{1}{2} (A + B)} = \frac{\tan \frac{1}{2} (a + b)}{\tan \frac{1}{2} c}, \qquad (5)$$

$$\frac{\sin \frac{1}{2} (a - b)}{\sin \frac{1}{2} (a + b)} = \frac{\tan \frac{1}{2} (A - B)}{\cot \frac{1}{2} C}, \qquad (6)$$

$$\frac{\cos \frac{1}{2} (a - b)}{\cos \frac{1}{2} (a + b)} = \frac{\tan \frac{1}{2} (A + B)}{\cot \frac{1}{2} C}. \qquad (7)$$

A Formula of Gauss:

$$\sin \tfrac{1}{2}(A - B) = \frac{\sin \tfrac{1}{2}(a - b)}{\sin \tfrac{1}{2} c} \cos \tfrac{1}{2} C. \tag{8}$$

Rule of Quadrants.

In any spherical triangle, one-half the sum of two angles is in the same quadrant as one-half the sum of the sides opposite.

As in the case of the right triangle we again recommend that the student adhere strictly to a model form. Again two model forms are presented, the Standard Model Form and the Navy Model Form, the most important difference in these two being the use of secant and cosecant tables in the latter. The reader should choose the form he considers more practical for his particular position. The Standard Model Form is used to illustrate Cases I and V, the Navy Model Form to illustrate Cases II and III. We also remark that the haversine law may be used wherever the cosine law is used.

Case I. *Given two sides and the included angle.*

First Method.

This method is particularly recommended when the unknown side alone is required.

Suppose a, b, C are given; the parts c, A, B are required.

1. Find the side c by the law of cosines, Formula (2).
2. Find the angles A and B by the law of sines, Formula (1). However, if we use the law of sines, two values are found for each angle. For example, for the angle A we will find an acute angle A_1 and an obtuse angle $A_2 = 180° - A_1$, each having the same sine. Since the triangle is uniquely determined by the given parts, only one of these values for A can be used. Generally the rule of quadrants given above will enable us to choose the correct value for A. If, however, $\frac{1}{2}(A_1 + C)$ and $\frac{1}{2}(A_2 + C)$ are in the same quadrant (and this may happen if A_1 is in the neighborhood of 90°) we cannot make the correct choice by using the rule. In that case, the triangle should be solved by the second method.
3. Check by Formula (8).

STANDARD MODEL SOLUTION. CASE I. (Second Method)

Given: a = 135°49′24″, c = 60°04′54″, B = 142°12′36″.

To find A and C:	a = 135°49′24″ c = 60°04′54 a + c = 195°54′18″ ½ (a + c) = 97°57′09″ a − c = 75°44′30″ ½ (a − c) = 37°52′15″ B = 142°12′36″ ½ B = 71°06′18″	
$\tan \frac{1}{2}(A + C) = \dfrac{\cot \frac{1}{2} B \cos \frac{1}{2}(a - c)}{\cos \frac{1}{2}(a + c)}$	l cot ½ B = 9.53438 − 10 l cos ½ (a − c) = 9.89730 − 10 sum = 9.43168 − 10 l cos ½ (a + c) = 9.14098 − 10 (n) l tan ½ (A + C) = 0.29070 (n)	½ (A + C) = 117°06′50″
$\tan \frac{1}{2}(A - C) = \dfrac{\cot \frac{1}{2} B \sin \frac{1}{2}(a - c)}{\sin \frac{1}{2}(a + c)}$	l cot ½ B = 9.53438 − 10 l sin ½ (a − c) = 9.78809 − 10 sum = 9.32247 − 10 l sin ½ (a + c) = 9.99581 − 10 l tan ½ (A − C) = 9.32266 − 10	½ (A − C) = 11°58′42″ A = 129°05′32″ C = 105°08′08″
To find b: $\tan \frac{1}{2} b = \dfrac{\tan \frac{1}{2}(a - c) \sin \frac{1}{2}(A + C)}{\sin \frac{1}{2}(A - C)}$	l tan ½ (a − c) = 9.89080 − 10 l sin ½ (A + C) = 9.94944 − 10 sum = 9.84024 − 10 l sin ½ (A − C) = 9.31710 − 10 l tan ½ b = 0.52314 ½ b = 73°18′36	b = 146°37′12″

Check:			
$\dfrac{\sin A}{\sin a} = \dfrac{\sin B}{\sin b} = \dfrac{\sin C}{\sin c}$	l sin A = 9.88944 − 10 l sin a = 9.84316 − 10 Diff. = 0.04678	l sin B = 9.78729 − 10 l sin b = 9.74051 − 10 Diff. = 0.04678	l sin C = 9.98467 − 10 l sin c = 9.93788 − 10 Diff. = 0.04679

Second Method.

If the angles are desired, this method is recommended since it avoids the difficulty which may arise in the First Method from the use of the law of sines. Suppose we have the same parts given as above.

1. Find the angles A and B from (6) and (7).
   From (6) we find $\frac{1}{2}(A - B)$;
   From (7) we find $\frac{1}{2}(A + B)$.
   Adding, we get A;
   subtracting, we get B.
2. Find the side c from equations (4) *or* (5).
3. Check the solution by the law of sines.

Case II. *Given two angles and the included side.*

Suppose A, B, and c are given; the parts a, b, and C are to be found.

Solution.

1. Find the sides a and b by equations (4) and (5).
   From (4) we find $\frac{1}{2}(a - b)$;
   From (5) we find $\frac{1}{2}(a + b)$.
   Adding, we get a;
   subtracting, we get b.
2. Find the angle C by equations (6) *or* (7).
3. Check the solutions by the law of sines.

Case III. *Given two sides and an angle opposite one of these sides.*

Suppose b, c, and C are given; a, A, and B are to be found.

Solution.

1. Find the angle B by the law of sines.
2. Find the angle A by Formula (7).
3. Find the side a by Formula (5).
4. Check the solution by Formula (8).

This is the ambiguous case. Since we use the law of sines to find B, we obtain two angles, B_1 and B_2, where B_1 is an acute angle and $B_2 = 180° - B_1$. If both B_1 and B_2 satisfy the *rule of quadrants* there are *two solutions;* that is, there are two different triangles containing the given parts. If only one of the values of B_1 and B_2 satisfy the *rule*, there is only *one solution.*

NAVY MODEL SOLUTION. CASE II.

Given: A = 57°56′53″, B = 137°20′33″, c = 94°48′06″.

Formulas:

$\tan \frac{1}{2}(b - a) = \sin \frac{1}{2}(B - A) \csc \frac{1}{2}(B + A) \tan \frac{1}{2} c$ .... from (4),

$\tan \frac{1}{2}(b + a) = \cos \frac{1}{2}(B - A) \sec \frac{1}{2}(B + A) \tan \frac{1}{2} c$ .... from (5),

$\cot \frac{1}{2} C = \tan \frac{1}{2}(B - A) \sin \frac{1}{2}(b + a) \csc \frac{1}{2}(b - a)$ .... from (6),

check by the Law of Sines.

Solution:

½ (B − A) = 39°41′50″	l sin 9.80532 − 10	l cos 9.88617 − 10	l tan 9.91915 − 10
½ (B + A) = 97°38′43″	l csc 0.00388	l sec 0.87602 (n)	....
½ c = 47°24′03″	l tan 0.03644	l tan 0.03644	....
½ (b − a) = 35°01′32″	l tan 9.84564 − 10	....	l csc 0.24113
½ (b + a) = 99°02′02″	....	l tan 0.79863 (n)	l sin 9.99458 − 10
½ C = 34°59′40″	....	....	l cot 0.15486
a = 64°00′30″			
b = 134°03′34″			
C = 69°59′20″			
Check:	l sin a 9.95369 − 10	l sin b 9.85650 − 10	l sin c 9.99847 − 10
	l sin A 9.92817 − 10	l sin B 9.83098 − 10	l sin C 9.97296 − 10
	Diff. 0.02552	Diff. 0.02552	Diff. 0.02551

NAVY MODEL SOLUTION. CASE III.

Given: b = 52°45′20″, c = 71°12′40″, B = 46°22′10″

Formulas:

(i) sin C = sin c sin B csc b . . . . from (1), Let $\frac{1}{2}(c-b)=\alpha$

(ii) cot $\frac{1}{2}$ A = cos $\frac{1}{2}$ (c + b) sec $\frac{1}{2}$ (c − b) tan $\frac{1}{2}$ (C + B) . . . . from (7), $\frac{1}{2}(c+b)=\beta$

(iii) tan $\frac{1}{2}$ a = sin $\frac{1}{2}$ (C + B) csc $\frac{1}{2}$ (C − B) tan $\frac{1}{2}$ (c − b) . . . . from (5), $\frac{1}{2}(C+B)=\gamma$

(iv) check: sin $\frac{1}{2}$ a = sin $\frac{1}{2}$ (c − b) csc $\frac{1}{2}$ (C − B) cos $\frac{1}{2}$ A . . . . from (8), $\frac{1}{2}(C-B)=\delta$

Solution:		(i)	(ii)₁	(ii)₂	(iii)₁	(iii)₂	(iv)₁	(iv)₂
c =	71°12′40″	l sin 9.97622 − 10						
b =	52°45′20″	l csc 0.09906						
α =	9°13′40″	. . . .	l sec 0.00565	l sec 0.00565	l tan 9.21075 − 10	l tan 9.21075 − 10	l sin 9.20509 − 10	l sin 9.30509 − 10
β =	61°59′00″	. . . .	l cos 9.67185 − 10	l cos 9.67185 − 10	. . . .	. . . .		
B =	46°22′10″	l sin 9.85962 − 10						
C_1 =	59°24′22″	l sin 9.93490 − 10						
C_2 =	120°35′38″							
γ_1 =	52°53′16″	. . . .	l tan 0.12112	. . . .	l sin 9.90171 − 10			
γ_2 =	83°28′54″	. . . .	. . . .	l tan 0.94211	. . . .	l sin 9.99718 − 10		
δ_1 =	6°31′06″	. . . .	. . . .	. . . .	l csc 0.94492	. . . .	l csc 0.94492	
δ_2 =	37°06′44″	. . . .	. . . .	. . . .	. . . .	l csc 0.21941	. . . .	l csc 0.21941
$\frac{1}{2}A_1$ =	57°49′56″	(A_1 = 115°39′52″)	l cot 9.79862 − 10	. . . .	. . . .	. . . .	l cos 9.72624 − 10	
$\frac{1}{2}A_2$ =	13°30′04″	(A_2 = 27°00′08″)	. . . .	l cot 0.61961	. . . .	. . . .	(sum 9.87625 − 10)	l cos 9.98783 − 10
$\frac{1}{2}a_1$ =	48°46′26″	(a_1 = 97°32′52″)	. . . .	. . . .	l tan 0.05738			
						. . . .	l sin 9.87628 − 10	(sum 9.41233 − 10)
$\frac{1}{2}a_2$ =	14°58′35″	(a_2 = 29°57′10″)	. . . .	. . . .	. . . .	l tan 9.42734 − 10	. . . .	l sin 9.41232 − 10

Case IV. *Given two angles and a side opposite one of these angles.*

Let A, B, and a be given; find C, b, and c.

Solution.

1. Find the side b by the law of sines.
2. Find the angle C by Formulas (6) *or* (7).
3. Find the side c by Formulas (4) *or* (5).
4. Check the solution by Gauss' formula (8).

This case is also *ambiguous* and the number of solutions is determined by a method analogous to the one given for Case III; that is, by the *rule of quadrants.*

Case V. *Given three sides.*

The three sides a, b, and c are given. The angles A, B, and C must be found.

Solution.

1. Find the three angles by the half-angle formulas (3).
2. Check the solution by the law of sines.

Case VI. *Given three angles.*

The angles A, B, and C are given; the sides a, b, and c must be found.

Solution.

1. Obtain a′, b′, c′ of the polar triangle by Theorem 12 of Chapter XII; $a' = 180° - A$, $b' = 180° - B$, $c' = 180° - C$.
2. Solve for A′, B′, C′ by Case V.
3. Then $a = 180° - A'$, $b = 180° - B'$, $c = 180° - C'$.
4. Check the solution by the law of sines.

Illustration.

Given: $A = 55°47'30''$, $B = 125°41'48''$, $C = 82°47'36''$.

Solution.

$$a' = 180° - 55°47'30'' = 124°12'30'',$$

$$b' = 180° - 125°41'48'' = 54°18'12'',$$

$$c' = 180° - 82°47'36'' = 97°12'24''.$$

Now it is necessary to find A′, B′, C′ by Case V. If the reader will refer to the illustration of Case V, he will see that

STANDARD MODEL SOLUTION. CASE V.

Given: a = 124°12′30″, b = 54°18′12″, c = 97°12′24″.

To find log p:	a = 124°12′30″ b = 54°18′12″ c = 97°12′24″ 2s = 275°43′06″ s = 137°51′33″ ← check →	s − a = 13°39′03″ s − b = 83°33′21″ s − c = 40°39′09″ s = 137°51′33″
$p = \sqrt{\dfrac{\sin(s-a)\sin(s-b)\sin(s-c)}{\sin s}}$	l sin (s − a) = 9.37292 − 10 l sin (s − b) = 9.99725 − 10 l sin (s − c) = 9.81389 − 10 sum = 9.18406 − 10 l sin s = 9.82669 − 10 2 log p = 9.35737 − 10	log p = 9.67868 − 10
To find A: $\tan \frac{1}{2} A = \dfrac{p}{\sin(s-a)}$	log p = 9.67868 − 10 l sin (s − a) = 9.37292 − 10 l tan ½ A = 0.30576 ½ A = 63°41′02″	A = 127°22′04″
To find B: $\tan \frac{1}{2} B = \dfrac{p}{\sin(s-b)}$	log p = 9.67868 − 10 l sin (s − b) = 9.99725 − 10 l tan ½ B = 9.68143 − 10 ½ B = 25°39′02″	B = 51°18′04″
To find C: $\tan \frac{1}{2} C = \dfrac{p}{\sin(s-c)}$	log p = 9.67868 − 10 l sin (s − c) = 9.81389 − 10 l tan ½ C = 9.86479 − 10 ½ C = 36°13′18″	C = 72°26′36″

Check: $\dfrac{\sin A}{\sin a} = \dfrac{\sin B}{\sin b} = \dfrac{\sin C}{\sin c}$

l sin A = 9.90023 − 10	l sin B = 9.89234 − 10	l sin C = 9.97928 − 10
l sin a = 9.91751 − 10	l sin b = 9.90962 − 10	l sin c = 9.99656 − 10
Diff. = 9.98272 − 10	Diff. = 9.98272 − 10	Diff. = 9.98272 − 10

$$A' = 127°22'04'',$$
$$B' = 51°18'04'',$$
$$C' = 72°26'36''.$$

Then

$$a = 180° - 127°22'04'' = 52°37'56'',$$
$$b = 180° - 51°18'04'' = 128°41'56'',$$
$$c = 180° - 72°26'36'' = 107°33'24''.$$

106. Exercise 21.

1. Solve the spherical triangle with $b = 39°25'11''$, $c = 109°46'19''$, and $A = 54°54'42''$.
2. Solve the spherical triangle with $a = 62°31'$, $b = 103°15'$, and $A = 63°21'$.
3. A spherical triangle whose sides are 300 ft. lies on a sphere of radius 150 ft. Find its angles.

CHAPTER XV

THE TERRESTRIAL AND ASTRONOMICAL TRIANGLES

107. The Terrestrial Triangle.

For the sake of simplicity, we shall assume the earth to be a sphere, its radius being approximately 3959 miles. While it revolves about the sun, the earth also rotates about one of its own diameters which is called its **axis.** The two extremities of this axis are the **north** and **south poles.** In the sense of Chapter XI these poles are associated with a great circle arc—the **equator.**

A **meridian** is a great circle on the earth which passes through the north and south poles. Through each point on the earth passes a meridian. To distinguish one meridian from another, the meridian of Greenwich, England, is chosen as a reference, or **prime meridian;** and any other meridian is characterized by its angular distance, between 0° and 180°, measured along the equator to the east or west of the meridian of Greenwich. This angular distance is the **longitude,** λ, of a point on the given meridian. Since the earth makes one complete revolution in 24 hours, it turns through 15° in one hour. This means that one hour from now a point with longitude 15° west and the same latitude as Greenwich will be in the same position relative to the sun that Greenwich is now. This has led to the practice of measuring longitude not in degrees, minutes, and seconds, but rather in the units of time: hours, minutes, and seconds. The conversion units are

$$\mathbf{1\ hour = 15°;\ 1\ min. = 15';\ 1\ sec. = 15''.}$$

Note the distinction between minutes and seconds of time and the smaller minutes and seconds of angular measure. When referring to time units the notation "h, m, s" are used for hours, minutes, and seconds.

We observe now that a point is not uniquely determined by its meridian, since there are many other points which may have the same longitude. In addition to knowing the meridian of a point, we must also know how far north or south of the equator it lies on that meridian. The angular magnitude of an arc on the meridian of a point and extending from the equator to that point is called the **latitude,** L, of the point. Latitude is therefore measured from 0° to 90° north or south of the equator. The **colatitude** is the complement of the latitude; i.e., $90° - L$. All points on the earth that have the same latitude lie on a small circle parallel to the equator which is called a **parallel of latitude.**

The latitude and longitude of a point completely determine its position on the earth.

Illustration.

In Fig. 77, the position of a point P is given by $L = 40°\text{N}$ and $\lambda = 90°\text{W}$ (or 6^hW). This means that $\widehat{AP} = 40°$, P being to the north of the equator; and $\widehat{AB} = 90°$, P being to the west of Greenwich.

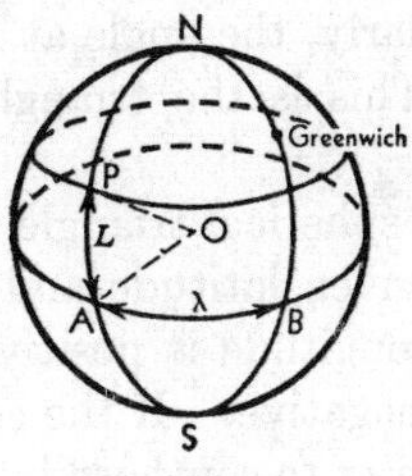

Fig. 77

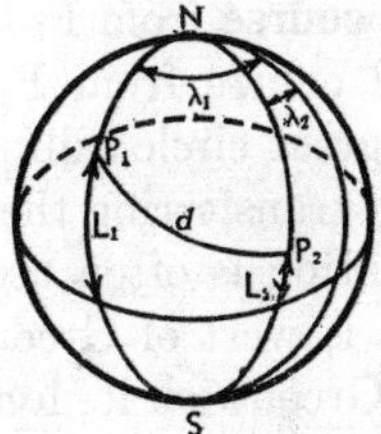

Fig. 78

If P_1 and P_2 are two points on the earth (see Fig. 78), the **distance** between P_1 and P_2 is the measure of the great circle arc between these points. If this distance, d, is measured in circular measure we shall call it the **angular distance.** In contrast we shall call the *length* of the great circle arc P_1P_2 the **linear distance** between the points P_1 and P_2. This length can be found by using the fact that P_1P_2 is an arc of a circle of radius 3959 miles and subtends an angle of d degrees at the center of the circle. The linear distance is commonly measured in nautical miles. A **nautical mile** is defined to be the length along a great circle on the earth of an arc whose angular distance is 1′. Since $360° = 21600'$

there are 21600 nautical miles in the circumference of a great circle. There are $(2\pi)(3959)$ statute miles. From these facts we obtain the relation

$$\textbf{1 nautical mile} = \frac{(2\pi)(3959)}{21600} = \textbf{1.1515 statute miles.}$$

The angular distance, d, can be found if we know the longitude and latitude of the two points P_1 and P_2. The arc d and the meridians of P_1 and P_2 form a spherical triangle with vertices P_1, P_2 and N, where N represents the north pole (see Fig. 78). (We note that we could also use the south pole and have the spherical triangle P_1P_2S.)

Let the latitudes L_1 and L_2 and the longitudes λ_1 and λ_2 of P_1 and P_2 be given. We then have a spherical triangle in which the known parts are two sides, the colatitudes $90° - L_1$ and $90° - L_2$, and the included angle, $\lambda_1 - \lambda_2$. To find d we may solve this triangle by Case I of the preceding chapter. We can also find the other two angles of this triangle. The angle at P_1 is called the **initial course** from P_1 to P_2; similarly, the angle at P_2 is called the initial course from P_2 to P_1. This is the triangle which is solved in great circle sailing.

When transferring the data to a spherical triangle the following convention is often used in the given latitude and longitude. If a point is west of Greenwich its longitude is positive and if it is east of Greenwich its longitude is negative. If the computor is in the northern hemisphere he may refer to southern latitude (i.e., south of the equator) as being negative. We note that if the longitude of P_1 is east and the longitude of P_2 is west, then by taking $\lambda_2 - \lambda_1$ we may obtain an included angle greater than 180°. This, however, is the larger spherical triangle formed by P_1P_2N and by Theorem 13, Chapter XII, is not the triangle we desire. In that case the included angle is $360° - (\lambda_2 - \lambda_1)$. The student should be able to visualize this by studying Fig. 78.

Illustration.

Find the distance and the initial course from Berkeley, California ($L_1 = +37°52'24''$, $\lambda_1 = 8^h9^m2.9^s$), to Rio de Janeiro, Brazil ($L_2 = -22°53'42''$, $\lambda_2 = 2^h52^m53.5^s$).

Solution:

Let B represent Berkeley; A, Rio de Janeiro; and C, the north pole. The spherical triangle then is

$$a = BC = 90° - 37°52'24'' = 52°07'36'',$$

$$b = AC = 90° - (-22°53'42'') = 112°53'42'',$$

$$C = \lambda_1 - \lambda_2 = 8^h9^m2.9^s - 2^h52^m53.5^s = 5^h16^m9.5^s = 79°02'22''.$$

Since we desire to find B and c we use the second method of Case I. If only the distance was required, we could use the first method of Case I. Solving the triangle we obtain:

$$A = 51°09'44'',\ B = 114°37'44'',\ c = 95°46'18''.$$

The distance, c, is then 5747′ = 5747 nautical miles
= 6617.67 statute miles.

The initial course is the angle B = 114°37′44″ east of north or E 24°38′ S.

108. The Celestial Sphere.

The **celestial sphere** is a fictitious sphere of infinite radius which is concentric with the earth and on which the heavenly bodies are considered to be projected. This conforms with the impression that we get when we observe the sky at night.

If the axis of the earth is extended in both directions through the north and south poles, it will determine on the celestial sphere two points, P_n and P_s, the **celestial north** and **south poles.** Refer to Fig. 79.

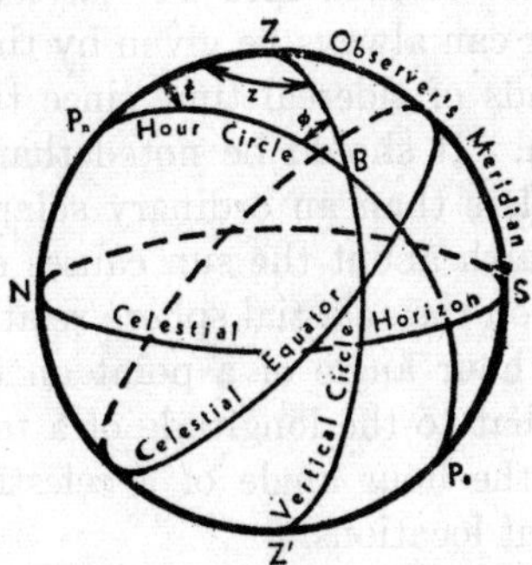

Fig. 79

The **celestial equator** is the great circle on the celestial sphere whose poles are the celestial north and south poles. The celestial equator and the earth's equator are therefore in the same plane.

The **zenith** of an observer is the point Z on the celestial sphere vertically above him. The point Z′ diametrically opposite the zenith is called the **nadir.**

The **horizon** of an observer is the great circle on the celestial sphere which has the zenith and nadir as its poles.

The **hour circle** of a point B on the celestial sphere is the great circle which passes through the poles and also through the point B. The hour circle on the sphere corresponds to the meridian on the earth. The hour circle of the zenith is called the **observer's meridian.**

The points in which the observer's meridian intersects the horizon are the **north point** and **south point**; an observer in northern latitudes finds the north point below the celestial north pole. The **east** and **west points** are points on the horizon to the *right* and *left*, respectively, of an observer facing north.

The **vertical circle** of a point B on the celestial sphere is the great circle which passes through zenith and nadir and also through the point B. This circle meets the horizon at right angles.

The **hour angle,** t, of a celestial point B is the angle at the pole between the observer's meridian and the hour circle of B. It is measured positively through the west from 0 hours to 24 hours. The time it takes the hour angle of a star to change from 0 hours to 24 hours; i.e., the time of one complete rotation of the star from the observer's meridian back to that meridian, is called a *sidereal day.* It is divided into 24 *sidereal hours.* Then the hour angle of a star can always be given by the number of hours, minutes, and seconds of sidereal time since the star crossed the observer's meridian. It should be noted that a sidereal hour is about ten seconds less than an ordinary solar hour, because the revolution of the earth about the sun causes daily change in the position of the sun on the celestial sphere relative to the positions of the stars. The hour angle of a point on the celestial sphere corresponds somewhat to the longitude of a terrestrial point; the two differ in that the hour angle of a celestial point varies for observers in different locations.

The **declination,** d, of a celestial point B is the angular distance of B from the celestial equator measured along the hour circle of B. For many well-known stars this quantity can be determined from tables in the *Nautical Almanac.* The declination is measured from 0° to 90° north or south of the equator, and therefore corresponds to the latitude of a point on the earth. In fact if L represents the latitude of an observer on the earth, then L is also the declination of the zenith point of that observer. This is true because the celestial equator and the earth's equator lie in the same plane. Consider Fig. 80. The latitude L of the ob-

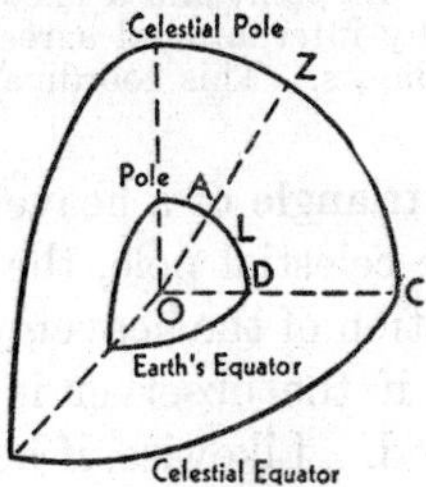

Fig. 80

server A on the earth is the angular measure of arc AD. Since this is the angular measure of the angle AOD at the center of the earth, it is also the measure of the arc CZ, along the observer's meridian from the celestial equator to the observer's zenith. The declination of the observer's zenith is the angular measure of the arc CZ, and is therefore L. If the observer is north of the earth's equator, d, should be taken positive for a celestial point having north declination and negative for a point having south declination. Conversely, if the observer is south of the equator.

The reader will note that a point on the celestial sphere can be designated by its declination and hour angle. These, however, as we have pointed out, are not quite analogous to the latitude and longitude of a point on the earth's surface. In the first place, the hour angle of a point depends on the position on the earth from which it is observed. Furthermore, the hour angle of a star varies constantly as the earth's rotation causes the star's position to move west. The time and place of observation, then, as well as the hour angle and declination, must be considered in discussing the position of a star. The same statement is true of another common coordinate system on the celestial sphere—the altitude-azimuth system—which we shall now discuss.

The **altitude,** h, of a celestial point B is the angular distance of B from the horizon measured along the vertical circle of B; see Fig. 79. It is considered positive if B is above the horizon; negative, if B is below. The altitude can be determined with the aid of a sextant.

The **azimuth,** z, of a celestial point B is the angle, at the zenith, between the observer's meridian and the vertical circle of B. It is measured from the north point around through the east from 0° to 360°.

There is a system of coordinates on the celestial sphere that does not depend on the time or place of observation. Astronomical bodies are often located by means of declination, which we have defined, and right ascension. The *right ascension* of a point on the celestial sphere is the angle at the pole between the hour circle of the point and a fixed hour circle determined, like the Greenwich meridian, by international agreement. It is measured eastward from 0 hours to 24 hours. This coordinate system is usually used in astronomy.

The **astronomical triangle** of a heavenly body is the spherical triangle formed by the celestial pole, the zenith Z, and the celestial point B, the projection of the heavenly body upon the celestial sphere. Customarily, if the observer is at a northern latitude, the north pole P_n is used. Likewise, if he is south of the equator, the south pole P_s is used. As indicated in Fig. 79, the astronomical triangle for an observer north of the equator has the following parts:

$$\textbf{sides:}\begin{cases}\text{arc } P_nZ = 90° - L = \text{colatitude of observer,}\\ \qquad\qquad\qquad\text{(codeclination of the zenith),}\\ \text{arc } P_nB = 90° - d = \text{codeclination of star,}\\ \text{arc } BZ = 90° - h = \text{co-altitude of star;}\end{cases}$$

$$\textbf{angles:}\begin{cases}\text{angle } ZP_nB = t = \text{hour angle of star,}\\ \text{angle } P_nZB = z = \text{azimuth of star,}\\ \text{angle } ZBP_n = \phi = \text{position angle of star.}\end{cases}$$

The **position angle,** ϕ, of a heavenly body is the angle at the heavenly body between its hour circle and its vertical circle.

109. Problems of Navigation and Astronomy.

We shall now present several of the important problems in astronomy and navigation that require the solution of spherical triangles.

I. *To find the time of day.*

We assume that the latitude and the date are known. The declination of the sun can be determined from the *Nautical Alma-*

nac for the given date, and its altitude can be measured by means of a sextant. Then, since the latitude of the observer is known, the three sides of the astronomical triangle are known and the hour angle can be found by the method of Case V, Chapter XIV. At noon the sun will be in the observer's meridian. Consequently, the hour angle, t, of the sun represents the time since noon. We shall generally disregard the small difference between a solar and a sidereal hour. The reader must understand that this time, t, which has been found is the observer's local time and may vary, but not more than an hour, from the standard time of the observer's time zone. [This statement, of course, ignores the fact that the local time will run from 0 to 24 hours, while standard time will be given in terms of two twelve hour periods.]

Solution: See p. 160.

Illustration.

A navigator at latitude 48°10′N measures the altitude of the sun to be 25°18′. The *Nautical Almanac* gives the declination of the sun at that date to be 5°20′S.

Note: In the above illustration we have assumed that the navigator saw the sun to the west of the meridian. If the sun were to the east, then the hour angle would be $24^h - 2^h46^m24^s$ and the time would be $21^h13^m36^s$ since noon.

II. *To find the observer's longitude.*

This problem is an extension of the preceding problem. We assume that the observer knows his own latitude. Fig. 81 represents the hour circles of Greenwich (G), of the observer's zenith (O), and of the sun (S). By the procedure of Problem I, the observer determines his local time t. Then by means of radio or telegraph he learns Greenwich local time t′. Considered as an angle, t′ is the hour angle of the sun as observed from Greenwich.

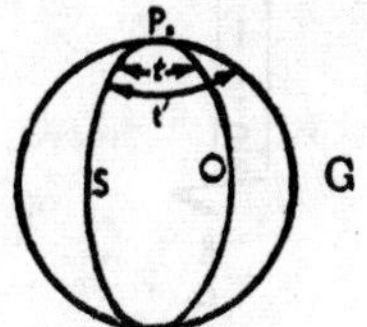

Fig. 81

Solution:

Given: $L = 48°10'N$, $d = 5°20'S = -5°20'$, $h = 25°18'$.

To find log p: $$p = \sqrt{\frac{\sin(s-a)\sin(s-b)\sin(s-c)}{\sin s}}$$	$a = 90° - L = 41°50'$ $s - a = 59°6'$ $b = 90° - d = 95°20'$ $s - b = 5°36'$ $c = 90° - h = 64°42'$ $s - c = 36°14'$ $2s = 201°52'$ $s = 100°56'$ ⟵ check ⟶ $s = 100°56'$
	$l \sin(s-a) = 9.93352 - 10$ $l \sin(s-b) = 8.98937 - 10$ $l \sin(s-c) = 9.77164 - 10$ sum $= 8.69453 - 10$ $l \sin s = 9.99204 - 10$ $2 \log p = 8.70249 - 10$ $\log p = 9.35125 - 10$
To find t: $$\tan\frac{t}{2} = \frac{p}{\sin(s-c)}$$	$\log p = 9.35125 - 10$ $l \sin(s-c) = 9.77164 - 10$ $l \tan t/2 = 9.57961 - 10$ $\frac{t}{2} = 20°47'57''$ $t = 41°35'54'' = 2^h46^m24^s$.

Then $t' - t$ is the angle between the Greenwich meridian and the observer's meridian, which angle is the longitude of the observer. In degrees the longitude is $\lambda = 15(t' - t)^\circ$.

Illustration.

In the preceding illustration the local time t of the navigator was found to be $2^h46^m24^s$ past noon. Suppose he receives by radio the Greenwich local time t' of $14^h40^m32^s$. Then the longitude of the navigator is $t' - t = 11^h54^m8^s$W $= 178°32'0''$ W

III. *To find the time of sunrise (or sunset).*

This problem is a special case of Problem I. At sunrise the sun is in the horizon; its altitude is therefore 0° and arc BZ of Fig. 79 is 90°. The spherical triangle has become a quadrantal triangle and can be solved by the method indicated in §99, Chapter XIII.

IV. *To find the observer's latitude.*

We assume that the observer knows the local time and therefore the hour angle of the sun. As in Problem I, he can determine the altitude and declination of the sun. Then, in the astronomical triangle of the sun (Fig. 79), the two sides BZ and BP_n and the angle t, opposite the side BZ, are known. The triangle can now be solved by the method of Case III, Chapter XIV, to determine the third side, PZ, which is the colatitude of the observer.

Illustration.

A navigator has determined the local time to be $2^h10^m15^s$. The declination of the sun is found to be 8°7′16″. By means of the sextant he finds that the sun's altitude is 65°20′32″. He desires to know his latitude.

Solution:

In the spherical triangle we have

$$a = \widehat{BZ} = 90° - 65°20'32'' = 24°39'28'',$$

$$b = \widehat{BP}_n = 90° - 8°7'16'' = 81°52'44'',$$

$$B = t = 2^h10^m15^s = 32°33'45''.$$

Using Case III we solve this triangle and obtain

$A = 13°06'32''$, $C = 147°59'12''$, $c = 102°48'46''$.

Now $c = \widehat{P_nZ}$ = colatitude of observer = $102°48'46''$.

The latitude of the observer $= 90° - 102°48'46''$

$= -12°48'46''$

$= 12°48'46''$ S.

IVa. *Second method for IV.*

A simple method by which an observer may determine his latitude is illustrated by means of Fig. 82. The latitude L of the observer is equal to the altitude of the pole P_n. This is true because the latitude L is equal to the declination of the observer's zenith (Z) north (or south) of the equator (Q). Since $\widehat{NZ} = \widehat{P_nQ} = 90°$, $L = \widehat{QZ} = \widehat{P_nQ} - \widehat{P_nZ} = \widehat{NZ} - \widehat{P_nZ} = \widehat{NP_n}$, the altitude

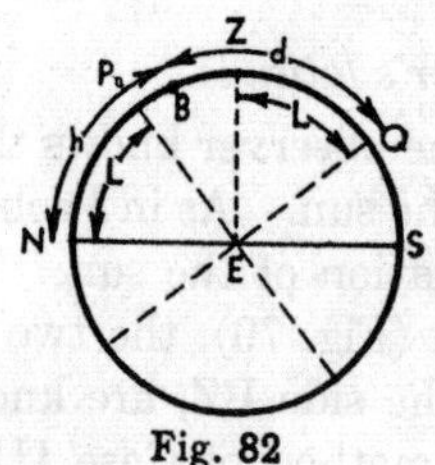

Fig. 82

of the pole. Suppose the observer determines the altitude h of a star B at the moment B crosses his meridian. The latitude $\widehat{NP_n} = h - \widehat{P_nB} = h -$ (codeclination of B). Therefore if the observer reads the declination d of B from the *Nautical Almanac*, he may obtain the desired latitude from the following relation:

$$L = h - (90° - d).$$

V. *To find the altitude and azimuth of a star.*

When the observer's latitude is known, the altitude and azimuth of any star whose declination and hour angle are known can be found. In the astronomical triangle of the star two sides P_nZ and P_nB and the included angle t are known. The triangle can be readily solved by the method of Case I, Chapter XIV.

110. Exercise 22.

1. Find the shortest air route and the initial course from Los Angeles (lat. 34°3′N, long. 118°15′W) to Tokyo (lat. 35°39′N, long. 139°45′E.)
2. Find the shortest time for an airplane flying at an average speed of 250 mi. per hours, to fly from New York (longitude, 73°58′W; latitude, 40°49′N) to Greenwich (longitude 0°; latitude, 51°29′N).
3. Find the solar time of a place on earth whose latitude is 38°55′N at a moment in the afternoon when the sun's declination is 18°20′, if the sun's altitude is observed to be 49°36′.
4. Find the time of sunset at a place with latitude 41°40′N on the date when the sun's declination is −20°15′.

CHAPTER XVI

PROJECTIONS

111. Introduction.

Of importance among the applications of spherical trigonometry are the various types of projections of the terrestrial sphere on a plane, which projections are the basis of map making.

A **map** may be considered the image of an object, or of a number of objects. As commonly thought of, however, a map represents the *dimensions* of an object or the distances between objects rather than any other characteristic of the objects themselves. Usually the dimensions of the image are smaller than those of the object. In order that the map or image may be used to give accurate information about the lengths or distances of an object, it is convenient that the dimensions of the map be a fixed fraction of the corresponding dimensions of the object, so that this fraction when multiplied by a dimension of the object will give the corresponding dimension of the image or map. This fraction is known as the **scale** of the map. It must be understood that the scale applies only to dimensions, i.e., lengths, and not to other relationships between object and image. In Fig. 83, triangle abc is the map of triangle ABC, the scale being $\frac{1}{2}$. That is, ab = $\frac{1}{2}$ AB, bc = $\frac{1}{2}$ BC. This does *not* mean, however, that the area of $\triangle$abc is one-half the area of $\triangle$ABC.

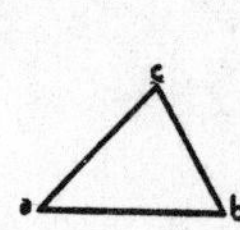

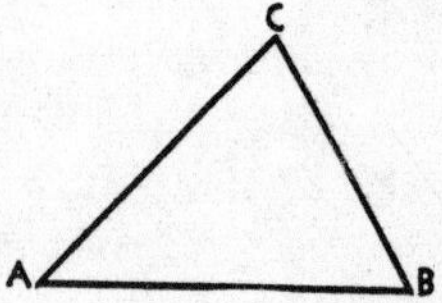

Fig. 83

Since dimension is measured in terms of length, we may give the following more careful definition of scale.

Definition: The **scale** of a map is the ratio of the length between two points on the map to the length between the corresponding two points on the object that is mapped.

Length is always considered positive; consequently the scale will always be a positive fraction.

112. Perspective Maps.

The simplest maps are those which are the images of plane objects; and the most common of these is the perspective map. In Fig. 84 triangle ABC is the object. The point O is chosen and the lines AO, OB, and OC drawn. The scale is chosen (in this case $\frac{1}{2}$) and a, b, and c are found so that $Oa = \frac{1}{2}\,OA$, $Ob = \frac{1}{2}\,OB$, and $Oc = \frac{1}{2}\,OC$. Then Δabc is the map of ΔABC, and it is seen from similarity of triangles that the sides of Δabc are one-half those of ΔABC. The triangles are said to be in **perspective** since all the lines joining pairs of corresponding points pass through one point—the **center of perspectivity** O.

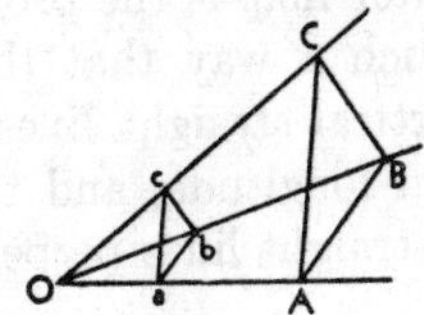

Fig. 84

Maps of this type are used in architecture and surveying, and also, with the point O above the plane of the object, in aerial photography.

Somewhat more complex than those of a plane are the perspective maps of **developable** surfaces. These are surfaces such as the cone and cylinder which can be constructed from a plane by merely bending it without stretching or contraction. To construct a perspective map of a cylinder, for example, the above method is first used to map the cylinder upon a smaller cylinder which is of the desired scale. The cylinder is then cut along a straight line parallel to its axis and laid out flat.

In the construction of many maps, distance is not preserved and, in fact, in many cases is not even represented according to a fixed scale. It is, however, important to keep *directions* unchanged. We will, therefore, concern ourselves with so-called **conformal** maps, particularly those mapping a sphere onto a plane.

Definition. The map of a sphere onto a plane is **conformal** if the angle between any two arcs intersecting at a point on the sphere is equal to the angle between the projections of the arcs on the plane.

Ordinarily, conformal projection does not preserve *size*; that is, neither distance nor area; nor does it set up a scale of lengths. But it does preserve *shape*.

113. Mercator Maps.

Definition. A **loxodrome,** or **rhumb line,** is a curve which intersects all meridians at a constant angle.

Consequently, a ship sailing a loxodrome is sailing a constant course. The great circle arcs, however, and not the loxodromes, are the shortest distances between points.

Definition. The **Mercator map** is the projection of a sphere on a plane in such a way that the meridians appear as parallel vertical straight lines, equidistant for equal distances in longitude, and the parallels of latitude appear as straight lines perpendicular to the first set of lines.

The projections of loxodromes on the Mercator map are straight lines which intersect the meridian projections at a constant angle. The **Mercator chart** is extensively used in navigation, since in practice the course of a ship is changed at regular intervals so that it follows a series of loxodromes which approximate a great circle arc and which can be plotted as straight lines.

In Fig. 85, we see that

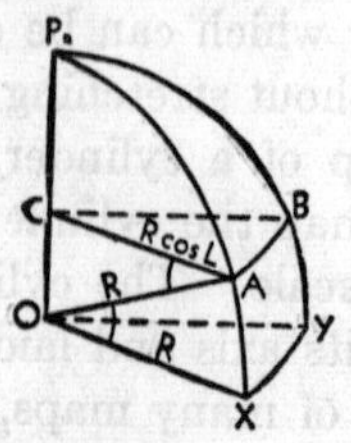

Fig. 85

XY = difference in longitude between A and B.
R = radius of the earth.
L = latitude of arc AB.
Suppose XY is measured in seconds,
R in nautical miles.

Then in the sector XOY, $\angle XOY = \frac{\widehat{XY}}{R}$ radians (see §13); $\angle ACB = \angle XOY$. From sector ACB, $\angle ACB = \frac{\widehat{AB}}{AC}$ radians. Then

$$\widehat{AB} = \angle ACB\ (\overline{AC}) = \frac{\widehat{XY}}{R}\ (R \cos L) = \widehat{XY} \cos L = \text{the number of nautical miles in the arc AB.}$$

On a Mercator chart, however, AB = XY. Therefore, the **distortion** of AB, defined as the ratio of the projection of AB to AB, is

$$\frac{\widehat{XY}}{\widehat{XY} \cos L} = \sec L.$$

To compensate for this distortion at A and B, the two meridians through these points must be expanded by equal amounts so that the scale along the meridians will be the same as the scale along the parallels. In Fig. 86, suppose XY represents 1′ along the

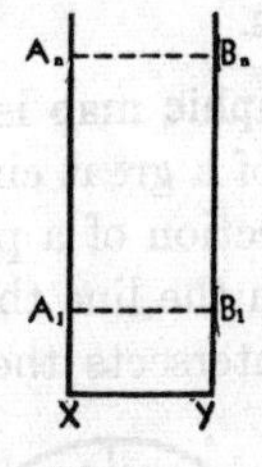

Fig. 86

equator and A_1B_1 is at 1′ north latitude. Then the distortion at A_1B_1 is sec 1′. So $B_1Y = A_1X = XY\ (\sec 1')$. Each minute of latitude along the meridian must be increased by the secant of the latitude. So, if A_nB_n is at n′ north latitude, then

$$B_nY = A_nX = XY\ (\sec 1' + \sec 2' + \cdot\ \cdot\ \cdot + \sec n').$$

The above remarks neglect the fact that the earth is ellipsoidal, not spherical. Consequently the last equation above is accurate only for small distances. In Bowditch's *The American Practical Navigator,* Table 3 is commonly used for constructing Mercator charts; in that table the figures take account of the elliptical shape of the earth. Since for small length of latitude L the ratio of expansion is approximately sec L, we see that for short distances on any loxodrome

$$d_m = d \sec L_m \quad \text{or} \quad d = d_m \cos L_m,$$

where d = actual distance in nautical miles; d_m = map distance in scale units; L_m = mid-latitude of the considered distance d_m.

114. Orthographic Maps.

Definition. An **orthographic map** is the perpendicular projection of a sphere on the plane of a meridian in such a way that the equator and the meridian 90° from the meridian which determines the projection plane are diameters. All other meridians appear as ellipses and all parallels of latitude appear as parallel straight lines.

115. Stereographic Maps.

Stereographic projection is commonly used in the construction of hemispherical maps.

Definition. The **stereographic map** is the projection of a sphere on the plane of a great circle (Fig. 87) in such a way that the projection of a point A on the sphere is the point in which the line through A and a pole of that great circle intersects the plane of the great circle.

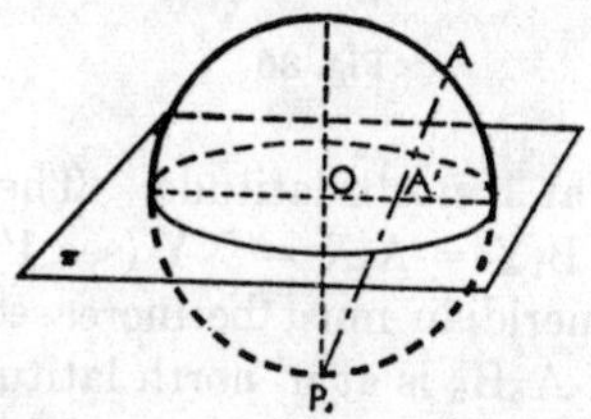

Fig. 87

The pole is called the **center of projection**; the plane, the **primitive plane**; and the great circle, the **primitive circle.** To each primitive plane correspond two centers of projection—a north pole and a south pole. The points of the hemisphere on the side of the primitive plane opposite to the center of projection will project inside the primitive circle. Consequently, to map the northern hemisphere inside a circle the south pole must be taken as the center of projection.

In stereographic projection of the northern or southern hemisphere all meridians will appear as diameters of the equator and all parallels of latitude will appear as circles, concentric with the equator. We state without proof a further important property of stereographic projection:

All other circles on the sphere will project as circles onto the primitive plane. Exceptionally, circles which pass through the center of projection (for example, the meridians mentioned above) *will project as straight lines.*

116. Graphical Constructions.

As a consequence of the last statement of the preceding section, it becomes possible to construct the stereographic projection of a spherical triangle and by means of it to determine the true spherical triangle. In the following pages various graphical constructions based on stereographic projection will be given by means of which many of the problems in the foregoing chapters can be solved. Most of these constructions will be described without proof, as the proofs require more study of solid geometry than has been undertaken in this book. For the proofs the reader may refer to Chapter XV of Bullard and Kiernan's: *Plane and Spherical Trigonometry*, Revised Edition.

I. To construct the angle between two great circles.

Construction. (See Fig. 88.)

a) Assume one great circle to be the primitive circle.
b) Let the second great circle intersect the first in the points M_1 and M_2, and let the angle between them be equal to θ degrees.
c) Draw M_1C so that $\angle OM_1C = \theta$. $M_1C = M_2C$ because we choose C to lie on the perpendicular bisector of M_1M_2.
d) Then the circle arc through M_1 and M_2 with radius equal to

M_1C and center at C will make an angle of θ degrees with the primitive circle.

This construction is based on the fact that the stereographic projection is conformal and that, therefore, the angle between the two great circles is equal to the angle between their projections, which, as indicated in Fig. 88, is the angle between their tangents.

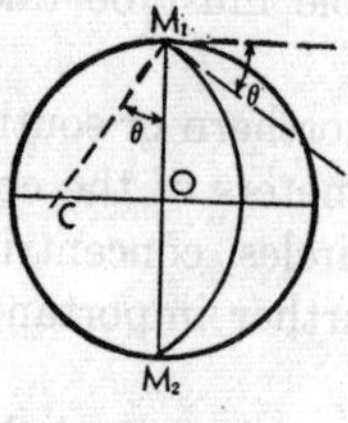

Fig. 88

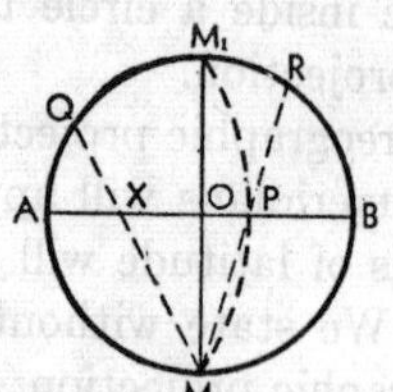

Fig. 89

II. To construct the projection of a great circle when the projection of its pole is given.

Construction. (See Fig. 89.)

a) Let X be the projection of the pole of the desired great circle. Suppose $\overline{P}$ represents the pole of which X is the projection.

b) Draw AB, the projection of the great circle determined by the given pole $\overline{P}$ and the center of projection O. The poles M_1 and M_2 of this great circle can be found by drawing the perpendicular bisector of AB. Furthermore, if M_1 and M_2 are poles of a great circle through $\overline{P}$, then there is a great circle through M_1 and M_2 of which $\overline{P}$ is a pole. In order to determine this circle we must find a third point.

c) Draw the line M_2X and extend it to cut the primitive circle at Q.

d) Construct on the primitive circle the arc QR equal to 90°, and connect M_2 with R. The point P at the intersection of AB with M_2R is the desired third point.

e) By the methods of elementary geometry construct the circle passing through M_1, M_2, and P. This is the desired circle.

III. To construct the projection of the pole of a great circle when the projection of the great circle is given.

Construction. This construction is the converse of the preceding one. (See Fig. 90.)

a) M_1PM_2 is a great circle projection.
b) Draw M_1P and extend it to intersect the primitive circle at A.
c) Construct M_1B so that arc AB = 90°.
d) The point X, the intersection of M_1B and OP, is then the pole of M_1PM_2.

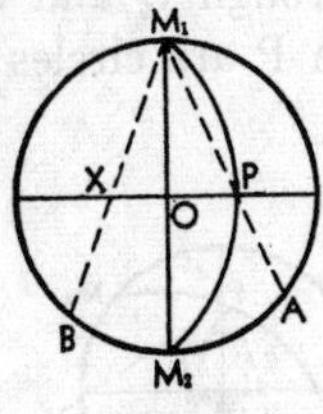

Fig. 90

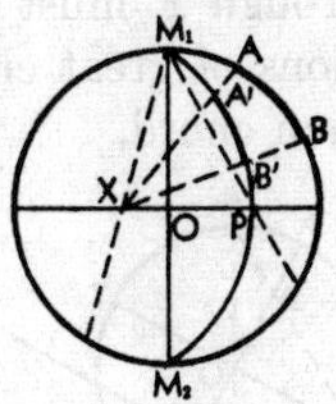

Fig. 91

IV. **To construct the projected length of a given arc.**

Construction. (See Fig. 91.)

a) Let M_1PM_2 be the projection of a great circle.
b) Let AB represent the true length of an arc on the great circle whose projection is M_1PM_2 and its true position with respect to M_1 and M_2. The problem is to find the projected length of AB (that is, the projection of A and B on the projected great circle arc M_1PM_2).
c) First construct X, the projection of the pole, by the method of Construction III. The lines joining X with A and B determine the projected arc A′B′.

Construction IV can be reversed to find the true length of an arc whose projection A′B′ is given.

V. **To construct the locus of the centers of the projections of all great circles which pass through a given point.**

Construction. (See Fig. 92.)

a) Let P be the projection of the given point. Call the given point $\overline{P}$.
b) Draw the extended diameter AB through P and the diameter ST perpendicular to it.
c) Extend SP to determine P′ on the primitive circle.
d) Extend P′O to determine Q′ on the primitive circle.
e) Extend SQ′ to determine Q on the line AB.
f) AB is the projection of the great circle through $\overline{P}$ perpendic-

ular to the primitive circle. Q is the projection of the point $\overline{Q}$, 180° from $\overline{P}$ on this great circle.

g) Now draw the perpendicular bisector of PQ. The centers of the projections of all great circles which pass through P lie on this perpendicular bisector of PQ; for all great circles that pass through $\overline{P}$ must also pass through $\overline{Q}$ and therefore all projections of great circles through P are circles with PQ as a common chord.

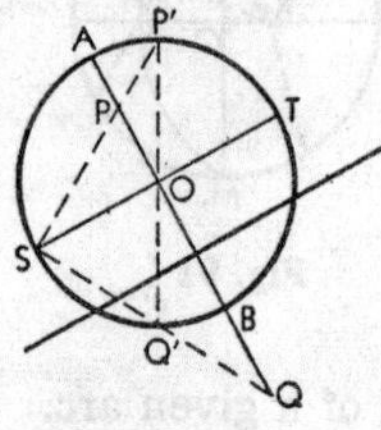

Fig. 92

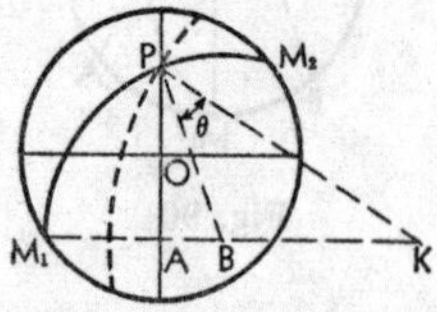

Fig. 93

VI. To construct the projection of a great circle which makes a given angle with a given great circle at a given point.

Construction. (See Fig. 93.)

a) Let M_1PM_2 be the projection of the given great circle; let P be the projection of the given point; and let θ be the given angle.

b) Construct the line M_1A of centers of all projections of great circles which pass through P. (Construction V.)

c) Construct the center B of M_1PM_2.

d) Construct K so that $\angle BPK$ is equal to the given angle θ.

e) K is then the center of the projection of the desired great circle through P which makes the angle θ with M_1PM_2 at P.

VII. To construct any circle of a sphere if its polar distance and the projection of its pole are given.

Construction. We shall consider here only the case when the projection of the pole lies on the primitive circle.

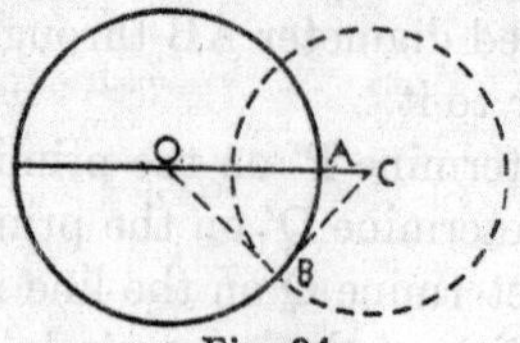

Fig. 94

a) Let A be the projection of the pole of a circle and let arc AB be the true measure of its polar distance.
b) Construct the extended diameter OA.
c) Construct the tangent to the primitive circle at B.
d) The intersection C of the tangent with OA is the center of the projected circle and its radius is CB.

117. Graphical Solutions of Spherical Triangles.

The preceding seven constructions enable us to project a spherical triangle if we have three parts given. The stereographic projections of the three unknown parts are then found. If one of these unknown parts is an angle then the true angle of the spherical triangle is equal to the projected angle. If the unknown part is a side, the true length of the side can be found from its projection by means of Construction IV. Since the graphical constructions propose to solve the same problems that were solved analytically in Chapter XIV, we would here also distinguish six cases. We shall, however, present only the graphical solutions to the following three cases:

Case 1. Given two sides and the included angle.
Case 2. Given two sides and the angle opposite one.
Case 3. Given three sides.

The remaining three cases of Chapter XIV can be solved similarly to the above three by means of polar triangles.

In all of the following constructions both sides and angles are measured by means of a protractor. The accuracy of the solution of a spherical triangle by the graphical method is consequently to be considered only as great as the accuracy of a draftsman's protractor.

Case 1. Given two sides and the included angle.

Illustration.

Given A, b, c. Find a, B, C.

Construction: (See Fig. 95.)

a) Construct arc AC = b on the primitive circle.
b) Construct angle A by the method of Construction I. The projection of arc c lies along the circle determined by this construction.
c) Measure off on this circle, by the method of Construction

IV, the arc AB which is the projected length of the true arc equal to c.

d) By the method of Construction V, draw the line of centers of all great circles passing through the point B.

e) Similarly construct the line of centers of all great circles passing through the point C.

f) The intersection, P, of the two lines constructed in (d) and (e) is the center of the projected arc BC, which can now be drawn.

g) By the converse of Construction IV, the true length of BC (which is a) can be found. The true magnitudes of angles B and C can be measured from the projected triangle, or by a converse of Construction I.

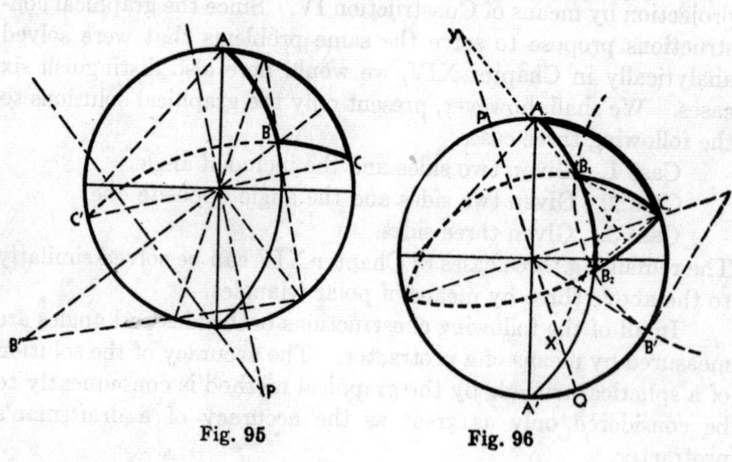

Fig. 95 Fig. 96

Case 2. Given two sides and the angle opposite one.

Illustration.

Given a, b, A. Find B, C, c.

Construction: (See Fig. 96.)

a) Construct arc AC = b on the primitive circle.

b) Construct $\angle A$ and the great circle AA' by the method of Construction I.

c) Construct the projection of a circle with pole at C and with a polar distance $CB' = a$ by the method of Construction VII.

d) The circle of step (c) determines two points B_1 and B_2 on AA' which have a as the true polar distance. As we have previously noted, in the discussion of Case 3 in Chapter XIV, this is the ambiguous case; here there may be two solutions, the triangles AB_1C and AB_2C.

e) The arcs B_1C and B_2C may be constructed by determining their poles X and Y. The pole X is the intersection of the perpendicular bisector of B_1C with the line PQ of centers of all projected great circles through C (Construction V). The pole Y is the intersection of PQ with the perpendicular bisector of B_2C.

f) The true lengths of AB_1 and AB_2 are again found by the method of Construction IV.

Case 3. Given the three sides; a, b, c. Find A, B, and C.

Construction:

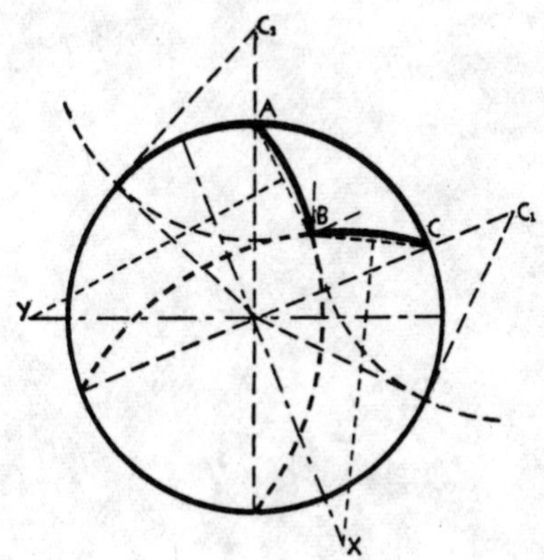

Fig. 97

a) On the primitive circle construct arc $AC = b$.

b) With A and C as poles and with polar distances respectively equal to c and a construct the two circles by the method of Construction VII. These circles intersect in B.

c) Construct the centers X and Y of arcs BC and AC by the method described in Step (2) of Case 2.

d) The magnitude of the angles, A, B, and C may be measured with a protractor. The magnitude of each angle, say A, is found best by measuring the angle between the radii YA and OA of the two projected great circles AB and AC which intersect at A.

FINAL EXAMINATIONS

Examination I. (*Three Hours.*)

1. Place the angle θ in the standard position and define the six trigonometric functions of θ. Using these definitions prove the following identities:

   a) $\cos \theta = \dfrac{\sin \theta}{\tan \theta}$, b) $\sec^2\theta \cot^2\theta = \cot^2\theta + 1$.

2. a) Express in radian measure: 270°, 21°, 780°.
   b) Express in degree measure: 2 radians, 1/3 radians, $\pi/9$ radians.
   c) Draw the graph for $y = 2 \sin x$.

3. Prove the following identities:

   a) $\dfrac{\sin x + \cot x}{\cos x \sin x} - 1 = \cot^2 x + \sec x$;

   b) $\tan 2x = \dfrac{2}{\cot x - \tan x}$;

   c) $\dfrac{\sin 4x - \sin 2x}{\sin 4x + \sin 2x} = \dfrac{\tan x}{\tan 3x}$.

4. Complete the following identities and prove one of them:

   a) $\sin x - \sin y =$ c) $\tan 2\theta =$

   b) $2 \sin x \cos y =$ d) $\cos x/2 =$

5. Find all the solutions between 0° and 360°, inclusive, for

   a) $[\cot x - 1][3 \cot x + \sqrt{3}] = 0$;

   b) $\tan (2x + 10°) = 1$;

   c) $\cot x + \tan x \csc x - \tan x = 0$.

6. Evaluate:

   a) $\sin \text{Arcsin } \frac{1}{2}$; b) $\tan (\frac{1}{2} \text{ Arccos } \frac{4}{5})$; c) $\text{Arccos } \frac{1}{2}$.

7. a) Find the area of the triangle ABC, if a = 12, c = 7, and B = 30°.

   b) Find the area of the triangle ABC, if a = 21, b = 9, and c = 20.

8. An aviator observes that the angle of depression of two points, A due N, B due S, are 30° and 75°, respectively. If A and B are 3/2 miles apart in the same horizontal plane, find the elevation of the aviator.

9. A surveyor's notebook reads: AB: 32 ft. due E; BC 13 ft., S23°E; CD: 19 ft., N52°30′E. Transfer the data to oblique triangles and find the distance BD.

10. Evaluate:

    a) $\sin(-17°19')$; c) $\cos(-115°49')$;

    b) $\tan 137°23'$; d) $\cot(-313°13')$.

Examination II. (*Three Hours.*)

1. Given the plane triangle ABC, in which a = 1002, b = 2338, $\gamma = 21°47'$. Find α, β, and c.

2. What angle is subtended at the eye by the Empire State Building (height 1248 ft.) when seen from a distance of 6390 ft.?

3. a) Evaluate cot [Arccot x − Arctan y].

   b) Given sin x = 3/5, cos y = −8/17, and x and y both in the second quadrant, find cos (x + y).

4. The sides of a triangle are 7, 15, and 20. Find the angle opposite the side of length 15.

5. An aviator finds that, after traveling 300 miles in an easterly direction and then 250 miles in a southerly direction, he is 400 miles due east of his starting point. Without logarithms, find the direction in which he traveled on the first leg of his journey.

6. Complete the following identities:

   a) $\sin x + \sin y =$ c) $\sin(\alpha + \beta) =$

   b) $\sin^2 \phi =$ d) $\tan x/2 =$

e) $\cos^2\alpha - \sin^2\alpha =$ g) $2 \cos x \cos y =$

f) $\cot (\alpha - \beta) =$ h) $\cos x/2 =$

7. Prove the following identities:

a) $\tan \frac{x}{2} = \cot \frac{x}{2} - 2 \cot x$;

b) $\cos 2\phi + \cos 2\theta = 2 \cos (\phi + \theta) \cos (\phi - \theta)$;

c) $\cot x + \tan x = \sec x \csc x$.

8. a) Draw the graph of $y = 2 \cos 3x$.

b) Solve for x: $\sin x = 2$; $\csc x = 1/2$; $\tan x = -\infty$.

9. From the tables find the following expressions:

a) $\log \tan 53°17'$; c) $\cot 135°13'$;

b) $\log \cos 47°38'$; d) $\sin (-135°24')$.

10. A pie has a radius of 6 inches. What is the area of a 33° section?

Examination III. (*Three Hours.*)

1. By use of four-place tables evaluate:

a) $\cos 35°12' + \tan 23°17' + \sin \pi/6$;

b) $\tan \pi/7$.

c) $\cot 133°13'$.

2. a) Given $\cot x = -1.0566$, find all values of x between 0° and 360°.

b) Given $\cos x = 5/12$ and $\sin x$ is negative. Find sec x, cot x, and csc x in the form of fractions.

3. If a force of 3 pounds pulls due east and another of 2 pounds pulls in a direction 30° west of north, what is the magnitude of the resultant force?

4. Find the area and the angles of a triangle whose sides are

$$a = 2.236, \quad b = 2.449, \quad c = 2.646.$$

5. a) Prove the identity: $\tan x = \frac{1 - \cos 2x}{\sin 2x}$.

b) Prove that

$$\cos (30° + x) \cos (30° - x) - \sin (30° + x) \sin (30° - x) = \tfrac{1}{2}.$$

6. An airplane is flying horizontally due east from us at a speed of 135 miles per hour. We observe the angle of elevation of the airplane to be 23°17′; twelve seconds later its angle of elevation is 21°39′. How high above us is the airplane flying?

7. At two points A and B, 1000 ft. apart on the same level and same side of a hill, observers find the angles of elevation of the top T of a tower on the hill, located in the same vertical plane as A and B, to be $\alpha = 10°22'$, $\beta = 15°51'$. Find the height of T above the line AB.

8. A wheel, 4 ft. in diameter, makes 420 revolutions per minute. Find the linear velocity of a point on the rim in feet per seconds.

9. Sketch the graph of $y = 2 \sin x/2$ over one period.

10. Solve the triangle ABC, if a = 13.46, b = 16.39, and $\alpha =$ 42°17′.

Examination IV.

You will need no tables or book for this examination. It should be completed in two hours.

1. Each of the expressions below is one member of an important trigonometric formula. Complete each of them by supplying the other member.

   (a) $1 - \sin^2 x =$ ....;

   (b) $\cos^2 x - \sin^2 x =$ ....;

   (c) $\sin (x - y) =$ ....;

   (d) $\cos x/2 =$ ....;

   (e) $\tan (x + y) =$ ....;

   (f) $\sin x - \sin y =$ ....

2. Fill out the table of values below.

θ	0°	30°	45°	90°	180°	270°
$\sin \theta$						
$\cos \theta$						
$\tan \theta$						

3. Express:
   (a) the angle 40° in radian measure . . . . ;
   (b) the angle 3 $\pi/2$ in both degrees and mils . . . . ;
   (c) the function csc (292°) in terms of a function of a positive acute angle . . . . ;
   (d) the function cot (−137°) in terms of a function of a positive acute angle . . . .

4. Sketch the graphs of
   (a) $y = \sin x$, from -2π to 2π; (b) $\rho = 6 \cos \theta$.

5. Simplify each of the expressions below by reducing it to a single trigonometric function of the angle x.

   (a) $\tan (90° - x)$; (b) $\sec (180° - x)$; (c) $\dfrac{\cos (-x)}{\sin x}$;

   (d) $\dfrac{1 + \cos 2x}{\sin 2x}$; (e) $\left[\sin \dfrac{x}{2} + \cos \dfrac{x}{2}\right]^2 - 1$;

   (f) $\sin 3x \cos 2x - \cos 3x \sin 2x$.

6. Find the (a) principal value of $\mathrm{Sin}^{-1} (\sin 27°)$;
   (b) principal value of Arc cos $(-\frac{1}{2})$;
   (c) value of tan (Arc tan $\frac{4}{7}$);
   (d) values of $\cot (\mathrm{Sec}^{-1}[-3])$;
   (e) value of $\sin (2\ \mathrm{Tan}^{-1}[1/3])$.

7. (a) Find side a in triangle ABC, where angle A = 45°, angle C = 30°, and side c = $5\sqrt{2}$.
   (b) Find side c in triangle ABC, where side a = 3, side b = 7, and angle C = 120°.
   (c) State the Law of Tangents for triangles;
   State a formula for $\tan \dfrac{\alpha}{2}$ in terms of s and the sides a, b, c of a triangle.

8. Find all values of x between 0° and 360° satisfying the equation $\cos^2 x - \sin x \cos x = 0$.

ANSWERS TO THE EXERCISES

In any problem which asks for the functions of an angle, the answers are given in the following order: sine, cosine, tangent, cotangent, secant, and cosecant.

Exercise 1.

1. **(a):** 13, α: $\left(\frac{5}{13}, \frac{12}{13}, \frac{5}{12}, \frac{12}{5}, \frac{13}{12}, \frac{13}{5}\right)$; β: $\left(\frac{12}{13}, \frac{5}{13}, \frac{12}{5}, \frac{5}{12}, \frac{13}{5}, \frac{13}{12}\right)$.

(b): 2, α: $\left(\frac{1}{2}, \frac{\sqrt{3}}{2}, \frac{1}{3}\sqrt{3}, \sqrt{3}, \frac{2}{3}\sqrt{3}, 2\right)$;

β: $\left(\frac{\sqrt{3}}{2}, \frac{1}{2}, \sqrt{3}, \frac{1}{3}\sqrt{3}, 2, \frac{2}{3}\sqrt{3}\right)$.

(c): 1, α: $\left(\frac{1}{2}\sqrt{2}, \frac{1}{2}\sqrt{2}, 1, 1, \sqrt{2}, \sqrt{2}\right)$; β: (same as α).

(d): $\sqrt{2}$, α, β: (same as [c]).

(e): 5, α: $\left(\frac{3}{5}, \frac{4}{5}, \frac{3}{4}, \frac{4}{3}, \frac{5}{4}, \frac{5}{3}\right)$; β: $\left(\frac{4}{5}, \frac{3}{5}, \frac{4}{3}, \frac{3}{4}, \frac{5}{3}, \frac{5}{4}\right)$.

(f): 7, α: $\left(\frac{4}{65}\sqrt{65}, \frac{7}{65}\sqrt{65}, \frac{4}{7}, \frac{7}{4}, \frac{\sqrt{65}}{7}, \frac{\sqrt{65}}{4}\right)$;

β: $\left(\frac{7}{65}\sqrt{65}, \frac{4}{65}\sqrt{65}, \frac{7}{4}, \frac{4}{7}, \frac{\sqrt{65}}{4}, \frac{\sqrt{65}}{7}\right)$.

2. **(a):** $\frac{3}{5}, \frac{4}{5}, \frac{3}{4}, \frac{5}{4}, \frac{5}{3}$. **(b):** $\frac{\sqrt{5}}{3}, \frac{2}{5}\sqrt{5}, \frac{\sqrt{5}}{2}, \frac{3}{5}\sqrt{5}, \frac{3}{2}$.

(c): $\frac{1}{2}\sqrt{2}, \frac{1}{2}\sqrt{2}, 1, \sqrt{2}, \sqrt{2}$. **(d):** $\frac{12}{13}, \frac{5}{13}, \frac{12}{5}, \frac{5}{12}, \frac{13}{12}$.

(e): $\frac{1}{2}\sqrt{2}, \frac{1}{2}\sqrt{2}, 1, 1, \sqrt{2}$. **(f):** $\frac{3}{5}, \frac{3}{4}, \frac{4}{3}, \frac{5}{4}, \frac{5}{3}$.

3. **(a):** cos 45°; **(b):** sin 22°; **(c):** cot 53°42′;
(d): tan 10°7′; **(e):** csc 53°; **(f):** sec 78°.

Exercise 2.

I. **1.** 0.7627. **2.** 0.4120. **3.** 0.9787. **4.** 0.9703.
5. 1.8165. **6.** 3.0523. **7.** 0.8847. **8.** 0.7944.
9. 0.9190. **10.** 0.7879. **11.** 3.3680. **12.** 7.0748.
13. 0.9323. **14.** 0.6085. **15.** 0.8556.

II. **1.** 13°20′. **2.** 71°20′. **3.** 23°10′. **4.** 55°20′.
5. 47°06′. **6.** 30°06′. **7.** 56°15.7′. **8.** 36°52.2′.

Exercise 3.

1. b = 40.706; c = 45.354; 63°50′.
2. b = 101.02; c = 112.715; 26°20′.
3. α = 34°46.5′; β = 55°13.5′; 219.1.
4. α = 25°32′; β = 64°28′; 5.23.
5. α = 70°32′; β = 19°28′; .57.
6. a = 21.84; b = 20.57; 46°43′.

Exercise 4.

1. 0.1745329; 1.3089969; 2.1852147; 4.4807305.
2. 90°; 330°; 135°3′28″; 0°56′2″.
3. 3.49066 ins.
4. 30π ft./sec.
5. 9π = 28.27 sq. in.; 2.23965 sq. in.

Exercise 5.

1. −.9744. **2.** .9272. **3.** 4.0108. **4.** −.9325.
5. −.5476. **6.** −.7280. **7.** −1.5311. **8.** 1.3481.
9. .9600.

Exercise 6.

1. −.9155. **2.** −.8420. **3.** .8632. **4.** 1.0416.
5. −.9336 **6.** −.8229.
7. 159°13.7′; 339°13.7′. **8.** 30°34′; 210°34′.
9. 200°54′; 339°06′. **10.** 50°37′; 309°23′.

Exercise 9.

1. 120°, 240°, 180°. **2.** 90°, 270°, 30°, 210°.
3. 153°26′, 333°26′, 26°34′, 206°34′.
4. 60°, 300°, 0°. **5.** 60°, 240°.

Exercise 10.

8. 60°, 300°. **9.** 30°, 90°, 150°, 210°, 270°, 330°.
10. 30°, 150°, 270°. **11.** 18°26′.
12. 30°, 150°, 90°, 270°, (22°30′ + n45°).

Exercise 11.

1. 1.92267.	**2.** 5.19312.	**3.** 0.22891.
4. 2.72198.	**5.** 7.59802 − 10.	**6.** 9.97621 − 10.
7. 8.45869 − 10.	**8.** 4.75587 − 10.	**9.** 4.88416.
10. 760.50.	**11.** 1.5950.	**12.** 596800.
13. 0.050553.	**14.** .00033347.	**15.** 0.27702.
16. 98.460.	**17.** 43870.	**18.** 8610.4.

Exercise 12.

1. 31.057. **2.** 4.3618. **3.** 1.7908. **4.** 22.066.

Exercise 13.

1. 9.55988 − 10.	**2.** 9.89820 − 10.	**3.** 9.55155 − 10.
4. 9.96662 − 10.	**5.** 8.62114 − 10.	**6.** 8.29486 − 10.
7. 8.46112 − 10.	**8.** 8.23157 − 10.	**9.** 9.95133 − 10.
10. 9.42026 − 10.	**11.** 0.83326.	**12.** 9.79938 − 10.
13. 39°17′08″.	**14.** 51°34′34″.	**15.** 50°33′21″.
16. 50°57′41″.	**17.** 1°22′00″.	**18.** 89°28′00″.
19. 2°43′01″.	**20.** 88°56′54″.	

Exercise 14.

1. b = 34.208, c = 36.865; 68°7′.
2. a = 9.9580, c = 11.836; 32°43′.
3. a = 0.23888, b = 0.28407; 49°56′19″.
4. α = 48°20′27″, β = 41°39′33″; 151.56.
5. α = 28°45′40″, β = 61°14′20″; 82.040.

Exercise 15.

1. 40°26′07″. **2.** 1334.5 ft. **3.** 889.92 ft., 688.23 ft.
4. W 7°38′51″ S, 23.67 mi./hr.

Exercise 16.

1. b = 610.5, c = 451.33; 43°49′50″; 126490.
2. α_1 = 60°12′17″, β_1 = 82°33′17″; 448.32; 53226;
α_2 = 119°47′43″, β_2 = 22°57′51″; 176.40; 20943.
3. γ = 69°17′22″, α = 43°30′22″; .52164; .095034.
4. α = 44°39′36″, β = 49°36′48″, γ = 85°43′40″; 51.70
5. $\sqrt{7}$; β = 130°53′36″, γ = 19°6′24″; $\sqrt{3}$

Exercise 17.

1. $2n\pi + \frac{\pi}{2}$. **2.** $n\pi + \frac{\pi}{6}$. **3.** $2n\pi \pm \frac{2}{3}\pi$.

4. $2n\pi \pm \frac{\pi}{3}$. **5.** $2n\pi + \frac{\pi}{4}$, $2n\pi + \frac{3}{4}\pi$. **6.** Impossible.

7. $2n\pi + 44°30'$, $2n\pi + 135°30'$.

8. $2n\pi + 72°$, $2n\pi + 288°$.

9. 2/3. **10.** $\pm 3/5$. **11.** -3.

12. $\pm 1/\sqrt{1 - x^2}$. **13.** $\sqrt{1 + y^2}/y$. **14.** $\sqrt{1 - z^2}/z$.

Exercise 18.

1. $-30°$, $-\frac{\pi}{6}$. **2.** $45°$, $\frac{\pi}{4}$. **3.** $60°$, $\frac{\pi}{3}$.

4. 18°10′, .31707. **5.** $\frac{120}{169}$. **6.** $\frac{1}{3}$.

7. $\sqrt{1 - y^2}\sqrt{1 - x^2} + xy$.

8. $\frac{y - x}{\sqrt{(1 + x^2)(1 + y^2)}}$.

Exercise 19.

1. A = B = C = 90°. **2.** A = B = C = 90°.
4. 53°5′ or 126°55′. **5.** 80°35′45″.

Exercise 20.

3. a = 57°13′8″, b = 53°39′16″, B = 58°15′30″.
4. A = 53°24′14″, B = 67°19′19″, c = 71°55′25″.
5. A_1 = 52°30′20″, C_1 = 143°33′46″, c_1 = 131°31′44″;
A_2 = 127°29′40″, C_2 = 36°26′14″, c_2 = 48°28′16″.

Exercise 21.

1. B = 31°25′26″, C = 129°24′30″, a = 85°17′12″.
2. B_1 = 78°42′30″, C_1 = 137°56′32″, c_1 = 138°19′28″;
B_2 = 101°17′30″, C_2 = 91°9′2″, c_2 = 97°4′14″
3. A = B = C = 135°28′20″.

Exercise 22.

1. 4755 nautical miles = 5478 statute miles; N54°W.
2. 13 hrs. 51 min.
3. 2:41 PM.
4. 4:43 PM.

Examination 1.

2. (a) 4.7123, 0.3665, 13.6136.
(b) 114°35′30″, 19°5′48″, 20°.

5. (a) **45°, 225°, 120°, 300°.**
(b) **17°30′, 107°30′, 197°30′, 287°30′.**
(c) **210°, 330°.**
6. (a) 1/2. (b) 1/3. (c) 60°.
7. (a) 21. (b) 89.4.
8. 3960 ft.
9. 20.16 ft.
10. (a) −.2977. (b) −.9201. (c) −.4355.
(d) .9396.

Examination 2.

1. $\alpha = 14°47'54''$; $\beta = 143°25'6''$; $c = 1455.8$.
2. $\alpha = 11°3'$.
3. (a) $\dfrac{x+y}{1-xy}$. (b) −13/85.
4. 36°52′.
5. N52°22′E.
8. (b) Impossible, Impossible, $(2n + 1)90°$.
9. (a) 0.12736. (b) 9.82858 − 10. (c) −1.0076. (d) −.7022.
10. 10.37 sq. in.

Examination 3.

1. (a) 1.7474. (b) .4816. (c) −.9396.
2. (a) 136°35′, 316°35′. (b) 12/5, $-\frac{5}{119}\sqrt{119}$, $-\frac{12}{119}\sqrt{119}$.
3. $\sqrt{7}$.
4. 2.5492; 51°53′12″, 59°30′44″, 68°36′4″.
6. 12157. ft. **7.** 514.31 ft. **8.** 87.96 ft./sec.
10. $\beta_1 = 55°0'34''$, $\gamma_1 = 82°42'26''$, $c_1 = 19.845$;
$\beta_2 = 124°59'26''$, $\gamma_2 = 12°43'34''$, $c_2 = 4.4073$.

INDEX

Numbers refer to pages

Abscissa, 22
Addition formulas, 46
Altitude, 157
Angle,
  complementary, 5
  of depression, 76
  of elevation, 76
  functions of an, 4, 23
  general definition of, 16
  negative, 16
  positive, 16
Angular measure, 16
Angular velocity, 19
Answers to exercises, 182
Antilogarithm, 58
Arc, length of, 19
Area,
  of a sector, 20
  of a segment, 20
  of a triangle, 92
Astronomical triangle, 158
Astronomy, problems of, 158
Auxiliaries S and T, 71

Base of logarithms, 56

Celestial sphere, 155
Characteristic of a logarithm, 57
  rules for, 58
Circle, segment of, 20
Cofunction, 5
Cologarithms, 66
Common logarithms, 56
Complementary angles, 5
  functions of, 6
Constructions, graphical, 169
Coordinates, polar, 29
  rectangular, 22
Cosecant,
  graph of, 38
  of an angle, 4, 23
Cosine,
  graph of, 36
  of an angle, 4, 23
  of a sum, 40
  of multiple angles, 49, 50
Cosines, law of, 81, 115
Cotangent, graph of, 37
  of an angle, 4, 23

Declination, 156
Decreasing function, 33
Degrees to radians, 17
Depression, angle of, 76
Difference of two angles,
  functions of the, 49
Dihedral angle, 105
Double angle formulas, 49

Elevation, angle of, 76
Equations, trigonometric, 29, 43, 54
Equator, 152
  celestial, 156
Examinations, 177
Extraneous solutions, 45

Functions,
  of an acute angle, 4
  of any angle, 23
  of complementary angles, 6
  of double angles, 49
  of half an angle, 50
  of negative angles, 28
  of $(\alpha + \beta)$, 46, 48
  of $(\alpha - \beta)$, 49
  of 30°, 45°, 60°, 7
  of the quadrantal angles, 24
  trigonometric, 3
Fundamental identities, 40

Gauss' formulas, 141, 144
General angle, 16
Geometry, spherical, 103

Graphical solutions
  of spherical triangles, 173
Graphs of the functions, 35

Half-angle formulas,
  for a plane triangle, 90
  for a spherical triangle, 136, 143
Haversine law, 118
Haversine nomogram, 119
Horizon, 156
Hour angle, 156
Hour circle, 156

Identities, 40, 52
Increasing functions, 33
Infinity, 35
Interpolation,
  for logarithms, 61
  for natural functions, 10
Inverse functions, 97
  principal values of, 98
Isosceles triangle, 134

Latitude, 153, 161
Law of cosines,
  for plane triangles, 81
  for spherical triangles, 115
Law of sines,
  for plane triangles, 84
  for spherical triangles, 114
Law of tangents, 89
Linear velocity, 19
Logarithms,
  characteristic of, 57
  common, 56
  computation using, 64
  definition of, 56
  of the functions, 67
  mantissa of, 57
  properties of, 62
Longitude, 152

Mantissa, 57
Maps, 164
  Mercator, 166
  orthographic, 168
  perspective, 165
  stereographic, 168
Meridian, 152
Mil, 17

Nadir, 156
Napier's analogies, 139, 143
Napier's rules, 127
Natural values of functions, 9
Nautical mile, 153
Navigation, problems of, 158
Negative angles, 16
  functions of, 28

Oblique plane triangles, 81
  ambiguous case, 86
  area of, 92
  summary of, 93
Oblique spherical triangles, 136
  solutions of, 143
Ordinate, 22
Origin, 16

Periodicity of functions, 34
Polar coordinates, 29
Polar triangles, 112
Position angle, 158
Position, standard, 23
Primitive circle, 169
Principal values, 98
Products of sines or cosines, 51
Projections, 164
Proportional parts, 61
Pythagorean Theorem, 5

Quadrantal spherical triangles, 134
Quadrants, 22

Radian measure, 17
Radius vector, 16, 22
Rectangular coordinates, 22
Reduction to acute angles, 26
Reference angle, 26
Resultant of two vectors, 78
Right triangles,
  applications of, 74
  formulas for spherical, 123
  solution of plane, 13, 74
  solution of spherical, 129
Rule for characteristic, 58
  of quadrants, 129, 143, 144
  for signs of functions, 24

Secant,
  of an angle, 4, 23
  graph of, 38
Segment of a circle, 20
Significant digits, 8

Signs of functions, 24
Sine,
  of an angle, 4, 23
  graph of, 36
  of a multiple angle, 49, 50
  of a sum, 40
Sphere, 107
Spherical triangles, 110
  oblique, 136, 143
  right, 122
Standard position of an angle, 23
Sunrise, 161
Surveying, 95

Tables, description for,
  of logarithms, 59
  of logarithms of functions, 67
  of natural functions, 9
Tangent,
  of an angle, 4, 23
  graph of, 37
Terrestrial triangle, 152
Terminal side of an angle, 16
Theorem, Pythagorean, 5
Trigonometric equations, 29, 43, 54
  functions, 3

Use of tables, 9, 59

Variation of the functions, 33
Vector, 78, 94
Velocity, 19

Zenith, 156

LOGARITHMIC
AND OTHER
TRIGONOMETRIC TABLES

TABLE OF THE COMMON
LOGARITHMS OF NUMBERS
WITH THE AUXILIARIES S AND T.

N	L 0	1	2	3	4	5	6	7	8	9
0	— ∞	00 000	30 103	47 712	60 206	69 897	77 815	84 510	90 309	95 424
1	00 000	04 139	07 918	11 394	14 613	17 609	20 412	23 045̄	25 527	27 875
2	30 103	32 222	34 242	36 173	38 021	39 794	41 497	43 136	44 716	46 240
3	47 712	49 136	50 515̄	51 851	53 148	54 407	55 630	56 820	57 978	59 106
4	60 206	61 278	62 325̄	63 347	64 345	65 321	66 276	67 210	68 124	69 020
5	69 897	70 757	71 600	72 428	73 239	74 036	74 819	75 587	76 343	77 085̄
6	77 815	78 533	79 239	79 934	80 618	81 291	81 954	82 607	83 251	83 885̄
7	84 510	85 126	85 733	86 332	86 923	87 506	88 081	88 649	89 209	89 763
8	90 309	90 849	91 381	91 908	92 428	92 942	93 450	93 952	94 448	94 939
9	95 424	95 904	96 379	96 848	97 313	97 772	98 227	98 677	99 123	99 564
10	00 000	00 432	00 860	01 284	01 703	02 119	02 531	02 938	03 342	03 743
11	04 139	04 532	04 922	05 308	05 690	06 070	06 446	06 819	07 188	07 555̄
12	07 918	08 279	08 636	08 991	09 342	09 691	10 037	10 380	10 721	11 059
13	11 394	11 727	12 057	12 385	12 710	13 033	13 354	13 672	13 988	14 301
14	14 613	14 922	15 229	15 534	15 836	16 137	16 435	16 732	17 026	17 319
15	17 609	17 898	18 184	18 469	18 752	19 033	19 312	19 590	19 866	20 140
16	20 412	20 683	20 952	21 219	21 484	21 748	22 011	22 272	22 531	22 789
17	23 045̄	23 300	23 553	23 805̄	24 055̄	24 304	24 551	24 797	25 042	25 285
18	25 527	25 768	26 007	26 245	26 482	26 717	26 951	27 184	27 416	27 646
19	27 875	28 103	28 330	28 556	28 780	29 003	29 226	29 447	29 667	29 885
20	30 103	30 320	30 535	30 750	30 963	31 175	31 387	31 597	31 806	32 015̄
21	32 222	32 428	32 634	32 838	33 041	33 244	33 445	33 646	33 846	34 044
22	34 242	34 439	34 635	34 830	35 025̄	35 218	35 411	35 603	35 793	35 984
23	36 173	36 361	36 549	36 736	36 922	37 107	37 291	37 475̄	37 658	37 840
24	38 021	38 202	38 382	38 561	38 739	38 917	39 094	39 270	39 445	39 620
25	39 794	39 967	40 140	40 312	40 483	40 654	40 824	40 993	41 162	41 330
26	41 497	41 664	41 830	41 996	42 160	42 325̄	42 488	42 651	42 813	42 975
27	43 136	43 297	43 457	43 616	43 775	43 933	44 091	44 248	44 404	44 560
28	44 716	44 871	45 025̄	45 179	45 332	45 484	45 637	45 788	45 939	46 090
29	46 240	46 389	46 538	46 687	46 835̄	46 982	47 129	47 276	47 422	47 567
30	47 712	47 857	48 001	48 144	48 287	48 430	48 572	48 714	48 855	48 996
31	49 136	49 276	49 415	49 554	49 693	49 831	49 969	50 106	50 243	50 379
32	50 515̄	50 651	50 786	50 920	51 055̄	51 188	51 322	51 455̄	51 587	51 720
33	51 851	51 983	52 114	52 244	52 375̄	52 504	52 634	52 763	52 892	53 020
34	53 148	53 275	53 403	53 529	53 656	53 782	53 908	54 033	54 158	54 283
35	54 407	54 531	54 654	54 777	54 900	55 023	55 145̄	55 267	55 388	55 509
36	55 630	55 751	55 871	55 991	56 110	56 229	56 348	56 467	56 585̄	56 703
37	56 820	56 937	57 054	57 171	57 287	57 403	57 519	57 634	57 749	57 864
38	57 978	58 092	58 206	58 320	58 433	58 546	58 659	58 771	58 883	58 995̄
39	59 106	59 218	59 329	59 439	59 550	59 660	59 770	59 879	59 988	60 097
40	60 206	60 314	60 423	60 531	60 638	60 746	60 853	60 959	61 066	61 172
41	61 278	61 384	61 490	61 595	61 700	61 805̄	61 909	62 014	62 118	62 221
42	62 325̄	62 428	62 531	62 634	62 737	62 839	62 941	63 043	63 144	63 246
43	63 347	63 448	63 548	63 649	63 749	63 849	63 949	64 048	64 147	64 246
44	64 345	64 444	64 542	64 640	64 738	64 836	64 933	65 031	65 128	65 225̄
45	65 321	65 418	65 514	65 610	65 706	65 801	65 896	65 992	66 087	66 181
46	66 276	66 370	66 464	66 558	66 652	66 745	66 839	66 932	67 025̄	67 117
47	67 210	67 302	67 394	67 486	67 578	67 669	67 761	67 852	67 943	68 034
48	68 124	68 215̄	68 305̄	68 395̄	68 485̄	68 574	68 664	68 753	68 842	68 931
49	69 020	69 108	69 197	69 285̄	69 373	69 461	69 548	69 636	69 723	69 810
50	69 897	69 984	70 070	70 157	70 243	70 329	70 415	70 501	70 586	70 672
N	L 0	1	2	3	4	5	6	7	8	9

		S	T			S	T
60″	= 0° 1′	4.68 557	4.68 557	300″	= 0° 5′	4.68 557	4.68 558
120	= 0 2	4.68 557	4.68 557	360	= 0 6	4.68 557	4.68 558
180	= 0 3	4.68 557	4.68 557	420	= 0 7	4.68 557	4.68 558
240	= 0 4	4.68 557	4.68 558	480	= 0 8	4.68 557	4.68 558

50—100

N	L 0	1	2	3	4	5	6	7	8	9
50	69 897	69 984	70 070	70 157	70 243	70 329	70 415	70 501	70 586	70 672
51	70 757	70 842	70 927	71 012	71 096	71 181	71 265̄	71 349	71 433	71 517
52	71 600	71 684	71 767	71 850	71 933	72 016	72 099	72 181	72 263	72 346
53	72 428	72 509	72 591	72 673	72 754	72 835	72 916	72 997	73 078	73 159
54	73 239	73 320	73 400	73 480	73 560	73 640	73 719	73 799	73 878	73 957
55	74 036	74 115	74 194	74 273	74 351	74 429	74 507	74 586	74 663	74 741
56	74 819	74 896	74 974	75 051	75 128	75 205̄	75 282	75 358	75 435̄	75 511
57	75 587	75 664	75 740	75 815	75 891	75 967	76 042	76 118	76 193	76 268
58	76 343	76 418	76 492	76 567	76 641	76 716	76 790	76 864	76 938	77 012
59	77 085	77 159	77 232	77 305	77 379	77 452	77 525̄	77 597	77 670	77 743
60	77 815	77 887	77 960	78 032	78 104	78 176	78 247	78 319	78 390	78 462
61	78 533	78 604	78 675	78 746	78 817	78 888	78 958	79 029	79 099	79 169
62	79 239	79 309	79 379	79 449	79 518	79 588	79 657	79 727	79 796	79 865
63	79 934	80 003	80 072	80 140	80 209	80 277	80 346	80 414	80 482	80 550
64	80 618	80 686	80 754	80 821	80 889	80 956	81 023	81 090	81 158	81 224
65	81 291	81 358	81 425̄	81 491	81 558	81 624	81 690	81 757	81 823	81 889
66	81 954	82 020	82 086	82 151	82 217	82 282	82 347	82 413	82 478	82 543
67	82 607	82 672	82 737	82 802	82 866	82 930	82 995̄	83 059	83 123	83 187
68	83 251	83 315̄	83 378	83 442	83 506	83 569	83 632	83 696	83 759	83 822
69	83 885̄	83 948	84 011	84 073	84 136	84 198	84 261	84 323	84 386	84 448
70	84 510	84 572	84 634	84 696	84 757	84 819	84 880	84 942	85 003	85 065̄
71	85 126	85 187	85 248	85 309	85 370	85 431	85 491	85 552	85 612	85 673
72	85 733	85 794	85 854	85 914	85 974	86 034	86 094	86 153	86 213	86 273
73	86 332	86 392	86 451	86 510	86 570	86 629	86 688	86 747	86 806	86 864
74	86 923	86 982	87 040	87 099	87 157	87 216	87 274	87 332	87 390	87 448
75	87 506	87 564	87 622	87 679	87 737	87 795̄	87 852	87 910	87 967	88 024
76	88 081	88 138	88 195	88 252	88 309	88 366	88 423	88 480	88 536	88 593
77	88 649	88 705	88 762	88 818	88 874	88 930	88 986	89 042	89 098	89 154
78	89 209	89 265	89 321	89 376	89 432	89 487	89 542	89 597	89 653	89 708
79	89 763	89 818	89 873	89 927	89 982	90 037	90 091	90 146	90 200	90 255̄
80	90 309	90 363	90 417	90 472	90 526	90 580	90 634	90 687	90 741	90 795̄
81	90 849	90 902	90 956	91 009	91 062	91 116	91 169	91 222	91 275	91 328
82	91 381	91 434	91 487	91 540	91 593	91 645	91 698	91 751	91 803	91 855
83	91 908	91 960	92 012	92 065̄	92 117	92 169	92 221	92 273	92 324	92 376
84	92 428	92 480	92 531	92 583	92 634	92 686	92 737	92 788	92 840	92 891
85	92 942	92 993	93 044	93 095̄	93 146	93 197	93 247	93 298	93 349	93 399
86	93 450̄	93 500	93 551	93 601	93 651	93 702	93 752	93 802	93 852	93 902
87	93 952	94 002	94 052	94 101	94 151	94 201	94 250	94 300	94 349	94 399
88	94 448	94 498	94 547	94 596	94 645	94 694	94 743	94 792	94 841	94 890
89	94 939	94 988	95 036	95 085	95 134	95 182	95 231	95 279	95 328	95 376
90	95 424	95 472	95 521	95 569	95 617	95 665̄	95 713	95 761	95 809	95 856
91	95 904	95 952	95 999	96 047	96 095̄	96 142	96 190	96 237	96 284	96 332
92	96 379	96 426	96 473	96 520	96 567	96 614	96 661	96 708	96 755̄	96 802
93	96 848	96 895̄	96 942	96 988	97 035̄	97 081	97 128	97 174	97 220	97 267
94	97 313	97 359	97 405	97 451	97 497	97 543	97 589	97 635̄	97 681	97 727
95	97 772	97 818	97 864	97 909	97 955̄	98 000	98 046	98 091	98 137	98 182
96	98 227	98 272	98 318	98 363	98 408	98 453	98 498	98 543	98 588	98 632
97	98 677	98 722	98 767	98 811	98 856	98 900	98 945̄	98 989	99 034	99 078
98	99 123	99 167	99 211	99 255	99 300	99 344	99 388	99 432	99 476	99 520
99	99 564	99 607	99 651	99 695̄	99 739	99 782	99 826	99 870	99 913	99 957
100	00 000	00 043	00 087	00 130	00 173	00 217	00 260	00 303	00 346	00 389
N	L 0	1	2	3	4	5	6	7	8	9

		S	T			S	T
540″	= 0° 9′	4.68 557	4.68 558	780″	= 0° 13′	4.68 557	4.68 558
600	= 0 10	4.68 557	4.68 558	840	= 0 14	4.68 557	4.68 558
660	= 0 11	4.68 557	4.68 558	900	= 0 15	4.68 557	4.68 558
720	= 0 12	4.68 557	4.68 558	960	= 0 16	4.68 557	4.68 558

100—150

N	L 0	1	2	3	4	5	6	7	8	9
100	00 000	043	087	130	173	217	260	303	346	389
101	432	475	518	561	604	647	689	732	775̄	817
102	860	903	945	988	*030	*072	*115̄	*157	*199	*242
103	01 284	326	368	410	452	494	536	578	620	662
104	703	745	787	828	870	912	953	995̄	*036	*078
105	02 119	160	202	243	284	325	366	407	449	490
106	531	572	612	653	694	735̄	776	816	857	898
107	938	979	*019	*060	*100	*141	*181	*222	*262	*302
108	03 342	383	423	463	503	543	583	623	663	703
109	743	782	822	862	902	941	981	*021	*060	*100
110	04 139	179	218	258	297	336	376	415̄	454	493
111	532	571	610	65̄0	689	727	766	805	844	883
112	922	961	999	*038	*077	*115	*154	*192	*231	*269
113	05 308	346	385̄	423	461	5̄00	538	576	614	652
114	690	729	767	805̄	843	881	918	956	994	*032
115	06 070	108	145	183	221	258	296	333	371	408
116	446	483	521	558	595	633	670	707	744	781
117	819	856	893	930	967	*004	*041	*078	*115̄	*151
118	07 188	225̄	262	298	335	372	408	445	482	518
119	555̄	591	628	664	700	737	773	809	846	882
120	918	954	990	*027	*063	*099	*135̄	*171	*207	*243
121	08 279	314	350	386	422	458	493	529	565̄	600
122	636	672	707	743	778	814	849	884	920	955
123	991	*026	*061	*096	*132	*167	*202	*237	*272	*307
124	09 342	377	412	447	482	517	552	587	621	656
125	691	726	760	795	830	864	899	934	968	*003
126	10 037	072	106	140	175̄	209	243	278	312	346
127	380	415̄	449	483	517	551	585	619	653	687
128	721	755̄	789	823	857	890	924	958	992	*025
129	11 059	093	126	160	193	227	261	294	327	361
130	394	428	461	494	528	561	594	628	661	694
131	727	760	793	826	860	893	926	959	992	*024
132	12 057	090	123	156	189	222	254	287	320	352
133	385	418	450	483	516	548	581	613	646	678
134	710	743	775	808	840	872	905̄	937	969	*001
135	13 033	066	098	130	162	194	226	258	290	322
136	354	386	418	45̄0	481	513	545	577	609	640
137	672	704	735	767	799	830	862	893	925̄	956
138	988	*019	*051	*082	*114	*145̄	*176	*208	*239	*270
139	14 301	333	364	395	426	457	489	520	551	582
140	613	644	675̄	706	737	768	799	829	860	891
141	922	953	983	*014	*045̄	*076	*106	*137	*168	*198
142	15 229	259	290	320	351	381	412	442	473	503
143	534	564	594	625̄	655̄	685	715	746	776	806
144	836	866	897	927	957	987	*017	*047	*077	*107
145	16 137	167	197	227	256	286	316	346	376	406
146	435	465	495̄	524	554	584	613	643	673	702
147	732	761	791	820	85̄0	879	909	938	967	997
148	17 026	056	085̄	114	143	173	202	231	260	289
149	319	348	377	406	435	464	493	522	551	580
150	17 609	638	667	696	725̄	754	782	811	840	869
N	L 0	1	2	3	4	5	6	7	8	9

P P

	44	43	42
1	4.4	4.3	4.2
2	8.8	8.6	8.4
3	13.2	12.9	12.6
4	17.6	17.2	16.8
5	22.0	21.5	21.0
6	26.4	25.8	25.2
7	30.8	30.1	29.4
8	35.2	34.4	33.6
9	39.6	38.7	37.8

	41	40	39
1	4.1	4.0	3.9
2	8.2	8.0	7.8
3	12.3	12.0	11.7
4	16.4	16.0	15.6
5	20.5	20.0	19.5
6	24.6	24.0	23.4
7	28.7	28.0	27.3
8	32.8	32.0	31.2
9	36.9	36.0	35.1

	38	37	36
1	3.8	3.7	3.6
2	7.6	7.4	7.2
3	11.4	11.1	10.8
4	15.2	14.8	14.4
5	19.0	18.5	18.0
6	22.8	22.2	21.6
7	26.6	25.9	25.2
8	30.4	29.6	28.8
9	34.2	33.3	32.4

	35	34	33
1	3.5	3.4	3.3
2	7.0	6.8	6.6
3	10.5	10.2	9.9
4	14.0	13.6	13.2
5	17.5	17.0	16.5
6	21.0	20.4	19.8
7	24.5	23.8	23.1
8	28.0	27.2	26.4
9	31.5	30.6	29.7

	32	31	30
1	3.2	3.1	3.0
2	6.4	6.2	6.0
3	9.6	9.3	9.0
4	12.8	12.4	12.0
5	16.0	15.5	15.0
6	19.2	18.6	18.0
7	22.4	21.7	21.0
8	25.6	24.8	24.0
9	28.8	27.9	27.0

P P

		S	T			S	T
960″	=0° 16′	4.68 557	4.68 558	1260″	=0° 21′	4.68 557	4.68 558
1020	=0 17	4.68 557	4.68 558	1320	=0 22	4.68 557	4.68 558
1080	=0 18	4.68 557	4.68 558	1380	=0 23	4.68 557	4.68 558
1140	=0 19	4.68 557	4.68 558	1440	=0 24	4.68 557	4.68 558
1200	=0 20	4.68 557	4.68 558	1500	=0 25	4.68 557	4.68 558

150—200

N	L 0	1	2	3	4	5	6	7	8	9
150	17 609	638	667	696	725̄	754	782	811	840	869
151	898	926	955	984	*013	*041	*070	*099	*127	*156
152	18 184	213	241	270	298	327	355	384	412	441
153	469	498	526	554	583	611	639	667	696	724
154	752	780	808	837	865̄	893	921	949	977	*005
155	19 033	061	089	117	145	173	201	229	257	285̄
156	312	340	368	396	424	451	479	507	535̄	562
157	590	618	645	673	700	728	756	783	811	838
158	866	893	921	948	976	*003	*030	*058	*085	*112
159	20 140	167	194	222	249	276	303	330	358	385̄
160	412	439	466	493	520	548	575̄	602	629	656
161	683	710	737	763	790	817	844	871	898	925̄
162	952	978	*005	*032	*059	*085	*112	*139	*165	*192
163	21 219	245	272	299	325	352	378	405̄	431	458
164	484	511	537	564	590	617	643	669	696	722
165	748	775̄	801	827	854	880	906	932	958	985̄
166	22 011	037	063	089	115	141	167	194	220	246
167	272	298	324	350	376	401	427	453	479	505
168	531	557	583	608	634	660	686	712	737	763
169	789	814	840	866	891	917	943	968	994	*019
170	23 045̄	070	096	121	147	172	198	223	249	274
171	300	325	350	376	401	426	452	477	502	528
172	553	578	603	629	654	679	704	729	754	779
173	805̄	830	855̄	880	905̄	930	955̄	980	*005̄	*030
174	24 055̄	080	105̄	130	155̄	180	204	229	254	279
175	304	329	353	378	403	428	452	477	502	527
176	551	576	601	625	650	674	699	724	748	773
177	797	822	846	871	895	920	944	969	993	*018
178	25 042	066	091	115	139	164	188	212	237	261
179	285	310	334	358	382	406	431	455̄	479	503
180	527	551	575	600	624	648	672	696	720	744
181	768	792	816	840	864	888	912	935	959	983
182	26 007	031	055̄	079	102	126	150	174	198	221
183	245	269	293	316	340	364	387	411	435̄	458
184	482	505	529	553	576	600	623	647	670	694
185	717	741	764	788	811	834	858	881	905̄	928
186	951	975̄	998	*021	*045̄	*068	*091	*114	*138	*161
187	27 184	207	231	254	277	300	323	346	370	393
188	416	439	462	485	508	531	554	577	600	623
189	646	669	692	715	738	761	784	807	830	852
190	875	898	921	944	967	989	*012	*035	*058	*081
191	28 103	126	149	171	194	217	240	262	285̄	307
192	330	353	375	398	421	443	466	488	511	533
193	556	578	601	623	646	668	691	713	735	758
194	780	803	825̄	847	870	892	914	937	959	981
195	29 003	026	048	070	092	115̄	137	159	181	203
196	226	248	270	292	314	336	358	380	403	425̄
197	447	469	491	513	535̄	557	579	601	623	645̄
198	667	688	710	732	754	776	798	820	842	863
199	885	907	929	951	973	994	*016	*038	*060	*081
200	30 103	125̄	146	168	190	211	233	255̄	276	298
N	L 0	1	2	3	4	5	6	7	8	9

P P

	29	28
1	2.9	2.8
2	5.8	5.6
3	8.7	8.4
4	11.6	11.2
5	14.5	14.0
6	17.4	16.8
7	20.3	19.6
8	23.2	22.4
9	26.1	25.2

	27	26
1	2.7	2.6
2	5.4	5.2
3	8.1	7.8
4	10.8	10.4
5	13.5	13.0
6	16.2	15.6
7	18.9	18.2
8	21.6	20.8
9	24.3	23.4

	25
1	2.5
2	5.0
3	7.5
4	10.0
5	12.5
6	15.0
7	17.5
8	20.0
9	22.5

	24	23
1	2.4	2.3
2	4.8	4.6
3	7.2	6.9
4	9.6	9.2
5	12.0	11.5
6	14.4	13.8
7	16.8	16.1
8	19.2	18.4
9	21.6	20.7

	22	21
1	2.2	2.1
2	4.4	4.2
3	6.6	6.3
4	8.8	8.4
5	11.0	10.5
6	13.2	12.6
7	15.4	14.7
8	17.6	16.8
9	19.8	18.9

P P

		S	T
1500″	=0° 25′	4.68 557	4.68 558
1560	=0 26	4.68 557	4.68 558
1620	=0 27	4.68 557	4.68 558
1680	=0 28	4.68 557	4.68 558
1740	=0 29	4.68 557	4.68 559

		S	T
1800″	=0° 30′	4.68 557	4.68 559
1860	=0 31	4.68 557	4.68 559
1920	=0 32	4.68 557	4.68 559
1980	=0 33	4.68 557	4.68 559
2040	=0 34	4.68 557	4.68 559

200—250

N	L 0	1	2	3	4	5	6	7	8	9
200	30 103	125̄	146	168	190	211	233	255̄	276	298
201	320	341	363	384	406	428	449	471	492	514
202	535	557	578	600	621	643	664	685	707	728
203	75̄0	771	792	814	835	856	878	899	920	942
204	963	984	*006	*027	*048	*069	*091	*112	*133	*154
205	31 175	197	218	239	260	281	302	323	345̄	366
206	387	408	429	45̄0	471	492	513	534	555	576
207	597	618	639	660	681	702	723	744	765̄	785
208	806	827	848	869	890	911	931	952	973	994
209	32 015̄	035	056	077	098	118	139	160	181	201
210	222	243	263	284	305̄	325	346	366	387	408
211	428	449	469	490	510	531	552	572	593	613
212	634	654	675̄	695̄	715	736	756	777	797	818
213	838	858	879	899	919	940	960	980	*001	*021
214	33 041	062	082	102	122	143	163	183	203	224
215	244	264	284	304	325̄	345̄	365̄	385	405	425
216	445	465	486	506	526	546	566	586	606	626
217	646	666	686	706	726	746	766	786	806	826
218	846	866	885	905	925	945	965	985̄	*005̄	*025̄
219	34 044	064	084	104	124	143	163	183	203	223
220	242	262	282	301	321	341	361	380	400	420
221	439	459	479	498	518	537	557	577	596	616
222	635	655̄	674	694	713	733	753	772	792	811
223	830	85̄0	869	889	908	928	947	967	986	*005
224	35 025̄	044	064	083	102	122	141	160	180	199
225	218	238	257	276	295	315̄	334	353	372	392
226	411	430	449	468	488	507	526	545	564	583
227	603	622	641	660	679	698	717	736	755	774
228	793	813	832	851	870	889	908	927	946	965̄
229	984	*003	*021	*040	*059	*078	*097	*116	*135	*154
230	36 173	192	211	229	248	267	286	305̄	324	342
231	361	380	399	418	436	455	474	493	511	530
232	549	568	586	605̄	624	642	661	680	698	717
233	736	754	773	791	810	829	847	866	884	903
234	922	940	959	977	996	*014	*033	*051	*070	*088
235	37 107	125	144	162	181	199	218	236	254	273
236	291	310	328	346	365̄	383	401	420	438	457
237	475̄	493	511	530	548	566	585̄	603	621	639
238	658	676	694	712	731	749	767	785	803	822
239	840	858	876	894	912	931	949	967	985̄	*003
240	38 021	039	057	075	093	112	130	148	166	184
241	202	220	238	256	274	292	310	328	346	364
242	382	399	417	435	453	471	489	507	525̄	543
243	561	578	596	614	632	65̄0	668	686	703	721
244	739	757	775̄	792	810	828	846	863	881	899
245	917	934	952	970	987	*005	*023	*041	*058	*076
246	39 094	111	129	146	164	182	199	217	235̄	252
247	270	287	305̄	322	340	358	375	393	410	428
248	445	463	480	498	515	533	550	568	585̄	602
249	620	637	655̄	672	690	707	724	742	759	777
250	794	811	829	846	863	881	898	915	933	950
N	L 0	1	2	3	4	5	6	7	8	9

P P

	22	21
1	2.2	2.1
2	4.4	4.2
3	6.6	6.3
4	8.8	8.4
5	11.0	10.5
6	13.2	12.6
7	15.4	14.7
8	17.6	16.8
9	19.8	18.9

	20
1	2.0
2	4.0
3	6.0
4	8.0
5	10.0
6	12.0
7	14.0
8	16.0
9	18.0

	19
1	1.9
2	3.8
3	5.7
4	7.6
5	9.5
6	11.4
7	13.3
8	15.2
9	17.1

	18
1	1.8
2	3.6
3	5.4
4	7.2
5	9.0
6	10.8
7	12.6
8	14.4
9	16.2

	17
1	1.7
2	3.4
3	5.1
4	6.8
5	8.5
6	10.2
7	11.9
8	13.6
9	15.3

P P

		S	T
1980″	=0° 33′	4.68 557	4.68 559
2040	=0 34	4.68 557	4.68 559
2100	=0 35	4 68 557	4.68 559
2160	=0 36	4.68 557	4.68 559
2220	=0 37	4.68 557	4.68 559
2280″	=0° 38′	4.68 557	4.68 559
2340	=0 39	4.68 557	4.68 559
2400	=0 40	4.68 557	4.68 559
2460	=0 41	4.68 556	4.68 560
2520	=0 42	4.68 556	4.68 560

250—300

N	L 0	1	2	3	4	5	6	7	8	9
250	39 794	811	829	846	863	881	898	915	933	950
251	967	985̄	*002	*019	*037	*054	*071	*088	*106	*123
252	40 140	157	175̄	192	209	226	243	261	278	295̄
253	312	329	346	364	381	398	415̄	432	449	466
254	483	500	518	535̄	552	569	586	603	620	637
255	654	671	688	705	722	739	756	773	790	807
256	824	841	858	875̄	892	909	926	943	960	976
257	993	*010	*027	*044	*061	*078	*095̄	*111	*128	*145
258	41 162	179	196	212	229	246	263	280	296	313
259	330	347	363	380	397	414	430	447	464	481
260	497	514	531	547	564	581	597	614	631	647
261	664	681	697	714	731	747	764	780	797	814
262	830	847	863	880	896	913	929	946	963	979
263	996	*012	*029	*045	*062	*078	*095̄	*111	*127	*144
264	42 160	177	193	210	226	243	259	275	292	308
265	325̄	341	357	374	390	406	423	439	455	472
266	488	504	521	537	553	570	586	602	619	635̄
267	651	667	684	700	716	732	749	765̄	781	797
268	813	830	846	862	878	894	911	927	943	959
269	975	991	*008	*024	*040	*056	*072	*088	*104	*120
270	43 136	152	169	185̄	201	217	233	249	265̄	281
271	297	313	329	345̄	361	377	393	409	425̄	441
272	457	473	489	505̄	521	537	553	569	584	600
273	616	632	648	664	680	696	712	727	743	759
274	775	791	807	823	838	854	870	886	902	917
275	933	949	965̄	981	996	*012	*028	*044	*059	*075
276	44 091	107	122	138	154	170	185	201	217	232
277	248	264	279	295̄	311	326	342	358	373	389
278	404	420	436	451	467	483	498	514	529	545̄
279	560	576	592	607	623	638	654	669	685̄	700
280	716	731	747	762	778	793	809	824	840	855
281	871	886	902	917	932	948	963	979	994	*010
282	45 025̄	040	056	071	086	102	117	133	148	163
283	179	194	209	225̄	240	255	271	286	301	317
284	332	347	362	378	393	408	423	439	454	469
285	484	500	515̄	530	545	561	576	591	606	621
286	637	652	667	682	697	712	728	743	758	773
287	788	803	818	834	849	864	879	894	909	924
288	939	954	969	984	*000	*015̄	*030	*045̄	*060	*075̄
289	46 090	105̄	120	135̄	150	165̄	180	195̄	210	225̄
290	240	255̄	270	285̄	300	315̄	330	345̄	359	374
291	389	404	419	434	449	464	479	494	509	523
292	538	553	568	583	598	613	627	642	657	672
293	687	702	716	731	746	761	776	790	805	820
294	835̄	850	864	879	894	909	923	938	953	967
295	982	997	*012	*026	*041	*056	*070	*085	*100	*114
296	47 129	144	159	173	188	202	217	232	246	261
297	276	290	305̄	319	334	349	363	378	392	407
298	422	436	451	465	480	494	509	524	538	553
299	567	582	596	611	625	640	654	669	683	698
300	712	727	741	756	770	784	799	813	828	842
N	L 0	1	2	3	4	5	6	7	8	9

P P

	18	17	16	15	14
1	1.8	1.7	1.6	1.5	1.4
2	3.6	3.4	3.2	3.0	2.8
3	5.4	5.1	4.8	4.5	4.2
4	7.2	6.8	6.4	6.0	5.6
5	9.0	8.5	8.0	7.5	7.0
6	10.8	10.2	9.6	9.0	8.4
7	12.6	11.9	11.2	10.5	9.8
8	14.4	13.6	12.8	12.0	11.2
9	16.2	15.3	14.4	13.5	12.6

		S	T			S	T
2460″	=0° 41′	4.68 556	4.68 560	2760″	=0° 46′	4.68 556	4.68 560
2520	=0 42	4.68 556	4.68 560	2820	=0 47	4.68 556	4.68 560
2580	=0 43	4.68 556	4.68 560	2880	=0 48	4.68 556	4.68 560
2640	=0 44	4.68 556	4.68 560	2940	=0 49	4.68 556	4.68 560
2700	=0 45	4.68 556	4.68 560	3000	=0 50	4.68 556	4.68 561

300—350

N	L 0	1	2	3	4	5	6	7	8	9
300	47 712	727	741	756	770	784	799	813	828	842
301	857	871	885	900	914	929	943	958	972	986
302	48 001	015	029	044	058	073	087	101	116	130
303	144	159	173	187	202	216	230	244	259	273
304	287	302	316	330	344	359	373	387	401	416
305	430	444	458	473	487	501	515	530	544	558
306	572	586	601	615̄	629	643	657	671	686	700
307	714	728	742	756	770	785̄	799	813	827	841
308	855	869	883	897	911	926	940	954	968	982
309	996	*010	*024	*038	*052	*066	*080	*094	*108	*122
310	49 136	150	164	178	192	206	220	234	248	262
311	276	290	304	318	332	346	360	374	388	402
312	415	429	443	457	471	485	499	513	527	541
313	554	568	582	596	610	624	638	651	665	679
314	693	707	721	734	748	762	776	790	803	817
315	831	845̄	859	872	886	900	914	927	941	955̄
316	969	982	996	*010	*024	*037	*051	*065̄	*079	*092
317	50 106	120	133	147	161	174	188	202	215	229
318	243	256	270	284	297	311	325̄	338	352	365
319	379	393	406	420	433	447	461	474	488	501
320	515̄	529	542	556	569	583	596	610	623	637
321	651	664	678	691	705̄	718	732	745	759	772
322	786	799	813	826	840	853	866	880	893	907
323	920	934	947	961	974	987	*001	*014	*028	*041
324	51 055̄	068	081	095̄	108	121	135̄	148	162	175̄
325	188	202	215	228	242	255	268	282	295	308
326	322	335	348	362	375	388	402	415̄	428	441
327	455̄	468	481	495̄	508	521	534	548	561	574
328	587	601	614	627	640	654	667	680	693	706
329	720	733	746	759	772	786	799	812	825	838
330	851	865̄	878	891	904	917	930	943	957	970
331	983	996	*009	*022	*035	*048	*061	*075̄	*088	*101
332	52 114	127	140	153	166	179	192	205	218	231
333	244	257	270	284	297	310	323	336	349	362
334	375̄	388	401	414	427	440	453	466	479	492
335	504	517	530	543	556	569	582	595	608	621
336	634	647	660	673	686	699	711	724	737	750
337	763	776	789	802	815̄	827	840	853	866	879
338	892	905̄	917	930	943	956	969	982	994	*007
339	53 020	033	046	058	071	084	097	110	122	135
340	148	161	173	186	199	212	224	237	250	263
341	275	288	301	314	326	339	352	364	377	390
342	403	415	428	441	453	466	479	491	504	517
343	529	542	555̄	567	580	593	605	618	631	643
344	656	668	681	694	706	719	732	744	757	769
345	782	794	807	820	832	845̄	857	870	882	895
346	908	920	933	945	958	970	983	995	*008	*020
347	54 033	045	058	070	083	095	108	120	133	145
348	158	170	183	195	208	220	233	245	258	270
349	283	295̄	307	320	332	345̄	357	370	382	394
350	407	419	432	444	456	469	481	494	506	518
N	L 0	1	2	3	4	5	6	7	8	9

P P

	15
1	1.5
2	3.0
3	4.5
4	6.0
5	7.5
6	9.0
7	10.5
8	12.0
9	13.5

	14
1	1.4
2	2.8
3	4.2
4	5.6
5	7.0
6	8.4
7	9.8
8	11.2
9	12.6

	13
1	1.3
2	2.6
3	3.9
4	5.2
5	6.5
6	7.8
7	9.1
8	10.4
9	11.7

	12
1	1.2
2	2.4
3	3.6
4	4.8
5	6.0
6	7.2
7	8.4
8	9.6
9	10.8

P P

	S	T		S	T
3000″ = 0° 50′	4.68 556	4.68 561	3300″ = 0° 55′	4.68 556	4.68 561
3060 = 0 51	4.68 556	4.68 561	3360 = 0 56	4.68 556	4.68 561
3120 = 0 52	4.68 556	4.68 561	3420 = 0 57	4.68 555	4.68 561
3180 = 0 53	4.68 556	4.68 561	3480 = 0 58	4.68 555	4.68 562
3240 = 0 54	4.68 556	4.68 561	3540 = 0 59	4.68 555	4.68 562

350—400

N	L 0	1	2	3	4	5	6	7	8	9
350	54 407	419	432	444	456	469	481	494	506	518
351	531	543	555	568	580	593	605̄	617	630	642
352	654	667	679	691	704	716	728	741	753	765
353	777	790	802	814	827	839	851	864	876	888
354	900	913	925̄	937	949	962	974	986	998	*011
355	55 023	035	047	060	072	084	096	108	121	133
356	145̄	157	169	182	194	206	218	230	242	255̄
357	267	279	291	303	315	328	340	352	364	376
358	388	400	413	425̄	437	449	461	473	485	497
359	509	522	534	546	558	570	582	594	606	618
360	630	642	654	666	678	691	703	715̄	727	739
361	751	763	775̄	787	799	811	823	835̄	847	859
362	871	883	895̄	907	919	931	943	955̄	967	979
363	991	*003	*015̄	*027	*038	*050	*062	*074	*086	*098
364	56 110	122	134	146	158	170	182	194	205	217
365	229	241	253	265̄	277	289	301	312	324	336
366	348	360	372	384	396	407	419	431	443	455̄
367	467	478	490	502	514	526	538	549	561	573
368	585̄	597	608	620	632	644	656	667	679	691
369	703	714	726	738	75̄0	761	773	785̄	797	808
370	820	832	844	855	867	879	891	902	914	926
371	937	949	961	972	984	996	*008	*019	*031	*043
372	57 054	066	078	089	101	113	124	136	148	159
373	171	183	194	206	217	229	241	252	264	276
374	287	299	310	322	334	345	357	368	380	392
375	403	415̄	426	438	449	461	473	484	496	507
376	519	530	542	553	565̄	576	588	600	611	623
377	634	646	657	669	680	692	703	715̄	726	738
378	749	761	772	784	795	807	818	830	841	852
379	864	875	887	898	910	921	933	944	955	967
380	978	990	*001	*013	*024	*035	*047	*058	*070	*081
381	58 092	104	115	127	138	149	161	172	184	195̄
382	206	218	229	240	252	263	274	286	297	309
383	320	331	343	354	365	377	388	399	410	422
384	433	444	456	467	478	490	501	512	524	535̄
385	546	557	569	580	591	602	614	625̄	636	647
386	659	670	681	692	704	715̄	726	737	749	760
387	771	782	794	805̄	816	827	838	85̄0	861	872
388	883	894	906	917	928	939	950	961	973	984
389	995̄	*006	*017	*028	*040	*051	*062	*073	*084	*095
390	59 106	118	129	140	151	162	173	184	195	207
391	218	229	240	251	262	273	284	295	306	318
392	329	340	351	362	373	384	395	406	417	428
393	439	450	461	472	483	494	506	517	528	539
394	55̄0	561	572	583	594	605̄	616	627	638	649
395	660	671	682	693	704	715̄	726	737	748	759
396	770	780	791	802	813	824	835	846	857	868
397	879	890	901	912	923	934	945̄	956	966	977
398	988	999	*010	*021	*032	*043	*054	*065̄	*076	*086
399	60 097	108	119	130	141	152	163	173	184	195
400	206	217	228	239	249	260	271	282	293	304
N	L 0	1	2	3	4	5	6	7	8	9

P P

	13	12	11	10
1	1.3	1.2	1.1	1.0
2	2.6	2.4	2.2	2.0
3	3.9	3.6	3.3	3.0
4	5.2	4.8	4.4	4.0
5	6.5	6.0	5.5	5.0
6	7.8	7.2	6.6	6.0
7	9.1	8.4	7.7	7.0
8	10.4	9.6	8.8	8.0
9	11.7	10.8	9.9	9.0

		S	T			S	T
3480″	= 0° 58′	4.68 555	4.68 562	3780″	= 1° 3′	4.68 555	4.68 562
3540	= 0 59	4.68 555	4.68 562	3840	= 1 4	4.68 555̄	4.68 563
3600	= 1 0	4.68 555	4.68 562	3900	= 1 5	4.68 555̄	4.68 563
3660	= 1 1	4.68 555	4.68 562	3960	= 1 6	4.68 555̄	4.68 563
3720	= 1 2	4.68 555	4.68 562	4020	= 1 7	4.68 555̄	4.68 563

400—450

N	L 0	1	2	3	4	5	6	7	8	9
400	60 206	217	228	239	249	260	271	282	293	304
401	314	325	336	347	358	369	379	390	401	412
402	423	433	444	455	466	477	487	498	509	520
403	531	541	552	563	574	584	595	606	617	627
404	638	649	660	670	681	692	703	713	724	735̄
405	746	756	767	778	788	799	810	821	831	842
406	853	863	874	885̄	895	906	917	927	938	949
407	959	970	981	991	*002	*013	*023	*034	*045̄	*055
408	61 066	077	087	098	109	119	130	140	151	162
409	172	183	194	204	215̄	225	236	247	257	268
410	278	289	300	310	321	331	342	352	363	374
411	384	395̄	405	416	426	437	448	458	469	479
412	490	500	511	521	532	542	553	563	574	584
413	595	606	616	627	637	648	658	669	679	690
414	700	711	721	731	742	752	763	773	784	794
415	805̄	815	826	836	847	857	868	878	888	899
416	909	920	930	941	951	962	972	982	993	*003
417	62 014	024	034	045̄	055	066	076	086	097	107
418	118	128	138	149	159	170	180	190	201	211
419	221	232	242	252	263	273	284	294	304	315̄
420	325̄	335	346	356	366	377	387	397	408	418
421	428	439	449	459	469	480	490	500	511	521
422	531	542	552	562	572	583	593	603	613	624
423	634	644	655̄	665̄	675	685	696	706	716	726
424	737	747	757	767	778	788	798	808	818	829
425	839	849	859	870	880	890	900	910	921	931
426	941	951	961	972	982	992	*002	*012	*022	*033
427	63 043	053	063	073	083	094	104	114	124	134
428	144	155̄	165̄	175̄	185̄	195	205	215	225	236
429	246	256	266	276	286	296	306	317	327	337
430	347	357	367	377	387	397	407	417	428	438
431	448	458	468	478	488	498	508	518	528	538
432	548	558	568	579	589	599	609	619	629	639
433	649	659	669	679	689	699	709	719	729	739
434	749	759	769	779	789	799	809	819	829	839
435	849	859	869	879	889	899	909	919	929	939
436	949	959	969	979	988	998	*008	*018	*028	*038
437	64 048	058	068	078	088	098	108	118	128	137
438	147	157	167	177	187	197	207	217	227	237
439	246	256	266	276	286	296	306	316	326	335
440	345	355	365	375̄	385̄	395̄	404	414	424	434
441	444	454	464	473	483	493	503	513	523	532
442	542	552	562	572	582	591	601	611	621	631
443	640	650	660	670	680	689	699	709	719	729
444	738	748	758	768	777	787	797	807	816	826
445	836	846	856	865	875	885̄	895̄	904	914	924
446	933	943	953	963	972	982	992	*002	*011	*021
447	65 031	040	050	060	070	079	089	099	108	118
448	128	137	147	157	167	176	186	196	205	215̄
449	225̄	234	244	254	263	273	283	292	302	312
450	321	331	341	350	360	369	379	389	398	408
N	L 0	1	2	3	4	5	6	7	8	9

P P

	11	10	9
1	1.1	1.0	0.9
2	2.2	2.0	1.8
3	3.3	3.0	2.7
4	4.4	4.0	3.6
5	5.5	5.0	4.5
6	6.6	6.0	5.4
7	7.7	7.0	6.3
8	8.8	8.0	7.2
9	9.9	9.0	8.1

		S	T			S	T
3960″	= 1° 6′	4.68 555̄	4.68 563	4260″	= 1° 11′	4.68 554	4.68 564
4020	= 1 7	4.68 555̄	4.68 563	4320	= 1 12	4.68 554	4.68 564
4080	= 1 8	4.68 555̄	4.68 563	4380	= 1 13	4.68 554	4.68 564
4140	= 1 9	4.68 555̄	4.68 563	4440	= 1 14	4.68 554	4.68 564
4200	= 1 10	4.68 554	4.68 563	4500	= 1 15	4.68 554	4.68 564

450—500

N	L 0	1	2	3	4	5	6	7	8	9
450	65 321	331	341	350	360	369	379	389	398	408
451	418	427	437	447	456	466	475	485	495	504
452	514	523	533	543	552	562	571	581	591	600
453	610	619	629	639	648	658	667	677	686	696
454	706	715	725	734	744	753	763	772	782	792
455	801	811	820	830	839	849	858	868	877	887
456	896	906	916	925	935	944	954	963	973	982
457	992	*001	*011	*020	*030	*039	*049	*058	*068	*077
458	66 087	096	106	115	124	134	143	153	162	172
459	181	191	200	210	219	229	238	247	257	266
460	276	285	295	304	314	323	332	342	351	361
461	370	380	389	398	408	417	427	436	445	455
462	464	474	483	492	502	511	521	530	539	549
463	558	567	577	586	596	605	614	624	633	642
464	652	661	671	680	689	699	708	717	727	736
465	745	755	764	773	783	792	801	811	820	829
466	839	848	857	867	876	885	894	904	913	922
467	932	941	950	960	969	978	987	997	*006	*015
468	67 025	034	043	052	062	071	080	089	099	108
469	117	127	136	145	154	164	173	182	191	201
470	210	219	228	237	247	256	265	274	284	293
471	302	311	321	330	339	348	357	367	376	385
472	394	403	413	422	431	440	449	459	468	477
473	486	495	504	514	523	532	541	550	560	569
474	578	587	596	605	614	624	633	642	651	660
475	669	679	688	697	706	715	724	733	742	752
476	761	770	779	788	797	806	815	825	834	843
477	852	861	870	879	888	897	906	916	925	934
478	943	952	961	970	979	988	997	*006	*015	*024
479	68 034	043	052	061	070	079	088	097	106	115
480	124	133	142	151	160	169	178	187	196	205
481	215	224	233	242	251	260	269	278	287	296
482	305	314	323	332	341	350	359	368	377	386
483	395	404	413	422	431	440	449	458	467	476
484	485	494	502	511	520	529	538	547	556	565
485	574	583	592	601	610	619	628	637	646	655
486	664	673	681	690	699	708	717	726	735	744
487	753	762	771	780	789	797	806	815	824	833
488	842	851	860	869	878	886	895	904	913	922
489	931	940	949	958	966	975	984	993	*002	*011
490	69 020	028	037	046	055	064	073	082	090	099
491	108	117	126	135	144	152	161	170	179	188
492	197	205	214	223	232	241	249	258	267	276
493	285	294	302	311	320	329	338	346	355	364
494	373	381	390	399	408	417	425	434	443	452
495	461	469	478	487	496	504	513	522	531	539
496	548	557	566	574	583	592	601	609	618	627
497	636	644	653	662	671	679	688	697	705	714
498	723	732	740	749	758	767	775	784	793	801
499	810	819	827	836	845	854	862	871	880	888
500	897	906	914	923	932	940	949	958	966	975
N	L 0	1	2	3	4	5	6	7	8	9

P P

	10	9	8
1	1.0	0.9	0.8
2	2.0	1.8	1.6
3	3.0	2.7	2.4
4	4.0	3.6	3.2
5	5.0	4.5	4.0
6	6.0	5.4	4.8
7	7.0	6.3	5.6
8	8.0	7.2	6.4
9	9.0	8.1	7.2

		S	T			S	T
4500″ = 1°	15′	4.68 554	4.68 564	4800″ = 1°	20′	4.68 554	4.68 565
4560 = 1	16	4.68 554	4.68 565	4860 = 1	21	4.68 553	4.68 566
4620 = 1	17	4.68 554	4.68 565	4920 = 1	22	4.68 553	4.68 566
4680 = 1	18	4.68 554	4.68 565	4980 = 1	23	4.68 553	4.68 566
4740 = 1	19	4.68 554	4.68 565	5040 = 1	24	4.68 553	4.68 566

500—550

N	L 0	1	2	3	4	5	6	7	8	9
500	69 897	906	914	923	932	940	949	958	966	975
501	984	992	*001	*010	*018	*027	*036	*044	*053	*062
502	70 070	079	088	096	105	114	122	131	140	148
503	157	165	174	183	191	200	209	217	226	234
504	243	252	260	269	278	286	295	303	312	321
505	329	338	346	355	364	372	381	389	398	406
506	415	424	432	441	449	458	467	475	484	492
507	501	509	518	526	535	544	552	561	569	578
508	586	595	603	612	621	629	638	646	655	663
509	672	680	689	697	706	714	723	731	740	749
510	757	766	774	783	791	800	808	817	825	834
511	842	851	859	868	876	885	893	902	910	919
512	927	935	944	952	961	969	978	986	995	*003
513	71 012	020	029	037	046	054	063	071	079	088
514	096	105	113	122	130	139	147	155	164	172
515	181	189	198	206	214	223	231	240	248	257
516	265	273	282	290	299	307	315	324	332	341
517	349	357	366	374	383	391	399	408	416	425
518	433	441	450	458	466	475	483	492	500	508
519	517	525	533	542	550	559	567	575	584	592
520	600	609	617	625	634	642	650	659	667	675
521	684	692	700	709	717	725	734	742	750	759
522	767	775	784	792	800	809	817	825	834	842
523	850	858	867	875	883	892	900	908	917	925
524	933	941	950	958	966	975	983	991	999	*008
525	72 016	024	032	041	049	057	066	074	082	090
526	099	107	115	123	132	140	148	156	165	173
527	181	189	198	206	214	222	230	239	247	255
528	263	272	280	288	296	304	313	321	329	337
529	346	354	362	370	378	387	395	403	411	419
530	428	436	444	452	460	469	477	485	493	501
531	509	518	526	534	542	550	558	567	575	583
532	591	599	607	616	624	632	640	648	656	665
533	673	681	689	697	705	713	722	730	738	746
534	754	762	770	779	787	795	803	811	819	827
535	835	843	852	860	868	876	884	892	900	908
536	916	925	933	941	949	957	965	973	981	989
537	997	*006	*014	*022	*030	*038	*046	*054	*062	*070
538	73 078	086	094	102	111	119	127	135	143	151
539	159	167	175	183	191	199	207	215	223	231
540	239	247	255	263	272	280	288	296	304	312
541	320	328	336	344	352	360	368	376	384	392
542	400	408	416	424	432	440	448	456	464	472
543	480	488	496	504	512	520	528	536	544	552
544	560	568	576	584	592	600	608	616	624	632
545	640	648	656	664	672	679	687	695	703	711
546	719	727	735	743	751	759	767	775	783	791
547	799	807	815	823	830	838	846	854	862	870
548	878	886	894	902	910	918	926	933	941	949
549	957	965	973	981	989	997	*005	*013	*020	*028
550	74 036	044	052	060	068	076	084	092	099	107
N	L 0	1	2	3	4	5	6	7	8	9

P P

	9	8	7
1	0.9	0.8	0.7
2	1.8	1.6	1.4
3	2.7	2.4	2.1
4	3.6	3.2	2.8
5	4.5	4.0	3.5
6	5.4	4.8	4.2
7	6.3	5.6	4.9
8	7.2	6.4	5.6
9	8.1	7.2	6.3

		S	T			S	T
4980″	= 1° 23′	4.68 553	4.68 566	5280″	= 1° 28′	4.68 553	4.68 567
5040	= 1 24	4.68 553	4.68 566	5340	= 1 29	4.68 553	4.68 567
5100	= 1 25	4.68 553	4.68 566	5400	= 1 30	4.68 553	4.68 567
5160	= 1 26	4.68 553	4.68 567	5460	= 1 31	4.68 552	4.68 568
5220	= 1 27	4.68 553	4.68 567	5520	= 1 32	4.68 552	4.68 568

550—600

N	L 0	1	2	3	4	5	6	7	8	9
550	74 036	044	052	060	068	076	084	092	099	107
551	115	123	131	139	147	155	162	170	178	186
552	194	202	210	218	225	233	241	249	257	265
553	273	280	288	296	304	312	320	327	335	343
554	351	359	367	374	382	390	398	406	414	421
555	429	437	445	453	461	468	476	484	492	500
556	507	515	523	531	539	547	554	562	570	578
557	586	593	601	609	617	624	632	640	648	656
558	663	671	679	687	695	702	710	718	726	733
559	741	749	757	764	772	780	788	796	803	811
560	819	827	834	842	850	858	865	873	881	889
561	896	904	912	920	927	935	943	950	958	966
562	974	981	989	997	*005	*012	*020	*028	*035	*043
563	75 051	059	066	074	082	089	097	105	113	120
564	128	136	143	151	159	166	174	182	189	197
565	205	213	220	228	236	243	251	259	266	274
566	282	289	297	305	312	320	328	335	343	351
567	358	366	374	381	389	397	404	412	420	427
568	435	442	450	458	465	473	481	488	496	504
569	511	519	526	534	542	549	557	565	572	580
570	587	595	603	610	618	626	633	641	648	656
571	664	671	679	686	694	702	709	717	724	732
572	740	747	755	762	770	778	785	793	800	808
573	815	823	831	838	846	853	861	868	876	884
574	891	899	906	914	921	929	937	944	952	959
575	967	974	982	989	997	*005	*012	*020	*027	*035
576	76 042	050	057	065	072	080	087	095	103	110
577	118	125	133	140	148	155	163	170	178	185
578	193	200	208	215	223	230	238	245	253	260
579	268	275	283	290	298	305	313	320	328	335
580	343	350	358	365	373	380	388	395	403	410
581	418	425	433	440	448	455	462	470	477	485
582	492	500	507	515	522	530	537	545	552	559
583	567	574	582	589	597	604	612	619	626	634
584	641	649	656	664	671	678	686	693	701	708
585	716	723	730	738	745	753	760	768	775	782
586	790	797	805	812	819	827	834	842	849	856
587	864	871	879	886	893	901	908	916	923	930
588	938	945	953	960	967	975	982	989	997	*004
589	77 012	019	026	034	041	048	056	063	070	078
590	085	093	100	107	115	122	129	137	144	151
591	159	166	173	181	188	195	203	210	217	225
592	232	240	247	254	262	269	276	283	291	298
593	305	313	320	327	335	342	349	357	364	371
594	379	386	393	401	408	415	422	430	437	444
595	452	459	466	474	481	488	495	503	510	517
596	525	532	539	546	554	561	568	576	583	590
597	597	605	612	619	627	634	641	648	656	663
598	670	677	685	692	699	706	714	721	728	735
599	743	750	757	764	772	779	786	793	801	808
600	815	822	830	837	844	851	859	866	873	880
N	L 0	1	2	3	4	5	6	7	8	9

P P

	8
1	0.8
2	1.6
3	2.4
4	3.2
5	4.0
6	4.8
7	5.6
8	6.4
9	7.2

	7
1	0.7
2	1.4
3	2.1
4	2.8
5	3.5
6	4.2
7	4.9
8	5.6
9	6.3

		S	T			S	T
5460″	= 1° 31′	4.68 552	4.68 568	5760″	= 1° 36′	4.68 552	4.68 569
5520	= 1 32	4.68 552	4.68 568	5820	= 1 37	4.68 552	4.68 569
5580	= 1 33	4.68 552	4.68 568	5880	= 1 38	4.68 552	4.68 569
5640	= 1 34	4.68 552	4.68 568	5940	= 1 39	4.68 551	4.68 569
5700	= 1 35	4.68 552	4.68 569	6000	= 1 40	4.68 551	4.68 570

N	L 0	1	2	3	4	5	6	7	8	9
600	77 815	822	830	837	844	851	859	866	873	880
601	887	895̄	902	909	916	924	931	938	945	952
602	960	967	974	981	988	996	*003	*010	*017	*025̄
603	78 032	039	046	053	061	068	075̄	082	089	097
604	104	111	118	125	132	140	147	154	161	168
605	176	183	190	197	204	211	219	226	233	240
606	247	254	262	269	276	283	290	297	305̄	312
607	319	326	333	340	347	355̄	362	369	376	383
608	390	398	405̄	412	419	426	433	440	447	455̄
609	462	469	476	483	490	497	504	512	519	526
610	533	540	547	554	561	569	576	583	590	597
611	604	611	618	625	633	640	647	654	661	668
612	675	682	689	696	704	711	718	725̄	732	739
613	746	753	760	767	774	781	789	796	803	810
614	817	824	831	838	845	852	859	866	873	880
615	888	895̄	902	909	916	923	930	937	944	951
616	958	965	972	979	986	993	*000	*007	*014	*021
617	79 029	036	043	050	057	064	071	078	085̄	092
618	099	106	113	120	127	134	141	148	155	162
619	169	176	183	190	197	204	211	218	225	232
620	239	246	253	260	267	274	281	288	295	302
621	309	316	323	330	337	344	351	358	365	372
622	379	386	393	400	407	414	421	428	435̄	442
623	449	456	463	470	477	484	491	498	505̄	511
624	518	525	532	539	546	553	560	567	574	581
625	588	595̄	602	609	616	623	630	637	644	650
626	657	664	671	678	685	692	699	706	713	720
627	727	734	741	748	754	761	768	775	782	789
628	796	803	810	817	824	831	837	844	851	858
629	865	872	879	886	893	900	906	913	920	927
630	934	941	948	955̄	962	969	975	982	989	996
631	80 003	010	017	024	030	037	044	051	058	065̄
632	072	079	085	092	099	106	113	120	127	134
633	140	147	154	161	168	175̄	182	188	195	202
634	209	216	223	229	236	243	250	257	264	271
635	277	284	291	298	305̄	312	318	325	332	339
636	346	353	359	366	373	380	387	393	400	407
637	414	421	428	434	441	448	455̄	462	468	475
638	482	489	496	502	509	516	523	530	536	543
639	550	557	564	570	577	584	591	598	604	611
640	618	625̄	632	638	645	652	659	665	672	679
641	686	693	699	706	713	720	726	733	740	747
642	754	760	767	774	781	787	794	801	808	814
643	821	828	835̄	841	848	855̄	862	868	875	882
644	889	895	902	909	916	922	929	936	943	949
645	956	963	969	976	983	990	996	*003	*010	*017
646	81 023	030	037	043	050	057	064	070	077	084
647	090	097	104	111	117	124	131	137	144	151
648	158	164	171	178	184	191	198	204	211	218
649	224	231	238	245̄	251	258	265̄	271	278	285̄
650	291	298	305̄	311	318	325̄	331	338	345̄	351
N	L 0	1	2	3	4	5	6	7	8	9

P P

8	
1	0.8
2	1.6
3	2.4
4	3.2
5	4.0
6	4.8
7	5.6
8	6.4
9	7.2

7	
1	0.7
2	1.4
3	2.1
4	2.8
5	3.5
6	4.2
7	4.9
8	5.6
9	6.3

6	
1	0.6
2	1.2
3	1.8
4	2.4
5	3.0
6	3.6
7	4.2
8	4.8
9	5.4

		S	T			S	T
6000″	= 1° 40	4.68 551	4.68 570	6300″	= 1° 45	4.68 551	4.68 571
6060	= 1 41	4.68 551	4.68 570	6360	= 1 46	4.68 551	4.68 571
6120	= 1 42	4.68 551	4.68 570	6420	= 1 47	4.68 550	4.68 572
6180	= 1 43	4.68 551	4.68 570	6480	= 1 48	4.68 550	4.68 572
6240	= 1 44	4.68 551	4.68 571	6540	= 1 49	4.68 550	4.68 572

650—700

N	L 0	1	2	3	4	5	6	7	8	9
650	81 291	298	305̄	311	318	325̄	331	338	345	351
651	358	365̄	371	378	385̄	391	398	405̄	411	418
652	425̄	431	438	445̄	451	458	465̄	471	478	485̄
653	491	498	505̄	511	518	525̄	531	538	544	551
654	558	564	571	578	584	591	598	604	611	617
655	624	631	637	644	651	657	664	671	677	684
656	690	697	704	710	717	723	730	737	743	75̄0
657	757	763	770	776	783	790	796	803	809	816
658	823	829	836	842	849	856	862	869	875	882
659	889	895	902	908	915̄	921	928	935̄	941	948
660	954	961	968	974	981	987	994	*000	*007	*014
661	82 020	027	033	040	046	053	060	066	073	079
662	086	092	099	105	112	119	125	132	138	145̄
663	151	158	164	171	178	184	191	197	204	210
664	217	223	230	236	243	249	256	263	269	276
665	282	289	295	302	308	315̄	321	328	334	341
666	347	354	360	367	373	380	387	393	400	406
667	413	419	426	432	439	445	452	458	465̄	471
668	478	484	491	497	504	510	517	523	530	536
669	543	549	556	562	569	575	582	588	595̄	601
670	607	614	620	627	633	640	646	653	659	666
671	672	679	685	692	698	705̄	711	718	724	730
672	737	743	75̄0	756	763	769	776	782	789	795
673	802	808	814	821	827	834	840	847	853	860
674	866	872	879	885	892	898	905̄	911	918	924
675	930	937	943	95̄0	956	963	969	975	982	988
676	995̄	*001	*008	*014	*020	*027	*033	*040	*046	*052
677	83 059	065	072	078	085̄	091	097	104	110	117
678	123	129	136	142	149	155̄	161	168	174	181
679	187	193	200	206	213	219	225	232	238	245̄
680	251	257	264	270	276	283	289	296	302	308
681	315̄	321	327	334	340	347	353	359	366	372
682	378	385̄	391	398	404	410	417	423	429	436
683	442	448	455̄	461	467	474	480	487	493	499
684	506	512	518	525̄	531	537	544	550	556	563
685	569	575	582	588	594	601	607	613	620	626
686	632	639	645	651	658	664	670	677	683	689
687	696	702	708	715̄	721	727	734	740	746	753
688	759	765	771	778	784	790	797	803	809	816
689	822	828	835̄	841	847	853	860	866	872	879
690	885̄	891	897	904	910	916	923	929	935	942
691	948	954	960	967	973	979	985	992	998	*004
692	84 011	017	023	029	036	042	048	055̄	061	067
693	073	080	086	092	098	105̄	111	117	123	130
694	136	142	148	155̄	161	167	173	180	186	192
695	198	205̄	211	217	223	230	236	242	248	255̄
696	261	267	273	280	286	292	298	305̄	311	317
697	323	330	336	342	348	354	361	367	373	379
698	386	392	398	404	410	417	423	429	435	442
699	448	454	460	466	473	479	485̄	491	497	504
700	510	516	522	528	535̄	541	547	553	559	566
N	L 0	1	2	3	4	5	6	7	8	9

P P

	7
1	0.7
2	1.4
3	2.1
4	2.8
5	3.5
6	4.2
7	4.9
8	5.6
9	6.3

	6
1	0.6
2	1.2
3	1.8
4	2.4
5	3.0
6	3.6
7	4.2
8	4.8
9	5.4

		S	T
6480″	= 1° 48′	4.68 550	4.68 572
6540	= 1 49	4.68 550	4.68 572
6600	= 1 50	4.68 550	4.68 572
6660	= 1 51	4.68 55̄0	4.68 573
6720	= 1 52	4.68 55̄0	4.68 573
6780″	= 1° 53′	4.68 55̄0	4.68 573
6840	= 1 54	4.68 55̄0	4.68 573
6900	= 1 55	4.68 549	4.68 574
6960	= 1 56	4.68 549	4.68 574
7020	= 1 57	4.68 549	4.68 574

N	L 0	1	2	3	4	5	6	7	8	9
700	84 510	516	522	528	535̄	541	547	553	559	566
701	572	578	584	590	597	603	609	615	621	628
702	634	640	646	652	658	665̄	671	677	683	689
703	696	702	708	714	720	726	733	739	745̄	751
704	757	763	770	776	782	788	794	800	807	813
705	819	825	831	837	844	85̄0	856	862	868	874
706	880	887	893	899	905	911	917	924	930	936
707	942	948	954	960	967	973	979	985̄	991	997
708	85 003	009	016	022	028	034	040	046	052	058
709	065̄	071	077	083	089	095	101	107	114	120
710	126	132	138	144	150	156	163	169	175̄	181
711	187	193	199	205	211	217	224	230	236	242
712	248	254	260	266	272	278	285̄	291	297	303
713	309	315	321	327	333	339	345	352	358	364
714	370	376	382	388	394	400	406	412	418	425̄
715	431	437	443	449	455̄	461	467	473	479	485
716	491	497	503	509	516	522	528	534	540	546
717	552	558	564	570	576	582	588	594	600	606
718	612	618	625̄	631	637	643	649	655̄	661	667
719	673	679	685̄	691	697	703	709	715	721	727
720	733	739	745	751	757	763	769	775	781	788
721	794	800	806	812	818	824	830	836	842	848
722	854	860	866	872	878	884	890	896	902	908
723	914	920	926	932	938	944	95̄0	956	962	968
724	974	980	986	992	998	*004	*010	*016	*022	*028
725	86 034	040	046	052	058	064	070	076	082	088
726	094	100	106	112	118	124	130	136	141	147
727	153	159	165	171	177	183	189	195	201	207
728	213	219	225	231	237	243	249	255̄	261	267
729	273	279	285̄	291	297	303	308	314	320	326
730	332	338	344	350	356	362	368	374	380	386
731	392	398	404	410	415	421	427	433	439	445
732	451	457	463	469	475̄	481	487	493	499	504
733	510	516	522	528	534	540	546	552	558	564
734	570	576	581	587	593	599	605	611	617	623
735	629	635̄	641	646	652	658	664	670	676	682
736	688	694	700	705	711	717	723	729	735̄	741
737	747	753	759	764	770	776	782	788	794	800
738	806	812	817	823	829	835	841	847	853	859
739	864	870	876	882	888	894	900	906	911	917
740	923	929	935̄	941	947	953	958	964	970	976
741	982	988	994	999	*005	*011	*017	*023	*029	*035̄
742	87 040	046	052	058	064	070	075	081	087	093
743	099	105̄	111	116	122	128	134	140	146	151
744	157	163	169	175̄	181	186	192	198	204	210
745	216	221	227	233	239	245̄	251	256	262	268
746	274	280	286	291	297	303	309	315̄	320	326
747	332	338	344	349	355	361	367	373	379	384
748	390	396	402	408	413	419	425̄	431	437	442
749	448	454	460	466	471	477	483	489	495̄	500
750	506	512	518	523	529	535	541	547	552	558
N	L 0	1	2	3	4	5	6	7	8	9

P P

	7	6	5
1	0.7	0.6	0.5
2	1.4	1.2	1.0
3	2.1	1.8	1.5
4	2.8	2.4	2.0
5	3.5	3.0	2.5
6	4.2	3.6	3.0
7	4.9	4.2	3.5
8	5.6	4.8	4.0
9	6.3	5.4	4.5

	S	T		S	T
6960′ = 1° 56′	4.68 549	4.68 574	7260′ = 2° 1′	4.68 549	4.68 575
7020 = 1 57	4.68 549	4.68 574	7320 = 2 2	4.68 548	4.68 576
7080 = 1 58	4.68 549	4.68 575̄	7380 = 2 3	4.68 548	4.68 576
7140 = 1 59	4.68 549	4.68 575̄	7440 = 2 4	4.68 548	4.68 576
7200 = 2 0	4.68 549	4.68 575	7500 = 2 5	4.68 548	4.68 577

750—800

N	L 0	1	2	3	4	5	6	7	8	9
750	87 506	512	518	523	529	535	541	547	552	558
751	564	570	576	581	587	593	599	604	610	616
752	622	628	633	639	645̄	651	656	662	668	674
753	679	685	691	697	703	708	714	720	726	731
754	737	743	749	754	760	766	772	777	783	789
755	795̄	800	806	812	818	823	829	835̄	841	846
756	852	858	864	869	875	881	887	892	898	904
757	910	915	921	927	933	938	944	950	955	961
758	967	973	978	984	990	996	*001	*007	*013	*018
759	88 024	030	036	041	047	053	058	064	070	076
760	081	087	093	098	104	110	116	121	127	133
761	138	144	150	156	161	167	173	178	184	190
762	195	201	207	213	218	224	230	235	241	247
763	252	258	264	270	275	281	287	292	298	304
764	309	315	321	326	332	338	343	349	355̄	360
765	366	372	377	383	389	395̄	400	406	412	417
766	423	429	434	440	446	451	457	463	468	474
767	480	485	491	497	502	508	513	519	525̄	530
768	536	542	547	553	559	564	570	576	581	587
769	593	598	604	610	615	621	627	632	638	643
770	649	655̄	660	666	672	677	683	689	694	700
771	705	711	717	722	728	734	739	745̄	750	756
772	762	767	773	779	784	790	795	801	807	812
773	818	824	829	835̄	840	846	852	857	863	868
774	874	880	885	891	897	902	908	913	919	925̄
775	930	936	941	947	953	958	964	969	975̄	981
776	986	992	997	*003	*009	*014	*020	*025	*031	*037
777	89 042	048	053	059	064	070	076	081	087	092
778	098	104	109	115̄	120	126	131	137	143	148
779	154	159	165̄	170	176	182	187	193	198	204
780	209	215	221	226	232	237	243	248	254	260
781	265	271	276	282	287	293	298	304	310	315
782	321	326	332	337	343	348	354	360	365	371
783	376	382	387	393	398	404	409	415̄	421	426
784	432	437	443	448	454	459	465̄	470	476	481
785	487	492	498	504	509	515̄	520	526	531	537
786	542	548	553	559	564	570	575	581	586	592
787	597	603	609	614	620	625	631	636	642	647
788	653	658	664	669	675̄	680	686	691	697	702
789	708	713	719	724	730	735	741	746	752	757
790	763	768	774	779	785̄	790	796	801	807	812
791	818	823	829	834	840	845	851	856	862	867
792	873	878	883	889	894	900	905	911	916	922
793	927	933	938	944	949	955̄	960	966	971	977
794	982	988	993	998	*004	*009	*015̄	*020	*026	*031
795	90 037	042	048	053	059	064	069	075̄	080	086
796	091	097	102	108	113	119	124	129	135̄	140
797	146	151	157	162	168	173	179	184	189	195̄
798	200	206	211	217	222	227	233	238	244	249
799	255̄	260	266	271	276	282	287	293	298	304
800	309	314	320	325	331	336	342	347	352	358
N	L 0	1	2	3	4	5	6	7	8	9

P P

	6
1	0.6
2	1.2
3	1.8
4	2.4
5	3.0
6	3.6
7	4.2
8	4.8
9	5.4

	5
1	0.5
2	1.0
3	1.5
4	2.0
5	2.5
6	3.0
7	3.5
8	4.0
9	4.5

		S	T			S	T
7500′	= 2° 5′	4.68 548	4.68 577	7800′	= 2° 10′	4.68 547	4.68 578
7560	= 2 6	4.68 548	4.68 577	7860	= 2 11	4.68 547	4.68 579
7620	= 2 7	4.68 548	4.68 577	7920	= 2 12	4.68 547	4.68 579
7680	= 2 8	4.68 547	4.68 578	7980	= 2 13	4.68 547	4.68 579
7740	= 2 9	4.68 547	4.68 578	8040	= 2 14	4.68 546	4.68 579

800—850

N	L 0	1	2	3	4	5	6	7	8	9
800	90 309	314	320	325	331	336	342	347	352	358
801	363	369	374	380	385̄	390	396	401	407	412
802	417	423	428	434	439	445̄	450	455	461	466
803	472	477	482	488	493	499	504	509	515̄	520
804	526	531	536	542	547	553	558	563	569	574
805	580	585̄	590	596	601	607	612	617	623	628
806	634	639	644	650	655	660	666	671	677	682
807	687	693	698	703	709	714	720	725	730	736
808	741	747	752	757	763	768	773	779	784	789
809	795̄	800	806	811	816	822	827	832	838	843
810	849	854	859	865̄	870	875	881	886	891	897
811	902	907	913	918	924	929	934	940	945̄	950
812	956	961	966	972	977	982	988	993	998	*004
813	91 009	014	020	025	030	036	041	046	052	057
814	062	068	073	078	084	089	094	100	105	110
815	116	121	126	132	137	142	148	153	158	164
816	169	174	180	185̄	190	196	201	206	212	217
817	222	228	233	238	243	249	254	259	265̄	270
818	275	281	286	291	297	302	307	312	318	323
819	328	334	339	344	350	355̄	360	365	371	376
820	381	387	392	397	403	408	413	418	424	429
821	434	440	445̄	450	455	461	466	471	477	482
822	487	492	498	503	508	514	519	524	529	535̄
823	540	545	551	556	561	566	572	577	582	587
824	593	598	603	609	614	619	624	630	635̄	640
825	645	651	656	661	666	672	677	682	687	693
826	698	703	709	714	719	724	730	735̄	740	745
827	751	756	761	766	772	777	782	787	793	798
828	803	808	814	819	824	829	834	840	845̄	850
829	855	861	866	871	876	882	887	892	897	903
830	908	913	918	924	929	934	939	944	950	955̄
831	960	965	971	976	981	986	991	997	*002	*007
832	92 012	018	023	028	033	038	044	049	054	059
833	065̄	070	075̄	080	085	091	096	101	106	111
834	117	122	127	132	137	143	148	153	158	163
835	169	174	179	184	189	195̄	200	205	210	215
836	221	226	231	236	241	247	252	257	262	267
837	273	278	283	288	293	298	304	309	314	319
838	324	330	335̄	340	345	350	355	361	366	371
839	376	381	387	392	397	402	407	412	418	423
840	428	433	438	443	449	454	459	464	469	474
841	480	485̄	490	495	500	505	511	516	521	526
842	531	536	542	547	552	557	562	567	572	578
843	583	588	593	598	603	609	614	619	624	629
844	634	639	645̄	650	655̄	660	665	670	675	681
845	686	691	696	701	706	711	716	722	727	732
846	737	742	747	752	758	763	768	773	778	783
847	788	793	799	804	809	814	819	824	829	834
848	840	845̄	850	855̄	860	865	870	875	881	886
849	891	896	901	906	911	916	921	927	932	937
850	942	947	952	957	962	967	973	978	983	988
N	L 0	1	2	3	4	5	6	7	8	9

P P

	6	5
1	0.6	0.5
2	1.2	1.0
3	1.8	1.5
4	2.4	2.0
5	3.0	2.5
6	3.6	3.0
7	4.2	3.5
8	4.8	4.0
9	5.4	4.5

		S	T			S	T
7980″	=2° 13′	4.68 547	4.68 579	8280″	=2° 18′	4.68 546	4.68 581
8040	=2 14	4.68 546	4.68 579	8340	=2 19	4.68 546	4.68 581
8100	=2 15	4.68 546	4.68 580	8400	=2 20	4.68 545	4.68 582
8160	=2 16	4.68 546	4.68 580	8460	=2 21	4.68 545	4.68 582
8220	=2 17	4.68 546	4.68 580	8520	=2 22	4.68 545	4.68 582

850—900

N	L 0	1	2	3	4	5	6	7	8	9
850	92 942	947	952	957	962	967	973	978	983	988
851	993	998	*003	*008	*013	*018	*024	*029	*034	*039
852	93 044	049	054	059	064	069	075̄	080	085̄	090
853	095̄	100	105	110	115	120	125	131	136	141
854	146	151	156	161	166	171	176	181	186	192
855	197	202	207	212	217	222	227	232	237	242
856	247	252	258	263	268	273	278	283	288	293
857	298	303	308	313	318	323	328	334	339	344
858	349	354	359	364	369	374	379	384	389	394
859	399	404	409	414	420	425̄	430	435̄	440	445̄
860	450̄	455̄	460	465̄	470	475	480	485	490	495
861	500	505	510	515	520	526	531	536	541	546
862	551	556	561	566	571	576	581	586	591	596
863	601	606	611	616	621	626	631	636	641	646
864	651	656	661	666	671	676	682	687	692	697
865	702	707	712	717	722	727	732	737	742	747
866	752	757	762	767	772	777	782	787	792	797
867	802	807	812	817	822	827	832	837	842	847
868	852	857	862	867	872	877	882	887	892	897
869	902	907	912	917	922	927	932	937	942	947
870	952	957	962	967	972	977	982	987	992	997
871	94 002	007	012	017	022	027	032	037	042	047
872	052	057	062	067	072	077	082	086	091	096
873	101	106	111	116	121	126	131	136	141	146
874	151	156	161	166	171	176	181	186	191	196
875	201	206	211	216	221	226	231	236	240	245
876	250	255	260	265	270	275	280	285	290	295
877	300	305̄	310	315̄	320	325̄	330	335̄	340	345̄
878	349	354	359	364	369	374	379	384	389	394
879	399	404	409	414	419	424	429	433	438	443
880	448	453	458	463	468	473	478	483	488	493
881	498	503	507	512	517	522	527	532	537	542
882	547	552	557	562	567	571	576	581	586	591
883	596	601	606	611	616	621	626	630	635	640
884	645	650	655	660	665̄	670	675̄	680	685̄	689
885	694	699	704	709	714	719	724	729	734	738
886	743	748	753	758	763	768	773	778	783	787
887	792	797	802	807	812	817	822	827	832	836
888	841	846	851	856	861	866	871	876	880	885
889	890	895	900	905̄	910	915̄	919	924	929	934
890	939	944	949	954	959	963	968	973	978	983
891	988	993	998	*002	*007	*012	*017	*022	*027	*032
892	95 036	041	046	051	056	061	066	071	075	080
893	085	090	095̄	100	105̄	109	114	119	124	129
894	134	139	143	148	153	158	163	168	173	177
895	182	187	192	197	202	207	211	216	221	226
896	231	236	240	245	250	255	260	265̄	270	274
897	279	284	289	294	299	303	308	313	318	323
898	328	332	337	342	347	352	357	361	366	371
899	376	381	386	390	395	400	405̄	410	415̄	419
900	424	429	434	439	444	448	453	458	463	468
N	L 0	1	2	3	4	5	6	7	8	9

P P

	6
1	0.6
2	1.2
3	1.8
4	2.4
5	3.0
6	3.6
7	4.2
8	4.8
9	5.4

	5
1	0.5
2	1.0
3	1.5
4	2.0
5	2.5
6	3.0
7	3.5
8	4.0
9	4.5

	4
1	0.4
2	0.8
3	1.2
4	1.6
5	2.0
6	2.4
7	2.8
8	3.2
9	3.6

		S	T			S	T
8460″	= 2° 21′	4.68 545	4.68 582	8760″	= 2° 26′	4.68 544	4.68 584
8520	= 2 22	4.68 545	4.68 582	8820	= 2 27	4.68 544	4.68 584
8580	= 2 23	4.68 545̄	4.68 583	8880	= 2 28	4.68 544	4.68 584
8640	= 2 24	4.68 545̄	4.68 583	8940	= 2 29	4.68 544	4.68 585̄
8700	= 2 25	4.68 545̄	4.68 583	9000	= 2 30	4.68 544	4.68 585

900—950

N	L 0	1	2	3	4	5	6	7	8	9
900	95 424	429	434	439	444	448	453	458	463	468
901	472	477	482	487	492	497	501	506	511	516
902	521	525	530	535	540	545	550	554	559	564
903	569	574	578	583	588	593	598	602	607	612
904	617	622	626	631	636	641	646	650	655	660
905	665	670	674	679	684	689	694	698	703	708
906	713	718	722	727	732	737	742	746	751	756
907	761	766	770	775	780	785	789	794	799	804
908	809	813	818	823	828	832	837	842	847	852
909	856	861	866	871	875	880	885	890	895	899
910	904	909	914	918	923	928	933	938	942	947
911	952	957	961	966	971	976	980	985	990	995
912	999	*004	*009	*014	*019	*023	*028	*033	*038	*042
913	96 047	052	057	061	066	071	076	080	085	090
914	095	099	104	109	114	118	123	128	133	137
915	142	147	152	156	161	166	171	175	180	185
916	190	194	199	204	209	213	218	223	227	232
917	237	242	246	251	256	261	265	270	275	280
918	284	289	294	298	303	308	313	317	322	327
919	332	336	341	346	350	355	360	365	369	374
920	379	384	388	393	398	402	407	412	417	421
921	426	431	435	440	445	450	454	459	464	468
922	473	478	483	487	492	497	501	506	511	515
923	520	525	530	534	539	544	548	553	558	562
924	567	572	577	581	586	591	595	600	605	609
925	614	619	624	628	633	638	642	647	652	656
926	661	666	670	675	680	685	689	694	699	703
927	708	713	717	722	727	731	736	741	745	750
928	755	759	764	769	774	778	783	788	792	797
929	802	806	811	816	820	825	830	834	839	844
930	848	853	858	862	867	872	876	881	886	890
931	895	900	904	909	914	918	923	928	932	937
932	942	946	951	956	960	965	970	974	979	984
933	988	993	997	*002	*007	*011	*016	*021	*025	*030
934	97 035	039	044	049	053	058	063	067	072	077
935	081	086	090	095	100	104	109	114	118	123
936	128	132	137	142	146	151	155	160	165	169
937	174	179	183	188	192	197	202	206	211	216
938	220	225	230	234	239	243	248	253	257	262
939	267	271	276	280	285	290	294	299	304	308
940	313	317	322	327	331	336	340	345	350	354
941	359	364	368	373	377	382	387	391	396	400
942	405	410	414	419	424	428	433	437	442	447
943	451	456	460	465	470	474	479	483	488	493
944	497	502	506	511	516	520	525	529	534	539
945	543	548	552	557	562	566	571	575	580	585
946	589	594	598	603	607	612	617	621	626	630
947	635	640	644	649	653	658	663	667	672	676
948	681	685	690	695	699	704	708	713	717	722
949	727	731	736	740	745	749	754	759	763	768
950	772	777	782	786	791	795	800	804	809	813
N	L 0	1	2	3	4	5	6	7	8	9

P P

	5
1	0.5
2	1.0
3	1.5
4	2.0
5	2.5
6	3.0
7	3.5
8	4.0
9	4.5

	4
1	0.4
2	0.8
3	1.2
4	1.6
5	2.0
6	2.4
7	2.8
8	3.2
9	3.6

P P

		S	T			S	T
9000″	=2° 30′	4.68 544	4.68 585	9300″	=2° 35′	4.68 543	4.68 587
9060	=2 31	4.68 544	4.68 585	9360	=2 36	4.68 543	4.68 587
9120	=2 32	4.68 543	4.68 586	9420	=2 37	4.68 542	4.68 588
9180	=2 33	4.68 543	4.68 586	9480	=2 38	4.68 542	4.68 588
9240	=2 34	4.68 543	4.68 587	9540	=2 39	4.68 542	4.68 588

950—1000

N	L 0	1	2	3	4	5	6	7	8	9
950	97 772	777	782	786	791	795	800	804	809	813
951	818	823	827	832	836	841	845	850	855	859
952	864	868	873	877	882	886	891	896	900	905
953	909	914	918	923	928	932	937	941	946	950
954	955	959	964	968	973	978	982	987	991	996
955	98 000	005	009	014	019	023	028	032	037	041
956	046	050	055	059	064	068	073	078	082	087
957	091	096	100	105	109	114	118	123	127	132
958	137	141	146	150	155	159	164	168	173	177
959	182	186	191	195	200	204	209	214	218	223
960	227	232	236	241	245	250	254	259	263	268
961	272	277	281	286	290	295	299	304	308	313
962	318	322	327	331	336	340	345	349	354	358
963	363	367	372	376	381	385	390	394	399	403
964	408	412	417	421	426	430	435	439	444	448
965	453	457	462	466	471	475	480	484	489	493
966	498	502	507	511	516	520	525	529	534	538
967	543	547	552	556	561	565	570	574	579	583
968	588	592	597	601	605	610	614	619	623	628
969	632	637	641	646	650	655	659	664	668	673
970	677	682	686	691	695	700	704	709	713	717
971	722	726	731	735	740	744	749	753	758	762
972	767	771	776	780	784	789	793	798	802	807
973	811	816	820	825	829	834	838	843	847	851
974	856	860	865	869	874	878	883	887	892	896
975	900	905	909	914	918	923	927	932	936	941
976	945	949	954	958	963	967	972	976	981	985
977	989	994	998	*003	*007	*012	*016	*021	*025	*029
978	99 034	038	043	047	052	056	061	065	069	074
979	078	083	087	092	096	100	105	109	114	118
980	123	127	131	136	140	145	149	154	158	162
981	167	171	176	180	185	189	193	198	202	207
982	211	216	220	224	229	233	238	242	247	251
983	255	260	264	269	273	277	282	286	291	295
984	300	304	308	313	317	322	326	330	335	339
985	344	348	352	357	361	366	370	374	379	383
986	388	392	396	401	405	410	414	419	423	427
987	432	436	441	445	449	454	458	463	467	471
988	476	480	484	489	493	498	502	506	511	515
989	520	524	528	533	537	542	546	550	555	559
990	564	568	572	577	581	585	590	594	599	603
991	607	612	616	621	625	629	634	638	642	647
992	651	656	660	664	669	673	677	682	686	691
993	695	699	704	708	712	717	721	726	730	734
994	739	743	747	752	756	760	765	769	774	778
995	782	787	791	795	800	804	808	813	817	822
996	826	830	835	839	843	848	852	856	861	865
997	870	874	878	883	887	891	896	900	904	909
998	913	917	922	926	930	935	939	944	948	952
999	957	961	965	970	974	978	983	987	991	996
1000	00 000	004	009	013	017	022	026	030	035	039
N	L 0	1	2	3	4	5	6	7	8	9

P P

	5
1	0.5
2	1.0
3	1.5
4	2.0
5	2.5
6	3.0
7	3.5
8	4.0
9	4.5

	4
1	0.4
2	0.8
3	1.2
4	1.6
5	2.0
6	2.4
7	2.8
8	3.2
9	3.6

		S	T			S	T
9480″	= 2° 38′	4.68 542	4.68 588	9780″	= 2° 43′	4.68 541	4.68 590
9540	= 2 39	4.68 542	4.68 588	9840	= 2 44	4.68 541	4.68 590
9600	= 2 40	4.68 542	4.68 589	9900	= 2 45	4.68 541	4.68 591
9660	= 2 41	4.68 542	4.68 589	9960	= 2 46	4.68 541	4.68 591
9720	= 2 42	4.68 541	4.68 590	10020	= 2 47	4.68 540	4.68 592

THE NATURAL LOGARITHMS
OF
WHOLE NUMBERS FROM 1 TO 200

Common logarithms may be converted into natural logarithms by multiplying them by 2.3025850930.

Natural logarithms may be converted into common logarithms by multiplying them by 0.4342944819.

N	Nat Log	N	Nat Log	N	Nat Log	N	Nat Log	N	Nat Log
0	$-\infty$	40	3.68 888	80	4.38 203	120	4.78 749	160	5.07 517
1	0.00 000	41	3.71 357	81	4.39 44$\bar{5}$	121	4.79 579	161	5.08 140
2	0.69 31$\bar{5}$	42	3.73 767	82	4.40 672	122	4.80 402	162	5.08 760
3	1.09 861	43	3.76 120	83	4.41 884	123	4.81 218	163	5.09 375
4	1.38 629	44	3.78 419	84	4.43 082	124	4.82 028	164	5.09 987
5	1.60 944	45	3.80 666	85	4.44 265	125	4.82 831	165	5.10 59$\bar{5}$
6	1.79 176	46	3.82 864	86	4.45 43$\bar{5}$	126	4.83 628	166	5.11 199
7	1.94 591	47	3.85 01$\bar{5}$	87	4.46 591	127	4.84 419	167	5.11 799
8	2.07 944	48	3.87 120	88	4.47 734	128	4.85 203	168	5.12 396
9	2.19 722	49	3.89 182	89	4.48 864	129	4.85 981	169	5.12 990
10	2.30 259	50	3.91 202	90	4.49 981	130	4.86 753	170	5.13 580
11	2.39 790	51	3.93 183	91	4.51 086	131	4.87 520	171	5.14 166
12	2.48 491	52	3.95 124	92	4.52 179	132	4.88 280	172	5.14 749
13	2.56 49$\bar{5}$	53	3.97 029	93	4.53 260	133	4.89 03$\bar{5}$	173	5.15 329
14	2.63 906	54	3.98 898	94	4.54 329	134	4.89 784	174	5.15 906
15	2.70 805	55	4.00 733	95	4.55 388	135	4.90 527	175	5.16 479
16	2.77 259	56	4.02 535	96	4.56 43$\bar{5}$	136	4.91 265	176	5.17 048
17	2.83 321	57	4.04 305	97	4.57 471	137	4.91 998	177	5.17 61$\bar{5}$
18	2.89 037	58	4.06 044	98	4.58 497	138	4.92 725	178	5.18 178
19	2.94 444	59	4.07 754	99	4.59 512	139	4.93 447	179	5.18 739
20	2.99 573	60	4.09 434	100	4.60 517	140	4.94 164	180	5.19 296
21	3.04 452	61	4.11 087	101	4.61 512	141	4.94 876	181	5.19 8$\bar{5}$0
22	3.09 104	62	4.12 713	102	4.62 497	142	4.95 583	182	5.20 401
23	3.13 549	63	4.14 313	103	4.63 473	143	4.96 284	183	5.20 949
24	3.17 805	64	4.15 888	104	4.64 439	144	4.96 981	184	5.21 494
25	3.21 888	65	4.17 439	105	4.65 396	145	4.97 673	185	5.22 036
26	3.25 810	66	4.18 965	106	4.66 344	146	4.98 361	186	5.22 57$\bar{5}$
27	3.29 584	67	4.20 469	107	4.67 283	147	4.99 043	187	5.23 111
28	3.33 220	68	4.21 951	108	4.68 213	148	4.99 721	188	5.23 644
29	3.36 730	69	4.23 411	109	4.69 13$\bar{5}$	149	5.00 39$\bar{5}$	189	5.24 17$\bar{5}$
30	3.40 120	70	4.24 8$\bar{5}$0	110	4.70 048	150	5.01 064	190	5.24 702
31	3.43 399	71	4.26 268	111	4.70 953	151	5.01 728	191	5.25 227
32	3.46 574	72	4.27 667	112	4.71 8$\bar{5}$0	152	5.02 388	192	5.25 7$\bar{5}$0
33	3.49 651	73	4.29 046	113	4.72 739	153	5.03 044	193	5.26 269
34	3.52 636	74	4.30 407	114	4.73 620	154	5.03 695	194	5.26 786
35	3.55 53$\bar{5}$	75	4.31 749	115	4.74 493	155	5.04 343	195	5.27 300
36	3.58 352	76	4.33 073	116	4.75 359	156	5.04 986	196	5.27 811
37	3.61 092	77	4.34 381	117	4.76 217	157	5.05 62$\bar{5}$	197	5.28 320
38	3.63 759	78	4.35 671	118	4.77 068	158	5.06 260	198	5.28 827
39	3.66 356	79	4.36 94$\bar{5}$	119	4.77 912	159	5.06 890	199	5.29 330
40	3.68 888	80	4.38 203	120	4.78 749	160	5.07 517	200	5.29 832

TABLE OF THE LOGARITHMS

OF THE

TRIGONOMETRIC FUNCTIONS

FROM 0° TO 1° AND 89° TO 90° FOR EVERY SECOND,

AND

FROM 1° TO 6° AND 84° TO 89° FOR EVERY TEN SECONDS.

L Cos *90 L Sin 0° L Tan 180° *270°

0.00	′ ″	0″	1″	2″	3″	4″	5″	6″	7″	8″	9″	10″	
000	0 0	4. —	68557	98660	*16270	*28763	*38454	*46373	*53067	*58866	*63982	*68557	50
000	10	5. 68557	72697	76476	79952	83170	86167	88969	91602	94085̄	96433	98660	40
000	20	98660	*00779	*02800	*04730	*06579	*08351	*10055̄	*11694	*13273	*14797	*16270	30
000	30	6. 16270	17694	19072	20409	21705	22964	24188	25378	26536	27664	28763	20
000	40	28763	29836	30882	31904	32903	33879	34833	35767	36682	37577	38454	10
000	50	38454	39315̄	40158	40985	41797	42594	43376	44145̄	44900	45643	46373	0 59
000	1 0	6.4 6373	7090	7797	8492	9175	9849	*0512	*1165̄	*1808	*2442	*3067	50
000	10	6.5 3067	3683	4291	4890	5481	6064	6639	7207	7767	8320	8866	40
000	20	8866	9406	9939	*0465	*0985	*1499	*2007	*2509	*3006	*3496	*3982	30
000	30	6.6 3982	4462	4936	5406	5870	6330	6785̄	7235̄	7680	8121	8557	20
000	40	8557	8990	9418	9841	*0261	*0676	*1088	*1496	*1900	*2300	*2697	10
000	50	6.7 2697	3090	3479	3865	4248	4627	5003	5376	5746	6112	6476	0 58
000	2 0	6476	6836	7193	7548	7900	8248	8595̄	8938	9278	9616	9952	50
000	10	9952	*0285̄	*0615̄	*0943	*1268	*1591	*1911	*2230	*2545	*2859	*3170	40
000	20	6.8 3170	3479	3786	4091	4394	4694	4993	5289	5584	5876	6167	30
000	30	6167	6455	6742	7027	7310	7591	7870	8147	8423	8697	8969	20
000	40	8969	9240	9509	9776	*0042	*0306	*0568	*0829	*1088	*1346	*1602	10
000	50	6.9 1602	1857	2110	2362	2612	2861	3109	3355̄	3599	3843	4085̄	0 57
000	3 0	4085̄	4325	4565̄	4803	5039	5275̄	5509	5742	5973	6204	6433	50
000	10	6433	6661	6888	7113	7338	7561	7783	8004	8224	8443	8660	40
000	20	8660	8877	9093	9307	9520	9733	9944	*0155̄	*0364	*0572	*0779	30
000	30	7.0 0779	0986	1191	1395	1599	1801	2003	2203	2403	2602	2800	20
000	40	2800	2997	3193	3388	3582	3776	3968	4160	4351	4541	4730	10
000	50	4730	4919	5106	5293	5479	5664	5849	6032	6215	6397	6579	0 56
000	4 0	6579	6759	6939	7118	7296	7474	7651	7827	8003	8177	8351	50
000	10	8351	8525̄	8698	8870	9041	9211	9381	9551	9719	9887	*0055̄	40
000	20	7.1 0055̄	0222	0388	0553	0718	0882	1046	1209	1371	1533	1694	30
000	30	1694	1854	2014	2174	2333	2491	2648	2805	2962	3118	3273	20
000	40	3273	3428	3582	3736	3889	4042	4194	4346	4497	4647	4797	10
000	50	4797	4947	5096	5244	5392	5540	5687	5833	5979	6125̄	6270	0 55
0.00		10″	9″	8″	7″	6″	5″	4″	3″	2″	1″	0″	″ ′

L Sin L Cos 89° L Cot *179° 269° *359°

L Cos | L Sin | **0°** | *90° 180° *270°

	144	143	142	141	140	139		138	137	136	135	134	133	
1	14.4	14.3	14.2	14.1	14.0	13.9	1	13.8	13.7	13.6	13.5	13.4	13.3	1
2	28.8	28.6	28.4	28.2	28.0	27.8	2	27.6	27.4	27.2	27.0	26.8	26.6	2
3	43.2	42.9	42.6	42.3	42.0	41.7	3	41.4	41.1	40.8	40.5	40.2	39.9	3
4	57.6	57.2	56.8	56.4	56.0	55.6	4	55.2	54.8	54.4	54.0	53.6	53.2	4
5	72.0	71.5	71.0	70.5	70.0	69.5	5	69.0	68.5	68.0	67.5	67.0	66.5	5
6	86.4	85.8	85.2	84.6	84.0	83.4	6	82.8	82.2	81.6	81.0	80.4	79.8	6
7	100.8	100.1	99.4	98.7	98.0	97.3	7	96.6	95.9	95.2	94.5	93.8	93.1	7
8	115.2	114.4	113.6	112.8	112.0	111.2	8	110.4	109.6	108.8	108.0	107.2	106.4	8
9	129.6	128.7	127.8	126.9	126.0	125.1	9	124.2	123.3	122.4	121.5	120.6	119.7	9
	132	**131**	**130**	**129**	**128**	**127**		**126**	**125**	**124**	**123**	**122**	**121**	
1	13.2	13.1	13.0	12.9	12.8	12.7	1	12.6	12.5	12.4	12.3	12.2	12.1	1
2	26.4	26.2	26.0	25.8	25.6	25.4	2	25.2	25.0	24.8	24.6	24.4	24.2	2
3	39.6	39.3	39.0	38.7	38.4	38.1	3	37.8	37.5	37.2	36.9	36.6	36.3	3
4	52.8	52.4	52.0	51.6	51.2	50.8	4	50.4	50.0	49.6	49.2	48.8	48.4	4
5	66.0	65.5	65.0	64.5	64.0	63.5	5	63.0	62.5	62.0	61.5	61.0	60.5	5
6	79.2	78.6	78.0	77.4	76.8	76.2	6	75.6	75.0	74.4	73.8	73.2	72.6	6
7	92.4	91.7	91.0	90.3	89.6	88.9	7	88.2	87.5	86.8	86.1	85.4	84.7	7
8	105.6	104.8	104.0	103.2	102.4	101.6	8	100.8	100.0	99.2	98.4	97.6	96.8	8
9	118.8	117.9	117.0	116.1	115.2	114.3	9	113.4	112.5	111.6	110.7	109.8	108.9	9
	120	**119**	**118**	**117**	**116**	**115**		**114**	**113**	**112**	**111**	**110**	**109**	
1	12.0	11.9	11.8	11.7	11.6	11.5	1	11.4	11.3	11.2	11.1	11.0	10.9	1
2	24.0	23.8	23.6	23.4	23.2	23.0	2	22.8	22.6	22.4	22.2	22.0	21.8	2
3	36.0	35.7	35.4	35.1	34.8	34.5	3	34.2	33.9	33.6	33.3	33.0	32.7	3
4	48.0	47.6	47.2	46.8	46.4	46.0	4	45.6	45.2	44.8	44.4	44.0	43.6	4
5	60.0	59.5	59.0	58.5	58.0	57.5	5	57.0	56.5	56.0	55.5	55.0	54.5	5
6	72.0	71.4	70.8	70.2	69.6	69.0	6	68.4	67.8	67.2	66.6	66.0	65.4	6
7	84.0	83.3	82.6	81.9	81.2	80.5	7	79.8	79.1	78.4	77.7	77.0	76.3	7
8	96.0	95.2	94.4	93.6	92.8	92.0	8	91.2	90.4	89.6	88.8	88.0	87.2	8
9	108.0	107.1	106.2	105.3	104.4	103.5	9	102.6	101.7	100.8	99.9	99.0	98.1	9

0.00	′	″	0″	1″	2″	3″	4″	5″	6″	7″	8″	9″	10″		
000	5	0	7.1 6270	6414	6558	6702	6845	6987	7130	7271	7413	7553	7694	50	
000		10	7694	7834	7973	8112	8250	8389	8526	8663	8800	8937	9072	40	
000		20	9072	9208	9343	9478	9612	9746	9879	*0012	*0145	*0277	*0409	30	
000		30	7.2 0409	0540	0671	0802	0932	1062	1191	1320	1449	1577	1705	20	
000		40	1705	1833	1960	2087	2213	2339	2465	2590	2715	2840	2964	10	
000		50	2964	3088	3212	3335	3458	3580	3702	3824	3946	4067	4188	0	**54**
000	6	0	4188	4308	4428	4548	4668	4787	4906	5024	5142	5260	5378	50	
000		10	5378	5495	5612	5728	5845	5961	6076	6192	6307	6421	6536	40	
000		20	6536	6650	6764	6877	6991	7104	7216	7329	7441	7552	7664	30	
000		30	7664	7775	7886	7997	8107	8217	8327	8437	8546	8655	8763	20	
000		40	8763	8872	8980	9088	9196	9303	9410	9517	9623	9730	9836	10	
000		50	9836	9942	*0047	*0152	*0257	*0362	*0467	*0571	*0675	*0779	*0882	0	**53**
000	7	0	7.3 0882	0986	1089	1191	1294	1396	1498	1600	1702	1803	1904	50	
000		10	1904	2005	2106	2206	2306	2406	2506	2606	2705	2804	2903	40	
000		20	2903	3001	3100	3198	3296	3393	3491	3588	3685	3782	3879	30	
000		30	3879	3975	4071	4167	4263	4359	4454	4549	4644	4739	4833	20	
000		40	4833	4928	5022	5116	5209	5303	5396	5489	5582	5675	5767	10	
000		50	5767	5860	5952	6044	6135	6227	6318	6409	6500	6591	6682	0	**52**
000	8	0	6682	6772	6862	6952	7042	7132	7221	7310	7399	7488	7577	50	
000		10	7577	7666	7754	7842	7930	8018	8106	8193	8280	8367	8454	40	
000		20	8454	8541	8628	8714	8800	8887	8972	9058	9144	9229	9314	30	
000		30	9314	9400	9484	9569	9654	9738	9822	9906	9990	*0074	*0158	20	
000		40	7.4 0158	0241	0324	0408	0491	0573	0656	0739	0821	0903	0985	10	
000		50	0985	1067	1149	1230	1312	1393	1474	1555	1636	1716	1797	0	**51**
000	9	0	1797	1877	1957	2037	2117	2197	2277	2356	2435	2515	2594	50	
000		10	2594	2673	2751	2830	2908	2987	3065	3143	3221	3299	3376	40	
000		20	3376	3454	3531	3608	3685	3762	3839	3916	3992	4069	4145	30	
000		30	4145	4221	4297	4373	4449	4524	4600	4675	4750	4825	4900	20	
000		40	4900	4975	5050	5124	5199	5273	5347	5421	5495	5569	5643	10	
000		50	5643	5716	5790	5863	5936	6009	6082	6155	6228	6300	6373	0	**50**
0.00			10″	9″	8″	7″	6″	5″	4″	3″	2″	1″	0″	″	′

L Sin | *179° 269° *359° | **89°** | L Cos

	108	107	106	105	104	103		102	101	99	98	97	96	
1	10.8	10.7	10.6	10.5	10.4	10.3	1	10.2	10.1	9.9	9.8	9.7	9.6	1
2	21.6	21.4	21.2	21.0	20.8	20.6	2	20.4	20.2	19.8	19.6	19.4	19.2	2
3	32.4	32.1	31.8	31.5	31.2	30.9	3	30.6	30.3	29.7	29.4	29.1	28.8	3
4	43.2	42.8	42.4	42.0	41.6	41.2	4	40.8	40.4	39.6	39.2	38.8	38.4	4
5	54.0	53.5	53.0	52.5	52.0	51.5	5	51.0	50.5	49.5	49.0	48.5	48.0	5
6	64.8	64.2	63.6	63.0	62.4	61.8	6	61.2	60.6	59.4	58.8	58.2	57.6	6
7	75.6	74.9	74.2	73.5	72.8	72.1	7	71.4	70.7	69.3	68.6	67.9	67.2	7
8	86.4	85.6	84.8	84.0	83.2	82.4	8	81.6	80.8	79.2	78.4	77.6	76.8	8
9	97.2	96.3	95.4	94.5	93.6	92.7	9	91.8	90.9	89.1	88.2	87.3	86.4	9

	95	94	93	92	91	90		89	88	87	86	85	84	
1	9.5	9.4	9.3	9.2	9.1	9.0	1	8.9	8.8	8.7	8.6	8.5	8.4	1
2	19.0	18.8	18.6	18.4	18.2	18.0	2	17.8	17.6	17.4	17.2	17.0	16.8	2
3	28.5	28.2	27.9	27.6	27.3	27.0	3	26.7	26.4	26.1	25.8	25.5	25.2	3
4	38.0	37.6	37.2	36.8	36.4	36.0	4	35.6	35.2	34.8	34.4	34.0	33.6	4
5	47.5	47.0	46.5	46.0	45.5	45.0	5	44.5	44.0	43.5	43.0	42.5	42.0	5
6	57.0	56.4	55.8	55.2	54.6	54.0	6	53.4	52.8	52.2	51.6	51.0	50.4	6
7	66.5	65.8	65.1	64.4	63.7	63.0	7	62.3	61.6	60.9	60.2	59.5	58.8	7
8	76.0	75.2	74.4	73.6	72.8	72.0	8	71.2	70.4	69.6	68.8	68.0	67.2	8
9	85.5	84.6	83.7	82.8	81.9	81.0	9	80.1	79.2	78.3	77.4	76.5	75.6	9

	83	82	81	80	79	78		77	76	75	74	73	72	
1	8.3	8.2	8.1	8.0	7.9	7.8	1	7.7	7.6	7.5	7.4	7.3	7.2	1
2	16.6	16.4	16.2	16.0	15.8	15.6	2	15.4	15.2	15.0	14.8	14.6	14.4	2
3	24.9	24.6	24.3	24.0	23.7	23.4	3	23.1	22.8	22.5	22.2	21.9	21.6	3
4	33.2	32.8	32.4	32.0	31.6	31.2	4	30.8	30.4	30.0	29.6	29.2	28.8	4
5	41.5	41.0	40.5	40.0	39.5	39.0	5	38.5	38.0	37.5	37.0	36.5	36.0	5
6	49.8	49.2	48.6	48.0	47.4	46.8	6	46.2	45.6	45.0	44.4	43.8	43.2	6
7	58.1	57.4	56.7	56.0	55.3	54.6	7	53.9	53.2	52.5	51.8	51.1	50.4	7
8	66.4	65.6	64.8	64.0	63.2	62.4	8	61.6	60.8	60.0	59.2	58.4	57.6	8
9	74.7	73.8	72.9	72.0	71.1	70.2	9	69.3	68.4	67.5	66.6	65.7	64.8	9

′	″	0″	1″	2″	3″	4″	5″	6″	7″	8″	9″	10″	″	′
5	0	7.1 6270	6414	6558	6702	6845̄	6988	7130	7271	7413	7553	7694	50	
	10	7694	7834	7973	8112	8250	8389	8526	8663	8800	8937	9073	40	
	20	9073	9208	9343	9478	9612	9746	9879	*0012	*0145̄	*0277	*0409	30	
	30	7.2 0409	0540	0671	0802	0932	1062	1191	1321	1449	1577	1703	20	
	40	1705	1833	1960	2087	2213	2339	2465	2590	2715	2840	2964	10	
	50	2964	3088	3212	3335̄	3458	3580	3703	3824	3946	4067	4188	0	54
6	0	4188	4308	4428	4548	4668	4787	4906	5024	5142	5260	5378	50	
	10	5378	5495̄	5612	5728	5845̄	5961	6076	6192	6307	6421	6536	40	
	20	6536	6650	6764	6877	6991	7104	7216	7329	7441	7552	7664	30	
	30	7664	7775	7886	7997	8107	8217	8327	8437	8546	8655̄	8764	20	
	40	8764	8872	8980	9088	9196	9303	9410	9517	9624	9730	9836	10	
	50	9836	9942	*0047	*0153	*0258	*0362	*0467	*0571	*0675	*0779	*0882	0	53
7	0	7.3 0882	0986	1089	1192	1294	1396	1499	1600	1702	1803	1904	50	
	10	1904	2005	2106	2206	2307	2406	2506	2606	2705̄	2804	2903	40	
	20	2903	3001	3100	3198	3296	3394	3491	3588	3685	3782	3879	30	
	30	3879	3975	4071	4167	4263	4359	4454	4549	4644	4739	4833	20	
	40	4833	4928	5022	5116	5209	5303	5396	5489	5582	5675̄	5767	10	
	50	5767	5860	5952	6044	6135	6227	6318	6409	6500	6591	6682	0	52
8	0	6682	6772	6862	6952	7042	7132	7221	7310	7400	7488	7577	50	
	10	7577	7666	7754	7842	7930	8018	8106	8193	8281	8368	8455̄	40	
	20	8455̄	8541	8628	8714	8801	8887	8973	9058	9144	9229	9315̄	30	
	30	9315̄	9400	9485̄	9569	9654	9738	9823	9907	9991	*0074	*0158	20	
	40	7.4 0158	0241	0325̄	0408	0491	0574	0656	0739	0821	0903	0985	10	
	50	0985	1067	1149	1230	1312	1393	1474	1555	1636	1716	1797	0	51
9	0	1797	1877	1958	2038	2117	2197	2277	2356	2436	2515̄	2594	50	
	10	2594	2673	2751	2830	2909	2987	3065	3143	3221	3299	3376	40	
	20	3376	3454	3531	3608	3686	3762	3839	3916	3992	4069	4145	30	
	30	4145	4221	4297	4373	4449	4524	4600	4675	4750	4825	4900	20	
	40	4900	4975	5050̄	5124	5199	5273	5347	5421	5495	5569	5643	10	
	50	5643	5716	5790	5863	5936	6009	6082	6155	6228	6300	6373	0	50
		10″	9″	8″	7″	6″	5″	4″	3″	2″	1″	0″	″	′

L Cos L Sin **0°** *90° 180° *270°

0.00	′	″	0″	1″	2″	3″	4″	5″	6″	7″	8″	9″	10″	″	′
000	10	0	7.46 373	445̄	517	589	661	733	805̄	876	948	*019	*090	50	
000		10	7.47 090	162	233	303	374	445̄	515	586	656	726	797	40	
000		20	797	867	936	*006	*076	*145	*215̄	*284	*353	*422	*491	30	
000		30	7.48 491	560	629	698	766	835̄	903	971	*039	*108	*175	20	
000		40	7.49 175	243	311	379	446	513	581	648	715̄	782	849	10	
000		50	849	916	982	*049	*115	*182	*248	*314	*380	*446	*512	0	49
000	11	0	7.50 512	578	643	709	774	840	905̄	970	*035	*100	*165̄	50	
000		10	7.51 165̄	230	294	359	423	488	552	616	680	744	808	40	
000		20	808	872	936	999	*063	*126	*190	*253	*316	*379	*442	30	
000		30	7.52 442	505	568	631	693	756	818	881	943	*005	*067	20	
000		40	7.53 067	129	191	253	315̄	376	438	499	561	622	683	10	
000		50	683	744	805	866	927	988	*049	*109	*170	*230	*291	0	48
000	12	0	7.54 291	351	411	471	531	591	651	711	771	830	890	50	
000		10	890	949	*009	*068	*127	*186	*245	*304	*363	*422	*481	40	
000		20	7.55 481	539	598	656	715̄	773	831	889	948	*006	*064	30	
000		30	7.56 064	121	179	237	295̄	352	410	467	524	582	639	20	
000		40	639	696	753	810	867	924	980	*037	*094	*150	*206	10	
000		50	7.57 206	263	319	375	431	488	544	599	655	711	767	0	47
000	13	0	767	822	878	934	989	*044	*100	*155̄	*210	*265	*320	50	
000		10	7.58 320	375	430	485̄	539	594	649	703	758	812	866	40	
000		20	866	921	975̄	*029	*083	*137	*191	*245̄	*299	*352	*406	30	
000		30	7.59 406	459	513	566	620	673	726	780	833	886	939	20	
000		40	939	992	*045̄	*097	*150	*203	*255	*308	*360	*413	*465	10	
000		50	7.60 465	517	570	622	674	726	778	830	882	934	985	0	46
000	14	0	985	*037	*089	*140	*192	*243	*294	*346	*397	*448	*499	50	
000		10	7.61 499	550	601	652	703	754	805̄	855	906	957	*007	40	
000		20	7.62 007	058	108	158	209	259	309	359	409	459	509	30	
000		30	509	559	609	659	708	758	808	857	907	956	*006	20	
000		40	7.63 006	055̄	104	153	203	252	301	350	399	448	496	10	
000		50	496	545	594	642	691	740	788	837	885̄	933	982	0	45
000	15	0	982	*030	*078	*126	*174	*222	*270	*318	*366	*414	*461	50	
000		10	7.64 461	509	557	604	652	699	747	794	842	889	936	40	
000		20	936	983	*030	*078	*125̄	*172	*218	*265	*312	*359	*406	30	
000		30	7.65 406	452	499	546	592	638	685̄	731	778	824	870	20	
000		40	870	916	962	*009	*055̄	*101	*146	*192	*238	*284	*330	10	
000		50	7.66 330	375	421	467	512	558	603	649	694	739	784	0	44
000	16	0	784	830	875̄	920	965	*010	*055	*100	*145̄	*190	*235̄	50	
000		10	7.67 235̄	279	324	369	413	458	502	547	591	636	680	40	
*000		20	680	724	768	813	857	901	945	989	*033	*077	*121	30	
*999		30	7.68 121	165̄	208	252	296	340	383	427	470	514	557	20	
999		40	557	601	644	687	731	774	817	860	903	946	989	10	
999		50	989	*032	*075	*118	*161	*204	*247	*289	*332	*375̄	*417	0	43
999	17	0	7.69 417	460	502	545̄	587	630	672	714	757	799	841	50	
999		10	841	883	925	967	*009	*051	*093	*135	*177	*219	*261	40	
999		20	7.70 261	302	344	386	427	469	510	552	593	635̄	676	30	
999		30	676	718	759	800	841	883	924	965̄	*006	*047	*088	20	
999		40	7.71 088	129	170	211	251	292	333	374	414	455	496	10	
999		50	496	536	577	617	658	698	739	779	819	859	900	0	42
999	18	0	900	940	980	*020	*060	*100	*140	*180	*220	*260	*300	50	
999		10	7.72 300	340	380	419	459	499	538	578	618	657	697	40	
999		20	697	736	775	815̄	854	894	933	972	*011	*050	*090	30	
999		30	7.73 090	129	168	207	246	285̄	324	363	401	440	479	20	
999		40	479	518	557	595	634	673	711	750	788	827	865	10	
999		50	865	904	942	980	*019	*057	*095	*133	*171	*210	*248	0	41
999	19	0	7.74 248	286	324	362	400	438	476	514	551	589	627	50	
999		10	627	665̄	703	740	778	815	853	891	928	966	*003	40	
999		20	7.75 003	040	078	115	153	190	227	264	302	339	376	30	
999		30	376	413	450	487	524	561	598	635̄	672	709	745	20	
999		40	745	782	819	856	892	929	966	*002	*039	*075	*112	10	
999		50	7.76 112	148	185̄	221	258	294	330	367	403	439	475	0	40
9.99			10″	9″	8″	7″	6″	5″	4″	3″	2″	1″	0″	″	′

P P	72	70	68	66	64	62	61	60
1	7.2	7.0	6.8	6.6	6.4	6.2	6.1	6.0
2	14.4	14.0	13.6	13.2	12.8	12.4	12.2	12.0
3	21.6	21.0	20.4	19.8	19.2	18.6	18.3	18.0
4	28.8	28.0	27.2	26.4	25.6	24.8	24.4	24.0
5	36.0	35.0	34.0	33.0	32.0	31.0	30.5	30.0
6	43.2	42.0	40.8	39.6	38.4	37.2	36.6	36.0
7	50.4	49.0	47.6	46.2	44.8	43.4	42.7	42.0
8	57.6	56.0	54.4	52.8	51.2	49.6	48.8	48.0
9	64.8	63.0	61.2	59.4	57.6	55.8	54.9	54.0

L Sin *179° 269° *359 **89°** L Cos

L Tan 0° *90° 180° *270°

′ ″	0″	1″	2″	3″	4″	5″	6″	7″	8″	9″	10″	
10 0	7.46 373	445	517	589	661	733	805̄	876	948	*019	*091	50
10	7.47 091	162	233	304	374	445	516	586	656	727	797	40
20	797	867	937	*006	*076	*146	*215	*284	*354	*423	*492	30
30	7.48 492	561	629	698	767	835̄	903	972	*040	*108	*176	20
40	7.49 176	243	311	379	446	514	581	648	715	782	849	10
50	849	916	982	*049	*115	*182	*248	*314	*380	*446	*512	0 49
11 0	7.50 512	578	643	709	774	840	905	970	*035	*100	*165	50
10	7.51 165	230	295̄	359	424	488	552	617	681	745̄	809	40
20	809	872	936	*000	*063	*127	*190	*253	*316	*380	*443	30
30	7.52 443	505	568	631	694	756	819	881	943	*005	*067	20
40	7.53 067	129	191	253	315̄	377	438	5̄00	561	622	683	10
50	683	745̄	806	867	927	988	*049	*110	*170	*231	*291	0 48
12 0	7.54 291	351	411	471	532	591	651	711	771	830	890	50
10	890	949	*009	*068	*127	*186	*245	*304	*363	*422	*481	40
20	7.55 481	539	598	657	715̄	773	832	890	948	*006	*064	30
30	7.56 064	122	179	237	295̄	352	410	467	525̄	582	639	20
40	639	696	753	810	867	924	981	*037	*094	*150	*207	10
50	7.57 207	263	319	376	432	488	544	600	656	711	767	0 47
13 0	767	823	878	934	989	*045̄	*100	*155	*210	*265	*320	50
10	7.58 320	375	430	485	540	594	649	704	758	812	867	40
20	867	921	975	*029	*083	*137	*191	*245	*299	*353	*406	30
30	7.59 406	460	513	567	620	673	727	780	833	886	939	20
40	939	992	*045̄	*098	*150	*203	*256	*308	*36[illegible]	*413	*466	10
50	7.60 466	518	570	622	674	726	778	830	882	934	986	0 46
14 0	986	*037	*089	*140	*192	*243	*295̄	*346	*397	*449	*5̄00	50
10	7.61 5̄00	551	602	653	704	754	805	856	906	957	*008	40
20	7.62 008	058	108	159	209	259	310	360	410	460	510	30
30	510	560	609	659	709	759	808	858	907	957	*006	20
40	7.63 006	055	105̄	154	203	252	301	350	399	448	497	10
50	497	546	594	643	692	740	789	837	885	934	982	0 45
15 0	982	*030	*078	*127	*175̄	*223	*271	*318	*366	*414	*462	50
10	7.64 462	510	557	605̄	652	700	747	795̄	842	889	937	40
20	937	984	*031	*078	*125̄	*172	*219	*266	*313	*359	*406	30
30	7.65 406	453	499	546	592	639	685	732	778	824	871	20
40	871	917	963	*009	*055̄	*101	*147	*193	*239	*284	*330	10
50	7.66 330	376	421	467	513	558	604	649	694	740	785̄	0 44
16 0	785̄	830	875	920	966	*011	*056	*100	*145	*190	*235̄	50
10	7.67 235̄	280	324	369	414	458	503	547	592	636	680	40
20	680	725̄	769	813	857	901	946	990	*034	*077	*121	30
30	7.68 121	165	209	253	296	340	384	427	471	514	558	20
40	558	601	645̄	688	731	774	818	861	904	947	990	10
50	990	*033	*076	*119	*162	*204	*247	*290	*333	*375	*418	0 43
17 0	7.69 418	460	503	545	588	630	673	715̄	757	799	842	50
10	842	884	926	968	*010	*052	*094	*136	*178	*219	*261	40
20	7.70 261	303	345̄	386	428	469	511	553	594	635	677	30
30	677	718	759	801	842	883	924	965	*006	*047	*088	20
40	7.71 088	129	170	211	252	293	334	374	415	456	496	10
50	496	537	577	618	658	699	739	779	820	860	900	0 42
18 0	900	940	981	*021	*061	*101	*141	*181	*221	*261	*301	50
10	7.72 301	340	380	420	460	499	539	579	618	658	697	40
20	697	737	776	815	855̄	894	933	973	*012	*051	*090	30
30	7.73 090	129	168	207	246	285	324	363	402	441	480	20
40	480	518	557	596	635̄	673	712	750	789	827	866	10
50	866	904	943	981	*019	*058	*096	*134	*172	*210	*248	0 41
19 0	7.74 248	286	325̄	363	401	438	476	514	552	590	628	50
10	628	665	703	741	779	816	854	891	929	966	*004	40
20	7.75 004	041	079	116	153	191	228	265	302	339	377	30
30	377	414	451	488	525̄	562	599	636	672	709	746	20
40	746	783	820	856	893	930	966	*003	*040	*076	*113	10
50	7.76 113	149	186	222	258	295̄	331	367	404	440	476	0 40
	10″	9″	8″	7″	6″	5″	4″	3″	2″	1″	0″	″ ′

P P

	59	58	57
1	5.9	5.8	5.7
2	11.8	11.6	11.4
3	17.7	17.4	17.1
4	23.6	23.2	22.8
5	29.5	29.0	28.5
6	35.4	34.8	34.2
7	41.3	40.6	39.9
8	47.2	46.4	45.6
9	53.1	52.2	51.3

	56	55	54
1	5.6	5.5	5.4
2	11.2	11.0	10.8
3	16.8	16.5	16.2
4	22.4	22.0	21.6
5	28.0	27.5	27.0
6	33.6	33.0	32.4
7	39.2	38.5	37.8
8	44.8	44.0	43.2
9	50.4	49.5	48.6

	53	52	51
1	5.3	5.2	5.1
2	10.6	10.4	10.2
3	15.9	15.6	15.3
4	21.2	20.8	20.4
5	26.5	26.0	25.5
6	31.8	31.2	30.6
7	37.1	36.4	35.7
8	42.4	41.6	40.8
9	47.7	46.8	45.9

	50	49	48
1	5.0	4.9	4.8
2	10.0	9.8	9.6
3	15.0	14.7	14.4
4	20.0	19.6	19.2
5	25.0	24.5	24.0
6	30.0	29.4	28.8
7	35.0	34.3	33.6
8	40.0	39.2	38.4
9	45.0	44.1	43.2

	47	46	45
1	4.7	4.6	4.5
2	9.4	9.2	9.0
3	14.1	13.8	13.5
4	18.8	18.4	18.0
5	23.5	23.0	22.5
6	28.2	27.6	27.0
7	32.9	32.2	31.5
8	37.6	36.8	36.0
9	42.3	41.4	40.5

	44	43	42
1	4.4	4.3	4.2
2	8.8	8.6	8.4
3	13.2	12.9	12.6
4	17.6	17.2	16.8
5	22.0	21.5	21.0
6	26.4	25.8	25.2
7	30.8	30.1	29.4
8	35.2	34.4	33.6
9	39.6	38.7	37.8

	41	40	39
1	4.1	4.0	3.9
2	8.2	8.0	7.8
3	12.3	12.0	11.7
4	16.4	16.0	15.6
5	20.5	20.0	19.5
6	24.6	24.0	23.4
7	28.7	28.0	27.3
8	32.8	32.0	31.2
9	36.9	36.0	35.1

	38	37	36
1	3.8	3.7	3.6
2	7.6	7.4	7.2
3	11.4	11.1	10.8
4	15.2	14.8	14.4
5	19.0	18.5	18.0
6	22.8	22.2	21.6
7	26.6	25.9	25.2
8	30.4	29.6	28.8
9	34.2	33.3	32.4

P P

*179° 269° *359 89° L Cot

L Cos　　L Sin　　**0°**　　*90°　180°　*270°

9·99	′	″	0″	1″	2″	3″	4″	5″	6″	7″	8″	9″	10″	″	′
999	20	0	7·76 475	512	548	584	620	656	692	728	764	800	836	50	
999		10	836	872	907	943	979	*015̄	*051	*086	*122	*158	*193	40	
999		20	7·77 193	229	264	300	335	371	406	442	477	512	548	30	
999		30	548	583	618	654	689	724	759	794	829	864	899	20	
999		40	899	934	969	*004	*039	*074	*109	*144	*179	*213	*248	10	
999		50	7·78 248	283	318	352	387	422	456	491	525	560	594	0	39
999	21	0	594	629	663	698	732	766	801	835̄	869	903	938	50	
999		10	938	972	*006	*040	*074	*108	*142	*176	*210	*244	*278	40	
999		20	7·79 278	312	346	380	414	448	481	515	549	582	616	30	
999		30	616	650	683	717	751	784	818	851	885̄	918	952	20	
999		40	952	985̄	*018	*052	*085̄	*118	*152	*185̄	*218	*251	*284	10	
999		50	7·80 284	317	351	384	417	450	483	516	549	582	615̄	0	38
999	22	0	615̄	647	680	713	746	779	812	844	877	910	942	50	
999		10	942	975̄	*008	*040	*073	*105	*138	*170	*203	*235	*268	40	
999		20	7·81 268	300	332	365̄	397	429	462	494	526	558	591	30	
999		30	591	623	655̄	687	719	751	783	815	847	879	911	20	
999		40	911	943	975̄	*007	*039	*070	*102	*134	*166	*198	*229	10	
999		50	7·82 229	261	293	324	356	387	419	451	482	514	545	0	37
999	23	0	545	577	608	639	671	702	733	765̄	796	827	859	50	
999		10	859	890	921	952	983	*015̄	*046	*077	*108	*139	*170	40	
999		20	7·83 170	201	232	263	294	325̄	356	387	417	448	479	30	
999		30	479	510	541	571	602	633	663	694	725̄	755	786	20	
999		40	786	817	847	878	908	939	969	*000	*030	*060	*091	10	
999		50	7·84 091	121	151	182	212	242	273	303	333	363	393	0	36
999	24	0	393	424	454	484	514	544	574	604	634	664	694	50	
999		10	694	724	754	784	814	843	873	903	933	963	992	40	
999		20	992	*022	*052	*082	*111	*141	*171	*200	*230	*259	*289	30	
999		30	7·85 289	318	348	377	407	436	466	495	525̄	554	583	20	
999		40	583	613	642	671	701	730	759	788	817	847	876	10	
999		50	876	905̄	934	963	992	*021	*050	*079	*108	*137	*166	0	35
999	25	0	7·86 166	195	224	253	282	311	340	368	397	426	455̄	50	
999		10	455̄	484	512	541	570	598	627	656	684	713	741	40	
999		20	741	770	799	827	856	884	913	941	969	998	*026	30	
999		30	7·87 026	055̄	083	111	140	168	196	224	253	281	309	20	
999		40	309	337	366	394	422	450	478	506	534	562	590	10	
999		50	590	618	646	674	702	730	758	786	814	842	870	0	34
999	26	0	870	897	925	953	981	*009	*036	*064	*092	*119	*147	50	
999		10	7·88 147	175̄	202	230	258	285	313	340	368	395	423	40	
999		20	423	450	478	505	533	560	587	615̄	642	669	697	30	
999		30	697	724	751	779	806	833	860	888	915̄	942	969	20	
999		40	969	996	*023	*050	*077	*105̄	*132	*159	*186	*213	*240	10	
999		50	7·89 240	267	294	320	347	374	401	428	455̄	482	509	0	33
999	27	0	509	535	562	589	616	642	669	696	722	749	776	50	
999		10	776	802	829	856	882	909	935	962	988	*015̄	*041	40	
999		20	7·90 041	068	094	121	147	174	200	226	253	279	305	30	
999		30	305	332	358	384	411	437	463	489	515	542	568	20	
999		40	568	594	620	646	672	698	725̄	751	777	803	829	10	
999		50	829	855̄	881	907	933	958	984	*010	*036	*062	*088	0	32
999	28	0	7·91 088	114	140	165	191	217	243	269	294	320	346	50	
999		10	346	371	397	423	448	474	500	525	551	576	602	40	
999		20	602	627	653	678	704	729	755̄	780	806	831	857	30	
999		30	857	882	907	933	958	983	*009	*034	*059	*085̄	*110	20	
998		40	7·92 110	135	160	186	211	236	261	286	311	336	362	10	
998		50	362	387	412	437	462	487	512	537	562	587	612	0	31
998	29	0	612	637	662	687	712	737	761	786	811	836	861	50	
998		10	861	886	910	935	960	985̄	*009	*034	*059	*084	*108	40	
998		20	7·93 108	133	158	182	207	231	256	281	305	330	354	30	
998		30	354	379	403	428	452	477	501	526	550	575̄	599	20	
998		40	599	623	648	672	696	721	745	769	794	818	842	10	
998		50	842	866	891	915̄	939	963	988	*012	*036	*060	*084	0	30
9·99			10″	9″	8″	7″	6″	5″	4″	3″	2″	1″	0″	″	′

L Sin　　*179°　269°　*359°　　**89°**　　L Cos

L Tan 0° *90° 180° *270°

′ ″	0″	1″	2″	3″	4″	5″	6″	7″	8″	9″	10″	″	′
20 0	7.76 476	512	548	585̄	621	657	693	729	765̄	801	837	50	
10	837	872	908	944	980	*016	*051	*087	*123	*158	*194	40	
20	7.77 194	230	265	301	336	372	407	442	478	513	549	30	
30	549	584	619	654	690	725̄	760	795̄	830	865	900	20	
40	900	935	970	*005	*040	*075̄	*110	*145̄	*179	*214	*249	10	
50	7.78 249	284	318	353	388	422	457	492	526	561	595	0	**39**
21 0	595	630	664	698	733	767	801	836	870	904	938	50	
10	938	973	*007	*041	*075̄	*109	*143	*177	*211	*245	*279	40	
20	7.79 279	313	347	381	415̄	448	482	516	550̄	583	617	30	
30	617	651	684	718	751	785	819	852	886	919	952	20	
40	952	986	*019	*053	*086	*119	*152	*186	*219	*252	*285	10	
50	7.80 285	318	351	385̄	418	451	484	517	550̄	583	615	0	**38**
22 0	615	648	681	714	747	780	812	345	878	911	943	50	
10	943	976	*009	*041	*074	*106	*139	*171	*204	*236	*269	40	
20	7.81 269	301	333	366	398	430	463	495̄	527	559	591	30	
30	591	624	656	688	720	752	784	816	848	880	912	20	
40	912	944	976	*008	*040	*071	*103	*135̄	*167	*198	*230	10	
50	7.82 230	262	294	325	357	388	420	452	483	515̄	546	0	**37**
23 0	546	578	609	640	672	703	734	766	797	828	860	50	
10	860	891	922	953	984	*016	*047	*078	*109	*140	*171	40	
20	7.83 171	202	233	264	295̄	326	357	388	418	449	480	30	
30	480	511	542	572	603	634	664	695	726	756	787	20	
40	787	818	848	879	909	940	970	*001	*031	*061	*092	10	
50	7.84 092	122	152	183	213	243	274	304	334	364	394	0	**36**
24 0	394	425̄	455̄	485̄	515̄	545̄	575	605	635	665	695	50	
10	695	725̄	755̄	785̄	815̄	845̄	874	904	934	964	993	40	
20	993	*023	*053	*083	*112	*142	*172	*201	*231	*260	*290	30	
30	7.85 290	319	349	378	408	437	467	496	526	555	584	20	
40	584	614	643	672	702	731	760	789	819	848	877	10	
50	877	906	935	964	993	*022	*051	*080	*109	*138	*167	0	**35**
25 0	7.86 167	196	225	254	283	312	341	370	398	427	456	50	
10	456	485̄	513	542	571	600	628	657	685	714	743	40	
20	743	771	800	828	857	885	914	942	971	999	*027	30	
30	7.87 027	056	084	113	141	169	197	226	254	282	310	20	
40	310	339	367	395̄	423	451	479	507	535	563	591	10	
50	591	619	647	675	703	731	759	787	815	843	871	0	**34**
26 0	871	899	926	954	982	*010	*037	*065	*093	*121	*148	50	
10	7.88 148	176	204	231	259	286	314	342	369	397	424	40	
20	424	452	479	506	534	561	589	616	643	671	698	30	
30	698	725	753	780	807	834	862	889	916	943	970	20	
40	970	997	*025̄	*052	*079	*106	*133	*160	*187	*214	*241	10	
50	7.89 241	268	295̄	322	349	376	403	429	456	483	510	0	**33**
27 0	510	537	563	590	617	644	670	697	724	750	777	50	
10	777	804	830	857	884	910	937	963	990	*016	*043	40	
20	7.90 043	069	096	122	149	175̄	201	228	254	280	307	30	
30	307	333	359	386	412	438	464	491	517	543	569	20	
40	569	595	622	648	674	700	726	752	778	804	830	10	
50	830	856	882	908	934	960	986	*012	*038	*064	*089	0	**32**
28 0	7.91 089	115	141	167	193	218	244	270	296	321	347	50	
10	347	373	398	424	450̄	475	501	527	552	578	603	40	
20	603	629	654	680	705	731	756	782	807	833	858	30	
30	858	883	909	934	960	985̄	*010	*036	*061	*086	*111	20	
40	7.92 111	137	162	187	212	237	263	288	313	338	363	10	
50	363	388	413	438	463	488	513	538	563	588	613	0	**31**
29 0	613	638	663	688	713	738	763	788	813	838	862	50	
10	862	887	912	937	961	986	*011	*036	*060	*085	*110	40	
20	7.93 110	134	159	184	208	233	258	282	307	331	356	30	
30	356	380	405̄	429	454	478	503	527	552	576	601	20	
40	601	625̄	649	674	698	722	747	771	795	820	844	10	
50	844	868	892	917	941	965	989	*013	*038	*062	*086	0	**30**
	10″	9″	8″	7″	6″	5″	4″	3″	2″	1″	0″	″	′

P P	37	36	35	34	33	32	31	30	29	28	27	26	25	24
1	3.7	3.6	3.5	3.4	3.3	3.2	3.1	3.0	2.9	2.8	2.7	2.6	2.5	2.4
2	7.4	7.2	7.0	6.8	6.6	6.4	6.2	6.0	5.8	5.6	5.4	5.2	5.0	4.8
3	11.1	10.8	10.5	10.2	9.9	9.6	9.3	9.0	8.7	8.4	8.1	7.8	7.5	7.2
4	14.8	14.4	14.0	13.6	13.2	12.8	12.4	12.0	11.6	11.2	10.8	10.4	10.0	9.6
5	18.5	18.0	17.5	17.0	16.5	16.0	15.5	15.0	14.5	14.0	13.5	13.0	12.5	12.0
6	22.2	21.6	21.0	20.4	19.8	19.2	18.6	18.0	17.4	16.8	16.2	15.6	15.0	14.4
7	25.9	25.2	24.5	23.8	23.1	22.4	21.7	21.0	20.3	19.6	18.9	18.2	17.5	16.8
8	29.6	28.8	28.0	27.2	26.4	25.6	24.8	24.0	23.2	22.4	21.6	20.8	20.0	19.2
9	33.3	32.4	31.5	30.6	29.7	28.8	27.9	27.0	26.1	25.2	24.3	23.4	22.5	21.6

*179° 269° *359° 89° L Cot

9·99	′	″	0″	1″	2″	3″	4″	5″	6″	7″	8″	9″	10″		
998	30	0	7.94 084	108	132	157	181	205̄	229	253	277	301	325̄	50	
998		10	325̄	349	373	397	421	445̄	469	492	516	540	564	40	
998		20	564	588	612	636	659	683	707	731	755̄	778	802	30	
998		30	802	826	849	873	897	921	944	968	991	*015	*039	20	
998		40	7.95 039	062	086	109	133	157	180	204	227	251	274	10	
998		50	274	298	321	344	368	391	415̄	438	461	485̄	508	0	29
998	31	0	508	532	555̄	578	601	625̄	648	671	695̄	718	741	50	
998		10	741	764	787	811	834	857	880	903	926	95̄0	973	40	
998		20	973	996	*019	*042	*065̄	*088	*111	*134	*157	*180	*203	30	
998		30	7.96 203	226	249	272	295̄	318	341	364	386	409	432	20	
998		40	432	455	478	501	524	546	569	592	615̄	637	660	10	
998		50	660	683	706	728	751	774	796	819	842	864	887	0	28
998	32	0	887	910	932	955̄	977	*000	*022	*045	*068	*090	*113	50	
998		10	7.97 113	135	158	180	202	225̄	247	270	292	315̄	337	40	
998		20	337	359	382	404	426	449	471	493	516	538	560	30	
998		30	560	583	605̄	627	649	672	694	716	738	760	782	20	
998		40	782	805̄	827	849	871	893	915	937	959	981	*003	10	
998		50	7.98 003	025	048	070	092	114	136	157	179	201	223	0	27
998	33	0	223	245	267	289	311	333	355̄	377	398	420	442	50	
998		10	442	464	486	508	529	551	573	595̄	616	638	660	40	
998		20	660	682	703	725̄	747	768	790	812	833	855̄	876	30	
998		30	876	898	920	941	963	984	*006	*027	*049	*070	*092	20	
998		40	7.99 092	113	135̄	156	178	199	221	242	264	285̄	306	10	
998		50	306	328	349	371	392	413	435̄	456	477	499	520	0	26
998	34	0	520	541	562	584	605̄	626	647	669	690	711	732	50	
998		10	732	753	775̄	796	817	838	859	880	901	922	943	40	
998		20	943	965̄	986	*007	*028	*049	*070	*091	*112	*133	*154	30	
998		30	8.00 154	175̄	196	217	238	259	279	300	321	342	363	20	
998		40	363	384	405̄	426	447	467	488	509	530	551	571	10	
998		50	571	592	613	634	654	675	696	717	737	758	779	0	25
998	35	0	779	799	820	841	861	882	903	923	944	964	985̄	50	
998		10	985̄	*006	*026	*047	*067	*088	*108	*129	*149	*170	*190	40	
998		20	8.01 190	211	231	252	272	293	313	333	354	374	395̄	30	
998		30	395̄	415	435	456	476	496	517	537	557	578	598	20	
998		40	598	618	639	659	679	699	720	740	760	780	801	10	
998		50	801	821	841	861	881	901	922	942	962	982	*002	0	24
998	36	0	8.02 002	022	042	062	082	102	123	143	163	183	203	50	
998		10	203	223	243	263	283	303	323	343	362	382	402	40	
998		20	402	422	442	462	482	502	522	542	561	581	601	30	
998		30	601	621	641	661	680	700	720	740	759	779	799	20	
998		40	799	819	838	858	878	898	917	937	957	976	996	10	
998		50	996	*016	*035	*055̄	*074	*094	*114	*133	*153	*172	*192	0	23
997	37	0	8.03 192	212	231	251	270	290	309	329	348	368	387	50	
997		10	387	407	426	446	465̄	484	504	523	543	562	581	40	
997		20	581	601	620	640	659	678	698	717	736	756	775̄	30	
997		30	775̄	794	813	833	852	871	891	910	929	948	967	20	
997		40	967	987	*006	*025	*044	*063	*083	*102	*121	*140	*159	10	
997		50	8.04 159	178	197	217	236	255̄	274	293	312	331	350	0	22
997	38	0	350	369	388	407	426	445	464	483	502	521	540	50	
997		10	540	559	578	597	616	635̄	654	673	692	710	729	40	
997		20	729	748	767	786	805̄	824	843	861	880	899	918	30	
997		30	918	937	955	974	993	*012	*030	*049	*068	*087	*105	20	
997		40	8.05 105	124	143	161	180	199	218	236	255̄	274	292	10	
997		50	292	311	329	348	367	385	404	422	441	460	478	0	21
997	39	0	478	497	515	534	552	571	589	608	626	645̄	663	50	
997		10	663	682	700	719	737	756	774	792	811	829	848	40	
997		20	848	866	885̄	903	921	940	958	976	995̄	*013	*031	30	
997		30	8.06 031	050	068	086	105̄	123	141	159	178	196	214	20	
997		40	214	232	251	269	287	305	324	342	360	378	396	10	
997		50	396	414	433	451	469	487	505	523	541	560	578	0	20
9·99			10″	9″	8″	7″	6″	5″	4″	3″	2″	1″	0″	″	′

L Tan **0°** *90° 180° *270°

′ ″	0″	1″	2″	3″	4″	5″	6″	7″	8″	9″	10″		
30 0	7.94 086	110	134	158	182	206	230	254	278	302	326	50	
10	326	350	374	398	422	446	470	494	518	542	566	40	
20	566	590	613	637	661	685̄	709	732	756	780	804	30	
30	804	827	851	875̄	899	922	946	970	993	*017	*040	20	
40	7.95 040	064	088	111	135̄	158	182	205	229	252	276	10	
50	276	299	323	346	370	393	416	440	463	487	510	0	29
31 0	510	533	557	580	603	627	650	673	696	720	743	50	
10	743	766	789	812	836	859	882	905	928	951	974	40	
20	974	998	*021	*044	*067	*090	*113	*136	*159	*182	*205̄	30	
30	7.96 205̄	228	251	274	297	320	343	365	388	411	434	20	
40	434	457	480	503	525	548	571	594	617	639	662	10	
50	662	685̄	708	730	753	776	798	821	844	866	889	0	28
32 0	889	911	934	957	979	*002	*024	*047	*069	*092	*114	50	
10	7.97 114	137	159	182	204	227	249	272	294	317	339	40	
20	339	361	384	406	428	451	473	495	518	540	562	30	
30	562	585̄	607	629	651	673	696	718	740	762	784	20	
40	784	807	829	851	873	895	917	939	961	983	*005	10	
50	7.98 005	027	050	072	094	116	138	159	181	203	225	0	27
33 0	225	247	269	291	313	335̄	357	379	400	422	444	50	
10	444	466	488	510	531	553	575̄	597	618	640	662	40	
20	662	684	705	727	749	770	792	814	835	857	878	30	
30	878	900	922	943	965̄	986	*008	*029	*051	*073	*094	20	
40	7.99 094	116	137	158	180	201	223	244	266	287	308	10	
50	308	330	351	373	394	415	437	458	479	501	522	0	26
34 0	522	543	564	586	607	628	649	671	692	713	734	50	
10	734	755	777	798	819	840	861	882	903	925̄	946	40	
20	946	967	988	*009	*030	*051	*072	*093	*114	*135̄	*156	30	
30	8.00 156	177	198	219	240	261	282	303	324	344	365	20	
40	365	386	407	428	449	470	490	511	532	553	574	10	
50	574	594	615	636	657	677	698	719	740	760	781	0	25
35 0	781	802	822	843	964	884	905̄	925	946	967	987	50	
10	987	*008	*028	*049	*070	*090	*111	*131	*152	*172	*193	40	
20	8.01 193	213	234	254	274	295̄	315	336	356	377	397	30	
30	397	417	438	458	478	499	519	539	560	580	600	20	
40	600	621	641	661	682	702	722	742	762	783	803	10	
50	803	823	843	863	884	904	924	944	964	984	*004	0	24
36 0	8.02 004	025̄	045̄	065̄	085̄	105̄	125̄	145̄	165	185	205	50	
10	205	225	245	265	285	305	325̄	345̄	365̄	385̄	405̄	40	
20	405̄	425̄	445̄	464	484	504	524	544	564	584	604	30	
30	604	623	643	663	683	703	722	742	762	782	801	20	
40	801	821	841	861	880	900	920	939	959	979	998	10	
50	998	*018	*038	*057	*077	*097	*116	*136	*155	*175̄	*194	0	23
37 0	8.03 194	214	234	253	273	292	312	331	351	370	390	50	
10	390	409	429	448	468	487	506	526	545	565̄	584	40	
20	584	603	623	642	661	681	700	720	739	758	777	30	
30	777	797	816	835	855̄	874	893	912	932	951	970	20	
40	970	989	*008	*028	*047	*066	*085	*104	*124	*143	*162	10	
50	8.04 162	181	200	219	238	257	276	296	315̄	334	353	0	22
38 0	353	372	391	410	429	448	467	486	505̄	524	543	50	
10	543	562	581	600	619	638	656	675	694	713	732	40	
20	732	751	770	789	808	826	845	864	883	902	921	30	
30	921	939	958	977	996	*014	*033	*052	*071	*089	*108	20	
40	8.05 108	127	146	164	183	202	220	239	258	276	295̄	10	
50	295̄	314	332	351	369	388	407	425	444	462	481	0	21
39 0	481	499	518	537	555	574	592	611	629	648	666	50	
10	666	685̄	703	722	740	758	777	795	814	832	851	40	
20	851	869	887	906	924	943	961	979	998	*016	*034	30	
30	8.06 034	053	071	089	107	126	144	162	181	199	217	20	
40	217	235	254	272	290	308	326	345̄	363	381	399	10	
50	399	417	436	454	472	490	508	526	544	562	581	0	20
	10″	9″	8″	7″	6″	5″	4″	3″	2″	1″	0″	″	′

P P

	25
1	2.5
2	5.0
3	7.5
4	10.0
5	12.5
6	15.0
7	17.5
8	20.0
9	22.5

	24	23
1	2.4	2.3
2	4.8	4.6
3	7.2	6.9
4	9.6	9.2
5	12.0	11.5
6	14.4	13.8
7	16.8	16.1
8	19.2	18.4
9	21.6	20.7

	22
1	2.2
2	4.4
3	6.6
4	8.8
5	11.0
6	13.2
7	15.4
8	17.6
9	19.8

	21
1	2.1
2	4.2
3	6.3
4	8.4
5	10.5
6	12.6
7	14.7
8	16.8
9	18.9

	20	19
1	2.0	1.9
2	4.0	3.8
3	6.0	5.7
4	8.0	7.6
5	10.0	9.5
6	12.0	11.4
7	14.0	13.3
8	16.0	15.2
9	18.0	17.1

	18
1	1.8
2	3.6
3	5.4
4	7.2
5	9.0
6	10.8
7	12.6
8	14.4
9	16.2

P P

*179° 269° *359° **89°** L Cot

L Cos **L Sin** **0°** *90° 180° *270°

9·99	′	″	0″	1″	2″	3″	4″	5″	6″	7″	8″	9″	10″	″	′
997	40	0	8.06 578	596	614	632	650	668	686	704	722	740	758	50	
997		10	758	776	794	812	830	848	866	884	902	920	938	40	
997		20	938	956	974	992	*010	*028	*046	*063	*081	*099	*117	30	
997		30	8.07 117	135	153	171	189	206	224	242	260	278	295	20	
997		40	295	313	331	349	367	384	402	420	438	455	473	10	
997		50	473	491	509	526	544	562	579	597	615	632	650	0	19
997	41	0	650	668	685	703	721	738	756	773	791	809	826	50	
997		10	826	844	861	879	896	914	932	949	967	984	*002	40	
997		20	8.08 002	019	037	054	072	089	107	124	141	159	176	30	
997		30	176	194	211	229	246	263	281	298	316	333	350	20	
997		40	350	368	385	403	420	437	455	472	489	506	524	10	
997		50	524	541	558	576	593	610	627	645	662	679	696	0	18
997	42	0	696	714	731	748	765	783	800	817	834	851	868	50	
997		10	868	886	903	920	937	954	971	988	*006	*023	*040	40	
997		20	8.09 040	057	074	091	108	125	142	159	176	193	210	30	
997		30	210	227	244	261	278	295	312	329	346	363	380	20	
997		40	380	397	414	431	448	465	482	499	516	533	550	10	
997		50	550	567	583	600	617	634	651	668	685	701	718	0	17
997	43	0	718	735	752	769	786	802	819	836	853	870	886	50	
997		10	886	903	920	937	953	970	987	*004	*020	*037	*054	40	
997		20	8.10 054	070	087	104	120	137	154	170	187	204	220	30	
997		30	220	237	254	270	287	303	320	337	353	370	386	20	
996		40	386	403	420	436	453	469	486	502	519	535	552	10	
996		50	552	568	585	601	618	634	651	667	684	700	717	0	16
996	44	0	717	733	750	766	782	799	815	832	848	864	881	50	
996		10	881	897	914	930	946	963	979	995	*012	*028	*044	40	
996		20	8.11 044	061	077	093	110	126	142	159	175	191	207	30	
996		30	207	224	240	256	272	289	305	321	337	354	370	20	
996		40	370	386	402	418	435	451	467	483	499	515	531	10	
996		50	531	548	564	580	596	612	628	644	660	677	693	0	15
996	45	0	693	709	725	741	757	773	789	805	821	837	853	50	
996		10	853	869	885	901	917	933	949	965	981	997	*013	40	
996		20	8.12 013	029	045	061	077	093	109	125	141	157	172	30	
996		30	172	188	204	220	236	252	268	284	300	315	331	20	
996		40	331	347	363	379	395	410	426	442	458	474	489	10	
996		50	489	505	521	537	553	568	584	600	616	631	647	0	14
996	46	0	647	663	679	694	710	726	741	757	773	788	804	50	
996		10	804	820	836	851	867	882	898	914	929	945	961	40	
996		20	961	976	992	*007	*023	*039	*054	*070	*085	*101	*117	30	
996		30	8.13 117	132	148	163	179	194	210	225	241	256	272	20	
996		40	272	287	303	318	334	349	365	380	396	411	427	10	
996		50	427	442	458	473	489	504	519	535	550	566	581	0	13
996	47	0	581	596	612	627	643	658	673	689	704	719	735	50	
996		10	735	750	765	781	796	811	827	842	857	873	888	40	
996		20	888	903	919	934	949	964	980	995	*010	*025	*041	30	
996		30	8.14 041	056	071	086	101	117	132	147	162	178	193	20	
996		40	193	208	223	238	253	269	284	299	314	329	344	10	
996		50	344	359	375	390	405	420	435	450	465	480	495	0	12
996	48	0	495	510	525	541	556	571	586	601	616	631	646	50	
996		10	646	661	676	691	706	721	736	751	766	781	796	40	
996		20	796	811	826	841	856	871	886	901	915	930	945	30	
996		30	945	960	975	990	*005	*020	*035	*050	*065	*079	*094	20	
996		40	8.15 094	109	124	139	154	169	183	198	213	228	243	10	
996		50	243	258	272	287	302	317	332	346	361	376	391	0	11
996	49	0	391	406	420	435	450	465	479	494	509	523	538	50	
996		10	538	553	568	582	597	612	626	641	656	670	685	40	
996		20	685	700	714	729	744	758	773	788	802	817	832	30	
995		30	832	846	861	875	890	905	919	934	948	963	978	20	
995		40	978	992	*007	*021	*036	*050	*065	*079	*094	*109	*123	10	
995		50	8.16 123	138	152	167	181	196	210	225	239	254	268	0	10
9·99			10″	9″	8″	7″	6″	5″	4″	3″	2″	1″	0″	″	′

L Sin *179° 269° *359° **89°** **L Cos**

L Tan **0°** *90° 180° *270°

′ ″	0″	1″	2″	3″	4″	5″	6″	7″	8″	9″	10″	
40 0	8.06 581	599	617	635̄	653	671	689	707	725	743	761	50
10	761	779	797	815	833	851	869	887	905	923	941	40
20	941	959	977	995̄	*013	*031	*049	*066	*084	*102	*120	30
30	8.07 120	138	156	174	192	209	227	245	263	281	298	20
40	298	316	334	352	370	387	405	423	441	458	476	10
50	476	494	512	529	547	565̄	582	600	618	635	653	0 **19**
41 0	653	671	688	706	724	741	759	776	794	812	829	50
10	829	847	864	882	900	917	935̄	952	970	987	*005̄	40
20	8.08 005̄	022	040	057	075̄	092	110	127	145̄	162	180	30
30	180	197	214	232	249	267	284	301	319	336	354	20
40	354	371	388	406	423	440	458	475	492	510	527	10
50	527	544	562	579	596	613	631	648	665	682	700	0 **18**
42 0	700	717	734	751	769	786	803	820	837	855̄	872	50
10	872	889	906	923	940	957	975̄	992	*009	*026	*043	40
20	8.09 043	060	077	094	111	128	146	163	180	197	214	30
30	214	231	248	265̄	282	299	316	333	35̄0	367	384	20
40	384	401	418	435̄	452	468	485	502	519	536	553	10
50	553	570	587	604	621	637	654	671	688	705̄	722	0 **17**
43 0	722	739	755	772	789	806	823	839	856	873	890	50
10	890	907	923	940	957	974	990	*007	*024	*040	*057	40
20	8.10 057	074	091	107	124	141	157	174	191	207	224	30
30	224	240	257	274	290	307	324	340	357	373	390	20
40	390	407	423	440	456	473	489	506	522	539	555	10
50	555	572	588	605̄	621	638	654	671	687	704	720	0 **16**
44 0	720	737	753	770	786	802	819	835	852	868	884	50
10	884	901	917	934	95̄0	966	983	999	*015	*032	*048	40
20	8.11 048	064	081	097	113	130	146	162	178	195̄	211	30
30	211	227	244	260	276	292	309	325̄	341	357	373	20
40	373	390	406	422	438	454	471	487	503	519	535	10
50	535	551	567	584	600	616	632	648	664	680	696	0 **15**
45 0	696	712	729	745̄	761	777	793	809	825̄	841	857	50
10	857	873	889	905̄	921	937	953	969	985̄	*001	*017	40
20	8.12 017	033	049	065̄	081	097	113	129	144	160	176	30
30	176	192	208	224	240	256	272	288	303	319	335	20
40	335	351	367	383	398	414	430	446	462	478	493	10
50	493	509	525̄	541	556	572	588	604	620	635	651	0 **14**
46 0	651	667	682	698	714	730	745	761	777	792	808	50
10	808	824	839	855	871	886	902	918	933	949	965̄	40
20	965̄	980	996	*011	*027	*043	*058	*074	*089	*105̄	*121	30
30	8.13 121	136	152	167	183	198	214	229	245̄	260	276	20
40	276	291	307	322	338	353	369	384	400	415	431	10
50	431	446	462	477	493	508	523	539	554	570	585	0 **13**
47 0	585	601	616	631	647	662	677	693	708	724	739	50
10	739	754	770	785̄	800	816	831	846	861	877	892	40
20	892	907	923	938	953	968	984	999	*014	*029	*045̄	30
30	8.14 045̄	060	075	090	106	121	136	151	166	182	197	20
40	197	212	227	242	258	273	288	303	318	333	348	10
50	348	364	379	394	409	424	439	454	469	484	5̄00	0 **12**
48 0	5̄00	515̄	530	545̄	560	575̄	590	605	620	635	650	50
10	650	665	680	695	710	725̄	740	755	770	785	800	40
20	800	815	830	845	860	875̄	890	905̄	920	935̄	95̄0	30
30	95̄0	965̄	980	994	*009	*024	*039	*054	*069	*084	*099	20
40	8.15 099	114	128	143	158	173	188	203	218	232	247	10
50	247	262	277	292	306	321	336	351	366	380	395	0 **11**
49 0	395	410	425̄	439	454	469	484	498	513	528	543	50
10	543	557	572	587	602	616	631	646	660	675̄	690	40
20	690	704	719	734	748	763	778	792	807	822	836	30
30	836	851	865	880	895̄	909	924	938	953	968	982	20
40	982	997	*011	*026	*040	*055̄	*070	*084	*099	*113	*128	10
50	8.16 128	142	157	171	186	200	215̄	229	244	258	273	0 **10**
	10″	9″	8″	7″	6″	5″	4″	3″	2″	1″	0″	″ ′

P	P
	18
1	1.8
2	3.6
3	5.4
4	7.2
5	9.0
6	10.8
7	12.6
8	14.4
9	16.2
	17
1	1.7
2	3.4
3	5.1
4	6.8
5	8.5
6	10.2
7	11.9
8	13.6
9	15.3
	16
1	1.6
2	3.2
3	4.8
4	6.4
5	8.0
6	9.6
7	11.2
8	12.8
9	14.4
	15
1	1.5
2	3.0
3	4.5
4	6.0
5	7.5
6	9.0
7	10.5
8	12.0
9	13.5
	14
1	1.4
2	2.8
3	4.2
4	5.6
5	7.0
6	8.4
7	9.8
8	11.2
9	12.6

*179° 269° *359° **89°** L Cot

L Cos L Sin **0°** *90° 180° *270°

9·99	′	″	0″	1″	2″	3″	4″	5″	6″	7″	8″	9″	10″	″	
995	50	0	8.16 268	283	297	311	326	340	355̄	369	384	398	413	50	
995		10	413	427	441	456	470	485̄	499	513	528	542	557	40	
995		20	557	571	585	600	614	628	643	657	672	686	700	30	
995		30	700	715̄	729	743	757	772	786	800	815̄	829	843	20	
995		40	843	858	872	886	900	915̄	929	943	957	972	986	10	
995		50	986	*000	*014	*029	*043	*057	*071	*085	*100	*114	*128	0	9
995	51	0	8.17 128	142	156	171	185̄	199	213	227	241	256	270	50	
995		10	270	284	298	312	326	340	355̄	369	383	397	411	40	
995		20	411	425	439	453	467	481	495	510	524	538	552	30	
995		30	552	566	580	594	608	622	636	650	664	678	692	20	
995		40	692	706	720	734	748	762	776	790	804	818	832	10	
995		50	832	846	860	874	888	902	916	930	943	957	971	0	8
995	52	0	971	985	999	*013	*027	*041	*055̄	*069	*082	*096	*110	50	
995̄		10	8.18 110	124	138	152	166	180	193	207	221	235̄	249	40	
995̄		20	249	263	276	290	304	318	332	345	359	373	387	30	
995̄		30	387	401	414	428	442	456	469	483	497	511	524	20	
995̄		40	524	538	552	566	579	593	607	621	634	648	662	10	
995̄		50	662	675	689	703	716	730	744	757	771	785̄	798	0	7
995̄	53	0	798	812	826	839	853	867	880	894	908	921	935̄	50	
995̄		10	935̄	948	962	976	989	*003	*016	*030	*044	*057	*071	40	
995̄		20	8.19 071	084	098	111	125̄	139	152	166	179	193	206	30	
995̄		30	206	220	233	247	260	274	287	301	314	328	341	20	
995̄		40	341	355̄	368	382	395	409	422	436	449	463	476	10	
995̄		50	476	489	503	516	530	543	557	570	583	597	610	0	6
995̄	54	0	610	624	637	650	664	677	691	704	717	731	744	50	
995̄		10	744	757	771	784	797	811	824	837	851	864	877	40	
995̄		20	877	891	904	917	931	944	957	971	984	997	*010	30	
995̄		30	8.20 010	024	037	050	064	077	090	103	117	130	143	20	
995̄		40	143	156	170	183	196	209	222	236	249	262	275	10	
994		50	275	288	302	315̄	328	341	354	368	381	394	407	0	5
994	55	0	407	420	433	446	460	473	486	499	512	525	538	50	
994		10	538	552	565̄	578	591	604	617	630	643	656	669	40	
994		20	669	682	696	709	722	735̄	748	761	774	787	800	30	
994		30	800	813	826	839	852	865	878	891	904	917	930	20	
994		40	930	943	956	969	982	995	*008	*021	*034	*047	*060	10	
994		50	8.21 060	073	086	099	112	125̄	138	151	164	177	189	0	4
994	56	0	189	202	215	228	241	254	267	280	293	306	319	50	
994		10	319	331	344	357	370	383	396	409	422	434	447	40	
994		20	447	460	473	486	499	511	524	537	550	563	576	30	
994		30	576	588	601	614	627	640	652	665	678	691	703	20	
994		40	703	716	729	742	754	767	780	793	805	818	831	10	
994		50	831	844	856	869	882	895̄	907	920	933	945	958	0	3
994	57	0	958	971	983	996	*009	*022	*034	*047	*060	*072	*085̄	50	
994		10	8.22 085̄	098	110	123	136	148	161	173	186	199	211	40	
994		20	211	224	237	249	262	274	287	300	312	325̄	337	30	
994		30	337	350	363	375	388	400	413	425	438	451	463	20	
994		40	463	476	488	501	513	526	538	551	563	576	588	10	
994		50	588	601	613	626	638	651	663	676	688	701	713	0	2
994	58	0	713	726	738	751	763	776	788	801	813	826	838	50	
994		10	838	850	863	875	888	900	913	925̄	937	950	962	40	
994		20	962	975̄	987	999	*012	*024	*037	*049	*061	*074	*086	30	
994		30	8.23 086	098	111	123	136	148	160	173	185̄	197	210	20	
994		40	210	222	234	247	259	271	284	296	308	321	333	10	
994		50	333	345	357	370	382	394	407	419	431	443	456	0	1
994	59	0	456	468	480	492	505̄	517	529	541	554	566	578	50	
994		10	578	590	603	615̄	627	639	652	664	676	688	700	40	
994		20	700	713	725̄	737	749	761	773	786	798	810	822	30	
993		30	822	834	846	859	871	883	895	907	919	931	944	20	
993		40	944	956	968	980	992	*004	*016	*028	*041	*053	*065̄	10	
993		50	8.24 065̄	077	089	101	113	125	137	149	161	173	186	0	0
9·99			10″	9″	8″	7″	6″	5″	4″	3″	2″	1″	0″	″	′

L Sin *179° 269° *359 **89°** L Cos

L Tan 0° *90° 180° *270°

′ ″	0″	1″	2″	3″	4″	5″	6″	7″	8″	9″	10″	
50 0	8.16 273	287	302	316	331	345	359	374	388	403	417	50
10	417	432	446	460	475̄	489	504	518	533	547	561	40
20	561	576	590	604	619	633	647	662	676	691	705̄	30
30	705̄	719	734	748	762	776	791	805	819	834	848	20
40	848	862	877	891	905	919	934	948	962	976	991	10
50	991	*005̄	*019	*033	*048	*062	*076	*090	*104	*119	*133	0 **9**
51 0	8.17 133	147	161	175	190	204	218	232	246	260	275̄	50
10	275̄	289	303	317	331	345	359	373	388	402	416	40
20	416	430	444	458	472	486	500	514	528	543	557	30
30	557	571	585̄	599	613	627	641	655̄	669	683	697	20
40	697	711	725̄	739	753	767	781	795̄	809	823	837	10
50	837	851	865̄	879	893	907	921	934	948	962	976	0 **8**
52 0	976	990	*004	*018	*032	*046	*060	*074	*087	*101	*115	50
10	8.18 115	129	143	157	171	185̄	198	212	226	240	254	40
20	254	268	281	295	309	323	337	351	364	378	392	30
30	392	406	419	433	447	461	475̄	488	502	516	530	20
40	530	543	557	571	585̄	598	612	626	639	653	667	10
50	667	681	694	708	722	735	749	763	776	790	804	0 **7**
53 0	804	817	831	845̄	858	872	886	899	913	926	940	50
10	940	954	967	981	994	*008	*022	*035	*049	*062	*076	40
20	8.19 076	090	103	117	130	144	157	171	184	198	211	30
30	211	225	239	252	266	279	293	306	320	333	347	20
40	347	360	374	387	401	414	427	441	454	468	481	10
50	481	495̄	508	522	535	548	562	575	589	602	616	0 **6**
54 0	616	629	642	656	669	683	696	709	723	736	749	50
10	749	763	776	789	803	816	830	843	856	870	883	40
20	883	896	910	923	936	949	963	976	989	*003	*016	30
30	8.20 016	029	042	056	069	082	096	109	122	135	149	20
40	149	162	175	188	201	215̄	228	241	254	268	281	10
50	281	294	307	320	334	347	360	373	386	399	413	0 **5**
55 0	413	426	439	452	465	478	491	505̄	518	531	544	50
10	544	557	570	583	596	610	623	636	649	662	675	40
20	675	688	701	714	727	740	753	767	780	793	806	30
30	806	819	832	845̄	858	871	884	897	910	923	936	20
40	936	949	962	975̄	988	*001	*014	*027	*040	*053	*066	10
50	8.21 066	079	092	105̄	118	131	144	156	169	182	195	0 **4**
56 0	195	208	221	234	247	260	273	286	299	311	324	50
10	324	337	350	363	376	389	402	414	427	440	453	40
20	453	466	479	492	504	517	530	543	556	569	581	30
30	581	594	607	620	633	645	658	671	684	697	709	20
40	709	722	735̄	748	760	773	786	799	811	824	837	10
50	837	850	862	875	888	901	913	926	939	951	964	0 **3**
57 0	964	977	989	*002	*015̄	*028	*040	*053	*066	*078	*091	50
10	8.22 091	104	116	129	142	154	167	179	192	205̄	217	40
20	217	230	243	255	268	280	293	306	318	331	343	30
30	343	356	369	381	394	406	419	431	444	457	469	20
40	469	482	494	507	519	532	544	557	569	582	595̄	10
50	595̄	607	620	632	645̄	657	670	682	695̄	707	720	0 **2**
58 0	720	732	744	757	769	782	794	807	819	832	844	50
10	844	857	869	881	894	906	919	931	944	956	968	40
20	968	981	993	*006	*018	030	*043	*055	*068	*080	*092	30
30	8.23 092	105̄	117	130	142	154	167	179	191	204	216	20
40	216	228	241	253	265	278	290	302	315̄	327	339	10
50	339	352	364	376	388	401	413	425	438	450	462	0 **1**
59 0	462	474	487	499	511	523	536	548	560	572	585̄	50
10	585̄	597	609	621	634	646	658	670	682	695̄	707	40
20	707	719	731	743	756	768	780	792	804	816	829	30
30	829	841	853	865	877	889	902	914	926	938	950	20
40	950	962	974	987	999	*011	*023	*035	*047	*059	*071	10
50	8.24 071	083	096	108	120	132	144	156	168	180	192	0 **0**
	10″	9″	8″	7″	6″	5″	4″	3″	2″	1″	0″	′ ″

P P

	15
1	1.5
2	3.0
3	4.5
4	6.0
5	7.5
6	9.0
7	10.5
8	12.0
9	13.5

	14
1	1.4
2	2.8
3	4.2
4	5.6
5	7.0
6	8.4
7	9.8
8	11.2
9	12.6

	13
1	1.3
2	2.6
3	3.9
4	5.2
5	6.5
6	7.8
7	9.1
8	10.4
9	11.7

	12
1	1.2
2	2.4
3	3.6
4	4.8
5	6.0
6	7.2
7	8.4
8	9.6
9	10.8

P P

*179° 269° *359° 89° L Cot

L Cos L Sin **1°** *91° 181° *271°

9.99	′	0″	10″	20″	30″	40″	50″	60″	
993	0	8.24 186	306	426	546	665	785̄	903	59
993	1	903	*022	*140	*258	*375	*493	*609	58
993	2	8.25 609	726	842	958	*074	*189	*304	57
993	3	8.26 304	419	533	648	761	875̄	988	56
992	4	988	*101	*214	*326	*438	*55̄0	*661	55
992	5	8.27 661	773	883	994	*104	*215̄	*324	54
992	6	8.28 324	434	543	652	761	869	977	53
992	7	977	*085	*193	*300	*407	*514	*621	52
992	8	8.29 621	727	833	939	*044	*15̄0	*255̄	51
991	9	8.30 255̄	359	464	568	672	776	879	50
991	10	879	983	*086	*188	*291	*393	*495	49
991	11	8.31 495	597	699	800	901	*002	*103	48
990	12	8.32 103	203	303	403	503	602	702	47
990	13	702	801	899	998	*096	*195̄	*292	46
990	14	8.33 292	390	488	585̄	682	779	875	45
990	15	875	972	*068	*164	*260	*355	*450	44
989	16	8.34 450	546	640	735	830	924	*018	43
989	17	8.35 018	112	206	299	392	485	578	42
989	18	578	671	764	856	948	*040	*131	41
989	19	8.36 131	223	314	405	496	587	678	40
988	20	678	768	858	948	*038	*128	*217	39
988	21	8.37 217	306	395	484	573	662	75̄0	38
988	22	75̄0	838	926	*014	*101	*189	*276	37
987	23	8.38 276	363	450	537	624	710	796	36
987	24	796	882	968	*054	*139	*225̄	*310	35
987	25	8.39 310	395	480	565̄	649	734	818	34
986	26	818	902	986	*070	*153	*237	*320	33
986	27	8.40 320	403	486	569	651	734	816	32
986	28	816	898	980	*062	*144	*225	*307	31
985	29	8.41 307	388	469	550	631	711	792	30
985	30	792	872	952	*032	*112	*192	*272	29
985̄	31	8.42 272	351	430	510	589	667	746	28
984	32	746	825̄	903	982	*060	*138	*216	27
984	33	8.43 216	293	371	448	526	603	680	26
984	34	680	757	834	910	987	*063	*139	25
983	35	8.44 139	216	292	367	443	519	594	24
983	36	594	669	745̄	820	895̄	969	*044	23
983	37	8.45 044	119	193	267	341	415	489	22
982	38	489	563	637	710	784	857	930	21
982	39	930	*003	*076	*149	*222	*294	*366	20
982	40	8.46 366	439	511	583	655̄	727	799	19
981	41	799	870	942	*013	*084	*155	*226	18
981	42	8.47 226	297	368	439	509	580	65̄0	17
981	43	65̄0	720	790	860	930	*000	*069	16
980	44	8.48 069	139	208	278	347	416	485̄	15
980	45	485̄	554	622	691	760	828	896	14
979	46	896	965̄	*033	*101	*169	*236	*304	13
979	47	8.49 304	372	439	506	574	641	708	12
979	48	708	775̄	842	908	975	*042	*108	11
978	49	8.50 108	174	241	307	373	439	504	10
978	50	504	570	636	701	767	832	897	9
977	51	897	963	*028	*092	*157	*222	*287	8
977	52	8.51 287	351	416	480	544	609	673	7
977	53	673	737	801	864	928	992	*055	6
976	54	8.52 055	119	182	245	308	371	434	5
976	55	434	497	560	623	685	748	810	4
975	56	810	872	935̄	997	*059	*121	*183	3
975̄	57	8.53 183	245̄	306	368	429	491	552	2
974	58	552	614	675̄	736	797	858	919	1
974	59	919	979	*040	*101	*161	*222	*282	0
9.99		60″	50″	40″	30″	20″	10″	0″	′

P P

	120	119	118
1	12.0	11.9	11.8
2	24.0	23.8	23.6
3	36.0	35.7	35.4
4	48.0	47.6	47.2
5	60.0	59.5	59.0
6	72.0	71.4	70.8
7	84.0	83.3	82.6
8	96.0	95.2	94.4
9	108.0	107.1	106.2

	117	116	115
1	11.7	11.6	11.5
2	23.4	23.2	23.0
3	35.1	34.8	34.5
4	46.8	46.4	46.0
5	58.5	58.0	57.5
6	70.2	69.6	69.0
7	81.9	81.2	80.5
8	93.6	92.8	92.0
9	105.3	104.4	103.5

	114	113	112	111
1	11.4	11.3	11.2	11.1
2	22.8	22.6	22.4	22.2
3	34.2	33.9	33.6	33.3
4	45.6	45.2	44.8	44.4
5	57.0	56.5	56.0	55.5
6	68.4	67.8	67.2	66.6
7	79.8	79.1	78.4	77.7
8	91.2	90.4	89.6	88.8
9	102.6	101.7	100.8	99.9

	110	109	108	107
1	11.0	10.9	10.8	10.7
2	22.0	21.8	21.6	21.4
3	33.0	32.7	32.4	32.1
4	44.0	43.6	43.2	42.8
5	55.0	54.5	54.0	53.5
6	66.0	65.4	64.8	64.2
7	77.0	76.3	75.6	74.9
8	88.0	87.2	86.4	85.6
9	99.0	98.1	97.2	96.3

	106	105	104	103
1	10.6	10.5	10.4	10.3
2	21.2	21.0	20.8	20.6
3	31.8	31.5	31.2	30.9
4	42.4	42.0	41.6	41.2
5	53.0	52.5	52.0	51.5
6	63.6	63.0	62.4	61.8
7	74.2	73.5	72.8	72.1
8	84.8	84.0	83.2	82.4
9	95.4	94.5	93.6	92.7

	102	101	100	99
1	10.2	10.1	10.0	9.9
2	20.4	20.2	20.0	19.8
3	30.6	30.3	30.0	29.7
4	40.8	40.4	40.0	39.6
5	51.0	50.5	50.0	49.5
6	61.2	60.6	60.0	59.4
7	71.4	70.7	70.0	69.3
8	81.6	80.8	80.0	79.2
9	91.8	90.9	90.0	89.1

	98	97	96	95
1	9.8	9.7	9.6	9.5
2	19.6	19.4	19.2	19.0
3	29.4	29.1	28.8	28.5
4	39.2	38.8	38.4	38.0
5	49.0	48.5	48.0	47.5
6	58.8	58.2	57.6	57.0
7	68.6	67.9	67.2	66.5
8	78.4	77.6	76.8	76.0
9	88.2	87.3	86.4	85.5

P P

L Sin *178° 268° *358° **88°** L Cos

L Tan 1° *91° 181° *271°

′	0″	10″	20″	30″	40″	50″	60″	
0	8.24 192	313	433	553	672	791	910	59
1	910	*029	*147	*265̄	*382	*5̄00	*616	58
2	8.25 616	733	849	965	*081	*196	*312	57
3	8.26 312	426	541	655̄	769	882	996	56
4	996	*109	*221	*334	*446	*558	*669	55
5	8.27 669	780	891	*002	*112	*223	*332	54
6	8.28 332	442	551	660	769	877	986	53
7	986	*094	*201	*309	*416	*523	*629	52
8	8.29 629	736	842	947	*053	*158	*203	51
9	8.30 263	368	473	577	681	785̄	888	50
10	888	992	*095	*198	*300	*403	*505̄	49
11	8.31 505̄	606	708	809	911	*012	*112	48
12	8.32 112	213	313	413	513	612	711	47
13	711	810	909	*008	*106	*205̄	*302	46
14	8.33 302	400	498	595	692	789	886	45
15	886	982	*078	*174	*270	*366	*461	44
16	8.34 461	556	651	746	840	935̄	*029	43
17	8.35 029	123	217	310	403	497	590	42
18	590	682	775̄	867	959	*051	*143	41
19	8.36 143	235̄	326	417	508	599	689	40
20	689	780	870	960	*05̄0	*140	*229	39
21	8.37 229	318	408	497	585	674	762	38
22	762	850	938	*026	*114	*202	*289	37
23	8.38 289	376	463	55̄0	636	723	809	36
24	809	895	981	*067	*153	*238	*323	35
25	8.39 323	408	493	578	663	747	832	34
26	832	916	*000	*083	*167	*250	*334	33
27	8.40 334	417	5̄00	583	665	748	830	32
28	830	913	995̄	*077	*158	*240	*321	31
29	8.41 321	403	484	565̄	646	726	807	30
30	807	887	967	*048	*127	*207	*287	29
31	8.42 287	366	446	525̄	604	683	762	28
32	762	840	919	997	*073	*154	*232	27
33	8.43 232	309	387	464	542	619	696	26
34	696	773	850	927	*003	*080	*156	25
35	8.44 156	232	308	384	460	536	611	24
36	611	686	762	837	912	987	*061	23
37	8.45 061	136	210	285̄	359	433	507	22
38	507	581	655̄	728	802	875̄	948	21
39	948	*021	*094	*167	*240	*312	*385̄	20
40	8.46 385̄	457	529	602	674	745	817	19
41	817	889	960	*032	*103	*174	*245	18
42	8.47 245	316	387	458	528	599	669	17
43	669	740	810	880	95̄0	*020	*089	16
44	8.48 089	159	228	298	367	436	505	15
45	505	574	643	711	780	849	917	14
46	917	985	*053	*121	*189	*257	*325	13
47	8.49 325	393	460	528	595̄	662	729	12
48	729	796	863	930	997	*063	*130	11
49	8.50 130	196	263	329	395̄	461	527	10
50	527	593	658	724	789	855̄	920	9
51	920	985	*050	*115	*180	*245	*310	8
52	8.51 310	374	439	503	568	632	696	7
53	696	760	824	888	952	*015	*079	6
54	8.52 079	143	206	269	332	396	459	5
55	459	522	584	647	710	772	835̄	4
56	835̄	897	960	*022	*084	*146	*208	3
57	8.53 208	270	332	393	455̄	516	578	2
58	578	639	700	762	823	884	945̄	1
59	945̄	*005	*066	*127	*187	*248	*308	0
	60″	50″	40″	30″	20″	10″	0″	′

P P

	94	93	92	91	90
1	9.4	9.3	9.2	9.1	9.0
2	18.8	18.6	18.4	18.2	18.0
3	28.2	27.9	27.6	27.3	27.0
4	37.6	37.2	36.8	36.4	36.0
5	47.0	46.5	46.0	45.5	45.0
6	56.4	55.8	55.2	54.6	54.0
7	65.8	65.1	64.4	63.7	63.0
8	75.2	74.4	73.6	72.8	72.0
9	84.6	83.7	82.8	81.9	81.0

	89	88	87	86	85
1	8.9	8.8	8.7	8.6	8.5
2	17.8	17.6	17.4	17.2	17.0
3	26.7	26.4	26.1	25.8	25.5
4	35.6	35.2	34.8	34.4	34.0
5	44.5	44.0	43.5	43.0	42.5
6	53.4	52.8	52.2	51.6	51.0
7	62.3	61.6	60.9	60.2	59.5
8	71.2	70.4	69.6	68.8	68.0
9	80.1	79.2	78.3	77.4	76.5

	84	83	82	81	80
1	8.4	8.3	8.2	8.1	8.0
2	16.8	16.6	16.4	16.2	16.0
3	25.2	24.9	24.6	24.3	24.0
4	33.6	33.2	32.8	32.4	32.0
5	42.0	41.5	41.0	40.5	40.0
6	50.4	49.8	49.2	48.6	48.0
7	58.8	58.1	57.4	56.7	56.0
8	67.2	66.4	65.6	64.8	64.0
9	75.6	74.7	73.8	72.9	72.0

	79	78	77	76	75
1	7.9	7.8	7.7	7.6	7.5
2	15.8	15.6	15.4	15.2	15.0
3	23.7	23.4	23.1	22.8	22.5
4	31.6	31.2	30.8	30.4	30.0
5	39.5	39.0	38.5	38.0	37.5
6	47.4	46.8	46.2	45.6	45.0
7	55.3	54.6	53.9	53.2	52.5
8	63.2	62.4	61.6	60.8	60.0
9	71.1	70.2	69.3	68.4	67.5

	74	73	72	71	70
1	7.4	7.3	7.2	7.1	7.0
2	14.8	14.6	14.4	14.2	14.0
3	22.2	21.9	21.6	21.3	21.0
4	29.6	29.2	28.8	28.4	28.0
5	37.0	36.5	36.0	35.5	35.0
6	44.4	43.8	43.2	42.6	42.0
7	51.8	51.1	50.4	49.7	49.0
8	59.2	58.4	57.6	56.8	56.0
9	66.6	65.7	64.8	63.9	63.0

	69	68	67	66	65
1	6.9	6.8	6.7	6.6	6.5
2	13.8	13.6	13.4	13.2	13.0
3	20.7	20.4	20.1	19.8	19.5
4	27.6	27.2	26.8	26.4	26.0
5	34.5	34.0	33.5	33.0	32.5
6	41.4	40.8	40.2	39.6	39.0
7	48.3	47.6	46.9	46.2	45.5
8	55.2	54.4	53.6	52.8	52.0
9	62.1	61.2	60.3	59.4	58.5

	64	63	62	61	60
1	6.4	6.3	6.2	6.1	6.0
2	12.8	12.6	12.4	12.2	12.0
3	19.2	18.9	18.6	18.3	18.0
4	25.6	25.2	24.8	24.4	24.0
5	32.0	31.5	31.0	30.5	30.0
6	38.4	37.8	37.2	36.6	36.0
7	44.8	44.1	43.4	42.7	42.0
8	51.2	50.4	49.6	48.8	48.0
9	57.6	56.7	55.8	54.9	54.0

P P

*178° 268° *358° **88°** L Cot

9·99	′	0″	10″	20″	30″	40″	50″	60″		
974	0	8.54 282	342	402	462	522	582	642	59	973
973	1	642	702	762	821	881	940	999	58	973
973	2	999	*059	*118	*177	*236	*295̄	*354	57	972
972	3	8.55 354	413	471	530	589	647	705	56	972
972	4	705	764	822	880	938	996	*054	55	971
971	5	8.56 054	112	170	227	285̄	342	400	54	971
971	6	400	457	515̄	572	629	686	743	53	970
970	7	743	800	857	914	970	*027	*084	52	970
970	8	8.57 084	140	196	253	309	365	421	51	969
969	9	421	477	533	589	645	701	757	50	969
969	10	757	812	868	923	979	*034	*089	49	968
968	11	8.58 089	144	200	255̄	310	364	419	48	968
968	12	419	474	529	583	638	693	747	47	967
967	13	747	801	856	910	964	*018	*072	46	967
967	14	8.59 072	126	180	234	288	341	395̄	45	967
967	15	395̄	448	502	555	609	662	715	44	966
966	16	715	768	821	874	927	980	*033	43	966
966	17	8.60 033	086	139	191	244	296	349	42	965̄
965̄	18	349	401	454	506	558	610	662	41	964
964	19	662	714	766	818	870	922	973	40	964
964	20	973	*025	*077	*128	*180	*231	*282	39	963
963	21	8.61 282	334	385̄	436	487	538	589	38	963
963	22	589	640	691	742	792	843	894	37	962
962	23	894	944	995̄	*045	*096	*146	*196	36	962
962	24	8.62 196	246	297	347	397	447	497	35	961
961	25	497	546	596	646	696	745	795̄	34	961
961	26	795̄	844	894	943	993	*042	*091	33	960
960	27	8.63 091	140	189	238	288	336	385	32	960
960	28	385	434	483	532	580	629	678	31	959
959	29	678	726	775̄	823	871	920	968	30	959
959	30	968	*016	*064	*112	*160	*208	*256	29	958
958	31	8.64 256	304	352	400	448	495	543	28	958
958	32	543	590	638	685	733	780	827	27	957
957	33	827	875̄	922	969	*016	*063	*110	26	956
956	34	8.65 110	157	204	251	298	344	391	25	956
956	35	391	438	484	531	577	624	670	24	955
955	36	670	717	763	809	855	901	947	23	955̄
955̄	37	947	994	*040	*085	*131	*177	*223	22	954
954	38	8.66 223	269	314	360	406	451	497	21	954
954	39	497	542	588	633	678	724	769	20	953
953	40	769	814	859	904	949	994	*039	19	952
952	41	8.67 039	084	129	174	219	263	308	18	952
952	42	308	353	397	442	486	531	575	17	951
951	43	575	619	664	708	752	796	841	16	951
951	44	841	885̄	929	973	*017	*060	*104	15	950
950	45	8.68 104	148	192	236	279	323	367	14	949
949	46	367	410	454	497	540	584	627	13	949
949	47	627	670	714	757	800	843	886	12	948
948	48	886	929	972	*015	*058	*101	*144	11	948
948	49	8.69 144	187	229	272	315̄	357	400	10	947
947	50	400	442	485̄	527	570	612	654	9	946
946	51	654	697	739	781	823	865	907	8	946
946	52	907	949	991	*033	*075	*117	*159	7	945̄
945̄	53	8.70 159	201	242	284	326	367	409	6	944
944	54	409	451	492	534	575̄	616	658	5	944
944	55	658	699	740	781	823	864	905̄	4	943
943	56	905̄	946	987	*028	*069	*110	*151	3	942
942	57	8.71 151	192	232	273	314	355̄	395	2	942
942	58	395	436	476	517	557	598	638	1	941
941	59	638	679	719	759	800	840	880	0	940
		60″	50″	40″	30″	20″	10″	0″	′	9·99

P P	61	60	59	58	57	56
1	6.1	6.0	5.9	5.8	5.7	5.6
2	12.2	12.0	11.8	11.6	11.4	11.2
3	18.3	18.0	17.7	17.4	17.1	16.8
4	24.4	24.0	23.6	23.2	22.8	22.4
5	30.5	30.0	29.5	29.0	28.5	28.0
6	36.6	36.0	35.4	34.8	34.2	33.6
7	42.7	42.0	41.3	40.6	39.9	39.2
8	48.8	48.0	47.2	46.4	45.6	44.8
9	54.9	54.0	53.1	52.2	51.3	50.4

L Tan 2° *92° 182° *272°

′	0″	10″	20″	30″	40″	50″	60″	
0	8.54 308	369	429	489	549	609	669	59
1	669	729	789	848	908	967	*027	58
2	8.55 027	086	145	205̄	264	323	382	57
3	382	441	499	558	617	675	734	56
4	734	792	850	909	967	*025̄	*083	55
5	8.56 083	141	199	256	314	372	429	54
6	429	487	544	601	659	716	773	53
7	773	830	887	944	*000	*057	*114	52
8	8.57 114	170	227	283	340	396	452	51
9	452	508	564	620	676	732	788	50
10	788	843	899	955̄	*010	*065	*121	49
11	8.58 121	176	231	286	341	396	451	48
12	451	506	561	616	670	725̄	779	47
13	779	834	888	943	997	*051	*105	46
14	8.59 105	159	213	267	321	375̄	428	45
15	428	482	536	589	642	696	749	44
16	749	802	856	909	962	*015̄	*068	43
17	8.60 068	121	173	226	279	331	384	42
18	384	436	489	541	593	646	698	41
19	698	750	802	854	906	958	*009	40
20	8.61 009	061	113	164	216	267	319	39
21	319	370	422	473	524	575	626	38
22	626	677	728	779	830	881	931	37
23	931	982	*033	*083	*134	*184	*234	36
24	8.62 234	285̄	335̄	385̄	435	485	535	35
25	535	585	635	685̄	735̄	784	834	34
26	834	884	933	983	*032	*081	*131	33
27	8.63 131	180	229	278	328	377	426	32
28	426	475̄	523	572	621	670	718	31
29	718	767	816	864	913	961	*009	30
30	8.64 009	058	106	154	202	250	298	29
31	298	346	394	442	490	538	585	28
32	585	633	681	728	776	823	870	27
33	870	918	965	*012	*060	*107	*154	26
34	8.65 154	201	248	295̄	342	388	435	25
35	435	482	529	575	622	668	715̄	24
36	715̄	761	808	854	900	947	993	23
37	993	*039	*085	*131	*177	*223	*269	22
38	8.66 269	315̄	361	406	452	498	543	21
39	543	589	634	680	725	771	816	20
40	816	861	906	952	997	*042	*087	19
41	8.67 087	132	177	222	267	312	356	18
42	356	401	446	490	535̄	579	624	17
43	624	668	713	757	801	846	890	16
44	890	934	978	*022	*066	*110	*154	15
45	8.68 154	198	242	286	330	373	417	14
46	417	461	504	548	592	635	678	13
47	678	722	765	808	852	895̄	938	12
48	938	981	*024	*067	*110	*153	*196	11
49	8.69 196	239	282	325̄	368	410	453	10
50	453	496	538	581	623	666	708	9
51	708	750	793	835	877	920	962	8
52	962	*004	*046	*088	*130	*172	*214	7
53	8.70 214	256	298	339	381	423	465̄	6
54	465̄	506	548	589	631	673	714	5
55	714	755	797	838	879	921	962	4
56	962	*003	*044	*085	*126	*167	*208	3
57	8.71 208	249	290	331	372	413	453	2
58	453	494	535̄	575	616	657	697	1
59	697	738	778	819	859	899	940	0
	60″	50″	40″	30″	20″	10″	0″	′

*177° 267° *357° 87° L Cot

P P

	55	54	53
1	5.5	5.4	5.3
2	11.0	10.8	10.6
3	16.5	16.2	15.9
4	22.0	21.6	21.2
5	27.5	27.0	26.5
6	33.0	32.4	31.8
7	38.5	37.8	37.1
8	44.0	43.2	42.4
9	49.5	48.6	47.7

	52	51
1	5.2	5.1
2	10.4	10.2
3	15.6	15.3
4	20.8	20.4
5	26.0	25.5
6	31.2	30.6
7	36.4	35.7
8	41.6	40.8
9	46.8	45.9

	50	49	48
1	5.0	4.9	4.8
2	10.0	9.8	9.6
3	15.0	14.7	14.4
4	20.0	19.6	19.2
5	25.0	24.5	24.0
6	30.0	29.4	28.8
7	35.0	34.3	33.6
8	40.0	39.2	38.4
9	45.0	44.1	43.2

	47	46	45
1	4.7	4.6	4.5
2	9.4	9.2	9.0
3	14.1	13.8	13.5
4	18.8	18.4	18.0
5	23.5	23.0	22.5
6	28.2	27.6	27.0
7	32.9	32.2	31.5
8	37.6	36.8	36.0
9	42.3	41.4	40.5

	44	43
1	4.4	4.3
2	8.8	8.6
3	13.2	12.9
4	17.6	17.2
5	22.0	21.5
6	26.4	25.8
7	30.8	30.1
8	35.2	34.4
9	39.6	38.7

	42	41	40
1	4.2	4.1	4.0
2	8.4	8.2	8.0
3	12.6	12.3	12.0
4	16.8	16.4	16.0
5	21.0	20.5	20.0
6	25.2	24.6	24.0
7	29.4	28.7	28.0
8	33.6	32.8	32.0
9	37.8	36.9	36.0

P P

L Cos L Sin **3°** *93° 183° *273°

9.99	′	0″	10″	20″	30″	40″	50″	60″	′	9.99
940	0	8.71 880	920	960	*000	*040	*080	*120	59	940
940	1	8.72 120	160	200	240	280	320	359	58	939
939	2	359	399	439	478	518	558	597	57	938
938	3	597	637	676	716	755̄	794	834	56	938
938	4	834	873	912	951	991	*030	*069	55	937
937	5	8.73 069	108	147	186	225̄	264	303	54	936
936	6	303	342	380	419	458	497	535	53	936
936	7	535	574	613	651	690	728	767	52	935
935	8	767	805	844	882	920	959	997	51	934
934	9	997	*035	*073	*112	*150	*188	*226	50	934
934	10	8.74 226	264	302	340	378	416	454	49	933
933	11	454	491	529	567	605̄	642	680	48	932
932	12	680	718	755	793	831	868	906	47	932
932	13	906	943	980	*018	*055	*092	*130	46	931
931	14	8.75 130	167	204	241	279	316	353	45	930
930	15	353	390	427	464	501	538	575̄	44	929
929	16	575̄	612	648	685	722	759	795	43	929
929	17	795	832	869	905	942	979	*015	42	928
928	18	8.76 015	052	088	125̄	161	197	234	41	927
927	19	234	270	306	343	379	415̄	451	40	926
926	20	451	487	523	559	595	631	667	39	926
926	21	667	703	739	775	811	847	883	38	925̄
925̄	22	883	919	954	990	*026	*061	*097	37	924
924	23	8.77 097	133	168	204	239	275̄	310	36	923
923	24	310	346	381	416	452	487	522	35	923
923	25	522	558	593	628	663	698	733	34	922
922	26	733	768	803	838	873	908	943	33	921
921	27	943	978	*013	*048	*083	*118	*152	32	920
920	28	8.78 152	187	222	257	291	326	360	31	920
920	29	360	395	430	464	499	533	568	30	919
919	30	568	602	636	671	705	739	774	29	918
918	31	774	808	842	876	910	945̄	979	28	917
917	32	979	*013	*047	*081	*115̄	*149	*183	27	917
917	33	8.79 183	217	251	284	318	352	386	26	916
916	34	386	420	453	487	521	555̄	588	25	915
915	35	588	622	655	689	722	756	789	24	914
914	36	789	823	856	890	923	956	990	23	913
913	37	990	*023	*056	*090	*123	*156	*189	22	913
913	38	8.80 189	222	255	289	322	355̄	388	21	912
912	39	388	421	454	487	519	552	585	20	911
911	40	585	618	651	684	716	749	782	19	910
910	41	782	815̄	847	880	913	945	978	18	909
909	42	978	*010	*043	*075	*108	*140	*173	17	909
909	43	8.81 173	205	237	270	302	334	367	16	908
908	44	367	399	431	463	496	528	560	15	907
907	45	560	592	624	656	688	720	752	14	906
906	46	752	784	816	848	880	912	944	13	905
905	47	944	975	*007	*039	*071	*103	*134	12	904
904	48	8.82 134	166	198	229	261	292	324	11	904
904	49	324	356	387	419	450	482	513	10	903
903	50	513	544	576	607	639	670	701	9	902
902	51	701	732	764	795̄	826	857	888	8	901
901	52	888	920	951	982	*013	*044	*075̄	7	900
900	53	8.83 075̄	106	137	168	199	230	261	6	899
899	54	261	292	322	353	384	415̄	446	5	898
898	55	446	476	507	538	568	599	630	4	898
898	56	630	660	691	721	752	783	813	3	897
897	57	813	844	874	904	935̄	965	996	2	896
896	58	996	*026	*056	*087	*117	*147	*177	1	895̄
895̄	59	8.84 177	208	238	268	298	328	358	0	894
		60″	50″	40″	30″	20″	10″	0″	′	9.99

P P	40	39
1	4.0	3.9
2	8.0	7.8
3	12.0	11.7
4	16.0	15.6
5	20.0	19.5
6	24.0	23.4
7	28.0	27.3
8	32.0	31.2
9	36.0	35.1

	38	37
1	3.8	3.7
2	7.6	7.4
3	11.4	11.1
4	15.2	14.8
5	19.0	18.5
6	22.8	22.2
7	26.6	25.9
8	30.4	29.6
9	34.2	33.3

	36
1	3.6
2	7.2
3	10.8
4	14.4
5	18.0
6	21.6
7	25.2
8	28.8
9	32.4

	35	34
1	3.5	3.4
2	7.0	6.8
3	10.5	10.2
4	14.0	13.6
5	17.5	17.0
6	21.0	20.4
7	24.5	23.8
8	28.0	27.2
9	31.5	30.6

	33	32
1	3.3	3.2
2	6.6	6.4
3	9.9	9.6
4	13.2	12.8
5	16.5	16.0
6	19.8	19.2
7	23.1	22.4
8	26.4	25.6
9	29.7	28.8

	31	30
1	3.1	3.0
2	6.2	6.0
3	9.3	9.0
4	12.4	12.0
5	15.5	15.0
6	18.6	18.0
7	21.7	21.0
8	24.8	24.0
9	27.9	27.0
P P		

*176° 266° *356° **86°** L Cos L Sin

L Tan 3° *93° 183° *273°

′	0″	10″	20″	30″	40″	50″	60″	
0	8.71 940	980	*020	*060	*100	*141	*181	59
1	8.72 181	221	261	301	341	380	420	58
2	420	460	500	540	579	619	659	57
3	659	698	738	777	817	856	896	56
4	896	935	975	*014	*053	*093	*132	55
5	8.73 132	171	210	249	288	327	366	54
6	366	405	444	483	522	561	600	53
7	600	638	677	716	754	793	832	52
8	832	870	909	947	986	*024	*063	51
9	8.74 063	101	139	178	216	254	292	50
10	292	330	369	407	445	483	521	49
11	521	559	597	634	672	710	748	48
12	748	786	823	861	899	936	974	47
13	974	*012	*049	*087	*124	*162	*199	46
14	8.75 199	236	274	311	348	385	423	45
15	423	460	497	534	571	608	645	44
16	645	682	719	756	793	830	867	43
17	867	904	940	977	*014	*051	*087	42
18	8.76 087	124	160	197	233	270	306	41
19	306	343	379	416	452	488	525	40
20	525	561	597	633	669	706	742	39
21	742	778	814	850	886	922	958	38
22	958	994	*030	*065	*101	*137	*173	37
23	8.77 173	208	244	280	315	351	387	36
24	387	422	458	493	529	564	600	35
25	600	635	670	706	741	776	811	34
26	811	847	882	917	952	987	*022	33
27	8.78 022	057	092	127	162	197	232	32
28	232	267	302	337	371	406	441	31
29	441	475	510	545	579	614	649	30
30	649	683	718	752	787	821	855	29
31	855	890	924	958	993	*027	*061	28
32	8.79 061	096	130	164	198	232	266	27
33	266	300	334	368	402	436	470	26
34	470	504	538	572	606	639	673	25
35	673	707	741	774	808	842	875	24
36	875	909	942	976	*009	*043	*076	23
37	8.80 076	110	143	177	210	243	277	22
38	277	310	343	376	409	443	476	21
39	476	509	542	575	608	641	674	20
40	674	707	740	773	806	839	872	19
41	872	905	937	970	*003	*036	*068	18
42	8.81 068	101	134	166	199	232	264	17
43	264	297	329	362	394	427	459	16
44	459	491	524	556	588	621	653	15
45	653	685	717	750	782	814	846	14
46	846	878	910	942	974	*006	*038	13
47	8.82 038	070	102	134	166	198	230	12
48	230	262	293	325	357	389	420	11
49	420	452	484	515	547	579	610	10
50	610	642	673	705	736	768	799	9
51	799	831	862	893	925	956	987	8
52	987	*019	*050	*081	*112	*144	*175	7
53	8.83 175	206	237	268	299	330	361	6
54	361	392	423	454	485	516	547	5
55	547	578	609	640	671	701	732	4
56	732	763	794	824	855	886	916	3
57	916	947	978	*008	*039	*069	*100	2
58	8.84 100	130	161	191	222	252	282	1
59	282	313	343	374	404	434	464	0
	60″	50″	40″	30″	20″	10″	0″	′

P P

	41	40
1	4.1	4.0
2	8.2	8.0
3	12.3	12.0
4	16.4	16.0
5	20.5	20.0
6	24.6	24.0
7	28.7	28.0
8	32.8	32.0
9	36.9	36.0

	39	38
1	3.9	3.8
2	7.8	7.6
3	11.7	11.4
4	15.6	15.2
5	19.5	19.0
6	23.4	22.8
7	27.3	26.6
8	31.2	30.4
9	35.1	34.2

	37	36
1	3.7	3.6
2	7.4	7.2
3	11.1	10.8
4	14.8	14.4
5	18.5	18.0
6	22.2	21.6
7	25.9	25.2
8	29.6	28.8
9	33.3	32.4

	35	34
1	3.5	3.4
2	7.0	6.8
3	10.5	10.2
4	14.0	13.6
5	17.5	17.0
6	21.0	20.4
7	24.5	23.8
8	28.0	27.2
9	31.5	30.6

	33	32
1	3.3	3.2
2	6.6	6.4
3	9.9	9.6
4	13.2	12.8
5	16.5	16.0
6	19.8	19.2
7	23.1	22.4
8	26.4	25.6
9	29.7	28.8

	31	30
1	3.1	3.0
2	6.2	6.0
3	9.3	9.0
4	12.4	12.0
5	15.5	15.0
6	18.6	18.0
7	21.7	21.0
8	24.8	24.0
9	27.9	27.0

P P

*176° 266° *356° 86° L Cot

L Cos L Sin **4°** *94° 184° *274°

9.99	′	0″	10″	20″	30″	40″	50″	60″	′	
894	0	8.84 358	389	419	449	479	509	539	59	893
893	1	539	569	599	629	659	688	718	58	892
892	2	718	748	778	808	838	867	897	57	891
891	3	897	927	957	986	*016	*045	*075	56	891
891	4	8.85 075	105̄	134	164	193	223	252	55	890
890	5	252	282	311	341	370	400	429	54	889
889	6	429	458	488	517	546	576	605̄	53	888
888	7	605̄	634	663	693	722	751	780	52	887
887	8	780	809	838	867	896	926	955̄	51	886
886	9	955̄	984	*013	*042	*070	*099	*128	50	885
885	10	8.86 128	157	186	215̄	244	273	301	49	884
884	11	301	330	359	388	416	445	474	48	883
883	12	474	502	531	560	588	617	645	47	882
882	13	645	674	703	731	760	788	816	46	881
881	14	816	845̄	873	902	930	958	987	45	880
880	15	987	*015	*043	*072	*100	*128	*156	44	879
879	16	8.87 156	185̄	213	241	269	297	325	43	879
879	17	325	354	382	410	438	466	494	42	878
878	18	494	522	55̄0	578	606	634	661	41	877
877	19	661	689	717	745	773	801	829	40	876
876	20	829	856	884	912	940	967	995̄	39	875̄
875̄	21	995̄	*023	*050	*078	*106	*133	*161	38	874
874	22	8.88 161	188	216	243	271	298	326	37	873
873	23	326	353	381	408	436	463	490	36	872
872	24	490	518	545	572	600	627	654	35	871
871	25	654	681	709	736	763	790	817	34	870
870	26	817	845̄	872	899	926	953	980	33	869
869	27	980	*007	*034	*061	*088	*115	*142	32	868
868	28	8.89 142	169	196	223	25̄0	277	304	31	867
867	29	304	330	357	384	411	438	464	30	866
866	30	464	491	518	545̄	571	598	625̄	29	865̄
865̄	31	625̄	651	678	704	731	758	784	28	864
864	32	784	811	837	864	890	917	943	27	863
863	33	943	970	996	*023	*049	*075	*102	26	862
862	34	8.90 102	128	154	181	207	233	260	25	861
861	35	260	286	312	338	364	391	417	24	860
860	36	417	443	469	495	521	548	574	23	859
859	37	574	600	626	652	678	704	730	22	858
858	38	730	756	782	808	834	859	885	21	857
857	39	885	911	937	963	989	*015̄	*040	20	856
856	40	8.91 040	066	092	118	143	169	195̄	19	855̄
855̄	41	195̄	221	246	272	298	323	349	18	854
854	42	349	374	400	426	451	477	502	17	853
853	43	502	528	553	579	604	630	655	16	852
852	44	655	680	706	731	757	782	807	15	851
851	45	807	833	858	883	909	934	959	14	85̄0
85̄0	46	959	984	*010	*035̄	*060	*085	*110	13	848
848	47	8.92 110	135	161	186	211	236	261	12	847
847	48	261	286	311	336	361	386	411	11	846
846	49	411	436	461	486	511	536	561	10	845
845	50	561	586	611	636	660	685	710	9	844
844	51	710	735̄	760	784	809	834	859	8	843
843	52	859	883	908	933	957	982	*007	7	842
842	53	8.93 007	031	056	081	105	130	154	6	841
841	54	154	179	203	228	253	277	301	5	840
840	55	301	326	350	375̄	399	424	448	4	839
839	56	448	472	497	521	546	570	594	3	838
838	57	594	619	643	667	691	716	740	2	837
837	58	740	764	788	812	837	861	885̄	1	836
836	59	885̄	909	933	957	981	*006	*030	0	834
		60″	50″	40″	30″	20″	10″	0″	′	9.99

P P

	31	30
1	3.1	3.0
2	6.2	6.0
3	9.3	9.0
4	12.4	12.0
5	15.5	15.0
6	18.6	18.0
7	21.7	21.0
8	24.8	24.0
9	27.9	27.0

	29
1	2.9
2	5.8
3	8.7
4	11.6
5	14.5
6	17.4
7	20.3
8	23.2
9	26.1

	28	27
1	2.8	2.7
2	5.6	5.4
3	8.4	8.1
4	11.2	10.8
5	14.0	13.5
6	16.8	16.2
7	19.6	18.9
8	22.4	21.6
9	25.2	24.3

	26
1	2.6
2	5.2
3	7.8
4	10.4
5	13.0
6	15.6
7	18.2
8	20.8
9	23.4

	25	24
1	2.5	2.4
2	5.0	4.8
3	7.5	7.2
4	10.0	9.6
5	12.5	12.0
6	15.0	14.4
7	17.5	16.8
8	20.0	19.2
9	22.5	21.6

P P

*175° 265° *355° **85°** L Cos L Sin

L Tan **4°** ***94° 184° *274°**

′	0″	10″	20″	30″	40″	50″	60″	′
0	8.84 464	495̄	525̄	555	585	615	646	59
1	646	676	706	736	766	796	826	58
2	826	856	886	916	946	976	*006	57
3	8.85 006	036	065	095	125	155̄	185̄	56
4	185̄	214	244	274	304	333	363	55
5	363	392	422	452	481	511	540	54
6	540	570	599	629	658	688	717	53
7	717	747	776	805	835̄	864	893	52
8	893	922	952	981	*010	*039	*069	51
9	8.86 069	098	127	156	185	214	243	50
10	243	272	301	330	359	388	417	49
11	417	447	475	504	533	562	591	48
12	591	619	648	677	706	734	763	47
13	763	792	821	849	878	907	935	46
14	935	964	992	*021	*049	*078	*106	45
15	8.87 106	135̄	163	192	220	249	277	44
16	277	305	334	362	390	419	447	43
17	447	475	503	532	560	588	616	42
18	616	644	673	701	729	757	785̄	41
19	785̄	813	841	869	897	925̄	953	40
20	953	981	*009	*037	*065̄	*092	*120	39
21	8.88 120	148	176	204	231	259	287	38
22	287	315̄	342	370	398	425	453	37
23	453	481	508	536	563	591	618	36
24	618	646	674	701	728	756	783	35
25	783	811	838	866	893	920	948	34
26	948	975̄	*002	*029	*057	*084	111	33
27	8.89 111	138	166	193	220	247	274	32
28	274	301	328	355	383	410	437	31
29	437	464	491	518	545̄	571	598	30
30	598	625	652	679	706	733	760	29
31	760	786	813	840	867	894	920	28
32	920	947	974	*000	*027	*054	*080	27
33	8.90 080	107	134	160	187	213	240	26
34	240	266	293	319	346	372	399	25
35	399	425	451	478	504	531	557	24
36	557	583	610	636	662	688	715̄	23
37	715̄	741	767	793	820	846	872	22
38	872	898	924	950	976	*002	*029	21
39	8.91 029	055̄	081	107	133	159	185̄	20
40	185̄	211	236	262	288	314	340	19
41	340	366	392	418	443	469	495	18
42	495	521	547	572	598	624	650	17
43	650	675	701	727	752	778	803	16
44	803	829	855̄	880	906	931	957	15
45	957	982	*008	*033	*059	*084	*110	14
46	8.92 110	135̄	160	186	211	237	262	13
47	262	287	313	338	363	388	414	12
48	414	439	464	489	515̄	540	565̄	11
49	565̄	590	615	640	665	691	716	10
50	716	741	766	791	816	841	866	9
51	866	891	916	941	966	991	*016	8
52	8.93 016	040	065	090	115	140	165̄	7
53	165̄	190	214	239	264	289	313	6
54	313	338	363	388	412	437	462	5
55	462	486	511	536	560	585̄	609	4
56	609	634	658	683	707	732	756	3
57	756	781	805	830	854	879	903	2
58	903	928	952	976	*001	*025	*049	1
59	8.94 049	074	098	122	147	171	195	0
	60″	50″	40″	30″	20″	10″	0″	′

P P

	31	30
1	3.1	3.0
2	6.2	6.0
3	9.3	9.0
4	12.4	12.0
5	15.5	15.0
6	18.6	18.0
7	21.7	21.0
8	24.8	24.0
9	27.9	27.0

	29
1	2.9
2	5.8
3	8.7
4	11.6
5	14.5
6	17.4
7	20.3
8	23.2
9	26.1

	28	27
1	2.8	2.7
2	5.6	5.4
3	8.4	8.1
4	11.2	10.8
5	14.0	13.5
6	16.8	16.2
7	19.6	18.9
8	22.4	21.6
9	25.2	24.3

	26
1	2.6
2	5.2
3	7.8
4	10.4
5	13.0
6	15.6
7	18.2
8	20.8
9	23.4

	25	24
1	2.5	2.4
2	5.0	4.8
3	7.5	7.2
4	10.0	9.6
5	12.5	12.0
6	15.0	14.4
7	17.5	16.8
8	20.0	19.2
9	22.5	21.6

P P

***175° 265° *355°** **85°** **L Cot**

L Cos L Sin **5°** *95° 185° *275°

9.99	′	0″	10″	20″	30″	40″	50″	60″		
834	0	8.94 030	054	078	102	126	150	174	59	833
833	1	174	198	222	246	270	294	317	58	832
832	2	317	341	365	389	413	437	461	57	831
831	3	461	484	508	532	556	580	603	56	830
830	4	603	627	651	675̄	698	722	746	55	829
829	5	746	769	793	817	840	864	887	54	828
828	6	887	911	935̄	958	982	*005	*029	53	827
827	7	8.95 029	052	076	099	123	146	170	52	825
825	8	170	193	216	240	263	287	310	51	824
824	9	310	333	357	380	403	427	450̄	50	823
823	10	450̄	473	496	520	543	566	589	49	822
822	11	589	613	636	659	682	705	728	48	821
821	12	728	752	775̄	798	821	844	867	47	820
820	13	867	890	913	936	959	982	*005	46	819
819	14	8.96 005	028	051	074	097	120	143	45	817
817	15	143	166	189	212	234	257	280	44	816
816	16	280	303	326	349	371	394	417	43	815
815	17	417	440	462	485	508	531	553	42	814
814	18	553	576	599	621	644	667	689	41	813
813	19	689	712	735̄	757	780	802	825̄	40	812
812	20	825̄	847	870	892	915̄	937	960	39	810
810	21	960	982	*005̄	*027	*050̄	*072	*095̄	38	809
809	22	8.97 095̄	117	139	162	184	207	229	37	808
808	23	229	251	274	296	318	341	363	36	807
807	24	363	385	407	430	452	474	496	35	806
806	25	496	518	541	563	585̄	607	629	34	804
804	26	629	651	674	696	718	740	762	33	803
803	27	762	784	806	828	850	872	894	32	802
802	28	894	916	938	960	982	*004	*026	31	801
801	29	8.98 026	048	070	092	114	135	157	30	800
800	30	157	179	201	223	245̄	266	288	29	798
798	31	288	310	332	354	375	397	419	28	797
797	32	419	441	462	484	506	527	549	27	796
796	33	549	571	592	614	636	657	679	26	795̄
795̄	34	679	701	722	744	765	787	808	25	793
793	35	808	830	851	873	894	916	937	24	792
792	36	937	959	980	*002	*023	*045̄	*066	23	791
791	37	8.99 066	087	109	130	152	173	194	22	790
790	38	194	216	237	258	280	301	322	21	788
788	39	322	343	365̄	386	407	428	450̄	20	787
787	40	450̄	471	492	513	534	556	577	19	786
786	41	577	598	619	640	661	682	704	18	785̄
785̄	42	704	725̄	746	767	788	809	830	17	783
783	43	830	851	872	893	914	935̄	956	16	782
782	44	956	977	998	*019	*040	*061	*082	15	781
781	45	9.00 082	103	123	144	165	186	207	14	780
780	46	207	228	249	269	290	311	332	13	778
778	47	332	353	373	394	415̄	436	456	12	777
777	48	456	477	498	518	539	560	581	11	776
776	49	581	601	622	642	663	684	704	10	775̄
775̄	50	704	725̄	746	766	787	807	828	9	773
773	51	828	848	869	889	910	930	951	8	772
772	52	951	971	992	*012	*033	*053	*074	7	771
771	53	9.01 074	094	115̄	135̄	155	176	196	6	769
769	54	196	217	237	257	278	298	318	5	768
768	55	318	339	359	379	399	420	440	4	767
767	56	440	460	480	501	521	541	561	3	765
765	57	561	582	602	622	642	662	682	2	764
764	58	682	703	723	743	763	783	803	1	763
763	59	803	823	843	863	883	903	923	0	761
		60″	50″	40″	30″	20″	10″	0″	′	9.99

P P

	24
1	2.4
2	4.8
3	7.2
4	9.6
5	12.0
6	14.4
7	16.8
8	19.2
9	21.6

	23
1	2.3
2	4.6
3	6.9
4	9.2
5	11.5
6	13.8
7	16.1
8	18.4
9	20.7

	22
1	2.2
2	4.4
3	6.6
4	8.8
5	11.0
6	13.2
7	15.4
8	17.6
9	19.8

	21
1	2.1
2	4.2
3	6.3
4	8.4
5	10.5
6	12.6
7	14.7
8	16.8
9	18.9

	20
1	2.0
2	4.0
3	6.0
4	8.0
5	10.0
6	12.0
7	14.0
8	16.0
9	18.0

P P

*174° 264° *354° **84°** L Cos L Sin

L Tan **5°** *95° 185° *275°

′	0″	10″	20″	30″	40″	50″	60″	
0	8.94 195	219	244	268	292	316	340	59
1	340	365̄	389	413	437	461	485	58
2	485	509	533	557	581	606	630	57
3	630	654	678	702	725	749	773	56
4	773	797	821	845	869	893	917	55
5	917	941	964	988	*012	*036	*060	54
6	8.95 060	083	107	131	155̄	178	202	53
7	202	226	249	273	297	320	344	52
8	344	368	391	415̄	439	462	486	51
9	486	509	533	556	580	603	627	50
10	627	650	674	697	721	744	767	49
11	767	791	814	838	861	884	908	48
12	908	931	954	977	*001	*024	*047	47
13	8.96 047	071	094	117	140	163	187	46
14	187	210	233	256	279	302	325	45
15	325	349	372	395̄	418	441	464	44
16	464	487	510	533	556	579	602	43
17	602	625̄	648	671	694	717	739	42
18	739	762	785	808	831	854	877	41
19	877	899	922	945̄	968	991	*013	40
20	8.97 013	036	059	081	104	127	15̄0	39
21	15̄0	172	195̄	218	240	263	285	38
22	285	308	331	353	376	398	421	37
23	421	443	466	488	511	533	556	36
24	556	578	601	623	646	668	691	35
25	691	713	735	758	780	802	825̄	34
26	825̄	847	869	892	914	936	959	33
27	959	981	*003	*025	*048	*070	*092	32
28	8.98 092	114	136	159	181	203	225	31
29	225	247	269	291	314	336	358	30
30	358	380	402	424	446	468	490	29
31	490	512	534	556	578	600	622	28
32	622	644	666	687	709	731	753	27
33	753	775	797	819	841	862	884	26
34	884	906	928	95̄0	971	993	*015̄	25
35	8.99 015̄	037	058	080	102	123	145	24
36	145	167	188	210	232	253	275	23
37	275	297	318	340	361	383	405̄	22
38	405̄	426	448	469	491	512	534	21
39	534	555	577	598	620	641	662	20
40	662	684	705	727	748	769	791	19
41	791	812	834	855̄	876	898	919	18
42	919	940	961	983	*004	*025	*046	17
43	9.00 046	068	089	110	131	153	174	16
44	174	195̄	216	237	258	280	301	15
45	301	322	343	364	385	406	427	14
46	427	448	469	490	511	532	553	13
47	553	574	595	616	637	658	679	12
48	679	700	721	742	763	784	805̄	11
49	805̄	826	846	867	888	909	930	10
50	930	951	971	992	*013	*034	*055̄	9
51	9.01 055̄	075	096	117	138	158	179	8
52	179	200	220	241	262	282	303	7
53	303	324	344	365	386	406	427	6
54	427	447	468	489	509	530	550	5
55	550	571	591	612	632	653	673	4
56	673	694	714	735̄	755	776	796	3
57	796	816	837	857	878	898	918	2
58	918	939	959	979	*000	*020	*040	1
59	9.02 040	061	081	101	121	142	162	0
	60″	50″	40″	30″	20″	10″	0″	

P P	25	24	23	22	21	20
1	2.5	2.4	2.3	2.2	2.1	2.0
2	5.0	4.8	4.6	4.4	4.2	4.0
3	7.5	7.2	6.9	6.6	6.3	6.0
4	10.0	9.6	9.2	8.8	8.4	8.0
5	12.5	12.0	11.5	11.0	10.5	10.0
6	15.0	14.4	13.8	13.2	12.6	12.0
7	17.5	16.8	16.1	15.4	14.7	14.0
8	20.0	19.2	18.4	17.6	16.8	16.0
9	22.5	21.6	20.7	19.8	18.9	18.0

*174° 264° *354° **84°** L Cot

L Cos L Sin **6°** *96° 186° *276°

9.99	′	0″	10″	20″	30″	40″	50″	60″	′	9.99
761	0	9.01 923	943	964	984	*004	*024	*043	59	760
760	1	9.02 043	063	083	103	123	143	163	58	759
759	2	163	183	203	223	243	263	283	57	757
757	3	283	302	322	342	362	382	402	56	756
756	4	402	421	441	461	481	501	520	55	755̄
755̄	5	520	540	560	579	599	619	639	54	753
753	6	639	658	678	698	717	737	757	53	752
752	7	757	776	796	816	835	855̄	874	52	751
751	8	874	894	914	933	953	972	992	51	749
749	9	992	*011	*031	*050	*070	*089	*109	50	748
748	10	9.03 109	128	148	167	187	206	226	49	747
747	11	226	245	265̄	284	303	323	342	48	745
745	12	342	361	381	400	420	439	458	47	744
744	13	458	478	497	516	535	555̄	574	46	742
742	14	574	593	613	632	651	670	690	45	741
741	15	690	709	728	747	766	786	805̄	44	740
740	16	805̄	824	843	862	881	901	920	43	738
738	17	920	939	958	977	996	*015	*034	42	737
737	18	9.04 034	053	072	091	110	129	149	41	736
736	19	149	168	187	206	225̄	244	262	40	734
734	20	262	281	300	319	338	357	376	39	733
733	21	376	395	414	433	452	471	490	38	731
731	22	490	508	527	546	565̄	584	603	37	730
730	23	603	621	640	659	678	697	715	36	728
728	24	715	734	753	772	790	809	828	35	727
727	25	828	847	8 65	884	903	921	940	34	726
726	26	940	959	977	996	*015̄	*033	*052	33	724
724	27	9.05 052	071	089	108	126	145̄	164	32	723
723	28	164	182	201	219	238	256	275̄	31	721
721	29	275̄	293	312	330	349	367	386	30	720
720	30	386	404	423	441	460	478	497	29	718
718	31	497	515	533	552	570	589	607	28	717
717	32	607	625	644	662	681	699	717	27	716
716	33	717	736	754	772	791	809	827	26	714
714	34	827	845	864	882	900	918	937	25	713
713	35	937	955̄	973	991	*010	*028	*046	24	711
711	36	9.06 046	064	082	101	119	137	155	23	710
710	37	155	173	191	210	228	246	264	22	708
708	38	264	282	300	318	336	354	372	21	707
707	39	372	390	408	426	445̄	463	481	20	705
705	40	481	499	517	535̄	553	571	589	19	704
704	41	589	606	624	642	660	678	696	18	702
702	42	696	714	732	750̄	768	786	804	17	701
701	43	804	821	839	857	875	893	911	16	699
699	44	911	929	946	964	982	*000	*018	15	698
698	45	9.07 018	035	053	071	089	106	124	14	696
696	46	124	142	160	177	195	213	231	13	695̄
695̄	47	231	248	266	284	301	319	337	12	693
693	48	337	354	372	390	407	425̄	442	11	692
692	49	442	460	478	495	513	530	548	10	690
690	50	548	566	583	601	618	636	653	9	689
689	51	653	671	688	706	723	741	758	8	687
687	52	758	776	793	811	828	846	863	7	686
686	53	863	881	898	915	933	950	968	6	684
684	54	968	985	*002	*020	*037	*055̄	*072	5	683
683	55	9.08 072	089	107	124	141	159	176	4	681
681	56	176	193	211	228	245	262	280	3	680
680	57	280	297	314	331	349	366	383	2	678
678	58	383	400	418	435̄	452	469	486	1	677
677	59	486	504	521	538	555	572	589	0	675
		60″	50″	40″	30″	20″	10″	0″	′	9.99

P	21	20	19	18	17
1	2.1	2.0	1.9	1.8	1.7
2	4.2	4.0	3.8	3.6	3.4
3	6.3	6.0	5.7	5.4	5.1
4	8.4	8.0	7.6	7.2	6.8
5	10.5	10.0	9.5	9.0	8.5
6	12.6	12.0	11.4	10.8	10.2
7	14.7	14.0	13.3	12.6	11.9
8	16.8	16.0	15.2	14.4	13.6
9	18.9	18.0	17.1	16.2	15.3

*173° 263° *353° **83°** L Cos L Sin

TABLE OF THE LOGARITHMS

OF THE

TRIGONOMETRIC FUNCTIONS

FROM MINUTE TO MINUTE

0°

*90° 180° *270°

″	′	L Sin	d	C S	C T	L Tan	c d	L Cot	L Cos	
0	0	—∞				—∞		∞	0.00 000	60
60	1	6.46 373	30103	5.31 443	5.31 443	6.46 373	30103	3.53 627	0.00 000	59
120	2	6.76 476	17609	5.31 443	5.31 443	6.76 476	17609	3.23 524	0.00 000	58
180	3	6.94 08$\bar{5}$	12494	5.31 443	5.31 443	6.94 08$\bar{5}$	12494	3.05 915	0.00 000	57
240	4	7.06 579	9691	5.31 443	5.31 442	7.06 579	9691	2.93 421	0.00 000	56
300	5	7.16 270	7918	5.31 443	5.31 442	7.16 270	7918	2.83 730	0.00 000	55
360	6	7.24 188	6694	5.31 443	5.31 442	7.24 188	6694	2.75 812	0.00 000	54
420	7	7.30 882	5800	5.31 443	5.31 442	7.30 882	5800	2.69 118	0.00 000	53
480	8	7.36 682	5115	5.31 443	5.31 442	7.36 682	5115	2.63 318	0.00 000	52
540	9	7.41 797	4576	5.31 443	5.31 442	7.41 797	4576	2.58 203	0.00 000	51
600	10	7.46 373	4139	5.31 443	5.31 442	7.46 373	4139	2.53 627	0.00 000	50
660	11	7.50 512	3779	5.31 443	5.31 442	7.50 512	3779	2.49 488	0.00 000	49
720	12	7.54 291	3476	5.31 443	5.31 442	7.54 291	3476	2.45 709	0.00 000	48
780	13	7.57 767	3218	5.31 443	5.31 442	7.57 767	3219	2.42 233	0.00 000	47
840	14	7.60 985	2997	5.31 443	5.31 442	7.60 986	2996	2.39 014	0.00 000	46
900	15	7.63 982	2802	5.31 443	5.31 442	7.63 982	2803	2.36 018	0.00 000	45
960	16	7.66 784	2633	5.31 443	5.31 442	7.66 78$\bar{5}$	2633	2.33 215	0.00 000	44
1020	17	7.69 417	2483	5.31 443	5.31 442	7.69 418	2482	2.30 582	9.99 999	43
1080	18	7.71 900	2348	5.31 443	5.31 442	7.71 900	2348	2.28 100	9.99 999	42
1140	19	7.74 248	2227	5.31 443	5.31 442	7.74 248	2228	2.25 752	9.99 999	41
1200	20	7.76 475	2119	5.31 443	5.31 442	7.76 476	2119	2.23 524	9.99 999	40
1260	21	7.78 594	2021	5.31 443	5.31 442	7.78 595	2020	2.21 40$\bar{5}$	9.99 999	39
1320	22	7.80 61$\bar{5}$	1930	5.31 443	5.31 442	7.80 615	1931	2.19 38$\bar{5}$	9.99 999	38
1380	23	7.82 545	1848	5.31 443	5.31 442	7.82 546	1848	2.17 454	9.99 999	37
1440	24	7.84 393	1773	5.31 443	5.31 442	7.84 394	1773	2.15 606	9.99 999	36
1500	25	7.86 166	1704	5.31 443	5.31 442	7.86 167	1704	2.13 833	9.99 999	35
1560	26	7.87 870	1639	5.31 443	5.31 442	7.87 871	1639	2.12 129	9.99 999	34
1620	27	7.89 509	1579	5.31 443	5.31 442	7.89 510	1579	2.10 490	9.99 999	33
1680	28	7.91 088	1524	5.31 443	5.31 442	7.91 089	1524	2.08 911	9.99 999	32
1740	29	7.92 612	1472	5.31 443	5.31 441	7.92 613	1473	2.07 387	9.99 998	31
1800	30	7.94 084	1424	5.31 443	5.31 441	7.94 086	1424	2.05 914	9.99 998	30
1860	31	7.95 508	1379	5.31 443	5.31 441	7.95 510	1379	2.04 490	9.99 998	29
1920	32	7.96 887	1336	5.31 443	5.31 441	7.96 889	1336	2.03 111	9.99 998	28
1980	33	7.98 223	1297	5.31 443	5.31 441	7.98 225	1297	2.01 77$\bar{5}$	9.99 998	27
2040	34	7.99 520	1259	5.31 443	5.31 441	7.99 522	1259	2.00 478	9.99 998	26
2100	35	8.00 779	1223	5.31 443	5.31 441	8.00 781	1223	1.99 219	9.99 998	25
2160	36	8.02 002	1190	5.31 443	5.31 441	8.02 004	1190	1.97 996	9.99 998	24
2220	37	8.03 192	1158	5.31 443	5.31 441	8.03 194	1159	1.96 806	9.99 997	23
2280	38	8.04 350	1128	5.31 443	5.31 441	8.04 353	1128	1.95 647	9.99 997	22
2340	39	8.05 478	1100	5.31 443	5.31 441	8.05 481	1100	1.94 519	9.99 997	21
2400	40	8.06 578	1072	5.31 443	5.31 441	8.06 581	1072	1.93 419	9.99 997	20
2460	41	8.07 650	1046	5.31 444	5.31 440	8.07 653	1047	1.92 347	9.99 997	19
2520	42	8.08 696	1022	5.31 444	5.31 440	8.08 700	1022	1.91 300	9.99 997	18
2580	43	8.09 718	999	5.31 444	5.31 440	8.09 722	998	1.90 278	9.99 997	17
2640	44	8.10 717	976	5.31 444	5.31 440	8.10 720	976	1.89 280	9.99 996	16
2700	45	8.11 693	954	5.31 444	5.31 440	8.11 696	955	1.88 304	9.99 996	15
2760	46	8.12 647	934	5.31 444	5.31 440	8.12 651	934	1.87 349	9.99 996	14
2820	47	8.13 581	914	5.31 444	5.31 440	8.13 585	915	1 86 41$\bar{5}$	9.99 996	13
2880	48	8.14 495	896	5.31 444	5.31 440	8.14 $\bar{5}$00	895	1.85 500	9.99 996	12
2940	49	8.15 391	877	5.31 444	5.31 440	8.15 395	878	1.84 60$\bar{5}$	9.99 996	11
3000	50	8.16 268	860	5.31 444	5.31 439	8.16 273	860	1.83 727	9.99 995	10
3060	51	8.17 128	843	5.31 444	5.31 439	8.17 133	843	1.82 867	9.99 995	9
3120	52	8.17 971	827	5.31 444	5.31 439	8.17 976	828	1.82 024	9.99 995	8
3180	53	8.18 798	812	5.31 444	5.31 439	8.18 804	812	1.81 196	9.99 99$\bar{5}$	7
3240	54	8.19 610	797	5.31 444	5.31 439	8.19 616	797	1.80 384	9.99 99$\bar{5}$	6
3300	55	8.20 407	782	5.31 444	5.31 439	8.20 413	782	1.79 587	9.99 994	5
3360	56	8.21 189	769	5.31 444	5.31 439	8.21 195	769	1.78 80$\bar{5}$	9.99 994	4
3420	57	8.21 958	755	5.31 44$\bar{5}$	5.31 439	8.21 964	756	1.78 036	9.99 994	3
3480	58	8.22 713	743	5.31 44$\bar{5}$	5.31 438	8.22 720	742	1.77 280	9.99 994	2
3540	59	8.23 456	730	5.31 44$\bar{5}$	5.31 438	8.23 462	730	1.76 538	9.99 994	1
3600	60	8.24 186		5.31 44$\bar{5}$	5.31 438	8.24 192		1.75 808	9.99 993	0
		L Cos	d			L Cot	c d	L Tan	L Sin	′

*179° 269° *359°

89°

1° *91° 181° *271°

″	′	L Sin	d	C S	C T	L Tan	c d	L Cot	L Cos	
3600	0	8.24 186	717	5.31 445̄	5.31 438	8.24 192	718	1.75 808	9.99 993	60
3660	1	8.24 903	706	5.31 445̄	5.31 438	8.24 910	706	1.75 090	9.99 993	59
3720	2	8.25 609	695	5.31 445̄	5.31 438	8.25 616	696	1.74 384	9.99 993	58
3780	3	8.26 304	684	5.31 445̄	5.31 438	8.26 312	684	1.73 688	9.99 993	57
3840	4	8.26 988	673	5.31 445	5.31 437	8.26 996	673	1.73 004	9.99 992	56
3900	5	8.27 661	663	5.31 445	5.31 437	8.27 669	663	1.72 331	9.99 992	55
3960	6	8.28 324	653	5.31 445	5.31 437	8.28 332	654	1.71 668	9.99 992	54
4020	7	8.28 977	644	5.31 445	5.31 437	8.28 986	643	1.71 014	9.99 992	53
4080	8	8.29 621	634	5.31 445	5.31 437	8.29 629	634	1.70 371	9.99 992	52
4140	9	8.30 255̄	624	5.31 445	5.31 437	8.30 263	625	1.69 737	9.99 991	51
4200	10	8.30 879	616	5.31 446	5.31 437	8.30 888	617	1.69 112	9.99 991	50
4260	11	8.31 495	608	5.31 446	5.31 436	8.31 505̄	607	1.68 495	9.99 991	49
4320	12	8.32 103	599	5.31 446	5.31 436	8.32 112	599	1.67 888	9.99 990	48
4380	13	8.32 702	590	5.31 446	5.31 436	8.32 711	591	1.67 289	9.99 990	47
4440	14	8.33 292	583	5.31 446	5.31 436	8.33 302	584	1.66 698	9.99 990	46
4500	15	8.33 875	575	5.31 446	5.31 436	8.33 886	575	1.66 114	9.99 990	45
4560	16	8.34 450	568	5.31 446	5.31 435	8.34 461	568	1.65 539	9.99 989	44
4620	17	8.35 018	560	5.31 446	5.31 435	8.35 029	561	1.64 971	9.99 989	43
4680	18	8.35 578	553	5.31 446	5.31 435	8.35 590	553	1.64 410	9.99 989	42
4740	19	8.36 131	547	5.31 446	5.31 435̄	8.36 143	546	1.63 857	9.99 989	41
4800	20	8.36 678	539	5.31 446	5.31 435̄	8.36 689	540	1.63 311	9.99 988	40
4860	21	8.37 217	533	5.31 447	5.31 434	8.37 229	533	1.62 771	9.99 988	39
4920	22	8.37 750̄	526	5.31 447	5.31 434	8.37 762	527	1.62 238	9.99 988	38
4980	23	8.38 276	520	5.31 447	5.31 434	8.38 289	520	1.61 711	9.99 987	37
5040	24	8.38 796	514	5.31 447	5.31 434	8.38 809	514	1.61 191	9.99 987	36
5100	25	8.39 310	508	5.31 447	5.31 434	8.39 323	509	1.60 677	9.99 987	35
5160	26	8.39 818	502	5.31 447	5.31 433	8.39 832	502	1.60 168	9.99 986	34
5220	27	8.40 320	496	5.31 447	5.31 433	8.40 334	496	1.59 666	9.99 986	33
5280	28	8.40 816	491	5.31 447	5.31 433	8.40 830	491	1.59 170	9.99 986	32
5340	29	8.41 307	485	5.31 447	5.31 433	8.41 321	486	1.58 679	9.99 985	31
5400	30	8.41 792	480	5.31 447	5.31 433	8.41 807	480	1.58 193	9.99 985	30
5460	31	8.42 272	474	5.31 448	5.31 432	8.42 287	475	1.57 713	9.99 985̄	29
5520	32	8.42 746	470	5.31 448	5.31 432	8.42 762	470	1.57 238	9.99 984	28
5580	33	8.43 216	464	5.31 448	5.31 432	8.43 232	464	1.56 768	9.99 984	27
5640	34	8.43 680	459	5.31 448	5.31 432	8.43 696	460	1.56 304	9.99 984	26
5700	35	8.44 139	455	5.31 448	5.31 431	8.44 156	455	1.55 844	9.99 983	25
5760	36	8.44 594	450	5.31 448	5.31 431	8.44 611	450	1.55 389	9.99 983	24
5820	37	8.45 044	445	5.31 448	5.31 431	8.45 061	446	1.54 939	9.99 983	23
5880	38	8.45 489	441	5.31 448	5.31 431	8.45 507	441	1.54 493	9.99 982	22
5940	39	8.45 930	436	5.31 449	5.31 431	8.45 948	437	1.54 052	9.99 982	21
6000	40	8.46 366	433	5.31 449	5.31 430	8.46 385̄	432	1.53 615	9.99 982	20
6060	41	8.46 799	427	5.31 449	5.31 430	8.46 817	428	1.53 183	9.99 981	19
6120	42	8.47 226	424	5.31 449	5.31 430	8.47 245	424	1.52 755̄	9.99 981	18
6180	43	8.47 650̄	419	5.31 449	5.31 430	8.47 669	420	1.52 331	9.99 981	17
6240	44	8.48 069	416	5.31 449	5.31 429	8.48 089	416	1.51 911	9.99 980	16
6300	45	8.48 485̄	411	5.31 449	5.31 429	8.48 505	412	1.51 495̄	9.99 980	15
6360	46	8.48 896	408	5.31 449	5.31 429	8.48 917	408	1.51 083	9.99 979	14
6420	47	8.49 304	404	5.31 450̄	5.31 428	8.49 325	404	1.50 675̄	9.99 979	13
6480	48	8.49 708	400	5.31 450̄	5.31 428	8.49 729	401	1.50 271	9.99 979	12
6540	49	8.50 108	396	5.31 450̄	5.31 428	8.50 130	397	1.49 870	9.99 978	11
6600	50	8.50 504	393	5.31 450̄	5.31 428	8.50 527	393	1.49 473	9.99 978	10
6660	51	8.50 897	390	5.31 450	5.31 427	8.50 920	390	1.49 080	9.99 977	9
6720	52	8.51 287	386	5.31 450	5.31 427	8.51 310	386	1.48 690	9.99 977	8
6780	53	8.51 673	382	5.31 450	5.31 427	8.51 696	383	1.48 304	9.99 977	7
6840	54	8.52 055	379	5.31 450	5.31 427	8.52 079	380	1.47 921	9.99 976	6
6900	55	8.52 434	376	5.31 451	5.31 426	8.52 459	376	1.47 541	9.99 976	5
6960	56	8.52 810	373	5.31 451	5.31 426	8.52 835̄	373	1.47 165	9.99 975	4
7020	57	8.53 183	369	5.31 451	5.31 426	8.53 208	370	1.46 792	9.99 975̄	3
7080	58	8.53 552	367	5.31 451	5.31 425	8.53 578	367	1.46 422	9.99 974	2
7140	59	8.53 919	363	5.31 451	5.31 425	8.53 945̄	363	1.46 055	9.99 974	1
7200	60	8.54 282		5.31 451	5.31 425̄	8.54 308		1.45 692	9.99 974	0
		L Cos	d			L Cot	c d	L Tan	L Sin	′

*178° 268° *358° **88°**

2°

*92° 182° *272°

"	'	L Sin	d	C S	C T	L Tan	c d	L Cot	L Cos	
7200	0	8.54 282	360	5.31 451	5.31 425̄	8.54 308	361	1.45 692	9.99 974	60
7260	1	8.54 642	357	5.31 451	5.31 425̄	8.54 6[illegible]	358	1.45 331	9.99 973	59
7320	2	8.54 999	355	5.31 452	5.31 424	8.55 027	355	1.44 973	9.99 973	58
7380	3	8.55 354	351	5.31 452	5.31 424	8.55 382	352	1.44 618	9.99 972	57
7440	4	8.55 705	349	5.31 452	5.31 424	8.55 734	349	1.44 266	9.99 972	56
7500	5	8.56 054	346	5.31 452	5.31 423	8.56 083	346	1.43 917	9.99 971	55
7560	6	8.56 400	343	5.31 452	5.31 423	8.56 429	344	1.43 571	9.99 971	54
7620	7	8.56 743	341	5.31 452	5.31 423	8.56 773	341	1.43 227	9.99 970	53
7680	8	8.57 084	337	5 31 453	5.31 422	8.57 114	338	1.42 886	9.99 970	52
7740	9	8.57 421	336	5.31 453	5.31 422	8.57 452	336	1.42 548	9.99 969	51
7800	10	8.57 757	332	5.31 453	5.31 422	8.57 788	333	1.42 212	9.99 969	50
7860	11	8.58 089	330	5.31 453	5.31 421	8.58 121	330	1.41 879	9.99 968	49
7920	12	8.58 419	328	5.31 453	5.31 421	8.58 451	328	1.41 549	9.99 968	48
7980	13	8.58 747	325	5.31 453	5.31 421	8.58 779	326	1.41 221	9.99 967	47
8040	14	8.59 072	323	5.31 454	5.31 421	8.59 105	323	1.40 895̄	9.99 967	46
8100	15	8.59 395̄	320	5.31 454	5.31 420	8.59 428	321	1.40 572	9.99 967	45
8160	16	8.59 715	318	5.31 454	5.31 420	8.59 749	319	1.40 251	9.99 966	44
8220	17	8.60 033	316	5.31 454	5.31 420	8.60 068	316	1.39 932	9.99 966	43
8280	18	8.60 349	313	5.31 454	5.31 419	8.60 384	314	1.39 616	9.99 965̄	42
8340	19	8.60 662	311	5.31 454	5.31 419	8.60 698	311	1.39 302	9.99 964	41
8400	20	8.60 973	309	5.31 455̄	5.31 418	8.61 009	310	1.38 991	9.99 964	40
8460	21	8.61 282	307	5.31 455̄	5.31 418	8.61 319	307	1.38 681	9.99 963	39
8520	22	8.61 589	305	5.31 455̄	5.31 418	8.61 626	305	1.38 374	9.99 963	38
8580	23	8.61 894	302	5.31 455	5.31 417	8.61 931	303	1.38 069	9.99 962	37
8640	24	8.62 196	301	5.31 455	5.31 417	8.62 234	301	1.37 766	9.99 962	36
8700	25	8.62 497	298	5.31 455	5.31 417	8.62 535	299	1.37 465̄	9.99 961	35
8760	26	8.62 795̄	296	5.31 456	5.31 416	8.62 834	297	1.37 166	9.99 961	34
8820	27	8.63 091	294	5.31 456	5.31 416	8.63 131	295	1.36 869	9.99 960	33
8880	28	8.63 385	293	5.31 456	5.31 416	8.63 426	292	1.36 574	9.99 960	32
8940	29	8.63 678	290	5.31 456	5.31 415	8.63 718	291	1.36 282	9.99 959	31
9000	30	8.63 968	288	5.31 456	5.31 415̄	8.64 009	289	1.35 991	9.99 959	30
9060	31	8.64 256	287	5.31 456	5.31 415̄	8.64 298	287	1.35 702	9.99 958	29
9120	32	8.64 543	284	5.31 457	5.31 414	8.64 585	285	1.35 415̄	9.99 958	28
9180	33	8.64 827	283	5.31 457	5.31 414	8.64 870	284	1.35 130	9.99 957	27
9240	34	8.65 110	281	5.31 457	5.31 413	8.65 154	281	1.34 846	9.99 956	26
9300	35	8.65 391	279	5.31 457	5.31 413	8.65 435	280	1.34 565̄	9.99 956	25
9360	36	8.65 670	277	5.31 457	5.31 413	8.65 715̄	278	1.34 285	9.99 955	24
9420	37	8.65 947	276	5.31 458	5.31 412	8.65 993	276	1.34 007	9.99 955̄	23
9480	38	8.66 223	274	5.31 458	5.31 412	8.66 269	274	1.33 731	9.99 954	22
9540	39	8.66 497	272	5.31 458	5.31 412	8.66 543	273	1.33 457	9.99 954	21
9600	40	8.66 769	270	5.31 458	5.31 411	8.66 816	271	1.33 184	9.99 953	20
9660	41	8.67 039	269	5.31 458	5.31 411	8.67 087	269	1.32 913	9.99 952	19
9720	42	8.67 308	267	5.31 459	5.31 410	8.67 356	268	1.32 644	9.99 952	18
9780	43	8.67 575	266	5.31 459	5.31 410	8.67 624	266	1.32 376	9.99 951	17
9840	44	8.67 841	263	5.31 459	5.31 410	8.67 890	264	1.32 110	9.99 951	16
9900	45	8.68 104	263	5.31 459	5.31 409	8.68 154	263	1.31 846	9.99 950	15
9960	46	8.68 367	260	5.31 459	5.31 409	8.68 417	261	1.31 583	9.99 949	14
10020	47	8.68 627	259	5.31 460	5.31 408	8.68 678	260	1.31 322	9.99 949	13
10080	48	8.68 886	258	5.31 460	5.31 408	8.68 938	258	1.31 062	9.99 948	12
10140	49	8.69 144	256	5.31 460	5.31 408	8.69 196	257	1.30 804	9.99 948	11
10200	50	8.69 400	254	5.31 460	5.31 407	8.69 453	255	1.30 547	9.99 947	10
10260	51	8.69 654	253	5.31 460	5.31 407	8.69 708	254	1.30 292	9.99 946	9
10320	52	8.69 907	252	5.31 461	5.31 406	8.69 962	252	1.30 038	9.99 946	8
10380	53	8.70 159	250	5.31 461	5.31 406	8.70 214	251	1.29 786	9.99 945̄	7
10440	54	8.70 409	249	5.31 461	5.31 405	8.70 465̄	249	1.29 535	9.99 944	6
10500	55	8.70 658	247	5.31 461	5.31 405̄	8.70 714	248	1.29 286	9.99 944	5
10560	56	8.70 905̄	246	5.31 461	5.31 405̄	8.70 962	246	1.29 038	9.99 943	4
10620	57	8.71 151	244	5.31 462	5.31 404	8.71 208	245	1.28 792	9.99 942	3
10680	58	8.71 395	243	5.31 462	5.31 404	8.71 453	244	1.28 547	9.99 942	2
10740	59	8.71 638	242	5.31 462	5.31 403	8.71 697	243	1.28 303	9.99 941	1
10800	60	8.71 880		5.31 462	5.31 403	8.71 940		1.28 060	9.99 940	0
		L Cos	d			L Cot	c d	L Tan	L Sin	'

*177° 267° *357° 87°

′	L Sin	d	L Tan	c d	L Cot	L Cos	
0	8.71 880	240	8.71 940	241	1.28 060	9.99 940	60
1	8.72 120	239	8.72 181	239	1.27 819	9.99 940	59
2	8.72 359	238	8.72 420	239	1.27 580	9.99 939	58
3	8.72 597	237	8.72 659	237	1.27 341	9.99 938	57
4	8.72 834	235	8.72 896	236	1.27 104	9.99 938	56
5	8.73 069	234	8.73 132	234	1.26 868	9.99 937	55
6	8.73 303	232	8.73 366	234	1.26 634	9.99 936	54
7	8.73 535	232	8.73 600	232	1.26 400	9.99 936	53
8	8.73 767	230	8.73 832	231	1.26 168	9.99 935	52
9	8.73 997	229	8.74 063	229	1.25 937	9.99 934	51
10	8.74 226	228	8.74 292	229	1.25 708	9.99 934	50
11	8.74 454	226	8.74 521	227	1.25 479	9.99 933	49
12	8.74 680	226	8.74 748	226	1.25 252	9.99 932	48
13	8.74 906	224	8.74 974	225	1.25 026	9.99 932	47
14	8.75 130	223	8.75 199	224	1.24 801	9.99 931	46
15	8.75 353	222	8.75 423	222	1.24 577	9.99 930	45
16	8.75 575̄	220	8.75 645	222	1.24 355̄	9.99 929	44
17	8.75 795	220	8.75 867	220	1.24 133	9.99 929	43
18	8.76 015	219	8.76 087	219	1.23 913	9.99 928	42
19	8.76 234	217	8.76 306	219	1.23 694	9.99 927	41
20	8.76 451	216	8.76 525̄	217	1.23 475	9.99 926	40
21	8.76 667	216	8.76 742	216	1.23 258	9.99 926	39
22	8.76 883	214	8.76 958	215	1.23 042	9.99 925̄	38
23	8.77 097	213	8.77 173	214	1.22 827	9.99 924	37
24	8.77 310	212	8.77 387	213	1.22 613	9.99 923	36
25	8.77 522	211	8.77 600	211	1.22 400	9.99 923	35
26	8.77 733	210	8.77 811	211	1.22 189	9.99 922	34
27	8.77 943	209	8.78 022	210	1.21 978	9.99 921	33
28	8.78 152	208	8.78 232	209	1.21 768	9.99 920	32
29	8.78 360	208	8.78 441	208	1.21 559	9.99 920	31
30	8.78 568	206	8.78 649	206	1.21 351	9.99 919	30
31	8.78 774	205	8.78 855	206	1.21 145̄	9.99 918	29
32	8.78 979	204	8.79 061	205	1.20 939	9.99 917	28
33	8.79 183	203	8.79 266	204	1.20 734	9.99 917	27
34	8.79 386	202	8.79 470	203	1.20 530	9.99 916	26
35	8.79 588	201	8.79 673	202	1.20 327	9.99 915	25
36	8.79 789	201	8.79 875	201	1.20 125̄	9.99 914	24
37	8.79 990	199	8.80 076	201	1.19 924	9.99 913	23
38	8.80 189	199	8.80 277	199	1.19 723	9.99 913	22
39	8.80 388	197	8.80 476	198	1.19 524	9.99 912	21
40	8.80 585	197	8.80 674	198	1.19 326	9.99 911	20
41	8.80 782	196	8.80 872	196	1.19 128	9.99 910	19
42	8.80 978	195	8.81 068	196	1.18 932	9.99 909	18
43	8.81 173	194	8.81 264	195	1.18 736	9.99 909	17
44	8.81 367	193	8.81 459	194	1.18 541	9.99 908	16
45	8.81 560	192	8.81 653	193	1.18 347	9.99 907	15
46	8.81 752	192	8.81 846	192	1.18 154	9.99 906	14
47	8.81 944	190	8.82 038	192	1.17 962	9.99 905	13
48	8.82 134	190	8.82 230	190	1.17 770	9.99 904	12
49	8.82 324	189	8.82 420	190	1.17 580	9.99 904	11
50	8.82 513	188	8.82 610	189	1.17 390	9.99 903	10
51	8.82 701	187	8.82 799	188	1.17 201	9.99 902	9
52	8.82 888	187	8.82 987	188	1.17 013	9.99 901	8
53	8.83 075̄	186	8.83 175̄	186	1.16 825	9.99 900	7
54	8.83 261	185	8.83 361	186	1.16 639	9.99 899	6
55	8.83 446	184	8.83 547	185	1.16 453	9.99 898	5
56	8.83 630	183	8.83 732	184	1.16 268	9.99 898	4
57	8.83 813	183	8.83 916	184	1.16 084	9.99 897	3
58	8.83 996	181	8.84 100	182	1.15 900	9.99 896	2
59	8.84 177	181	8.84 282	182	1.15 718	9.99 895̄	1
60	8.84 358		8.84 464		1.15 536	9.99 894	0
	L Cos	d	L Cot	c d	L Tan	L Sin	′

P P

	241	239	237	235	234
1	4.0	4.0	4.0	3.9	3.9
2	8.0	8.0	7.9	7.8	7.8
3	12.0	12.0	11.8	11.8	11.7
4	16.1	15.9	15.8	15.7	15.6
5	20.1	19.9	19.8	19.6	19.5
6	24.1	23.9	23.7	23.5	23.4
7	28.1	27.9	27.6	27.4	27.3
8	32.1	31.9	31.6	31.3	31.2
9	36.2	35.8	35.6	35.2	35.1
10	40.2	39.8	39.5	39.2	39.0
20	80.3	79.7	79.0	78.3	78.0
30	120.5	119.5	118.5	117.5	117.0
40	160.7	159.3	158.0	156.7	156.0
50	200.8	199.2	197.5	195.8	195.0

	232	229	227	225	223
1	3.9	3.8	3.8	3.8	3.7
2	7.7	7.6	7.6	7.5	7.4
3	11.6	11.4	11.4	11.2	11.2
4	15.5̄	15.3	15.1	15.0	14.9
5	19.3	19.1	18.9	18.8	18.6
6	23.2	22.9	22.7	22.5	22.3
7	27.1	26.7	26.5̄	26.2	26.0
8	30.9	30.5	30.3	30.0	29.7
9	34.8	34.4	34.0	33.8	33.4
10	38.7	38.2	37.8	37.5	37.2
20	77.3	76.3	75.7	75.0	74.3
30	116.0	114.5	113.5	112.5	111.5
40	154.7	152.7	151.3	150.0	148.7
50	193.3	190.8	189.2	187.5	185.8

	222	220	217	215	213
1	3.7	3.7	3.6	3.6	3.6
2	7.4	7.3	7.2	7.2	7.1
3	11.1	11.0	10.8	10.8	10.6
4	14.8	14.7	14.5̄	14.3	14.2
5	18.5	18.3	18.1	17.9	17.8
6	22.2	22.0	21.7	21.5	21.3
7	25.9	25.7	25.3	25.1	24.8
8	29.6	29.3	28.9	28.7	28.4
9	33.3	33.0	32.6	32.2	32.0
10	37.0	36.7	36.2	35.8	35.5
20	74.0	73.3	72.3	71.7	71.0
30	111.0	110.0	108.5	107.5	106.5
40	148.0	146.7	144.7	143.3	142.0
50	185.0	183.3	180.8	179.2	177.5

	211	208	206	203	201
1	3.5	3.5̄	3.4	3.4	3.4
2	7.0	6.9	6.9	6.8	6.7
3	10.6	10.4	10.3	10.2	10.0
4	14.1	13.9	13.7	13.5	13.4
5	17.6	17.3	17.2	16.9	16.8
6	21.1	20.8	20.6	20.3	20.1
7	24.6	24.3	24.0	23.7	23.4
8	28.1	27.7	27.5̄	27.1	26.8
9	31.6	31.2	30.9	30.4	30.2
10	35.2	34.7	34.3	33.8	33.5
20	70.3	69.3	68.7	67.7	67.0
30	105.5	104.0	103.0	101.5	100.5
40	140.7	138.7	137.3	135.3	134.0
50	175.8	173.3	171.7	169.2	167.5

	199	197	195	193	192
1	3.3	3.3	3.2	3.2	3.2
2	6.6	6.6	6.5	6.4	6.4
3	10.0	9.8	9.8	9.6	9.6
4	13.3	13.1	13.0	12.9	12.8
5	16.6	16.4	16.2	16.1	16.0
6	19.9	19.7	19.5	19.3	19.2
7	23.2	23.0	22.8	22.5	22.4
8	26.5	26.3	26.0	25.7	25.6
9	29.8	29.6	29.2	29.0	28.8
10	33.2	32.8	32.5	32.2	32.0
20	66.3	65.7	65.0	64.3	64.0
30	99.5	98.5	97.5	96.5	96.0
40	132.7	131.3	130.0	128.7	128.0
50	165.8	164.2	162.5	160.8	160.0

	189	187	185	183	181
1	3.2	3.1	3.1	3.0	3.0
2	6.3	6.2	6.2	6.1	6.0
3	9.4	9.4	9.2	9.2	9.0
4	12.6	12.5̄	12.3	12.2	12.1
5	15.8	15.6	15.4	15.2	15.1
6	18.9	18.7	18.5	18.3	18.1
7	22.0	21.8	21.6	21.4	21.1
8	25.2	24.9	24.7	24.4	24.1
9	28.4	28.0	27.8	27.4	27.2
10	31.5	31.2	30.8	30.5	30.2
20	63.0	62.3	61.7	61.0	60.3
30	94.5	93.5	92.5	91.5	90.5
40	126.0	124.7	123.3	122.0	120.7
50	157.5	155.8	154.2	152.5	150.8

P P

′	L Sin	d	L Tan	c d	L Cot	L Cos	′
0	8.84 358	181	8.84 464	182	1.15 536	9.99 894	60
1	8.84 539	179	8.84 646	180	1.15 354	9.99 893	59
2	8.84 718	179	8.84 826	180	1.15 174	9.99 892	58
3	8.84 897	178	8.85 006	179	1.14 994	9.99 891	57
4	8.85 075	177	8.85 185	178	1.14 815	9.99 891	56
5	8.85 252	177	8.85 363	177	1.14 637	9.99 890	55
6	8.85 429	176	8.85 540	177	1.14 460	9.99 889	54
7	8.85 605	175	8.85 717	176	1.14 283	9.99 888	53
8	8.85 780	175	8.85 893	176	1.14 107	9.99 887	52
9	8.85 955	173	8.86 069	174	1.13 931	9.99 886	51
10	8.86 128	173	8.86 243	174	1.13 757	9.99 885	50
11	8.86 301	173	8.86 417	174	1.13 583	9.99 884	49
12	8.86 474	171	8.86 591	172	1.13 409	9.99 883	48
13	8.86 645	171	8.86 763	172	1.13 237	9.99 882	47
14	8.86 816	171	8.86 935	171	1.13 065	9.99 881	46
15	8.86 987	169	8.87 106	171	1.12 894	9.99 880	45
16	8.87 156	169	8.87 277	170	1.12 723	9.99 879	44
17	8.87 325	169	8.87 447	169	1.12 553	9.99 879	43
18	8.87 494	167	8.87 616	169	1.12 384	9.99 878	42
19	8.87 661	168	8.87 785	168	1.12 215	9.99 877	41
20	8.87 829	166	8.87 953	167	1.12 047	9.99 876	40
21	8.87 995	166	8.88 120	167	1.11 880	9.99 875	39
22	8.88 161	165	8.88 287	166	1.11 713	9.99 874	38
23	8.88 326	164	8.88 453	165	1.11 547	9.99 873	37
24	8.88 490	164	8.88 618	165	1.11 382	9.99 872	36
25	8.88 654	163	8.88 783	165	1.11 217	9.99 871	35
26	8.88 817	163	8.88 948	163	1.11 052	9.99 870	34
27	8.88 980	162	8.89 111	163	1.10 889	9.99 869	33
28	8.89 142	162	8.89 274	163	1.10 726	9.99 868	32
29	8.89 304	160	8.89 437	161	1.10 563	9.99 867	31
30	8.89 464	161	8.89 598	162	1.10 402	9.99 866	30
31	8.89 625	159	8.89 760	160	1.10 240	9.99 865	29
32	8.89 784	159	8.89 920	160	1.10 080	9.99 864	28
33	8.89 943	159	8.90 080	160	1.09 920	9.99 863	27
34	8.90 102	158	8.90 240	159	1.09 760	9.99 862	26
35	8.90 260	157	8.90 399	158	1.09 601	9.99 861	25
36	8.90 417	157	8.90 557	158	1.09 443	9.99 860	24
37	8.90 574	156	8.90 715	157	1.09 285	9.99 859	23
38	8.90 730	155	8.90 872	157	1.09 128	9.99 858	22
39	8.90 885	155	8.91 029	156	1.08 971	9.99 857	21
40	8.91 040	155	8.91 185	155	1.08 815	9.99 856	20
41	8.91 195	154	8.91 340	155	1.08 660	9.99 855	19
42	8.91 349	153	8.91 495	155	1.08 505	9.99 854	18
43	8.91 502	153	8.91 650	153	1.08 350	9.99 853	17
44	8.91 655	152	8.91 803	154	1.08 197	9.99 852	16
45	8.91 807	152	8.91 957	153	1.08 043	9.99 851	15
46	8.91 959	151	8.92 110	152	1.07 890	9.99 850	14
47	8.92 110	151	8.92 262	152	1.07 738	9.99 848	13
48	8.92 261	150	8.92 414	151	1.07 586	9.99 847	12
49	8.92 411	150	8.92 565	151	1.07 435	9.99 846	11
50	8.92 561	149	8.92 716	150	1.07 284	9.99 845	10
51	8.92 710	149	8.92 866	150	1.07 134	9.99 844	9
52	8.92 859	148	8.93 016	149	1.06 984	9.99 843	8
53	8.93 007	147	8.93 165	148	1.06 835	9.99 842	7
54	8.93 154	147	8.93 313	149	1.06 687	9.99 841	6
55	8.93 301	147	8.93 462	147	1.06 538	9.99 840	5
56	8.93 448	146	8.93 609	147	1.06 391	9.99 839	4
57	8.93 594	146	8.93 756	147	1.06 244	9.99 838	3
58	8.93 740	145	8.93 903	146	1.06 097	9.99 837	2
59	8.93 885	145	8.94 049	146	1.05 951	9.99 836	1
60	8.94 030		8.94 195		1.05 805	9.99 834	0
	L Cos	d	L Cot	c d	L Tan	L Sin	′

P P					
	182	181	179	178	177
1	3.0	3.0	3.0	3.0	3.0
2	6.1	6.0	6.0	5.9	5.9
3	9.1	9.0	9.0	8.9	8.8
4	12.1	12.1	11.9	11.9	11.8
5	15.2	15.1	14.9	14.8	14.8
6	18.2	18.1	17.9	17.8	17.7
7	21.2	21.1	20.9	20.8	20.6
8	24.3	24.1	23.9	23.7	23.6
9	27.3	27.2	26.8	26.7	26.6
10	30.3	30.2	29.8	29.7	29.5
20	60.7	60.3	59.7	59.3	59.0
30	91.0	90.5	89.5	89.0	88.5
40	121.3	120.7	119.3	118.7	118.0
50	151.7	150.8	149.2	148.3	147.5
	176	175	174	173	172
1	2.9	2.9	2.9	2.9	2.9
2	5.9	5.8	5.8	5.8	5.7
3	8.8	8.8	8.7	8.6	8.6
4	11.7	11.7	11.6	11.5	11.5
5	14.7	14.6	14.5	14.4	14.3
6	17.6	17.5	17.4	17.3	17.2
7	20.5	20.4	20.3	20.2	20.1
8	23.5	23.3	23.2	23.1	22.9
9	26.4	26.2	26.1	26.0	25.8
10	29.3	29.2	29.0	28.8	28.7
20	58.7	58.3	58.0	57.7	57.3
30	88.0	87.5	87.0	86.5	86.0
40	117.3	116.7	116.0	115.3	114.7
50	146.7	145.8	145.0	144.2	143.3
	171	170	169	168	167
1	2.8	2.8	2.8	2.8	2.8
2	5.7	5.7	5.6	5.6	5.6
3	8.6	8.5	8.4	8.4	8.4
4	11.4	11.3	11.3	11.2	11.1
5	14.2	14.2	14.1	14.0	13.9
6	17.1	17.0	16.9	16.8	16.7
7	20.0	19.8	19.7	19.6	19.5
8	22.8	22.7	22.5	22.4	22.3
9	25.6	25.5	25.4	25.2	25.0
10	28.5	28.3	28.2	28.0	27.8
20	57.0	56.7	56.3	56.0	55.7
30	85.5	85.0	84.5	84.0	83.5
40	114.0	113.3	112.7	112.0	111.3
50	142.5	141.7	140.8	140.0	139.2
	166	165	164	163	162
1	2.8	2.8	2.7	2.7	2.7
2	5.5	5.5	5.5	5.4	5.4
3	8.3	8.2	8.2	8.2	8.1
4	11.1	11.0	10.9	10.9	10.8
5	13.8	13.8	13.7	13.6	13.5
6	16.6	16.5	16.4	16.3	16.2
7	19.4	19.2	19.1	19.0	18.9
8	22.1	22.0	21.9	21.7	21.6
9	24.9	24.8	24.6	24.4	24.3
10	27.7	27.5	27.3	27.2	27.0
20	55.3	55.0	54.7	54.3	54.0
30	83.0	82.5	82.0	81.5	81.0
40	110.7	110.0	109.3	108.7	108.0
50	138.3	137.5	136.7	135.8	135.0
	161	160	159	158	157
1	2.7	2.7	2.6	2.6	2.6
2	5.4	5.3	5.3	5.3	5.2
3	8.0	8.0	8.0	7.9	7.8
4	10.7	10.7	10.6	10.5	10.5
5	13.4	13.3	13.2	13.2	13.1
6	16.1	16.0	15.9	15.8	15.7
7	18.8	18.7	18.6	18.4	18.3
8	21.5	21.3	21.2	21.1	20.9
9	24.2	24.0	23.8	23.7	23.6
10	26.8	26.7	26.5	26.3	26.2
20	53.7	53.3	53.0	52.7	52.3
30	80.5	80.0	79.5	79.0	78.5
40	107.3	106.7	106.0	105.3	104.7
50	134.2	133.3	132.5	131.7	130.8
	156	155	154	153	152
1	2.6	2.6	2.6	2.6	2.5
2	5.2	5.2	5.1	5.1	5.1
3	7.8	7.8	7.7	7.6	7.6
4	10.4	10.3	10.3	10.2	10.1
5	13.0	12.9	12.8	12.8	12.7
6	15.6	15.5	15.4	15.3	15.2
7	18.2	18.1	18.0	17.8	17.7
8	20.8	20.7	20.5	20.4	20.3
9	23.4	23.2	23.1	23.0	22.8
10	26.0	25.8	25.7	25.5	25.3
20	52.0	51.7	51.3	51.0	50.7
30	78.0	77.5	77.0	76.5	76.0
40	104.0	103.3	102.7	102.0	101.3
50	130.0	129.2	128.3	127.5	126.7
P P					

'	L Sin	d	L Tan	c d	L Cot	L Cos	'
0	8.94 030	144	8.94 195	145	1.05 805̄	9.99 834	60
1	8.94 174	143	8.94 340	145	1.05 660	9.99 833	59
2	8.94 317	144	8.94 485	145	1.05 515̄	9.99 832	58
3	8.94 461	142	8.94 630	143	1.05 370	9.99 831	57
4	8.94 603	143	8.94 773	144	1.05 227	9.99 830	56
5	8.94 746	141	8.94 917	143	1.05 083	9.99 829	55
6	8.94 887	142	8.95 060	142	1.04 940	9.99 828	54
7	8.95 029	141	8.95 202	142	1.04 798	9.99 827	53
8	8.95 170	140	8.95 344	142	1.04 656	9.99 825	52
9	8.95 310	140	8.95 486	141	1.04 514	9.99 824	51
10	8.95 450̄	139	8.95 627	140	1.04 373	9.99 823	50
11	8.95 589	139	8.95 767	141	1.04 233	9.99 822	49
12	8.95 728	139	8.95 908	139	1.04 092	9.99 821	48
13	8.95 867	138	8.96 047	140	1.03 953	9.99 820	47
14	8.96 005	138	8.96 187	138	1.03 813	9.99 819	46
15	8.96 143	137	8.96 325	139	1.03 675̄	9.99 817	45
16	8.96 280	137	8.96 464	138	1.03 536	9.99 816	44
17	8.96 417	136	8.96 602	137	1.03 398	9.99 815	43
18	8.96 553	136	8.96 739	138	1.03 261	9.99 814	42
19	8.96 689	136	8.96 877	136	1.03 123	9.99 813	41
20	8.96 825̄	135	8.97 013	137	1.02 987	9.99 812	40
21	8.96 960	135	8.97 150̄	135	1.02 850	9.99 810	39
22	8.97 095̄	134	8.97 285	136	1.02 715̄	9.99 809	38
23	8.97 229	134	8.97 421	135	1.02 579	9.99 808	37
24	8.97 363	133	8.97 556	135	1.02 444	9.99 807	36
25	8.97 496	133	8.97 691	134	1.02 309	9.99 806	35
26	8.97 629	133	8.97 825̄	134	1.02 175	9.99 804	34
27	8.97 762	132	8.97 959	133	1.02 041	9.99 803	33
28	8.97 894	132	8.98 092	133	1.01 908	9.99 802	32
29	8.98 026	131	8.98 225	133	1.01 775̄	9.99 801	31
30	8.98 157	131	8.98 358	132	1.01 642	9.99 800	30
31	8.98 288	131	8.98 490	132	1.01 510	9.99 798	29
32	8.98 419	130	8.98 622	131	1.01 378	9.99 797	28
33	8.98 549	130	8.98 753	131	1.01 247	9.99 796	27
34	8.98 679	129	8.98 884	131	1.01 116	9.99 795̄	26
35	8.98 808	129	8.99 015̄	130	1.00 985	9.99 793	25
36	8.98 937	129	8.99 145	130	1.00 855̄	9.99 792	24
37	8.99 066	128	8.99 275	130	1.00 725̄	9.99 791	23
38	8.99 194	128	8.99 405̄	129	1.00 595	9.99 790	22
39	8.99 322	128	8.99 534	128	1.00 466	9.99 788	21
40	8.99 450̄	127	8.99 662	129	1.00 338	9.99 787	20
41	8.99 577	127	8.99 791	128	1.00 209	9.99 786	19
42	8.99 704	126	8.99 919	127	1.00 081	9.99 785̄	18
43	8.99 830	126	9.00 046	128	0.99 954	9.99 783	17
44	8.99 956	126	9.00 174	127	0.99 826	9.99 782	16
45	9.00 082	125	9.00 301	126	0.99 699	9.99 781	15
46	9.00 207	125	9.00 427	126	0.99 573	9.99 780	14
47	9.00 332	124	9.00 553	126	0.99 447	9.99 778	13
48	9.00 456	125	9.00 679	126	0.99 321	9.99 777	12
49	9.00 581	123	9.00 805̄	125	0.99 195	9.99 776	11
50	9.00 704	124	9.00 930	125	0.99 070	9.99 775̄	10
51	9.00 828	123	9.01 055̄	124	0.98 945	9.99 773	9
52	9.00 951	123	9.01 179	124	0.98 821	9.99 772	8
53	9.01 074	122	9.01 303	124	0.98 697	9.99 771	7
54	9.01 196	122	9.01 427	123	0.98 573	9.99 769	6
55	9.01 318	122	9.01 550	123	0.98 450̄	9.99 768	5
56	9.01 440	121	9.01 673	123	0.98 327	9.99 767	4
57	9.01 561	121	9.01 796	122	0.98 204	9.99 765	3
58	9.01 682	121	9.01 918	122	0.98 082	9.99 764	2
59	9.01 803	120	9.02 040	122	0.97 960	9.99 763	1
60	9.01 923		9.02 162		0.97 838	9.99 761	0
	L Cos	d	L Cot	c d	L Tan	L Sin	'

P P

	151	149	148	147	146
1	2.5	2.5̄	2.5̄	2.4	2.4
2	5.0	5̄.0	4.9	4.9	4.9
3	7.6	7.4	7.4	7.4	7.3
4	10.1	9.9	9.9	9.8	9.7
5	12.6	12.4	12.3	12.2	12.2
6	15.1	14.9	14.8	14.7	14.6
7	17.6	17.4	17.3	17.2	17.0
8	20.1	19.9	19.7	19.6	19.5̄
9	22.6	22.4	22.2	22.0	21.9
10	25.2	24.8	24.7	24.5	24.3
20	50.3	49.7	49.3	49.0	48.7
30	75.5	74.5	74.0	73.5	73.0
40	100.7	99.3	98.7	98.0	97.3
50	125.8	124.2	123.3	122.5	121.7

	145	144	143	142	141
1	2.4	2.4	2.4	2.4	2.4
2	4.8	4.8	4.8	4.7	4.7
3	7.2	7.2	7.2	7.1	7.0
4	9.7	9.6	9.5	9.5̄	9.4
5	12.1	12.0	11.9	11.8	11.8
6	14.5	14.4	14.3	14.2	14.1
7	16.9	16.8	16.7	16.6	16.4
8	19.3	19.2	19.1	18.9	18.8
9	21.8	21.6	21.4	21.3	21.2
10	24.2	24.0	23.8	23.7	23.5
20	48.3	48.0	47.7	47.3	47.0
30	72.5	72.0	71.5	71.0	70.5
40	96.7	96.0	95.3	94.7	94.0
50	120.8	120.0	119.2	118.3	117.5

	140	139	138	137	136
1	2.3	2.3	2.3	2.3	2.3
2	4.7	4.6	4.6	4.6	4.5
3	7.0	7.0	6.9	6.8	6.8
4	9.3	9.3	9.2	9.1	9.1
5	11.7	11.6	11.5	11.4	11.3
6	14.0	13.9	13.8	13.7	13.6
7	16.3	16.2	16.1	16.0	15.9
8	18.7	18.5	18.4	18.3	18.1
9	21.0	20.8	20.7	20.6	20.4
10	23.3	23.2	23.0	22.8	22.7
20	46.7	46.3	46.0	45.7	45.3
30	70.0	69.5	69.0	68.5	68.0
40	93.3	92.7	92.0	91.3	90.7
50	116.7	115.8	115.0	114.2	113.3

	135	134	133	132	131
1	2.2	2.2	2.2	2.2	2.2
2	4.5	4.5̄	4.4	4.4	4.4
3	6.8	6.7	6.6	6.6	6.6
4	9.0	8.9	8.9	8.8	8.7
5	11.2	11.2	11.1	11.0	10.9
6	13.5	13.4	13.3	13.2	13.1
7	15.8	15.6	15.5	15.4	15.3
8	18.0	17.9	17.7	17.6	17.5̄
9	20.2	20.1	20.0	19.8	19.6
10	22.5	22.3	22.2	22.0	21.8
20	45.0	44.7	44.3	44.0	43.7
30	67.5	67.0	66.5	66.0	65.5
40	90.0	89.3	88.7	88.0	87.3
50	112.5	111.7	110.8	110.0	109.2

	130	129	128	127	126
1	2.2	2.2	2.1	2.1	2.1
2	4.3	4.3	4.3	4.2	4.2
3	6.5	6.4	6.4	6.4	6.3
4	8.7	8.6	8.5	8.5̄	8.4
5	10.8	10.8	10.7	10.6	10.5
6	13.0	12.9	12.8	12.7	12.6
7	15.2	15.0	14.9	14.8	14.7
8	17.3	17.2	17.1	16.9	16.8
9	19.5	19.4	19.2	19.0	18.9
10	21.7	21.5	21.3	21.2	21.0
20	43.3	43.0	42.7	42.3	42.0
30	65.0	64.5	64.0	63.5	63.0
40	86.7	86.0	85.3	84.7	84.0
50	108.3	107.5	106.7	105.8	105.0

	125	124	123	122	121
1	2.1	2.1	2.0	2.0	2.0
2	4.2	4.1	4.1	4.1	4.0
3	6.2	6.2	6.2	6.1	6.0
4	8.3	8.3	8.2	8.1	8.1
5	10.4	10.3	10.2	10.2	10.1
6	12.5	12.4	12.3	12.2	12.1
7	14.6	14.5̄	14.4	14.2	14.1
8	16.7	16.5	16.4	16.3	16.1
9	18.8	18.6	18.4	18.3	18.2
10	20.8	20.7	20.5	20.3	20.2
20	41.7	41.3	41.0	40.7	40.3
30	62.5	62.0	61.5	61.0	60.5
40	83.3	82.7	82.0	81.3	80.7
50	104.2	103.3	102.5	101.7	100.8

P P

′	L Sin	d	L Tan	c d	L Cot	L Cos	′
0	9.01 923	120	9.02 162	121	0.97 838	9.99 761	60
1	9.02 043	120	9.02 283	121	0.97 717	9.99 760	59
2	9.02 163	120	9.02 404	121	0.97 596	9.99 759	58
3	9.02 283	119	9.02 525	120	0.97 475̄	9.99 757	57
4	9.02 402	118	9.02 645	121	0.97 355̄	9.99 756	56
5	9.02 520	119	9.02 766	119	0.97 234	9.99 755̄	55
6	9.02 639	118	9.02 885	120	0.97 115̄	9.99 753	54
7	9.02 757	117	9.03 005̄	119	0.96 995	9.99 752	53
8	9.02 874	118	9.03 124	118	0.96 876	9.99 751	52
9	9.02 992	117	9.03 242	119	0.96 758	9.99 749	51
10	9.03 109	117	9.03 361	118	0.96 639	9.99 748	50
11	9.03 226	116	9.03 479	118	0.96 521	9.99 747	49
12	9.03 342	116	9.03 597	117	0.96 403	9.99 745	48
13	9.03 458	116	9.03 714	118	0.96 286	9.99 744	47
14	9.03 574	116	9.03 832	116	0.96 168	9.99 742	46
15	9.03 690	115	9.03 948	117	0.96 052	9.99 741	45
16	9.03 805̄	115	9.04 065	116	0.95 935̄	9.99 740	44
17	9.03 920	114	9.04 181	116	0.95 819	9.99 738	43
18	9.04 034	115	9.04 297	116	0.95 703	9.99 737	42
19	9.04 149	113	9.04 413	115	0.95 587	9.99 736	41
20	9.04 262	114	9.04 528	115	0.95 472	9.99 734	40
21	9.04 376	114	9.04 643	115	0.95 357	9.99 733	39
22	9.04 490	113	9.04 758	115	0.95 242	9.99 731	38
23	9.04 603	112	9.04 873	114	0.95 127	9.99 730	37
24	9.04 715	113	9.04 987	114	0.95 013	9.99 728	36
25	9.04 828	112	9.05 101	113	0.94 899	9.99 727	35
26	9.04 940	112	9.05 214	114	0.94 786	9.99 726	34
27	9.05 052	112	9.05 328	113	0.94 672	9.99 724	33
28	9.05 164	111	9.05 441	112	0.94 559	9.99 723	32
29	9.05 275̄	111	9.05 553	113	0.94 447	9.99 721	31
30	9.05 386	111	9.05 666	112	0.94 334	9.99 720	30
31	9.05 497	110	9.05 778	112	0.94 222	9.99 718	29
32	9.05 607	110	9.05 890	112	0.94 110	9.99 717	28
33	9.05 717	110	9.06 002	111	0.93 998	9.99 716	27
34	9.05 827	110	9.06 113	111	0.93 887	9.99 714	26
35	9.05 937	109	9.06 224	111	0.93 776	9.99 713	25
36	9.06 046	109	9.06 335̄	110	0.93 665	9.99 711	24
37	9.06 155	109	9.06 445	111	0.93 555̄	9.99 710	23
38	9.06 264	108	9.06 556	110	0.93 444	9.99 708	22
39	9.06 372	109	9.06 666	109	0.93 334	9.99 707	21
40	9.06 481	108	9.06 775	110	0.93 225̄	9.99 705	20
41	9.06 589	107	9.06 885̄	109	0.93 115	9.99 704	19
42	9.06 696	108	9.06 994	109	0.93 006	9.99 702	18
43	9.06 804	107	9.07 103	108	0.92 897	9.99 701	17
44	9.06 911	107	9.07 211	109	0.92 789	9.99 699	16
45	9.07 018	106	9.07 320	108	0.92 680	9.99 698	15
46	9.07 124	107	9.07 428	108	0.92 572	9.99 696	14
47	9.07 231	106	9.07 536	107	0.92 464	9.99 695̄	13
48	9.07 337	105	9.07 643	108	0.92 357	9.99 693	12
49	9.07 442	106	9.07 751	107	0.92 249	9.99 692	11
50	9.07 548	105	9.07 858	106	0.92 142	9.99 690	10
51	9.07 653	105	9.07 964	107	0.92 036	9.99 689	9
52	9.07 758	105	9.08 071	106	0.91 929	9.99 687	8
53	9.07 863	105	9.08 177	106	0.91 823	9.99 686	7
54	9.07 968	104	9.08 283	106	0.91 717	9.99 684	6
55	9.08 072	104	9.08 389	106	0.91 611	9.99 683	5
56	9.08 176	104	9.08 495̄	105	0.91 505	9.99 681	4
57	9.08 280	103	9.08 600	105	0.91 400	9.99 680	3
58	9.08 383	103	9.08 705	105	0.91 295̄	9.99 678	2
59	9.08 486	103	9.08 810	104	0.91 190	9.99 677	1
60	9.08 589		9.08 914		0.91 086	9.99 675	0
	L Cos	d	L Cot	c d	L Tan	L Sin	′

P P

	121	120	119	118
1	2.0	2.0	2.0	2.0
2	4.0	4.0	4.0	3.9
3	6.0	6.0	6.0	5.9
4	8.1	8.0	7.9	7.9
5	10.1	10.0	9.9	9.8
6	12.1	12.0	11.9	11.8
7	14.1	14.0	13.9	13.8
8	16.1	16.0	15.9	15.7
9	18.2	18.0	17.8	17.7
10	20.2	20.0	19.8	19.7
20	40.3	40.0	39.7	39.3
30	60.5	60.0	59.5	59.0
40	80.7	80.0	79.3	78.7
50	100.8	100.0	99.2	98.3

	117	116	115	114
1	2.0	1.9	1.9	1.9
2	3.9	3.9	3.8	3.8
3	5.8	5.8	5.8	5.7
4	7.8	7.7	7.7	7.6
5	9.8	9.7	9.6	9.5
6	11.7	11.6	11.5	11.4
7	13.6	13.5	13.4	13.3
8	15.6	15.5̄	15.3	15.2
9	17.6	17.4	17.2	17.1
10	19.5	19.3	19.2	19.0
20	39.0	38.7	38.3	38.0
30	58.5	58.0	57.5	57.0
40	78.0	77.3	76.7	76.0
50	97.5	96.7	95.8	95.0

	113	112	111	110
1	1.9	1.9	1.8	1.8
2	3.8	3.7	3.7	3.7
3	5.6	5.6	5.6	5.5
4	7.5	7.5̄	7.4	7.3
5	9.4	9.3	9.2	9.2
6	11.3	11.2	11.1	11.0
7	13.2	13.1	13.0	12.8
8	15.1	14.9	14.8	14.7
9	17.0	16.8	16.6	16.5
10	18.8	18.7	18.5	18.3
20	37.7	37.3	37.0	36.7
30	56.5	56.0	55.5	55.0
40	75.3	74.7	74.0	73.3
50	94.2	93.3	92.5	91.7

	109	108	107	106
1	1.8	1.8	1.8	1.8
2	3.6	3.6	3.6	3.5
3	5.4	5.4	5.4	5.3
4	7.3	7.2	7.1	7.1
5	9.1	9.0	8.9	8.8
6	10.9	10.8	10.7	10.6
7	12.7	12.6	12.5̄	12.4
8	14.5	14.4	14.3	14.1
9	16.4	16.2	16.0	15.9
10	18.2	18.0	17.8	17.7
20	36.3	36.0	35.7	35.3
30	54.5	54.0	53.5	53.0
40	72.7	72.0	71.3	70.7
50	90.8	90.0	89.2	88.3

P P

′	L Sin	d	L Tan	c d	L Cot	L Cos	
0	9.08 589	103	9.08 914	105	0.91 086	9.99 675	60
1	9.08 692	103	9.09 019	104	0.90 981	9.99 674	59
2	9.08 795̄	102	9.09 123	104	0.90 877	9.99 672	58
3	9.08 897	102	9.09 227	103	0.90 773	9.99 670	57
4	9.08 999	102	9.09 330	104	0.90 670	9.99 669	56
5	9.09 101	101	9.09 434	103	0.90 566	9.99 667	55
6	9.09 202	102	9.09 537	103	0.90 463	9.99 666	54
7	9.09 304	101	9.09 640	102	0.90 360	9.99 664	53
8	9.09 405̄	101	9.09 742	103	0.90 258	9.99 663	52
9	9.09 506	100	9.09 845̄	102	0.90 155	9.99 661	51
10	9.09 606	101	9.09 947	102	0.90 053	9.99 659	50
11	9.09 707	100	9.10 049	101	0.89 951	9.99 658	49
12	9.09 807	100	9.10 150	102	0.89 850	9.99 656	48
13	9.09 907	99	9.10 252	101	0.89 748	9.99 655̄	47
14	9.10 006	100	9.10 353	101	0.89 647	9.99 653	46
15	9.10 106	99	9.10 454	101	0.89 546	9.99 651	45
16	9.10 205̄	99	9.10 555̄	101	0.89 445	9.99 650	44
17	9.10 304	98	9.10 656	100	0.89 344	9.99 648	43
18	9.10 402	99	9.10 756	100	0.89 244	9.99 647	42
19	9.10 501	98	9.10 856	100	0.89 144	9.99 645̄	41
20	9.10 599	98	9.10 956	100	0.89 044	9.99 643	40
21	9.10 697	98	9.11 056	99	0.88 944	9.99 642	39
22	9.10 795	98	9.11 155	99	0.88 845̄	9.99 640	38
23	9.10 893	97	9.11 254	99	0.88 746	9.99 638	37
24	9.10 990	97	9.11 353	99	0.88 647	9.99 637	36
25	9.11 087	97	9.11 452	99	0.88 548	9.99 635	35
26	9.11 184	97	9.11 551	98	0.88 449	9.99 633	34
27	9.11 281	96	9.11 649	98	0.88 351	9.99 632	33
28	9.11 377	97	9.11 747	98	0.88 253	9.99 630	32
29	9.11 474	96	9.11 845	98	0.88 155̄	9.99 629	31
30	9.11 570	96	9.11 943	97	0.88 057	9.99 627	30
31	9.11 666	95	9.12 040	98	0.87 960	9.99 625	29
32	9.11 761	96	9.12 138	97	0.87 862	9.99 624	28
33	9.11 857	95	9.12 235̄	97	0.87 765	9.99 622	27
34	9.11 952	95	9.12 332	96	0.87 668	9.99 620	26
35	9.12 047	95	9.12 428	97	0.87 572	9.99 618	25
36	9.12 142	94	9.12 525̄	96	0.87 475	9.99 617	24
37	9.12 236	95	9.12 621	96	0.87 379	9.99 615	23
38	9.12 331	94	9.12 717	96	0.87 283	9.99 613	22
39	9.12 425̄	94	9.12 813	96	0.87 187	9.99 612	21
40	9.12 519	93	9.12 909	95	0.87 091	9.99 610	20
41	9.12 612	94	9.13 004	95	0.86 996	9.99 608	19
42	9.12 706	93	9.13 099	95	0.86 901	9.99 607	18
43	9.12 799	93	9.13 194	95	0.86 806	9.99 605̄	17
44	9.12 892	93	9.13 289	95	0.86 711	9.99 603	16
45	9.12 985	93	9.13 384	94	0.86 616	9.99 601	15
46	9.13 078	93	9.13 478	95	0.86 522	9.99 600	14
47	9.13 171	92	9.13 573	94	0.86 427	9.99 598	13
48	9.13 263	92	9.13 667	94	0.86 333	9.99 596	12
49	9.13 355	92	9.13 761	93	0.86 239	9.99 595̄	11
50	9.13 447	92	9.13 854	94	0.86 146	9.99 593	10
51	9.13 539	91	9.13 948	93	0.86 052	9.99 591	9
52	9.13 630	92	9.14 041	93	0.85 959	9.99 589	8
53	9.13 722	91	9.14 134	93	0.85 866	9.99 588	7
54	9.13 813	91	9.14 227	93	0.85 773	9.99 586	6
55	9.13 904	90	9.14 320	92	0.85 680	9.99 584	5
56	9.13 994	91	9.14 412	92	0.85 588	9.99 582	4
57	9.14 085	90	9.14 504	93	0.85 496	9.99 581	3
58	9.14 175	91	9.14 597	91	0.85 403	9.99 579	2
59	9.14 266	90	9.14 688	92	0.85 312	9.99 577	1
60	9.14 356		9.14 780		0.85 220	9.99 575	0
	L Cos	d	L Cot	c d	L Tan	L Sin	′

P P

	105	104	103	102
1	1.8	1.7	1.7	1.7
2	3.5	3.5̄	3.4	3.4
3	5.2	5.2	5.2	5.1
4	7.0	6.9	6.9	6.8
5	8.8	8.7	8.6	8.5
6	10.5	10.4	10.3	10.2
7	12.2	12.1	12.0	11.9
8	14.0	13.9	13.7	13.6
9	15.8	15.6	15.4	15.3
10	17.5	17.3	17.2	17.0
20	35.0	34.7	34.3	34.0
30	52.5	52.0	51.5	51.0
40	70.0	69.3	68.7	68.0
50	87.5	86.7	85.8	85.0

	101	100	99	98
1	1.7	1.7	1.6	1.6
2	3.4	3.3	3.3	3.3
3	5.0	5.0	5̄.0	4.9
4	6.7	6.7	6.6	6.5
5	8.4	8.3	8.2	8.2
6	10.1	10.0	9.9	9.8
7	11.8	11.7	11.6	11.4
8	13.5̄	13.3	13.2	13.1
9	15.2	15.0	14.8	14.7
10	16.8	16.7	16.5	16.3
20	33.7	33.3	33.0	32.7
30	50.5	50.0	49.5	49.0
40	67.3	66.7	66.0	65.3
50	84.2	83.3	82.5	81.7

	97	96	95	94
1	1.6	1.6	1.6	1.6
2	3.2	3.2	3.2	3.1
3	4.8	4.8	4.8	4.7
4	6.5̄	6.4	6.3	6.3
5	8.1	8.0	7.9	7.8
6	9.7	9.6	9.5	9.4
7	11.3	11.2	11.1	11.0
8	12.9	12.8	12.7	12.5
9	14.6	14.4	14.2	14.1
10	16.2	16.0	15.8	15.7
20	32.3	32.0	31.7	31.3
30	48.5	48.0	47.5	47.0
40	64.7	64.0	63.3	62.7
50	80.8	80.0	79.2	78.3

	93	92	91	90
1	1.6	1.5	1.5	1.5
2	3.1	3.1	3.0	3.0
3	4.6	4.6	4.6	4.5
4	6.2	6.1	6.1	6.0
5	7.8	7.7	7.6	7.5
6	9.3	9.2	9.1	9.0
7	10.8	10.7	10.6	10.5
8	12.4	12.3	12.1	12.0
9	14.0	13.8	13.6	13.5
10	15.5	15.3	15.2	15.0
20	31.0	30.7	30.3	30.0
30	46.5	46.0	45.5	45.0
40	62.0	61.3	60.7	60.0
50	77.5	76.7	75.8	75.0

P P

8°

*98° 188° *278°

′	L Sin	d	L Tan	c d	L Cot	L Cos	
0	9.14 356	89	9.14 780	92	0.85 220	9.99 575	60
1	9.14 445	90	9.14 872	91	0.85 128	9.99 574	59
2	9.14 535̄	89	9.14 963	91	0.85 037	9.99 572	58
3	9.14 624	90	9.15 054	91	0.84 946	9.99 570	57
4	9.14 714	89	9.15 145	91	0.84 855̄	9.99 568	56
5	9.14 803	88	9.15 236	91	0.84 764	9.99 566	55
6	9.14 891	89	9.15 327	90	0.84 673	9.99 565̄	54
7	9.14 980	89	9.15 417	91	0.84 583	9.99 563	53
8	9.15 069	88	9.15 508	90	0.84 492	9.99 561	52
9	9.15 157	88	9.15 598	90	0.84 402	9.99 559	51
10	9.15 245	88	9.15 688	89	0.84 312	9.99 557	50
11	9.15 333	88	9.15 777	90	0.84 223	9.99 556	49
12	9.15 421	87	9.15 867	89	0.84 133	9.99 554	48
13	9.15 508	88	9.15 956	90	0.84 044	9.99 552	47
14	9.15 596	87	9.16 046	89	0.83 954	9.99 550	46
15	9.15 683	87	9.16 135̄	89	0.83 865	9.99 548	45
16	9.15 770	87	9.16 224	88	0.83 776	9.99 546	44
17	9.15 857	87	9.16 312	89	0.83 688	9.99 545̄	43
18	9.15 944	86	9.16 401	88	0.83 599	9.99 543	42
19	9.16 030	86	9.16 489	88	0.83 511	9.99 541	41
20	9.16 116	87	9.16 577	88	0.83 423	9.99 539	40
21	9.16 203	86	9.16 665	88	0.83 335̄	9.99 537	39
22	9.16 289	85	9.16 753	88	0.83 247	9.99 535	38
23	9.16 374	86	9.16 841	87	0.83 159	9.99 533	37
24	9.16 460	85	9.16 928	88	0.83 072	9.99 532	36
25	9.16 545	86	9.17 016	87	0.82 984	9.99 530	35
26	9.16 631	85	9.17 103	87	0.82 897	9.99 528	34
27	9.16 716	85	9.17 190	87	0.82 810	9.99 526	33
28	9.16 801	85	9.17 277	86	0.82 723	9.99 524	32
29	9.16 886	84	9.17 363	87	0.82 637	9.99 522	31
30	9.16 970	85	9.17 450	86	0.82 550	9.99 520	30
31	9.17 055̄	84	9.17 536	86	0.82 464	9.99 518	29
32	9.17 139	84	9.17 622	86	0.82 378	9.99 517	28
33	9.17 223	84	9.17 708	86	0.82 292	9.99 515̄	27
34	9.17 307	84	9.17 794	86	0.82 206	9.99 513	26
35	9.17 391	83	9.17 880	85	0.82 120	9.99 511	25
36	9.17 474	84	9.17 965	86	0.82 035̄	9.99 509	24
37	9.17 558	83	9.18 051	85	0.81 949	9.99 507	23
38	9.17 641	83	9.18 136	85	0.81 864	9.99 505	22
39	9.17 724	83	9.18 221	85	0.81 779	9.99 503	21
40	9.17 807	83	9.18 306	85	0.81 694	9.99 501	20
41	9.17 890	83	9.18 391	84	0.81 609	9.99 499	19
42	9.17 973	82	9.18 475	85	0.81 525̄	9.99 497	18
43	9.18 055	82	9.18 560	84	0.81 440	9.99 495	17
44	9.18 137	83	9.18 644	84	0.81 356	9.99 494	16
45	9.18 220	82	9.18 728	84	0.81 272	9.99 492	15
46	9.18 302	81	9.18 812	84	0.81 188	9.99 490	14
47	9.18 383	82	9.18 896	83	0.81 104	9.99 488	13
48	9.18 465	82	9.18 979	84	0.81 021	9.99 486	12
49	9.18 547	81	9.19 063	83	0.80 937	9.99 484	11
50	9.18 628	81	9.19 146	83	0.80 854	9.99 482	10
51	9.18 709	81	9.19 229	83	0.80 771	9.99 480	9
52	9.18 790	81	9.19 312	83	0.80 688	9.99 478	8
53	9.18 871	81	9.19 395	83	0.80 605̄	9.99 476	7
54	9.18 952	81	9.19 478	83	0.80 522	9.99 474	6
55	9.19 033	80	9.19 561	82	0.80 439	9.99 472	5
56	9.19 113	80	9.19 643	82	0.80 357	9.99 470	4
57	9.19 193	80	9.19 725	82	0.80 275̄	9.99 468	3
58	9.19 273	80	9.19 807	82	0.80 193	9.99 466	2
59	9.19 353	80	9.19 889	82	0.80 111	9.99 464	1
60	9.19 433		9.19 971		0.80 029	9.99 462	0
	L Cos	d	L Cot	c d	L Tan	L Sin	′

P P	92	91	90
1	1.5	1.5	1.5
2	3.1	3.0	3.0
3	4.6	4.6	4.5
4	6.1	6.1	6.0
5	7.7	7.6	7.5
6	9.2	9.1	9.0
7	10.7	10.6	10.5
8	12.3	12.1	12.0
9	13.8	13.6	13.5
10	15.3	15.2	15.0
20	30.7	30.3	30.0
30	46.0	45.5	45.0
40	61.3	60.7	60.0
50	76.7	75.8	75.0
	89	**88**	**87**
1	1.5̄	1.5̄	1.4
2	3.0	2.9	2.9
3	4.4	4.4	4.4
4	5.9	5.9	5.8
5	7.4	7.3	7.2
6	8.9	8.8	8.7
7	10.4	10.3	10.2
8	11.9	11.7	11.6
9	13.4	13.2	13.0
10	14.8	14.7	14.5
20	29.7	29.3	29.0
30	44.5	44.0	43.5
40	59.3	58.7	58.0
50	74.2	73.3	72.5
	86	**85**	**84**
1	1.4	1.4	1.4
2	2.9	2.8	2.8
3	4.3	4.2	4.2
4	5.7	5.7	5.6
5	7.2	7.1	7.0
6	8.6	8.5	8.4
7	10.0	9.9	9.8
8	11.5̄	11.3	11.2
9	12.9	12.8	12.6
10	14.3	14.2	14.0
20	28.7	28.3	28.0
30	43.0	42.5	42.0
40	57.3	56.7	56.0
50	71.7	70.8	70.0
	83	**82**	**81**
1	1.4	1.4	1.4
2	2.8	2.7	2.7
3	4.2	4.1	4.0
4	5.5	5.5̄	5.4
5	6.9	6.8	6.8
6	8.3	8.2	8.1
7	9.7	9.6	9.4
8	11.1	10.9	10.8
9	12.4	12.3	12.2
10	13.8	13.7	13.5
20	27.7	27.3	27.0
30	41.5	41.0	40.5
40	55.3	54.7	54.0
50	69.2	68.3	67.5
P P			

*171° 261° *351°

81°

′	L Sin	d	L Tan	c d	L Cot	L Cos	
0	9.19 433	80	9.19 971	82	0.80 029	9.99 462	60
1	9.19 513	79	9.20 053	81	0.79 947	9.99 460	59
2	9.19 592	80	9.20 134	82	0.79 866	9.99 458	58
3	9.19 672	79	9.20 216	81	0.79 784	9.99 456	57
4	9.19 751	79	9.20 297	81	0.79 703	9.99 454	56
5	9.19 830	79	9.20 378	81	0.79 622	9.99 452	55
6	9.19 909	79	9.20 459	81	0.79 541	9.99 450	54
7	9.19 988	79	9.20 540	81	0.79 460	9.99 448	53
8	9.20 067	78	9.20 621	80	0.79 379	9.99 446	52
9	9.20 145	78	9.20 701	81	0.79 299	9.99 444	51
10	9.20 223	79	9.20 782	80	0.79 218	9.99 442	50
11	9.20 302	78	9.20 862	80	0.79 138	9.99 440	49
12	9.20 380	78	9.20 942	80	0.79 058	9.99 438	48
13	9.20 458	77	9.21 022	80	0.78 978	9.99 436	47
14	9.20 535	78	9.21 102	80	0.78 898	9.99 434	46
15	9.20 613	78	9.21 182	79	0.78 818	9.99 432	45
16	9.20 691	77	9.21 261	80	0.78 739	9.99 429	44
17	9.20 768	77	9.21 341	79	0.78 659	9.99 427	43
18	9.20 845	77	9.21 420	79	0.78 580	9.99 425	42
19	9.20 922	77	9.21 499	79	0.78 501	9.99 423	41
20	9.20 999	77	9.21 578	79	0.78 422	9.99 421	40
21	9.21 076	77	9.21 657	79	0.78 343	9.99 419	39
22	9.21 153	76	9.21 736	78	0.78 264	9.99 417	38
23	9.21 229	77	9.21 814	79	0.78 186	9.99 415̄	37
24	9.21 306	76	9.21 893	78	0.78 107	9.99 413	36
25	9.21 382	76	9.21 971	78	0.78 029	9.99 411	35
26	9.21 458	76	9.22 049	78	0.77 951	9.99 409	34
27	9.21 534	76	9.22 127	78	0.77 873	9.99 407	33
28	9.21 610	75	9.22 205	78	0.77 795̄	9.99 404	32
29	9.21 685	76	9.22 283	78	0.77 717	9.99 402	31
30	9.21 761	75	9.22 361	77	0.77 639	9.99 400	30
31	9.21 836	76	9.22 438	78	0.77 562	9.99 398	29
32	9.21 912	75	9.22 516	77	0.77 484	9.99 396	28
33	9.21 987	75	9.22 593	77	0.77 407	9.99 394	27
34	9.22 062	75	9.22 670	77	0.77 330	9.99 392	26
35	9.22 137	74	9.22 747	77	0.77 253	9.99 390	25
36	9.22 211	75	9.22 824	77	0.77 176	9.99 388	24
37	9.22 286	75	9.22 901	76	0.77 099	9.99 385	23
38	9.22 361	74	9.22 977	77	0.77 023	9.99 383	22
39	9.22 435̄	74	9.23 054	76	0.76 946	9.99 381	21
40	9.22 509	74	9.23 130	76	0.76 870	9.99 379	20
41	9.22 583	74	9.23 206	77	0.76 794	9.99 377	19
42	9.22 657	74	9.23 283	76	0.76 717	9.99 375̄	18
43	9.22 731	74	9.23 359	76	0.76 641	9.99 372	17
44	9.22 805̄	73	9.23 435̄	75	0.76 565	9.99 370	16
45	9.22 878	74	9.23 510	76	0.76 490	9.99 368	15
46	9.22 952	73	9.23 586	75	0.76 414	9.99 366	14
47	9.23 025	73	9.23 661	76	0.76 339	9.99 364	13
48	9.23 098	73	9.23 737	75	0.76 263	9.99 362	12
49	9.23 171	73	9.23 812	75	0.76 188	9.99 359	11
50	9.23 244	73	9.23 887	75	0.76 113	9.99 357	10
51	9.23 317	73	9.23 962	75	0.76 038	9.99 355	9
52	9.23 390	72	9.24 037	75	0.75 963	9.99 353	8
53	9.23 462	73	9.24 112	74	0.75 888	9.99 351	7
54	9.23 535̄	72	9.24 186	75	0.75 814	9.99 348	6
55	9.23 607	72	9.24 261	74	0.75 739	9.99 346	5
56	9.23 679	73	9.24 335	75	0.75 665̄	9.99 344	4
57	9.23 752	71	9.24 410	74	0.75 590	9.99 342	3
58	9.23 823	72	9.24 484	74	0.75 516	9.99 340	2
59	9.23 895	72	9.24 558	74	0.75 442	9.99 337	1
60	9.23 967		9.24 632		0.75 368	9.99 335	0
	L Cos	d	L Cot	c d	L Tan	L Sin	′

P P

	80	79	78	77
1	1.3	1.3	1.3	1.3
2	2.7	2.6	2.6	2.6
3	4.0	4.0	3.9	3.8
4	5.3	5.3	5.2	5.1
5	6.7	6.6	6.5	6.4
6	8.0	7.9	7.8	7.7
7	9.3	9.2	9.1	9.0
8	10.7	10.5	10.4	10.3
9	12.0	11.8	11.7	11.6
10	13.3	13.2	13.0	12.8
20	26.7	26.3	26.0	25.7
30	40.0	39.5	39.0	38.5
40	53.3	52.7	52.0	51.3
50	66.7	65.8	65.0	64.2

	76	75	74	73
1	1.3	1.2	1.2	1.2
2	2.5	2.5	2.5̄	2.4
3	3.8	3.8	3.7	3.6
4	5.1	5.0	4.9	4.9
5	6.3	6.2	6.2	6.1
6	7.6	7.5	7.4	7.3
7	8.9	8.8	8.6	8.5
8	10.1	10.0	9.9	9.7
9	11.4	11.2	11.1	11.0
10	12.7	12.5	12.3	12.2
20	25.3	25.0	24.7	24.3
30	38.0	37.5	37.0	36.5
40	50.7	50.0	49.3	48.7
50	63.3	62.5	61.7	60.8

	72	71	3	2
1	1.2	1.2	0.0	0.0
2	2.4	2.4	0.1	0.1
3	3.6	3.6	0.2	0.1
4	4.8	4.7	0.2	0.1
5	6.0	5.9	0.2	0.2
6	7.2	7.1	0.3	0.2
7	8.4	8.3	0.4	0.2
8	9.6	9.5̄	0.4	0.3
9	10.8	10.6	0.4	0.3
10	12.0	11.8	0.5	0.3
20	24.0	23.7	1.0	0.7
30	36.0	35.5	1.5	1.0
40	48.0	47.3	2.0	1.3
50	60.0	59.2	2.5	1.7

	3/79	3/78	3/77
0			
1	13.2	13.0	12.8
2	39.5	39.0	38.5
3	65.8	65.0	64.2

	3/76	3/75	3/74
0			
1	12.7	12.5	12.3
2	38.0	37.5	37.0
3	63.3	62.5	61.7

P P

10°

*100° 190° *280°

′	L Sin	d	L Tan	c d	L Cot	L Cos	d	′
0	9.23 967	72	9.24 632	74	0.75 368	9.99 335	2	60
1	9.24 039	71	9.24 706	73	0.75 294	9.99 333	2	59
2	9.24 110	71	9.24 779	74	0.75 221	9.99 331	3	58
3	9.24 181	72	9.24 853	73	0.75 147	9.99 328	2	57
4	9.24 253	71	9.24 926	74	0.75 074	9.99 326	2	56
5	9.24 324	71	9.25 000	73	0.75 000	9.99 324	2	55
6	9.24 395	71	9.25 073	73	0.74 927	9.99 322	3	54
7	9.24 466	70	9.25 146	73	0.74 854	9.99 319	2	53
8	9.24 536	71	9.25 219	73	0.74 781	9.99 317	2	52
9	9.24 607	70	9.25 292	73	0.74 708	9.99 315	2	51
10	9.24 677	71	9.25 365	72	0.74 635	9.99 313	3	50
11	9.24 748	70	9.25 437	73	0.74 563	9.99 310	2	49
12	9.24 818	70	9.25 510	72	0.74 490	9.99 308	2	48
13	9.24 888	70	9.25 582	73	0.74 418	9.99 306	2	47
14	9.24 958	70	9.25 655	72	0.74 345	9.99 304	3	46
15	9.25 028	70	9.25 727	72	0.74 273	9.99 301	2	45
16	9.25 098	70	9.25 799	72	0.74 201	9.99 299	2	44
17	9.25 168	69	9.25 871	72	0.74 129	9.99 297	3	43
18	9.25 237	70	9.25 943	72	0.74 057	9.99 294	2	42
19	9.25 307	69	9.26 015	71	0.73 985	9.99 292	2	41
20	9.25 376	69	9.26 086	72	0.73 914	9.99 290	2	40
21	9.25 445	69	9.26 158	71	0.73 842	9.99 288	3	39
22	9.25 514	69	9.26 229	72	0.73 771	9.99 285	2	38
23	9.25 583	69	9.26 301	71	0.73 699	9.99 283	2	37
24	9.25 652	69	9.26 372	71	0.73 628	9.99 281	3	36
25	9.25 721	69	9.26 443	71	0.73 557	9.99 278	2	35
26	9.25 790	68	9.26 514	71	0.73 486	9.99 276	2	34
27	9.25 858	69	9.26 585	70	0.73 415	9.99 274	3	33
28	9.25 927	68	9.26 655	71	0.73 345	9.99 271	2	32
29	9.25 995	68	9.26 726	71	0.73 274	9.99 269	2	31
30	9.26 063	68	9.26 797	70	0.73 203	9.99 267	3	30
31	9.26 131	68	9.26 867	70	0.73 133	9.99 264	2	29
32	9.26 199	68	9.26 937	71	0.73 063	9.99 262	2	28
33	9.26 267	68	9.27 008	70	0.72 992	9.99 260	3	27
34	9.26 335	68	9.27 078	70	0.72 922	9.99 257	2	26
35	9.26 403	67	9.27 148	70	0.72 852	9.99 255	3	25
36	9.26 470	68	9.27 218	70	0.72 782	9.99 252	2	24
37	9.26 538	67	9.27 288	69	0.72 712	9.99 250	2	23
38	9.26 605	67	9.27 357	70	0.72 643	9.99 248	3	22
39	9.26 672	67	9.27 427	69	0.72 573	9.99 245	2	21
40	9.26 739	67	9.27 496	70	0.72 504	9.99 243	2	20
41	9.26 806	67	9.27 566	69	0.72 434	9.99 241	3	19
42	9.26 873	67	9.27 635	69	0.72 365	9.99 238	2	18
43	9.26 940	67	9.27 704	69	0.72 296	9.99 236	3	17
44	9.27 007	66	9.27 773	69	0.72 227	9.99 233	2	16
45	9.27 073	67	9.27 842	69	0.72 158	9.99 231	2	15
46	9.27 140	66	9.27 911	69	0.72 089	9.99 229	3	14
47	9.27 206	67	9.27 980	69	0.72 020	9.99 226	2	13
48	9.27 273	66	9.28 049	68	0.71 951	9.99 224	3	12
49	9.27 339	66	9.28 117	69	0.71 883	9.99 221	2	11
50	9.27 405	66	9.28 186	68	0.71 814	9.99 219	2	10
51	9.27 471	66	9.28 254	69	0.71 746	9.99 217	3	9
52	9.27 537	65	9.28 323	68	0.71 677	9.99 214	2	8
53	9.27 602	66	9.28 391	68	0.71 609	9.99 212	3	7
54	9.27 668	66	9.28 459	68	0.71 541	9.99 209	2	6
55	9.27 734	65	9.28 527	68	0.71 473	9.99 207	3	5
56	9.27 799	65	9.28 595	67	0.71 405	9.99 204	2	4
57	9.27 864	66	9.28 662	68	0.71 338	9.99 202	2	3
58	9.27 930	65	9.28 730	68	0.71 270	9.99 200	3	2
59	9.27 995	65	9.28 798	67	0.71 202	9.99 197	2	1
60	9.28 060		9.28 865		0.71 135	9.99 195		0
	L Cos	d	L Cot	c d	L Tan	L Sin	d	′

P P

	74	73	72
1	1.2	1.2	1.2
2	2.5	2.4	2.4
3	3.7	3.6	3.6
4	4.9	4.9	4.8
5	6.2	6.1	6.0
6	7.4	7.3	7.2
7	8.6	8.5	8.4
8	9.9	9.7	9.6
9	11.1	11.0	10.8
10	12.3	12.2	12.0
20	24.7	24.3	24.0
30	37.0	36.5	36.0
40	49.3	48.7	48.0
50	61.7	60.8	60.0

	71	70	69
1	1.2	1.2	1.2
2	2.4	2.3	2.3
3	3.6	3.5	3.4
4	4.7	4.7	4.6
5	5.9	5.8	5.8
6	7.1	7.0	6.9
7	8.3	8.2	8.0
8	9.5	9.3	9.2
9	10.6	10.5	10.4
10	11.8	11.7	11.5
20	23.7	23.3	23.0
30	35.5	35.0	34.5
40	47.3	46.7	46.0
50	59.2	58.3	57.5

	68	67	66
1	1.1	1.1	1.1
2	2.3	2.2	2.2
3	3.4	3.4	3.3
4	4.5	4.5	4.4
5	5.7	5.6	5.5
6	6.8	6.7	6.6
7	7.9	7.8	7.7
8	9.1	8.9	8.8
9	10.2	10.0	9.9
10	11.3	11.2	11.0
20	22.7	22.3	22.0
30	34.0	33.5	33.0
40	45.3	44.7	44.0
50	56.7	55.8	55.0

	3/74	3/73	3/72
0	12.3	12.2	12.0
1	37.0	36.5	36.0
2	61.7	60.8	60.0
3			

	3/71	3/70	3/69	3/68
0	11.8	11.7	11.5	11.3
1	35.5	35.0	34.5	34.0
2	59.2	58.3	57.5	56.7
3				

P P

*169° 259° *349°

79°

11° *101° 191° *281°

′	L Sin	d	L Tan	c d	L Cot	L Cos	d	
0	9.28 060	65	9.28 865	68	0.71 135	9.99 195	3	60
1	9.28 125	65	9.28 933	67	0.71 067	9.99 192	2	59
2	9.28 190	64	9.29 000	67	0.71 000	9.99 190	3	58
3	9.28 254	65	9.29 067	67	0.70 933	9.99 187	2	57
4	9.28 319	65	9.29 134	67	0.70 866	9.99 185	3	56
5	9.28 384	64	9.29 201	67	0.70 799	9.99 182	2	55
6	9.28 448	64	9.29 268	67	0.70 732	9.99 180	3	54
7	9.28 512	65	9.29 335	67	0.70 665	9.99 177	2	53
8	9.28 577	64	9.29 402	66	0.70 598	9.99 175	3	52
9	9.28 641	64	9.29 468	67	0.70 532	9.99 172	2	51
10	9.28 705	64	9.29 535	66	0.70 465	9.99 170	3	50
11	9.28 769	64	9.29 601	67	0.70 399	9.99 167	2	49
12	9.28 833	63	9.29 668	66	0.70 332	9.99 165	3	48
13	9.28 896	64	9.29 734	66	0.70 266	9.99 162	2	47
14	9.28 960	64	9.29 800	66	0.70 200	9.99 160	3	46
15	9.29 024	63	9.29 866	66	0.70 134	9.99 157	2	45
16	9.29 087	63	9.29 932	66	0.70 068	9.99 155	3	44
17	9.29 150	64	9.29 998	66	0.70 002	9.99 152	2	43
18	9.29 214	63	9.30 064	66	0.69 936	9.99 150	3	42
19	9.29 277	63	9.30 130	65	0.69 870	9.99 147	2	41
20	9.29 340	63	9.30 195	66	0.69 805	9.99 145	3	40
21	9.29 403	63	9.30 261	65	0.69 739	9.99 142	2	39
22	9.29 466	63	9.30 326	65	0.69 674	9.99 140	3	38
23	9.29 529	62	9.30 391	66	0.69 609	9.99 137	2	37
24	9.29 591	63	9.30 457	65	0.69 543	9.99 135	3	36
25	9.29 654	62	9.30 522	65	0.69 478	9.99 132	2	35
26	9.29 716	63	9.30 587	65	0.69 413	9.99 130	3	34
27	9.29 779	62	9.30 652	65	0.69 348	9.99 127	3	33
28	9.29 841	62	9.30 717	65	0.69 283	9.99 124	2	32
29	9.29 903	63	9.30 782	64	0.69 218	9.99 122	3	31
30	9.29 966	62	9.30 846	65	0.69 154	9.99 119	2	30
31	9.30 028	62	9.30 911	64	0.69 089	9.99 117	3	29
32	9.30 090	61	9.30 975	65	0.69 025	9.99 114	2	28
33	9.30 151	62	9.31 040	64	0.68 960	9.99 112	3	27
34	9.30 213	62	9.31 104	64	0.68 896	9.99 109	3	26
35	9.30 275	61	9.31 168	65	0.68 832	9.99 106	2	25
36	9.30 336	62	9.31 233	64	0.68 767	9.99 104	3	24
37	9.30 398	61	9.31 297	64	0.68 703	9.99 101	2	23
38	9.30 459	62	9.31 361	64	0.68 639	9.99 099	3	22
39	9.30 521	61	9.31 425	64	0.68 575	9.99 096	3	21
40	9.30 582	61	9.31 489	63	0.68 511	9.99 093	2	20
41	9.30 643	61	9.31 552	64	0.68 448	9.99 091	3	19
42	9.30 704	61	9.31 616	63	0.68 384	9.99 088	2	18
43	9.30 765	61	9.31 679	64	0.68 321	9.99 086	3	17
44	9.30 826	61	9.31 743	63	0.68 257	9.99 083	3	16
45	9.30 887	60	9.31 806	64	0.68 194	9.99 080	2	15
46	9.30 947	61	9.31 870	63	0.68 130	9.99 078	3	14
47	9.31 008	60	9.31 933	63	0.68 067	9.99 075	3	13
48	9.31 068	61	9.31 996	63	0.68 004	9.99 072	2	12
49	9.31 129	60	9.32 059	63	0.67 941	9.99 070	3	11
50	9.31 189	61	9.32 122	63	0.67 878	9.99 067	3	10
51	9.31 250	60	9.32 185	63	0.67 815	9.99 064	2	9
52	9.31 310	60	9.32 248	63	0.67 752	9.99 062	3	8
53	9.31 370	60	9.32 311	62	0.67 689	9.99 059	3	7
54	9.31 430	60	9.32 373	63	0.67 627	9.99 056	2	6
55	9.31 490	59	9.32 436	62	0.67 564	9.99 054	3	5
56	9.31 549	60	9.32 498	63	0.67 502	9.99 051	3	4
57	9.31 609	60	9.32 561	62	0.67 439	9.99 048	2	3
58	9.31 669	59	9.32 623	62	0.67 377	9.99 046	3	2
59	9.31 728	60	9.32 685	62	0.67 315	9.99 043	3	1
60	9.31 788		9.32 747		0.67 253	9.99 040		0
	L Cos	d	L Cot	c d	L Tan	L Sin	d	′

P P

	65	64	63
1	1.1	1.1	1.0
2	2.2	2.1	2.1
3	3.2	3.2	3.2
4	4.3	4.3	4.2
5	5.4	5.3	5.2
6	6.5	6.4	6.3
7	7.6	7.5	7.4
8	8.7	8.5	8.4
9	9.8	9.6	9.4
10	10.8	10.7	10.5
20	21.7	21.3	21.0
30	32.5	32.0	31.5
40	43.3	42.7	42.0
50	54.2	53.3	52.5

	62	61	60
1	1.0	1.0	1.0
2	2.1	2.0	2.0
3	3.1	3.0	3.0
4	4.1	4.1	4.0
5	5.2	5.1	5.0
6	6.2	6.1	6.0
7	7.2	7.1	7.0
8	8.3	8.1	8.0
9	9.3	9.2	9.0
10	10.3	10.2	10.0
20	20.7	20.3	20.0
30	31.0	30.5	30.0
40	41.3	40.7	40.0
50	51.7	50.8	50.0

	59	3	2
1	1.0	0.0	0.0
2	2.0	0.1	0.1
3	3.0	0.2	0.1
4	3.9	0.2	0.1
5	4.9	0.2	0.2
6	5.9	0.3	0.2
7	6.9	0.4	0.2
8	7.9	0.4	0.3
9	8.8	0.4	0.3
10	9.8	0.5	0.3
20	19.7	1.0	0.7
30	29.5	1.5	1.0
40	39.3	2.0	1.3
50	49.2	2.5	1.7

	3/67	3/66	3/65
0	11.2	11.0	10.8
1	33.5	33.0	32.5
2	55.8	55.0	54.2
3			

	3/64	3/63	3/62
0	10.7	10.5	10.3
1	32.0	31.5	31.0
2	53.3	52.5	51.7
3			

P P

12°

′	L Sin	d	L Tan	c d	L Cot	L Cos	d	′
0	9.31 788	59	9.32 747	63	0.67 253	9.99 040	2	60
1	9.31 847	60	9.32 810	62	0.67 190	9.99 038	3	59
2	9.31 907	59	9.32 872	61	0.67 128	9.99 035	3	58
3	9.31 966	59	9.32 933	62	0.67 067	9.99 032	2	57
4	9.32 025̄	59	9.32 995	62	0.67 005̄	9.99 030	3	56
5	9.32 084	59	9.33 057	62	0.66 943	9.99 027	3	55
6	9.32 143	59	9.33 119	61	0.66 881	9.99 024	2	54
7	9.32 202	59	9.33 180	62	0.66 820	9.99 022	3	53
8	9.32 261	58	9.33 242	61	0.66 758	9.99 019	3	52
9	9.32 319	59	9.33 303	62	0.66 697	9.99 016	3	51
10	9.32 378	59	9.33 365̄	61	0.66 635	9.99 013	2	50
11	9.32 437	58	9.33 426	61	0.66 574	9.99 011	3	49
12	9.32 495	58	9.33 487	61	0.66 513	9.99 008	3	48
13	9.32 553	59	9.33 548	61	0.66 452	9.99 005	3	47
14	9.32 612	58	9.33 609	61	0.66 391	9.99 002	2	46
15	9.32 670	58	9.33 670	61	0.66 330	9.99 000	3	45
16	9.32 728	58	9.33 731	61	0.66 269	9.98 997	3	44
17	9.32 786	58	9.33 792	61	0.66 208	9.98 994	3	43
18	9.32 844	58	9.33 853	60	0.66 147	9.98 991	2	42
19	9.32 902	58	9.33 913	61	0.66 087	9.98 989	3	41
20	9.32 960	58	9.33 974	60	0.66 026	9.98 986	3	40
21	9.33 018	57	9.34 034	61	0.65 966	9.98 983	3	39
22	9.33 075	58	9.34 095̄	60	0.65 905	9.98 980	2	38
23	9.33 133	57	9.34 155	60	0.65 845̄	9.98 978	3	37
24	9.33 190	58	9.34 215	61	0.65 785̄	9.98 975̄	3	36
25	9.33 248	57	9.34 276	60	0.65 724	9.98 972	3	35
26	9.33 305	57	9.34 336	60	0.65 664	9.98 969	2	34
27	9.33 362	58	9.34 396	60	0.65 604	9.98 967	3	33
28	9.33 420	57	9.34 456	60	0.65 544	9.98 964	3	32
29	9.33 477	57	9.34 516	60	0.65 484	9.98 961	3	31
30	9.33 534	57	9.34 576	59	0.65 424	9.98 958	3	30
31	9.33 591	56	9.34 635	60	0.65 365̄	9.98 955	2	29
32	9.33 647	57	9.34 695̄	60	0.65 305	9.98 953	3	28
33	9.33 704	57	9.34 755̄	59	0.65 245	9.98 950	3	27
34	9.33 761	57	9.34 814	60	0.65 186	9.98 947	3	26
35	9.33 818	56	9.34 874	59	0.65 126	9.98 944	3	25
36	9.33 874	57	9.34 933	59	0.65 067	9.98 941	3	24
37	9.33 931	56	9.34 992	59	0.65 008	9.98 938	2	23
38	9.33 987	56	9.35 051	60	0.64 949	9.98 936	3	22
39	9.34 043	57	9.35 111	59	0.64 889	9.98 933	3	21
40	9.34 100	56	9.35 170	59	0.64 830	9.98 930	3	20
41	9.34 156	56	9.35 229	59	0.64 771	9.98 927	3	19
42	9.34 212	56	9.35 288	59	0.64 712	9.98 924	3	18
43	9.34 268	56	9.35 347	58	0.64 653	9.98 921	2	17
44	9.34 324	56	9.35 405	59	0.64 595̄	9.98 919	3	16
45	9.34 380	56	9.35 464	59	0.64 536	9.98 916	3	15
46	9.34 436	55	9.35 523	58	0.64 477	9.98 913	3	14
47	9.34 491	56	9.35 581	59	0.64 419	9.98 910	3	13
48	9.34 547	55	9.35 640	58	0.64 360	9.98 907	3	12
49	9.34 602	56	9.35 698	59	0.64 302	9.98 904	3	11
50	9.34 658	55	9.35 757	58	0.64 243	9.98 901	3	10
51	9.34 713	56	9.35 815̄	58	0.64 185	9.98 898	2	9
52	9.34 769	55	9.35 873	58	0.64 127	9.98 896	3	8
53	9.34 824	55	9.35 931	58	0.64 069	9.98 893	3	7
54	9.34 879	55	9.35 989	58	0.64 011	9.98 890	3	6
55	9.34 934	55	9.36 047	58	0.63 953	9.98 887	3	5
56	9.34 989	55	9.36 105	58	0.63 895̄	9.98 884	3	4
57	9.35 044	55	9.36 163	58	0.63 837	9.98 881	3	3
58	9.35 099	55	9.36 221	58	0.63 779	9.98 878	3	2
59	9.35 154	55	9.36 279	57	0.63 721	9.98 875	3	1
60	9.35 209		9.36 336		0.63 664	9.98 872		0
	L Cos	d	L Cot	c d	L Tan	L Sin	d	′

P P

	63	62	61
1	1.0	1.0	1.0
2	2.1	2.1	2.0
3	3.2	3.1	3.0
4	4.2	4.1	4.1
5	5.2	5.2	5.1
6	6.3	6.2	6.1
7	7.4	7.2	7.1
8	8.4	8.3	8.1
9	9.4	9.3	9.2
10	10.5	10.3	10.2
20	21.0	20.7	20.3
30	31.5	31.0	30.5
40	42.0	41.3	40.7
50	52.5	51.7	50.8

	60	59	58
1	1.0	1.0	1.0
2	2.0	2.0	1.9
3	3.0	3.0	2.9
4	4.0	3.9	3.9
5	5.0	4.9	4.8
6	6.0	5.9	5.8
7	7.0	6.9	6.8
8	8.0	7.9	7.7
9	9.0	8.8	8.7
10	10.0	9.8	9.7
20	20.0	19.7	19.3
30	30.0	29.5	29.0
40	40.0	39.3	38.7
50	50.0	49.2	48.3

	57	56	55
1	1.0	0.9	0.9
2	1.9	1.9	1.8
3	2.8	2.8	2.8
4	3.8	3.7	3.7
5	4.8	4.7	4.6
6	5.7	5.6	5.5
7	6.6	6.5	6.4
8	7.6	7.5̄	7.3
9	8.6	8.4	8.2
10	9.5	9.3	9.2
20	19.0	18.7	18.3
30	28.5	28.0	27.5
40	38.0	37.3	36.7
50	47.5	46.7	45.8

	3/62	3/61	3/60
0			
1	10.3	10.2	10.0
2	31.0	30.5	30.0
3	51.7	50.8	50.0

	3/59	3/58	3/57
0			
1	9.8	9.7	9.5
2	29.5	29.0	28.5
3	49.2	48.3	47.5

P P

77°

13°

*103° 193° *283°

′	L Sin	d	L Tan	c d	L Cot	L Cos	d	
0	9.35 209	54	9.36 336	58	0.63 664	9.98 872	3	60
1	9.35 263	55	9.36 394	58	0.63 606	9.98 869	2	59
2	9.35 318	55	9.36 452	57	0.63 548	9.98 867	3	58
3	9.35 373	54	9.36 509	57	0.63 491	9.98 864	3	57
4	9.35 427	54	9.36 566	58	0.63 434	9.98 861	3	56
5	9.35 481	55	9.36 624	57	0.63 376	9.98 858	3	55
6	9.35 536	54	9.36 681	57	0.63 319	9.98 855	3	54
7	9.35 590	54	9.36 738	57	0.63 262	9.98 852	3	53
8	9.35 644	54	9.36 795	57	0.63 205	9.98 849	3	52
9	9.35 698	54	9.36 852	57	0.63 148	9.98 846	3	51
10	9.35 752	54	9.36 909	57	0.63 091	9.98 843	3	50
11	9.35 806	54	9.36 966	57	0.63 034	9.98 840	3	49
12	9.35 860	54	9.37 023	57	0.62 977	9.98 837	3	48
13	9.35 914	54	9.37 080	57	0.62 920	9.98 834	3	47
14	9.35 968	54	9.37 137	56	0.62 863	9.98 831	3	46
15	9.36 022	53	9.37 193	57	0.62 807	9.98 828	3	45
16	9.36 075	54	9.37 250	56	0.62 750	9.98 825	3	44
17	9.36 129	53	9.37 306	57	0.62 694	9.98 822	3	43
18	9.36 182	54	9.37 363	56	0.62 637	9.98 819	3	42
19	9.36 236	53	9.37 419	57	0.62 581	9.98 816	3	41
20	9.36 289	53	9.37 476	56	0.62 524	9.98 813	3	40
21	9.36 342	53	9.37 532	56	0.62 468	9.98 810	3	39
22	9.36 395	54	9.37 588	56	0.62 412	9.98 807	3	38
23	9.36 449	53	9.37 644	56	0.62 356	9.98 804	3	37
24	9.36 502	53	9.37 700	56	0.62 300	9.98 801	3	36
25	9.36 555	53	9.37 756	56	0.62 244	9.98 798	3	35
26	9.36 608	52	9.37 812	56	0.62 188	9.98 795	3	34
27	9.36 660	53	9.37 868	56	0.62 132	9.98 792	3	33
28	9.36 713	53	9.37 924	56	0.62 076	9.98 789	3	32
29	9.36 766	53	9.37 980	55	0.62 020	9.98 786	3	31
30	9.36 819	52	9.38 035	56	0.61 965	9.98 783	3	30
31	9.36 871	53	9.38 091	56	0.61 909	9.98 780	3	29
32	9.36 924	52	9.38 147	55	0.61 853	9.98 777	3	28
33	9.36 976	52	9.38 202	55	0.61 798	9.98 774	3	27
34	9.37 028	53	9.38 257	56	0.61 743	9.98 771	3	26
35	9.37 081	52	9.38 313	55	0.61 687	9.98 768	3	25
36	9.37 133	52	9.38 368	55	0.61 632	9.98 765	3	24
37	9.37 185	52	9.38 423	56	0.61 577	9.98 762	3	23
38	9.37 237	52	9.38 479	55	0.61 521	9.98 759	3	22
39	9.37 289	52	9.38 534	55	0.61 466	9.98 756	3	21
40	9.37 341	52	9.38 589	55	0.61 411	9.98 753	3	20
41	9.37 393	52	9.38 644	55	0.61 356	9.98 750	4	19
42	9.37 445	52	9.38 699	55	0.61 301	9.98 746	3	18
43	9.37 497	52	9.38 754	54	0.61 246	9.98 743	3	17
44	9.37 549	51	9.38 808	55	0.61 192	9.98 740	3	16
45	9.37 600	52	9.38 863	55	0.61 137	9.98 737	3	15
46	9.37 652	51	9.38 918	54	0.61 082	9.98 734	3	14
47	9.37 703	52	9.38 972	55	0.61 028	9.98 731	3	13
48	9.37 755	51	9.39 027	55	0.60 973	9.98 728	3	12
49	9.37 806	52	9.39 082	54	0.60 918	9.98 725	3	11
50	9.37 858	51	9.39 136	54	0.60 864	9.98 722	3	10
51	9.37 909	51	9.39 190	55	0.60 810	9.98 719	4	9
52	9.37 960	51	9.39 245	54	0.60 755	9.98 715	3	8
53	9.38 011	51	9.39 299	54	0.60 701	9.98 712	3	7
54	9.38 062	51	9.39 353	54	0.60 647	9.98 709	3	6
55	9.38 113	51	9.39 407	54	0.60 593	9.98 706	3	5
56	9.38 164	51	9.39 461	54	0.60 539	9.98 703	3	4
57	9.38 215	51	9.39 515	54	0.60 485	9.98 700	3	3
58	9.38 266	51	9.39 569	54	0.60 431	9.98 697	3	2
59	9.38 317	51	9.39 623	54	0.60 377	9.98 694	4	1
60	9.38 368		9.39 677		0.60 323	9.98 690		0
	L Cos	d	L Cot	c d	L Tan	L Sin	d	′

P P

	57	56	55
1	1.0	0.9	0.9
2	1.9	1.9	1.8
3	2.8	2.8	2.8
4	3.8	3.7	3.7
5	4.8	4.7	4.6
6	5.7	5.6	5.5
7	6.6	6.5	6.4
8	7.6	7.5	7.3
9	8.6	8.4	8.2
10	9.5	9.3	9.2
20	19.0	18.7	18.3
30	28.5	28.0	27.5
40	38.0	37.3	36.7
50	47.5	46.7	45.8

	54	53	52
1	0.9	0.9	0.9
2	1.8	1.8	1.7
3	2.7	2.6	2.6
4	3.6	3.5	3.5
5	4.5	4.4	4.3
6	5.4	5.3	5.2
7	6.3	6.2	6.1
8	7.2	7.1	6.9
9	8.1	8.0	7.8
10	9.0	8.8	8.7
20	18.0	17.7	17.3
30	27.0	26.5	26.0
40	36.0	35.3	34.7
50	45.0	44.2	43.3

	51	4	3	2
1	0.8	0.1	0.0	0.0
2	1.7	0.1	0.1	0.1
3	2.6	0.2	0.2	0.1
4	3.4	0.3	0.2	0.1
5	4.2	0.3	0.2	0.2
6	5.1	0.4	0.3	0.2
7	6.0	0.5	0.4	0.2
8	6.8	0.5	0.4	0.3
9	7.6	0.6	0.4	0.3
10	8.5	0.7	0.5	0.3
20	17.0	1.3	1.0	0.7
30	25.5	2.0	1.5	1.0
40	34.0	2.7	2.0	1.3
50	42.5	3.3	2.5	1.7

	4/55	4/54	3/58	3/57
0	6.9	6.8	9.7	9.5
1	20.6	20.2	29.0	28.5
2	34.4	33.8	48.3	47.5
3	48.1	47.2	—	—
4				

	3/56	3/55	3/54
0	9.3	9.2	9.0
1	28.0	27.5	27.0
2	46.7	45.8	45.0
3			

P P

*166° 256° *346°

76°

14°

*104° 194° *284°

′	L Sin	d	L Tan	cd	L Cot	L Cos	d	
0	9.38 368	50	9.39 677	54	0.60 323	9.98 690	3	60
1	9.38 418	51	9.39 731	54	0.60 269	9.98 687	3	59
2	9.38 469	50	9.39 785	53	0.60 215	9.98 684	3	58
3	9.38 519	51	9.39 838	54	0.60 162	9.98 681	3	57
4	9.38 570	50	9.39 892	53	0.60 108	9.98 678	3	56
5	9.38 620	50	9.39 945	54	0.60 055	9.98 675	4	55
6	9.38 670	51	9.39 999	53	0.60 001	9.98 671	3	54
7	9.38 721	50	9.40 052	54	0.59 948	9.98 668	3	53
8	9.38 771	50	9.40 106	53	0.59 894	9.98 665	3	52
9	9.38 821	50	9.40 159	53	0.59 841	9.98 662	3	51
10	9.38 871	50	9.40 212	54	0.59 788	9.98 659	3	50
11	9.38 921	50	9.40 266	53	0.59 734	9.98 656	4	49
12	9.38 971	50	9.40 319	53	0.59 681	9.98 652	3	48
13	9.39 021	50	9.40 372	53	0.59 628	9.98 649	3	47
14	9.39 071	50	9.40 425	53	0.59 575	9.98 646	3	46
15	9.39 121	49	9.40 478	53	0.59 522	9.98 643	3	45
16	9.39 170	50	9.40 531	53	0.59 469	9.98 640	4	44
17	9.39 220	50	9.40 584	52	0.59 416	9.98 636	3	43
18	9.39 270	49	9.40 636	53	0.59 364	9.98 633	3	42
19	9.39 319	50	9.40 689	53	0.59 311	9.98 630	3	41
20	9.39 369	49	9.40 742	53	0.59 258	9.98 627	4	40
21	9.39 418	49	9.40 795	52	0.59 205	9.98 623	3	39
22	9.39 467	50	9.40 847	53	0.59 153	9.98 620	3	38
23	9.39 517	49	9.40 900	52	0.59 100	9.98 617	3	37
24	9.39 566	49	9.40 952	53	0.59 048	9.98 614	4	36
25	9.39 615	49	9.41 005	52	0.58 995	9.98 610	3	35
26	9.39 664	49	9.41 057	52	0.58 943	9.98 607	3	34
27	9.39 713	49	9.41 109	52	0.58 891	9.98 604	3	33
28	9.39 762	49	9.41 161	53	0.58 839	9.98 601	4	32
29	9.39 811	49	9.41 214	52	0.58 786	9.98 597	3	31
30	9.39 860	49	9.41 266	52	0.58 734	9.98 594	3	30
31	9.39 909	49	9.41 318	52	0.58 682	9.98 591	3	29
32	9.39 958	48	9.41 370	52	0.58 630	9.98 588	4	28
33	9.40 006	49	9.41 422	52	0.58 578	9.98 584	3	27
34	9.40 055	48	9.41 474	52	0.58 526	9.98 581	3	26
35	9.40 103	49	9.41 526	52	0.58 474	9.98 578	4	25
36	9.40 152	48	9.41 578	51	0.58 422	9.98 574	3	24
37	9.40 200	49	9.41 629	52	0.58 371	9.98 571	3	23
38	9.40 249	48	9.41 681	52	0.58 319	9.98 568	3	22
39	9.40 297	49	9.41 733	51	0.58 267	9.98 565	4	21
40	9.40 346	48	9.41 784	52	0.58 216	9.98 561	3	20
41	9.40 394	48	9.41 836	51	0.58 164	9.98 558	3	19
42	9.40 442	48	9.41 887	52	0.58 113	9.98 555	4	18
43	9.40 490	48	9.41 939	51	0.58 061	9.98 551	3	17
44	9.40 538	48	9.41 990	51	0.58 010	9.98 548	3	16
45	9.40 586	48	9.42 041	52	0.57 959	9.98 545	4	15
46	9.40 634	48	9.42 093	51	0.57 907	9.98 541	3	14
47	9.40 682	48	9.42 144	51	0.57 856	9.98 538	3	13
48	9.40 730	48	9.42 195	51	0.57 805	9.98 535	4	12
49	9.40 778	47	9.42 246	51	0.57 754	9.98 531	3	11
50	9.40 825	48	9.42 297	51	0.57 703	9.98 528	3	10
51	9.40 873	48	9.42 348	51	0.57 652	9.98 525	4	9
52	9.40 921	47	9.42 399	51	0.57 601	9.98 521	3	8
53	9.40 968	48	9.42 450	51	0.57 550	9.98 518	3	7
54	9.41 016	47	9.42 501	51	0.57 499	9.98 515	4	6
55	9.41 063	48	9.42 552	51	0.57 448	9.98 511	3	5
56	9.41 111	47	9.42 603	50	0.57 397	9.98 508	3	4
57	9.41 158	47	9.42 653	51	0.57 347	9.98 505	4	3
58	9.41 205	47	9.42 704	51	0.57 296	9.98 501	3	2
59	9.41 252	48	9.42 755	50	0.57 245	9.98 498	4	1
60	9.41 300		9.42 805		0.57 195	9.98 494		0
	L Cos	d	L Cot	cd	L Tan	L Sin	d	′

P P

	54	53	52
1	0.9	0.9	0.9
2	1.8	1.8	1.7
3	2.7	2.6	2.6
4	3.6	3.5	3.5
5	4.5	4.4	4.3
6	5.4	5.3	5.2
7	6.3	6.2	6.1
8	7.2	7.1	6.9
9	8.1	8.0	7.8
10	9.0	8.8	8.7
20	18.0	17.7	17.3
30	27.0	26.5	26.0
40	36.0	35.3	34.7
50	45.0	44.2	43.3

	51	50	49
1	0.8	0.8	0.8
2	1.7	1.7	1.6
3	2.6	2.5	2.4
4	3.4	3.3	3.3
5	4.2	4.2	4.1
6	5.1	5.0	4.9
7	6.0	5.8	5.7
8	6.8	6.7	6.5
9	7.6	7.5	7.4
10	8.5	8.3	8.2
20	17.0	16.7	16.3
30	25.5	25.0	24.5
40	34.0	33.3	32.7
50	42.5	41.7	40.8

	48	47	4	3
1	0.8	0.8	0.1	0.0
2	1.6	1.6	0.1	0.1
3	2.4	2.4	0.2	0.2
4	3.2	3.1	0.3	0.2
5	4.0	3.9	0.3	0.2
6	4.8	4.7	0.4	0.3
7	5.6	5.5	0.5	0.4
8	6.4	6.3	0.5	0.4
9	7.2	7.0	0.6	0.4
10	8.0	7.8	0.7	0.5
20	16.0	15.7	1.3	1.0
30	24.0	23.5	2.0	1.5
40	32.0	31.3	2.7	2.0
50	40.0	39.2	3.3	2.5

	4/54	4/53	4/52	4/51
0–1	6.8	6.6	6.5	6.4
1–2	20.2	19.9	19.5	19.1
2–3	33.8	33.1	32.5	31.9
3–4	47.2	46.4	45.5	44.6

	3/54	3/53	3/52	3/51
0–1	9.0	8.8	8.7	8.5
1–2	27.0	26.5	26.0	25.5
2–3	45.0	44.2	43.3	42.5

P P

*165° 255° *345°

75°

15°

*105° 195° *285°

'	L Sin	d	L Tan	c d	L Cot	L Cos	d	
0	9.41 300	47	9.42 805	51	0.57 195	9.98 494	3	60
1	9.41 347	47	9.42 856	50	0.57 144	9.98 491	3	59
2	9.41 394	47	9.42 906	51	0.57 094	9.98 488	4	58
3	9.41 441	47	9.42 957	50	0.57 043	9.98 484	3	57
4	9.41 488	47	9.43 007	50	0.56 993	9.98 481	4	56
5	9.41 535	47	9.43 057	51	0.56 943	9.98 477	3	55
6	9.41 582	46	9.43 108	50	0.56 892	9.98 474	3	54
7	9.41 628	47	9.43 158	50	0.56 842	9.98 471	4	53
8	9.41 675	47	9.43 208	50	0.56 792	9.98 467	3	52
9	9.41 722	46	9.43 258	50	0.56 742	9.98 464	4	51
10	9.41 768	47	9.43 308	50	0.56 692	9.98 460	3	50
11	9.41 815	46	9.43 358	50	0.56 642	9.98 457	4	49
12	9.41 861	47	9.43 408	50	0.56 592	9.98 453	3	48
13	9.41 908	46	9.43 458	50	0.56 542	9.98 450	3	47
14	9.41 954	47	9.43 508	50	0.56 492	9.98 447	4	46
15	9.42 001	46	9.43 558	49	0.56 442	9.98 443	3	45
16	9.42 047	46	9.43 607	50	0.56 393	9.98 440	4	44
17	9.42 093	47	9.43 657	50	0.56 343	9.98 436	3	43
18	9.42 140	46	9.43 707	49	0.56 293	9.98 433	4	42
19	9.42 186	46	9.43 756	50	0.56 244	9.98 429	3	41
20	9.42 232	46	9.43 806	49	0.56 194	9.98 426	4	40
21	9.42 278	46	9.43 855	50	0.56 145	9.98 422	3	39
22	9.42 324	46	9.43 905	49	0.56 095	9.98 419	4	38
23	9.42 370	46	9.43 954	50	0.56 046	9.98 415	3	37
24	9.42 416	45	9.44 004	49	0.55 996	9.98 412	3	36
25	9.42 461	46	9.44 053	49	0.55 947	9.98 409	4	35
26	9.42 507	46	9.44 102	49	0.55 898	9.98 405	3	34
27	9.42 553	46	9.44 151	50	0.55 849	9.98 402	4	33
28	9.42 599	45	9.44 201	49	0.55 799	9.98 398	3	32
29	9.42 644	46	9.44 250	49	0.55 750	9.98 395	4	31
30	9.42 690	45	9.44 299	49	0.55 701	9.98 391	3	30
31	9.42 735	46	9.44 348	49	0.55 652	9.98 388	4	29
32	9.42 781	45	9.44 397	49	0.55 603	9.98 384	3	28
33	9.42 826	46	9.44 446	49	0.55 554	9.98 381	4	27
34	9.42 872	45	9.44 495	49	0.55 505	9.98 377	4	26
35	9.42 917	45	9.44 544	48	0.55 456	9.98 373	3	25
36	9.42 962	46	9.44 592	49	0.55 408	9.98 370	4	24
37	9.43 008	45	9.44 641	49	0.55 359	9.98 366	3	23
38	9.43 053	45	9.44 690	48	0.55 310	9.98 363	4	22
39	9.43 098	45	9.44 738	49	0.55 262	9.98 359	3	21
40	9.43 143	45	9.44 787	49	0.55 213	9.98 356	4	20
41	9.43 188	45	9.44 836	48	0.55 164	9.98 352	3	19
42	9.43 233	45	9.44 884	49	0.55 116	9.98 349	4	18
43	9.43 278	45	9.44 933	48	0.55 067	9.98 345	3	17
44	9.43 323	44	9.44 981	48	0.55 019	9.98 342	4	16
45	9.43 367	45	9.45 029	49	0.54 971	9.98 338	4	15
46	9.43 412	45	9.45 078	48	0.54 922	9.98 334	3	14
47	9.43 457	45	9.45 126	48	0.54 874	9.98 331	4	13
48	9.43 502	44	9.45 174	48	0.54 826	9.98 327	3	12
49	9.43 546	45	9.45 222	49	0.54 778	9.98 324	4	11
50	9.43 591	44	9.45 271	48	0.54 729	9.98 320	3	10
51	9.43 635	45	9.45 319	48	0.54 681	9.98 317	4	9
52	9.43 680	44	9.45 367	48	0.54 633	9.98 313	4	8
53	9.43 724	45	9.45 415	48	0.54 585	9.98 309	3	7
54	9.43 769	44	9.45 463	48	0.54 537	9.98 306	4	6
55	9.43 813	44	9.45 511	48	0.54 489	9.98 302	3	5
56	9.43 857	44	9.45 559	47	0.54 441	9.98 299	4	4
57	9.43 901	45	9.45 606	48	0.54 394	9.98 295	4	3
58	9.43 946	44	9.45 654	48	0.54 346	9.98 291	3	2
59	9.43 990	44	9.45 702	48	0.54 298	9.98 288	4	1
60	9.44 034		9.45 750		0.54 250	9.98 284		0
	L Cos	d	L Cot	c d	L Tan	L Sin	d	'

P P

	51	50	49
1	0.8	0.8	0.8
2	1.7	1.7	1.6
3	2.6	2.5	2.4
4	3.4	3.3	3.3
5	4.2	4.2	4.1
6	5.1	5.0	4.9
7	6.0	5.8	5.7
8	6.8	6.7	6.5
9	7.6	7.5	7.4
10	8.5	8.3	8.2
20	17.0	16.7	16.3
30	25.5	25.0	24.5
40	34.0	33.3	32.7
50	42.5	41.7	40.8

	48	47	46
1	0.8	0.8	0.8
2	1.6	1.6	1.5
3	2.4	2.4	2.3
4	3.2	3.1	3.1
5	4.0	3.9	3.8
6	4.8	4.7	4.6
7	5.6	5.5	5.4
8	6.4	6.3	6.1
9	7.2	7.0	6.9
10	8.0	7.8	7.7
20	16.0	15.7	15.3
30	24.0	23.5	23.0
40	32.0	31.3	30.7
50	40.0	39.2	38.3

	45	44	4	3
1	0.8	0.7	0.1	0.0
2	1.5	1.5	0.1	0.1
3	2.2	2.2	0.2	0.2
4	3.0	2.9	0.3	0.2
5	3.8	3.7	0.3	0.2
6	4.5	4.4	0.4	0.3
7	5.2	5.1	0.5	0.4
8	6.0	5.9	0.5	0.4
9	6.8	6.6	0.6	0.4
10	7.5	7.3	0.7	0.5
20	15.0	14.7	1.3	1.0
30	22.5	22.0	2.0	1.5
40	30.0	29.3	2.7	2.0
50	37.5	36.7	3.3	2.5

	$\frac{4}{50}$	$\frac{4}{49}$	$\frac{4}{48}$	$\frac{4}{47}$
0				
	6.2	6.1	6.0	5.9
1				
	18.8	18.4	18.0	17.6
2				
	31.2	30.6	30.0	29.4
3				
	43.8	42.9	42.0	41.1
4				

	$\frac{3}{51}$	$\frac{3}{50}$	$\frac{3}{49}$	$\frac{3}{48}$
0				
	8.5	8.3	8.2	8.0
1				
	25.5	25.0	24.5	24.0
2				
	42.5	41.7	40.8	40.0
3				

P P

*164° 254° *344° **74°**

16°   *106°   196°   *286°

′	L Sin	d	L Tan	c d	L Cot	L Cos	d	′
0	9.44 034		9.45 75̄0		0.54 250	9.98 284		60
1	9.44 078	44	9.45 797	47	0.54 203	9.98 281	3	59
2	9.44 122	44	9.45 845̄	48	0.54 155	9.98 277	4	58
3	9.44 166	44	9.45 892	47	0.54 108	9.98 273	4	57
4	9.44 210	44	9.45 940	48	0.54 060	9.98 270	3	56
5	9.44 253	43	9.45 987	47	0.54 013	9.98 266	4	55
6	9.44 297	44	9.46 035̄	48	0.53 965	9.98 262	4	54
7	9.44 341	44	9.46 082	47	0.53 918	9.98 259	3	53
8	9.44 385̄	44	9.46 130	48	0.53 870	9.98 255	4	52
9	9.44 428	43	9.46 177	47	0.53 823	9.98 251	4	51
10	9.44 472	44	9.46 224	47	0.53 776	9.98 248	3	50
11	9.44 516	44	9.46 271	47	0.53 729	9.98 244	4	49
12	9.44 559	43	9.46 319	48	0.53 681	9.98 240	4	48
13	9.44 602	43	9.46 366	47	0.53 634	9.98 237	3	47
14	9.44 646	44	9.46 413	47	0.53 587	9.98 233	4	46
15	9.44 689	43	9.46 460	47	0.53 540	9.98 229	4	45
16	9.44 733	44	9.46 507	47	0.53 493	9.98 226	3	44
17	9.44 776	43	9.46 554	47	0.53 446	9.98 222	4	43
18	9.44 819	43	9.46 601	47	0.53 399	9.98 218	4	42
19	9.44 862	43	9.46 648	47	0.53 352	9.98 215̄	3	41
20	9.44 905	43	9.46 694	46	0.53 306	9.98 211	4	40
21	9.44 948	43	9.46 741	47	0.53 259	9.98 207	4	39
22	9.44 992	44	9.46 788	47	0.53 212	9.98 204	3	38
23	9.45 035̄	43	9.46 835̄	47	0.53 165	9.98 200	4	37
24	9.45 077	42	9.46 881	46	0.53 119	9.98 196	4	36
25	9.45 120	43	9.46 928	47	0.53 072	9.98 192	4	35
26	9.45 163	43	9.46 975̄	47	0.53 025	9.98 189	3	34
27	9.45 206	43	9.47 021	46	0.52 979	9.98 185̄	4	33
28	9.45 249	43	9.47 068	47	0.52 932	9.98 181	4	32
29	9.45 292	43	9.47 114	46	0.52 886	9.98 177	4	31
30	9.45 334	42	9.47 160	46	0.52 840	9.98 174	3	30
31	9.45 377	43	9.47 207	47	0.52 793	9.98 170	4	29
32	9.45 419	42	9.47 253	46	0.52 747	9.98 166	4	28
33	9.45 462	43	9.47 299	46	0.52 701	9.98 162	4	27
34	9.45 504	42	9.47 346	47	0.52 654	9.98 159	3	26
35	9.45 547	43	9.47 392	46	0.52 608	9.98 155̄	4	25
36	9.45 589	42	9.47 438	46	0.52 562	9.98 151	4	24
37	9.45 632	43	9.47 484	46	0.52 516	9.98 147	4	23
38	9.45 674	42	9.47 530	46	0.52 470	9.98 144	3	22
39	9.45 716	42	9.47 576	46	0.52 424	9.98 140	4	21
40	9.45 758	42	9.47 622	46	0.52 378	9.98 136	4	20
41	9.45 801	43	9.47 668	46	0.52 332	9.98 132	4	19
42	9.45 843	42	9.47 714	46	0.52 286	9.98 129	3	18
43	9.45 885̄	42	9.47 760	46	0.52 240	9.98 125̄	4	17
44	9.45 927	42	9.47 806	46	0.52 194	9.98 121	4	16
45	9.45 969	42	9.47 852	46	0.52 148	9.98 117	4	15
46	9.46 011	42	9.47 897	45	0.52 103	9.98 113	4	14
47	9.46 053	42	9.47 943	46	0.52 057	9.98 110	3	13
48	9.46 095̄	42	9.47 989	46	0.52 011	9.98 106	4	12
49	9.46 136	41	9.48 035̄	46	0.51 965	9.98 102	4	11
50	9.46 178	42	9.48 080	45	0.51 920	9.98 098	4	10
51	9.46 220	42	9.48 126	46	0.51 874	9.98 094	4	9
52	9.46 262	42	9.48 171	45	0.51 829	9.98 090	4	8
53	9.46 303	41	9.48 217	46	0.51 783	9.98 087	3	7
54	9.46 345̄	42	9.48 262	45	0.51 738	9.98 083	4	6
55	9.46 386	41	9.48 307	45	0.51 693	9.98 079	4	5
56	9.46 428	42	9.48 353	46	0.51 647	9.98 075	4	4
57	9.46 469	41	9.48 398	45	0.51 602	9.98 071	4	3
58	9.46 511	42	9.48 443	45	0.51 557	9.98 067	4	2
59	9.46 552	41	9.48 489	46	0.51 511	9.98 063	4	1
60	9.46 594	42	9.48 534	45	0.51 466	9.98 060	3	0
	L Cos	d	L Cot	c d	L Tan	L Sin	d	

P P

	48	47	46
1	0.8	0.8	0.8
2	1.6	1.6	1.5
3	2.4	2.4	2.3
4	3.2	3.1	3.1
5	4.0	3.9	3.8
6	4.8	4.7	4.6
7	5.6	5.5̄	5.4
8	6.4	6.3	6.1
9	7.2	7.0	6.9
10	8.0	7.8	7.7
20	16.0	15.7	15.3
30	24.0	23.5	23.0
40	32.0	31.3	30.7
50	40.0	39.2	38.3

	45	44	43
1	0.8	0.7	0.7
2	1.5	1.5̄	1.4
3	2.2	2.2	2.2
4	3.0	2.9	2.9
5	3.8	3.7	3.6
6	4.5	4.4	4.3
7	5.2	5.1	5.0
8	6.0	5.9	5.7
9	6.8	6.6	6.4
10	7.5	7.3	7.2
20	15.0	14.7	14.3
30	22.5	22.0	21.5
40	30.0	29.3	28.7
50	37.5	36.7	35.8

	42	41	4	3
1	0.7	0.7	0.1	0.0
2	1.4	1.4	0.1	0.1
3	2.1	2.0	0.2	0.2
4	2.8	2.7	0.3	0.2
5	3.5	3.4	0.3	0.2
6	4.2	4.1	0.4	0.3
7	4.9	4.8	0.5̄	0.4
8	5.6	5.5̄	0.5	0.4
9	6.3	6.2	0.6	0.4
10	7.0	6.8	0.7	0.5
20	14.0	13.7	1.3	1.0
30	21.0	20.5	2.0	1.5
40	28.0	27.3	2.7	2.0
50	35.0	34.2	3.3	2.5

	4/48	4/47	4/46	4/45
0	6.0	5.9	5.8	5.6
1	18.0	17.6	17.2	16.9
2	30.0	29.4	28.8	28.1
3	42.0	41.1	40.2	39.4
4				

	3/48	3/47	3/46	3/45
0	8.0	7.8	7.7	7.5
1	24.0	23.5	23.0	22.5
2	40.0	39.2	38.3	37.5
3				

P P

*163°   253°   *343°   73°

17°

′	L Sin	d	L Tan	c d	L Cot	L Cos	d	
0	9.46 594	41	9.48 534	45	0.51 466	9.98 060	4	60
1	9.46 635	41	9.48 579	45	0.51 421	9.98 056	4	59
2	9.46 676	41	9.48 624	45	0.51 376	9.98 052	4	58
3	9.46 717	41	9.48 669	45	0.51 331	9.98 048	4	57
4	9.46 758	42	9.48 714	45	0.51 286	9.98 044	4	56
5	9.46 800	41	9.48 759	45	0.51 241	9.98 040	4	55
6	9.46 841	41	9.48 804	45	0.51 196	9.98 036	4	54
7	9.46 882	41	9.48 849	45	0.51 151	9.98 032	3	53
8	9.46 923	41	9.48 894	45	0.51 106	9.98 029	4	52
9	9.46 964	41	9.48 939	45	0.51 061	9.98 025	4	51
10	9.47 005	40	9.48 984	45	0.51 016	9.98 021	4	50
11	9.47 045	41	9.49 029	44	0.50 971	9.98 017	4	49
12	9.47 086	41	9.49 073	45	0.50 927	9.98 013	4	48
13	9.47 127	41	9.49 118	45	0.50 882	9.98 009	4	47
14	9.47 168	41	9.49 163	44	0.50 837	9.98 005	4	46
15	9.47 209	40	9.49 207	45	0.50 793	9.98 001	4	45
16	9.47 249	41	9.49 252	44	0.50 748	9.97 997	4	44
17	9.47 290	40	9.49 296	45	0.50 704	9.97 993	4	43
18	9.47 330	41	9.49 341	44	0.50 659	9.97 989	3	42
19	9.47 371	40	9.49 385	45	0.50 615	9.97 986	4	41
20	9.47 411	41	9.49 430	44	0.50 570	9.97 982	4	40
21	9.47 452	40	9.49 474	45	0.50 526	9.97 978	4	39
22	9.47 492	41	9.49 519	44	0.50 481	9.97 974	4	38
23	9.47 533	40	9.49 563	44	0.50 437	9.97 970	4	37
24	9.47 573	40	9.49 607	45	0.50 393	9.97 966	4	36
25	9.47 613	41	9.49 652	44	0.50 348	9.97 962	4	35
26	9.47 654	40	9.49 696	44	0.50 304	9.97 958	4	34
27	9.47 694	40	9.49 740	44	0.50 260	9.97 954	4	33
28	9.47 734	40	9.49 784	44	0.50 216	9.97 950	4	32
29	9.47 774	40	9.49 828	44	0.50 172	9.97 946	4	31
30	9.47 814	40	9.49 872	44	0.50 128	9.97 942	4	30
31	9.47 854	40	9.49 916	44	0.50 084	9.97 938	4	29
32	9.47 894	40	9.49 960	44	0.50 040	9.97 934	4	28
33	9.47 934	40	9.50 004	44	0.49 996	9.97 930	4	27
34	9.47 974	40	9.50 048	44	0.49 952	9.97 926	4	26
35	9.48 014	40	9.50 092	44	0.49 908	9.97 922	4	25
36	9.48 054	40	9.50 136	44	0.49 864	9.97 918	4	24
37	9.48 094	39	9.50 180	43	0.49 820	9.97 914	4	23
38	9.48 133	40	9.50 223	44	0.49 777	9.97 910	4	22
39	9.48 173	40	9.50 267	44	0.49 733	9.97 906	4	21
40	9.48 213	39	9.50 311	44	0.49 689	9.97 902	4	20
41	9.48 252	40	9.50 355	43	0.49 645	9.97 898	4	19
42	9.48 292	40	9.50 398	44	0.49 602	9.97 894	4	18
43	9.48 332	39	9.50 442	43	0.49 558	9.97 890	4	17
44	9.48 371	40	9.50 485	44	0.49 515	9.97 886	4	16
45	9.48 411	39	9.50 529	43	0.49 471	9.97 882	4	15
46	9.48 450	40	9.50 572	44	0.49 428	9.97 878	4	14
47	9.48 490	39	9.50 616	43	0.49 384	9.97 874	4	13
48	9.48 529	39	9.50 659	44	0.49 341	9.97 870	4	12
49	9.48 568	39	9.50 703	43	0.49 297	9.97 866	5	11
50	9.48 607	40	9.50 746	43	0.49 254	9.97 861	4	10
51	9.48 647	39	9.50 789	44	0.49 211	9.97 857	4	9
52	9.48 686	39	9.50 833	43	0.49 167	9.97 853	4	8
53	9.48 725	39	9.50 876	43	0.49 124	9.97 849	4	7
54	9.48 764	39	9.50 919	43	0.49 081	9.97 845	4	6
55	9.48 803	39	9.50 962	43	0.49 038	9.97 841	4	5
56	9.48 842	39	9.51 005	43	0.48 995	9.97 837	4	4
57	9.48 881	39	9.51 048	44	0.48 952	9.97 833	4	3
58	9.48 920	39	9.51 092	43	0.48 908	9.97 829	4	2
59	9.48 959	39	9.51 135	43	0.48 865	9.97 825	4	1
60	9.48 998		9.51 178		0.48 822	9.97 821		0
	L Cos	d	L Cot	c d	L Tan	L Sin	d	′

P P

	45	44	43
1	0.8	0.7	0.7
2	1.5	1.5	1.4
3	2.2	2.2	2.2
4	3.0	2.9	2.9
5	3.8	3.7	3.6
6	4.5	4.4	4.3
7	5.2	5.1	5.0
8	6.0	5.9	5.7
9	6.8	6.6	6.4
10	7.5	7.3	7.2
20	15.0	14.7	14.3
30	22.5	22.0	21.5
40	30.0	29.3	28.7
50	37.5	36.7	35.8

	42	41	40
1	0.7	0.7	0.7
2	1.4	1.4	1.3
3	2.1	2.0	2.0
4	2.8	2.7	2.7
5	3.5	3.4	3.3
6	4.2	4.1	4.0
7	4.9	4.8	4.7
8	5.6	5.5	5.3
9	6.3	6.2	6.0
10	7.0	6.8	6.7
20	14.0	13.7	13.3
30	21.0	20.5	20.0
40	28.0	27.3	26.7
50	35.0	34.2	33.3

	39	5	4	3
1	0.6	0.1	0.1	0.0
2	1.3	0.2	0.1	0.1
3	2.0	0.2	0.2	0.2
4	2.6	0.3	0.3	0.2
5	3.2	0.4	0.3	0.2
6	3.9	0.5	0.4	0.3
7	4.6	0.6	0.5	0.4
8	5.2	0.7	0.5	0.4
9	5.8	0.8	0.6	0.4
10	6.5	0.8	0.7	0.5
20	13.0	1.7	1.3	1.0
30	19.5	2.5	2.0	1.5
40	26.0	3.3	2.7	2.0
50	32.5	4.2	3.3	2.5

	5/43	4/45	4/44
0	4.3	5.6	5.5
1	12.9	16.9	16.5
2	21.5	28.1	27.5
3	30.1	39.4	38.5
4	38.7	—	—
5			

	4/43	3/45	3/44
0	5.4	7.5	7.3
1	16.1	22.5	22.0
2	26.9	37.5	36.7
3	37.6	—	—
4			

P P

72°

18°

*108° 198° *288°

'	L Sin	d	L Tan	cd	L Cot	L Cos	d	'
0	9.48 998		9.51 178		0.48 822	9.97 821		60
1	9.49 037	39	9.51 221	43	0.48 779	9.97 817	4	59
2	9.49 076	39	9.51 264	43	0.48 736	9.97 812	5	58
3	9.49 115	39	9.51 306	42	0.48 694	9.97 808	4	57
4	9.49 153	38	9.51 349	43	0.48 651	9.97 804	4	56
5	9.49 192	39	9.51 392	43	0.48 608	9.97 800	4	55
6	9.49 231	39	9.51 435	43	0.48 565	9.97 796	4	54
7	9.49 269	38	9.51 478	43	0.48 522	9.97 792	4	53
8	9.49 308	39	9.51 520	42	0.48 480	9.97 788	4	52
9	9.49 347	39	9.51 563	43	0.48 437	9.97 784	4	51
10	9.49 385	38	9.51 606	43	0.48 394	9.97 779	5	50
11	9.49 424	39	9.51 648	42	0.48 352	9.97 775	4	49
12	9.49 462	38	9.51 691	43	0.48 309	9.97 771	4	48
13	9.49 500	38	9.51 734	43	0.48 266	9.97 767	4	47
14	9.49 539	39	9.51 776	42	0.48 224	9.97 763	4	46
15	9.49 577	38	9.51 819	43	0.48 181	9.97 759	4	45
16	9.49 615	38	9.51 861	42	0.48 139	9.97 754	5	44
17	9.49 654	39	9.51 903	42	0.48 097	9.97 750	4	43
18	9.49 692	38	9.51 946	43	0.48 054	9.97 746	4	42
19	9.49 730	38	9.51 988	42	0.48 012	9.97 742	4	41
20	9.49 768	38	9.52 031	43	0.47 969	9.97 738	4	40
21	9.49 806	38	9.52 073	42	0.47 927	9.97 734	4	39
22	9.49 844	38	9.52 115	42	0.47 885	9.97 729	5	38
23	9.49 882	38	9.52 157	42	0.47 843	9.97 725	4	37
24	9.49 920	38	9.52 200	43	0.47 800	9.97 721	4	36
25	9.49 958	38	9.52 242	42	0.47 758	9.97 717	4	35
26	9.49 996	38	9.52 284	42	0.47 716	9.97 713	4	34
27	9.50 034	38	9.52 326	42	0.47 674	9.97 708	5	33
28	9.50 072	38	9.52 368	42	0.47 632	9.97 704	4	32
29	9.50 110	38	9.52 410	42	0.47 590	9.97 700	4	31
30	9.50 148	38	9.52 452	42	0.47 548	9.97 696	4	30
31	9.50 185	37	9.52 494	42	0.47 506	9.97 691	5	29
32	9.50 223	38	9.52 536	42	0.47 464	9.97 687	4	28
33	9.50 261	38	9.52 578	42	0.47 422	9.97 683	4	27
34	9.50 298	37	9.52 620	42	0.47 380	9.97 679	4	26
35	9.50 336	38	9.52 661	41	0.47 339	9.97 674	5	25
36	9.50 374	38	9.52 703	42	0.47 297	9.97 670	4	24
37	9.50 411	37	9.52 745	42	0.47 255	9.97 666	4	23
38	9.50 449	38	9.52 787	42	0.47 213	9.97 662	4	22
39	9.50 486	37	9.52 829	42	0.47 171	9.97 657	5	21
40	9.50 523	37	9.52 870	41	0.47 130	9.97 653	4	20
41	9.50 561	38	9.52 912	42	0.47 088	9.97 649	4	19
42	9.50 598	37	9.52 953	41	0.47 047	9.97 645	4	18
43	9.50 635	37	9.52 995	42	0.47 005	9.97 640	5	17
44	9.50 673	38	9.53 037	42	0.46 963	9.97 636	4	16
45	9.50 710	37	9.53 078	41	0.46 922	9.97 632	4	15
46	9.50 747	37	9.53 120	42	0.46 880	9.97 628	4	14
47	9.50 784	37	9.53 161	41	0.46 839	9.97 623	5	13
48	9.50 821	37	9.53 202	41	0.46 798	9.97 619	4	12
49	9.50 858	37	9.53 244	42	0.46 756	9.97 615	4	11
50	9.50 896	38	9.53 285	41	0.46 715	9.97 610	5	10
51	9.50 933	37	9.53 327	42	0.46 673	9.97 606	4	9
52	9.50 970	37	9.53 368	41	0.46 632	9.97 602	4	8
53	9.51 007	37	9.53 409	41	0.46 591	9.97 597	5	7
54	9.51 043	36	9.53 450	41	0.46 550	9.97 593	4	6
55	9.51 080	37	9.53 492	42	0.46 508	9.97 589	4	5
56	9.51 117	37	9.53 533	41	0.46 467	9.97 584	5	4
57	9.51 154	37	9.53 574	41	0.46 426	9.97 580	4	3
58	9.51 191	37	9.53 615	41	0.46 385	9.97 576	4	2
59	9.51 227	36	9.53 656	41	0.46 344	9.97 571	5	1
60	9.51 264	37	9.53 697	41	0.46 303	9.97 567	4	0
	L Cos	d	L Cot	cd	L Tan	L Sin	d	'

P P

	43	42	41
1	0.7	0.7	0.7
2	1.4	1.4	1.4
3	2.2	2.1	2.0
4	2.9	2.8	2.7
5	3.6	3.5	3.4
6	4.3	4.2	4.1
7	5.0	4.9	4.8
8	5.7	5.6	5.5
9	6.4	6.3	6.2
10	7.2	7.0	6.8
20	14.3	14.0	13.7
30	21.5	21.0	20.5
40	28.7	28.0	27.3
50	35.8	35.0	34.2

	39	38	37
1	0.6	0.6	0.6
2	1.3	1.3	1.2
3	2.0	1.9	1.8
4	2.6	2.5	2.5
5	3.2	3.2	3.1
6	3.9	3.8	3.7
7	4.6	4.4	4.3
8	5.2	5.1	4.9
9	5.8	5.7	5.6
10	6.5	6.3	6.2
20	13.0	12.7	12.3
30	19.5	19.0	18.5
40	26.0	25.3	24.7
50	32.5	31.7	30.8

	36	5	4
1	0.6	0.1	0.1
2	1.2	0.2	0.1
3	1.8	0.2	0.2
4	2.4	0.3	0.3
5	3.0	0.4	0.3
6	3.6	0.5	0.4
7	4.2	0.6	0.5
8	4.8	0.7	0.5
9	5.4	0.8	0.6
10	6.0	0.8	0.7
20	12.0	1.7	1.3
30	18.0	2.5	2.0
40	24.0	3.3	2.7
50	30.0	4.2	3.3

	$\frac{5}{43}$	$\frac{5}{42}$	$\frac{5}{41}$
0			
	4.3	4.2	4.1
1			
	12.9	12.6	12.3
2			
	21.5	21.0	20.5
3			
	30.1	29.4	28.7
4			
	38.7	37.8	36.9
5			

	$\frac{4}{43}$	$\frac{4}{42}$	$\frac{4}{41}$
0			
	5.4	5.2	5.1
1			
	16.1	15.8	15.4
2			
	26.9	26.2	25.6
3			
	37.6	36.8	35.9
4			

P P

*161° 251° *341° **71°**

19° *109° 199° *289°

′	L Sin	d	L Tan	c d	L Cot	L Cos	d	′
0	9.51 264		9.53 697		0.46 303	9.97 567		60
1	9.51 301	37	9.53 738	41	0.46 262	9.97 563	4	59
2	9.51 338	37	9.53 779	41	0.46 221	9.97 558	5	58
3	9.51 374	36	9.53 820	41	0.46 180	9.97 554	4	57
4	9.51 411	37	9.53 861	41	0.46 139	9.97 550	4	56
5	9.51 447	36	9.53 902	41	0.46 098	9.97 545	5	55
6	9.51 484	37	9.53 943	41	0.46 057	9.97 541	4	54
7	9.51 520	36	9.53 984	41	0.46 016	9.97 536	5	53
8	9.51 557	37	9.54 025̄	41	0.45 975	9.97 532	4	52
9	9.51 593	36	9.54 065	40	0.45 935̄	9.97 528	4	51
10	9.51 629	36	9.54 106	41	0.45 894	9.97 523	5	50
11	9.51 666	37	9.54 147	41	0.45 853	9.97 519	4	49
12	9.51 702	36	9.54 187	40	0.45 813	9.97 515̄	4	48
13	9.51 738	36	9.54 228	41	0.45 772	9.97 510	5	47
14	9.51 774	36	9.54 269	41	0.45 731	9.97 506	4	46
15	9.51 811	37	9.54 309	40	0.45 691	9.97 501	5	45
16	9.51 847	36	9.54 35̄0	41	0.45 650	9.97 497	4	44
17	9.51 883	36	9.54 390	40	0.45 610	9.97 492	5	43
18	9.51 919	36	9.54 431	41	0.45 569	9.97 488	4	42
19	9.51 955	36	9.54 471	40	0.45 529	9.97 484	4	41
20	9.51 991	36	9.54 512	41	0.45 488	9.97 479	5	40
21	9.52 027	36	9.54 552	40	0.45 448	9.97 475̄	4	39
22	9.52 063	36	9.54 593	41	0.45 407	9.97 470	5	38
23	9.52 099	36	9.54 633	40	0.45 367	9.97 466	4	37
24	9.52 135̄	36	9.54 673	40	0.45 327	9.97 461	5	36
25	9.52 171	36	9.54 714	41	0.45 286	9.97 457	4	35
26	9.52 207	36	9.54 754	40	0.45 246	9.97 453	4	34
27	9.52 242	35	9.54 794	40	0.45 206	9.97 448	5	33
28	9.52 278	36	9.54 835̄	41	0.45 165	9.97 444	4	32
29	9.52 314	36	9.54 875̄	40	0.45 125	9.97 439	5	31
30	9.52 35̄0	36	9.54 915̄	40	0.45 085	9.97 435̄	4	30
31	9.52 385	35	9.54 955	40	0.45 045̄	9.97 430	5	29
32	9.52 421	36	9.54 995	40	0.45 005̄	9.97 426	4	28
33	9.52 456	35	9.55 035	40	0.44 965̄	9.97 421	5	27
34	9.52 492	36	9.55 075	40	0.44 925̄	9.97 417	4	26
35	9.52 527	35	9.55 115	40	0.44 885̄	9.97 412	5	25
36	9.52 563	36	9.55 155	40	0.44 845̄	9.97 408	4	24
37	9.52 598	35	9.55 195	40	0.44 805̄	9.97 403	5	23
38	9.52 634	36	9.55 235	40	0.44 765̄	9.97 399	4	22
39	9.52 669	35	9.55 275	40	0.44 725̄	9.97 394	5	21
40	9.52 705̄	36	9.55 315̄	40	0.44 685	9.97 390	4	20
41	9.52 740	35	9.55 355̄	40	0.44 645	9.97 385	5	19
42	9.52 775	35	9.55 395̄	40	0.44 605	9.97 381	4	18
43	9.52 811	36	9.55 434	39	0.44 566	9.97 376	5	17
44	9.52 846	35	9.55 474	40	0.44 526	9.97 372	4	16
45	9.52 881	35	9.55 514	40	0.44 486	9.97 367	5	15
46	9.52 916	35	9.55 554	40	0.44 446	9.97 363	4	14
47	9.52 951	35	9.55 593	39	0.44 407	9.97 358	5	13
48	9.52 986	35	9.55 633	40	0.44 367	9.97 353	5	12
49	9.53 021	35	9.55 673	40	0.44 327	9.97 349	4	11
50	9.53 056	35	9.55 712	39	0.44 288	9.97 344	5	10
51	9.53 092	36	9.55 752	40	0.44 248	9.97 340	4	9
52	9.53 126	34	9.55 791	39	0.44 209	9.97 335	5	8
53	9.53 161	35	9.55 831	40	0.44 169	9.97 331	4	7
54	9.53 196	35	9.55 870	39	0.44 130	9.97 326	5	6
55	9.53 231	35	9.55 910	40	0.44 090	9.97 322	4	5
56	9.53 266	35	9.55 949	39	0.44 051	9.97 317	5	4
57	9.53 301	35	9.55 989	40	0.44 011	9.97 312	5	3
58	9.53 336	35	9.56 028	39	0.43 972	9.97 308	4	2
59	9.53 370	34	9.56 067	39	0.43 933	9.97 303	5	1
60	9.53 405	35	9.56 107	40	0.43 893	9.97 299	4	0
	L Cos	d	L Cot	c d	L Tan	L Sin	d	′

P P

	41	40	39
1	0.7	0.7	0.6
2	1.4	1.3	1.3
3	2.0	2.0	2.0
4	2.7	2.7	2.6
5	3.4	3.3	3.2
6	4.1	4.0	3.9
7	4.8	4.7	4.6
8	5.5̄	5.3	5.2
9	6.2	6.0	5.8
10	6.8	6.7	6.5
20	13.7	13.3	13.0
30	20.5	20.0	19.5
40	27.3	26.7	26.0
50	34.2	33.3	32.5

	37	36	35
1	0.6	0.6	0.6
2	1.2	1.2	1.2
3	1.8	1.8	1.8
4	2.5̄	2.4	2.3
5	3.1	3.0	2.9
6	3.7	3.6	3.5
7	4.3	4.2	4.1
8	4.9	4.8	4.7
9	5.6	5.4	5.2
10	6.2	6.0	5.8
20	12.3	12.0	11.7
30	18.5	18.0	17.5
40	24.7	24.0	23.3
50	30.8	30.0	29.2

	34	5	4
1	0.6	0.1	0.1
2	1.1	0.2	0.1
3	1.7	0.2	0.2
4	2.3	0.3	0.3
5	2.8	0.4	0.3
6	3.4	0.5	0.4
7	4.0	0.6	0.5̄
8	4.5	0.7	0.5
9	5.1	0.8	0.6
10	5.7	0.8	0.7
20	11.3	1.7	1.3
30	17.0	2.5	2.0
40	22.7	3.3	2.7
50	28.3	4.2	3.3

	5/41	5/40	5/39
0–1	4.1	4.0	3.9
1–2	12.3	12.0	11.7
2–3	20.5	20.0	19.5
3–4	28.7	28.0	27.3
4–5	36.9	36.0	35.1

	4/41	4/40	4/39
0–1	5.1	5.0	4.9
1–2	15.4	15.0	14.6
2–3	25.6	25.0	24.4
3–4	35.9	35.0	34.1

P P

*160° 250° *340° 70°

20° *110° 200° *290°

′	L Sin	d	L Tan	c d	L Cot	L Cos	d	′
0	9.53 405		9.56 107		0.43 893	9.97 299		60
1	9.53 440	35	9.56 146	39	0.43 854	9.97 294	5	59
2	9.53 475̄	35	9.56 185	39	0.43 815̄	9.97 289	5	58
3	9.53 509	34	9.56 224	39	0.43 776	9.97 285̄	4	57
4	9.53 544	35	9.56 264	40	0.43 736	9.97 280	5	56
5	9.53 578	34	9.56 303	39	0.43 697	9.97 276	4	55
6	9.53 613	35	9.56 342	39	0.43 658	9.97 271	5	54
7	9.53 647	34	9.56 381	39	0.43 619	9.97 266	5	53
8	9.53 682	35	9.56 420	39	0.43 580	9.97 262	4	52
9	9.53 716	34	9.56 459	39	0.43 541	9.97 257	5	51
10	9.53 751	35	9.56 498	39	0.43 502	9.97 252	5	50
11	9.53 785	34	9.56 537	39	0.43 463	9.97 248	4	49
12	9.53 819	34	9.56 576	39	0.43 424	9.97 243	5	48
13	9.53 854	35	9.56 615	39	0.43 385̄	9.97 238	5	47
14	9.53 888	34	9.56 654	39	0.43 346	9.97 234	4	46
15	9.53 922	34	9.56 693	39	0.43 307	9.97 229	5	45
16	9.53 957	35	9.56 732	39	0.43 268	9.97 224	5	44
17	9.53 991	34	9.56 771	39	0.43 229	9.97 220	4	43
18	9.54 025̄	34	9.56 810	39	0.43 190	9.97 215	5	42
19	9.54 059	34	9.56 849	39	0.43 151	9.97 210	5	41
20	9.54 093	34	9.56 887	38	0.43 113	9.97 206	4	40
21	9.54 127	34	9.56 926	39	0.43 074	9.97 201	5	39
22	9.54 161	34	9.56 965̄	39	0.43 035	9.97 196	5	38
23	9.54 195	34	9.57 004	39	0.42 996	9.97 192	4	37
24	9.54 229	34	9.57 042	38	0.42 958	9.97 187	5	36
25	9.54 263	34	9.57 081	39	0.42 919	9.97 182	5	35
26	9.54 297	34	9.57 120	39	0.42 880	9.97 178	4	34
27	9.54 331	34	9.57 158	38	0.42 842	9.97 173	5	33
28	9.54 365̄	34	9.57 197	39	0.42 803	9.97 168	5	32
29	9.54 399	34	9.57 235	38	0.42 765̄	9.97 163	5	31
30	9.54 433	34	9.57 274	39	0.42 726	9.97 159	4	30
31	9.54 466	33	9.57 312	38	0.42 688	9.97 154	5	29
32	9.54 500	34	9.57 351	39	0.42 649	9.97 149	5	28
33	9.54 534	34	9.57 389	38	0.42 611	9.97 145̄	4	27
34	9.54 567	33	9.57 428	39	0.42 572	9.97 140	5	26
35	9.54 601	34	9.57 466	38	0.42 534	9.97 135	5	25
36	9.54 635̄	34	9.57 504	38	0.42 496	9.97 130	5	24
37	9.54 668	33	9.57 543	39	0.42 457	9.97 126	4	23
38	9.54 702	34	9.57 581	38	0.42 419	9.97 121	5	22
39	9.54 735	33	9.57 619	38	0.42 381	9.97 116	5	21
40	9.54 769	34	9.57 658	39	0.42 342	9.97 111	5	20
41	9.54 802	33	9.57 696	38	0.42 304	9.97 107	4	19
42	9.54 836	34	9.57 734	38	0.42 266	9.97 102	5	18
43	9.54 869	33	9.57 772	38	0.42 228	9.97 097	5	17
44	9.54 903	34	9.57 810	38	0.42 190	9.97 092	5	16
45	9.54 936	33	9.57 849	39	0.42 151	9.97 087	5	15
46	9.54 969	33	9.57 887	38	0.42 113	9.97 083	4	14
47	9.55 003	34	9.57 925̄	38	0.42 075	9.97 078	5	13
48	9.55 036	33	9.57 963	38	0.42 037	9.97 073	5	12
49	9.55 069	33	9.58 001	38	0.41 999	9.97 068	5	11
50	9.55 102	33	9.58 039	38	0.41 961	9.97 063	5	10
51	9.55 136	34	9.58 077	38	0.41 923	9.97 059	4	9
52	9.55 169	33	9.58 115̄	38	0.41 885	9.97 054	5	8
53	9.55 202	33	9.58 153	38	0.41 847	9.97 049	5	7
54	9.55 235̄	33	9.58 191	38	0.41 809	9.97 044	5	6
55	9.55 268	33	9.58 229	38	0.41 771	9.97 039	5	5
56	9.55 301	33	9.58 267	38	0.41 733	9.97 035̄	4	4
57	9.55 334	33	9.58 304	37	0.41 696	9.97 030	5	3
58	9.55 367	33	9.58 342	38	0.41 658	9.97 025̄	5	2
59	9.55 400	33	9.58 380	38	0.41 620	9.97 020	5	1
60	9.55 433	33	9.58 418	38	0.41 582	9.97 015	5	0
	L Cos	d	L Cot	c d	L Tan	L Sin	d	′

P P

	40	39	38
1	0.7	0.6	0.6
2	1.3	1.3	1.3
3	2.0	2.0	1.9
4	2.7	2.6	2.5
5	3.3	3.2	3.2
6	4.0	3.9	3.8
7	4.7	4.6	4.4
8	5.3	5.2	5.1
9	6.0	5.8	5.7
10	6.7	6.5	6.3
20	13.3	13.0	12.7
30	20.0	19.5	19.0
40	26.7	26.0	25.3
50	33.3	32.5	31.7

	37	35	34
1	0.6	0.6	0.6
2	1.2	1.2	1.1
3	1.8	1.8	1.7
4	2.5̄	2.3	2.3
5	3.1	2.9	2.8
6	3.7	3.5	3.4
7	4.3	4.1	4.0
8	4.9	4.7	4.5
9	5.6	5.2	5.1
10	6.2	5.8	5.7
20	12.3	11.7	11.3
30	18.5	17.5	17.0
40	24.7	23.3	22.7
50	30.8	29.2	28.3

	33	5	4
1	0.6	0.1	0.1
2	1.1	0.2	0.1
3	1.6	0.2	0.2
4	2.2	0.3	0.3
5	2.8	0.4	0.3
6	3.3	0.5	0.4
7	3.8	0.6	0.5̄
8	4.4	0.7	0.5
9	5.0	0.8	0.6
10	5.5	0.8	0.7
20	11.0	1.7	1.3
30	16.5	2.5	2.0
40	22.0	3.3	2.7
50	27.5	4.2	3.3

	5/40	5/39	5/38
0	4.0	3.9	3.8
1	12.0	11.7	11.4
2	20.0	19.5	19.0
3	28.0	27.3	26.6
4	36.0	35.1	34.2
5			

	5/37	4/39	4/38
0	3.7	4.9	4.8
1	11.1	14.6	14.2
2	18.5	24.4	23.8
3	25.9	34.1	33.2
4	33.3	—	—
5			

P P

*159° 249° *339° 69°

′	L Sin	d	L Tan	cd	L Cot	L Cos	d	
0	9.55 433	33	9.58 418	37	0.41 582	9.97 015	5	60
1	9.55 466	33	9.58 455	38	0.41 545̄	9.97 010	5	59
2	9.55 499	33	9.58 493	38	0.41 507	9.97 005	4	58
3	9.55 532	32	9.58 531	38	0.41 469	9.97 001	5	57
4	9.55 564	33	9.58 569	37	0.41 431	9.96 996	5	56
5	9.55 597	33	9.58 606	38	0.41 394	9.96 991	5	55
6	9.55 630	33	9.58 644	37	0.41 356	9.96 986	5	54
7	9.55 663	32	9.58 681	38	0.41 319	9.96 981	5	53
8	9.55 695	33	9.58 719	38	0.41 281	9.96 976	5	52
9	9.55 728	33	9.58 757	37	0.41 243	9.96 971	5	51
10	9.55 761	32	9.58 794	38	0.41 206	9.96 966	4	50
11	9.55 793	33	9.58 832	37	0.41 168	9.96 962	5	49
12	9.55 826	32	9.58 869	38	0.41 131	9.96 957	5	48
13	9.55 858	33	9.58 907	37	0.41 093	9.96 952	5	47
14	9.55 891	32	9.58 944	37	0.41 056	9.96 947	5	46
15	9.55 923	33	9.58 981	38	0.41 019	9.96 942	5	45
16	9.55 956	32	9.59 019	37	0.40 981	9.96 937	5	44
17	9.55 988	33	9.59 056	38	0.40 944	9.96 932	5	43
18	9.56 021	32	9.59 094	37	0.40 906	9.96 927	5	42
19	9.56 053	32	9.59 131	37	0.40 869	9.96 922	5	41
20	9.56 085	33	9.59 168	37	0.40 832	9.96 917	5	40
21	9.56 118	32	9.59 205	38	0.40 795̄	9.96 912	5	39
22	9.56 150	32	9.59 243	37	0.40 757	9.96 907	4	38
23	9.56 182	33	9.59 280	37	0.40 720	9.96 903	5	37
24	9.56 215̄	32	9.59 317	37	0.40 683	9.96 898	5	36
25	9.56 247	32	9.59 354	37	0.40 646	9.96 893	5	35
26	9.56 279	32	9.59 391	38	0.40 609	9.96 888	5	34
27	9.56 311	32	9.59 429	37	0.40 571	9.96 883	5	33
28	9.56 343	32	9.59 466	37	0.40 534	9.96 878	5	32
29	9.56 375	33	9.59 503	37	0.40 497	9.96 873	5	31
30	9.56 408	32	9.59 540	37	0.40 460	9.96 868	5	30
31	9.56 440	32	9.59 577	37	0.40 423	9.96 863	5	29
32	9.56 472	32	9.59 614	37	0.40 386	9.96 858	5	28
33	9.56 504	32	9.59 651	37	0.40 349	9.96 853	5	27
34	9.56 536	32	9.59 688	37	0.40 312	9.96 848	5	26
35	9.56 568	31	9.59 725̄	37	0.40 275	9.96 843	5	25
36	9.56 599	32	9.59 762	37	0.40 238	9.96 838	5	24
37	9.56 631	32	9.59 799	36	0.40 201	9.96 833	5	23
38	9.56 663	32	9.59 835	37	0.40 165̄	9.96 828	5	22
39	9.56 695	32	9.59 872	37	0.40 128	9.96 823	5	21
40	9.56 727	32	9.59 909	37	0.40 091	9.96 818	5	20
41	9.56 759	31	9.59 946	37	0.40 054	9.96 813	5	19
42	9.56 790	32	9.59 983	36	0.40 017	9.96 808	5	18
43	9.56 822	32	9.60 019	37	0.39 981	9.96 803	5	17
44	9.56 854	32	9.60 056	37	0.39 944	9.96 798	5	16
45	9.56 886	31	9.60 093	37	0.39 907	9.96 793	5	15
46	9.56 917	32	9.60 130	36	0.39 870	9.96 788	5	14
47	9.56 949	31	9.60 166	37	0.39 834	9.96 783	5	13
48	9.56 980	32	9.60 203	37	0.39 797	9.96 778	6	12
49	9.57 012	32	9.60 240	36	0.39 760	9.96 772	5	11
50	9.57 044	31	9.60 276	37	0.39 724	9.96 767	5	10
51	9.57 075	32	9.60 313	36	0.39 687	9.96 762	5	9
52	9.57 107	31	9.60 349	37	0.39 651	9.96 757	5	8
53	9.57 138	31	9.60 386	36	0.39 614	9.96 752	5	7
54	9.57 169	32	9.60 422	37	0.39 578	9.96 747	5	6
55	9.57 201	31	9.60 459	36	0.39 541	9.96 742	5	5
56	9.57 232	32	9.60 495	37	0.39 505̄	9.96 737	5	4
57	9.57 264	31	9.60 532	36	0.39 468	9.96 732	5	3
58	9.57 295̄	31	9.60 568	37	0.39 432	9.96 727	5	2
59	9.57 326	32	9.60 605̄	36	0.39 395	9.96 722	5	1
60	9.57 358		9.60 641		0.39 359	9.96 717		0
	L Cos	d	L Cot	cd	L Tan	L Sin	d	′

P P

	38	37	36
1	0.6	0.6	0.6
2	1.3	1.2	1.2
3	1.9	1.8	1.8
4	2.5	2.5̄	2.4
5	3.2	3.1	3.0
6	3.8	3.7	3.6
7	4.4	4.3	4.2
8	5.1	4.9	4.8
9	5.7	5.6	5.4
10	6.3	6.2	6.0
20	12.7	12.3	12.0
30	19.0	18.5	18.0
40	25.3	24.7	24.0
50	31.7	30.8	30.0

	33	32	31
1	0.6	0.5	0.5
2	1.1	1.1	1.0
3	1.6	1.6	1.6
4	2.2	2.1	2.1
5	2.8	2.7	2.6
6	3.3	3.2	3.1
7	3.8	3.7	3.6
8	4.4	4.3	4.1
9	5̄.0	4.8	4.6
10	5.5	5.3	5.2
20	11.0	10.7	10.3
30	16.5	16.0	15.5
40	22.0	21.3	20.7
50	27.5	26.7	25.8

	6	5	4
1	0.1	0.1	0.1
2	0.2	0.2	0.1
3	0.3	0.2	0.2
4	0.4	0.3	0.3
5	0.5	0.4	0.3
6	0.6	0.5	0.4
7	0.7	0.6	0.5̄
8	0.8	0.7	0.5
9	0.9	0.8	0.6
10	1.0	0.8	0.7
20	2.0	1.7	1.3
30	3.0	2.5	2.0
40	4.0	3.3	2.7
50	5.0	4.2	3.3

	6/37	5/38	5/37
0	3.1	3.8	3.7
1	9.2	11.4	11.1
2	15.4	19.0	18.5
3	21.6	26.6	25.9
4	27.8	34.2	33.3
5	33.9	—	—
6			

	5/36	4/38	4/37
0	3.6	4.8	4.6
1	10.8	14.2	13.9
2	18.0	23.8	23.1
3	25.2	33.2	32.4
4	32.4	—	—
5			

P P

22°

*112° 202° *292°

′	L Sin	d	L Tan	c d	L Cot	L Cos	d	′
0	9.57 358		9.60 641		0.39 359	9.96 717		60
1	9.57 389	31	9.60 677	36	0.39 323	9.96 711	6	59
2	9.57 420	31	9.60 714	37	0.39 286	9.96 706	5	58
3	9.57 451	31	9.60 750	36	0.39 250	9.96 701	5	57
4	9.57 482	31	9.60 786	36	0.39 214	9.96 696	5	56
5	9.57 514	32	9.60 823	37	0.39 177	9.96 691	5	55
6	9.57 545	31	9.60 859	36	0.39 141	9.96 686	5	54
7	9.57 576	31	9.60 895	36	0.39 105	9.96 681	5	53
8	9.57 607	31	9.60 931	36	0.39 069	9.96 676	5	52
9	9.57 638	31	9.60 967	36	0.39 033	9.96 670	6	51
10	9.57 669	31	9.61 004	37	0.38 996	9.96 665	5	50
11	9.57 700	31	9.61 040	36	0.38 960	9.96 660	5	49
12	9.57 731	31	9.61 076	35	0.38 924	9.96 655	5	48
13	9.57 762	31	9.61 112	36	0.38 888	9.96 650	5	47
14	9.57 793	31	9.61 148	36	0.38 852	9.96 645	5	46
15	9.57 824	31	9.61 184	36	0.38 816	9.96 640	5	45
16	9.57 855	31	9.61 220	36	0.38 780	9.96 634	6	44
17	9.57 885	30	9.61 256	36	0.38 744	9.96 629	5	43
18	9.57 916	31	9.61 292	36	0.38 708	9.96 624	5	42
19	9.57 947	31	9.61 328	36	0.38 672	9.96 619	5	41
20	9.57 978	31	9.61 364	36	0.38 636	9.96 614	5	40
21	9.58 008	30	9.61 400	36	0.38 600	9.96 608	6	39
22	9.58 039	31	9.61 436	36	0.38 564	9.96 603	5	38
23	9.58 070	31	9.61 472	36	0.38 528	9.96 598	5	37
24	9.58 101	31	9.61 508	36	0.38 492	9.96 593	5	36
25	9.58 131	30	9.61 544	36	0.38 456	9.96 588	5	35
26	9.58 162	31	9.61 579	35	0.38 421	9.96 582	6	34
27	9.58 192	30	9.61 615	36	0.38 385	9.96 577	5	33
28	9.58 223	31	9.61 651	36	0.38 349	9.96 572	5	32
29	9.58 253	30	9.61 687	36	0.38 313	9.96 567	5	31
30	9.58 284	31	9.61 722	35	0.38 278	9.96 562	5	30
31	9.58 314	30	9.61 758	36	0.38 242	9.96 556	6	29
32	9.58 345	31	9.61 794	36	0.38 206	9.96 551	5	28
33	9.58 375	30	9.61 830	36	0.38 170	9.96 546	5	27
34	9.58 406	31	9.61 865	35	0.38 135	9.96 541	5	26
35	9.58 436	30	9.61 901	36	0.38 099	9.96 535	6	25
36	9.58 467	31	9.61 936	35	0.38 064	9.96 530	5	24
37	9.58 497	30	9.61 972	36	0.38 028	9.96 525	5	23
38	9.58 527	30	9.62 008	36	0.37 992	9.96 520	5	22
39	9.58 557	30	9.62 043	35	0.37 957	9.96 514	6	21
40	9.58 588	31	9.62 079	36	0.37 921	9.96 509	5	20
41	9.58 618	30	9.62 114	35	0.37 886	9.96 504	5	19
42	9.58 648	30	9.62 150	36	0.37 850	9.96 498	6	18
43	9.58 678	30	9.62 185	35	0.37 815	9.96 493	5	17
44	9.58 709	31	9.62 221	36	0.37 779	9.96 488	5	16
45	9.58 739	30	9.62 256	35	0.37 744	9.96 483	5	15
46	9.58 769	30	9.62 292	36	0.37 708	9.96 477	6	14
47	9.58 799	30	9.62 327	35	0.37 673	9.96 472	5	13
48	9.58 829	30	9.62 362	35	0.37 638	9.96 467	5	12
49	9.58 859	30	9.62 398	36	0.37 602	9.96 461	6	11
50	9.58 889	30	9.62 433	35	0.37 567	9.96 456	5	10
51	9.58 919	30	9.62 468	35	0.37 532	9.96 451	5	9
52	9.58 949	30	9.62 504	36	0.37 496	9.96 445	6	8
53	9.58 979	30	9.62 539	35	0.37 461	9.96 440	5	7
54	9.59 009	30	9.62 574	35	0.37 426	9.96 435	5	6
55	9.59 039	30	9.62 609	35	0.37 391	9.96 429	6	5
56	9.59 069	30	9.62 645	36	0.37 355	9.96 424	5	4
57	9.59 098	29	9.62 680	35	0.37 320	9.96 419	5	3
58	9.59 128	30	9.62 715	35	0.37 285	9.96 413	6	2
59	9.59 158	30	9.62 750	35	0.37 250	9.96 408	5	1
60	9.59 188	30	9.62 785	35	0.37 215	9.96 403	5	0
	L Cos	d	L Cot	c d	L Tan	L Sin	d	′

P P

	37	36	35
1	0.6	0.6	0.6
2	1.2	1.2	1.2
3	1.8	1.8	1.8
4	2.5	2.4	2.3
5	3.1	3.0	2.9
6	3.7	3.6	3.5
7	4.3	4.2	4.1
8	4.9	4.8	4.7
9	5.6	5.4	5.2
10	6.2	6.0	5.8
20	12.3	12.0	11.7
30	18.5	18.0	17.5
40	24.7	24.0	23.3
50	30.8	30.0	29.2

	32	31	30
1	0.5	0.5	0.5
2	1.1	1.0	1.0
3	1.6	1.6	1.5
4	2.1	2.1	2.0
5	2.7	2.6	2.5
6	3.2	3.1	3.0
7	3.7	3.6	3.5
8	4.3	4.1	4.0
9	4.8	4.6	4.5
10	5.3	5.2	5.0
20	10.7	10.3	10.0
30	16.0	15.5	15.0
40	21.3	20.7	20.0
50	26.7	25.8	25.0

	29	6	5
1	0.5	0.1	0.1
2	1.0	0.2	0.2
3	1.4	0.3	0.2
4	1.9	0.4	0.3
5	2.4	0.5	0.4
6	2.9	0.6	0.5
7	3.4	0.7	0.6
8	3.9	0.8	0.7
9	4.4	0.9	0.8
10	4.8	1.0	0.8
20	9.7	2.0	1.7
30	14.5	3.0	2.5
40	19.3	4.0	3.3
50	24.2	5.0	4.2

	6/36	6/35
0		
	3.0	2.9
1		
	9.0	8.8
2		
	15.0	14.6
3		
	21.0	20.4
4		
	27.0	26.2
5		
	33.0	32.1
6		

	5/37	5/36	5/35
0			
	3.7	3.6	3.5
1			
	11.1	10.8	10.5
2			
	18.5	18.0	17.5
3			
	25.9	25.2	24.5
4			
	33.3	32.4	31.5
5			

P P

*157° 247° *337°

67°

23°

*113° 203° *293°

′	L Sin	d	L Tan	c d	L Cot	L Cos	d	
0	9.59 188	30	9.62 785	35	0.37 215̄	9.96 403	6	60
1	9.59 218	29	9.62 820	35	0.37 180	9.96 397	5	59
2	9.59 247	30	9.62 855	35	0.37 145̄	9.96 392	5	58
3	9.59 277	30	9.62 890	36	0.37 110	9.96 387	6	57
4	9.59 307	29	9.62 926	35	0.37 074	9.96 381	5	56
5	9.59 336	30	9.62 961	35	0.37 039	9.96 376	6	55
6	9.59 366	30	9.62 996	35	0.37 004	9.96 370	5	54
7	9.59 396	29	9.63 031	35	0.36 969	9.96 365̄	5	53
8	9.59 425	30	9.63 066	35	0.36 934	9.96 360	6	52
9	9.59 455̄	29	9.63 101	34	0.36 899	9.96 354	5	51
10	9.59 484	30	9.63 135	35	0.36 865̄	9.96 349	6	50
11	9.59 514	29	9.63 170	35	0.36 830	9.96 343	5	49
12	9.59 543	30	9.63 205	35	0.36 795̄	9.96 338	5	48
13	9.59 573	29	9.63 240	35	0.36 760	9.96 333	6	47
14	9.59 602	30	9.63 275	35	0.36 725̄	9.96 327	5	46
15	9.59 632	29	9.63 310	35	0.36 690	9.96 322	6	45
16	9.59 661	29	9.63 345̄	34	0.36 655	9.96 316	5	44
17	9.59 690	30	9.63 379	35	0.36 621	9.96 311	6	43
18	9.59 720	29	9.63 414	35	0.36 586	9.96 305	5	42
19	9.59 749	29	9.63 449	35	0.36 551	9.96 300	6	41
20	9.59 778	30	9.63 484	35	0.36 516	9.96 294	5	40
21	9.59 808	29	9.63 519	34	0.36 481	9.96 289	5	39
22	9.59 837	29	9.63 553	35	0.36 447	9.96 284	6	38
23	9.59 866	29	9.63 588	35	0.36 412	9.96 278	5	37
24	9.59 895	29	9.63 623	34	0.36 377	9.96 273	6	36
25	9.59 924	30	9.63 657	35	0.36 343	9.96 267	5	35
26	9.59 954	29	9.63 692	34	0.36 308	9.96 262	6	34
27	9.59 983	29	9.63 726	35	0.36 274	9.96 256	5	33
28	9.60 012	29	9.63 761	35	0.36 239	9.96 251	6	32
29	9.60 041	29	9.63 796	34	0.36 204	9.96 245	5	31
30	9.60 070	29	9.63 830	35	0.36 170	9.96 240	6	30
31	9.60 099	29	9.63 865̄	34	0.36 135	9.96 234	5	29
32	9.60 128	29	9.63 899	35	0.36 101	9.96 229	6	28
33	9.60 157	29	9.63 934	34	0.36 066	9.96 223	5	27
34	9.60 186	29	9.63 968	35	0.36 032	9.96 218	6	26
35	9.60 215̄	29	9.64 003	34	0.35 997	9.96 212	5	25
36	9.60 244	29	9.64 037	35	0.35 963	9.96 207	6	24
37	9.60 273	29	9.64 072	34	0.35 928	9.96 201	5	23
38	9.60 302	29	9.64 106	34	0.35 894	9.96 196	6	22
39	9.60 331	28	9.64 140	35	0.35 860	9.96 190	5	21
40	9.60 359	29	9.64 175̄	34	0.35 825	9.96 185̄	6	20
41	9.60 388	29	9.64 209	34	0.35 791	9.96 179	5	19
42	9.60 417	29	9.64 243	35	0.35 757	9.90 174	6	18
43	9.60 446	28	9.64 278	34	0.35 722	9.96 168	6	17
44	9.60 474	29	9.64 312	34	0.35 688	9.96 162	5	16
45	9.60 503	29	9.64 346	35	0.35 654	9.96 157	6	15
46	9.60 532	29	9.64 381	34	0.35 619	9.96 151	5	14
47	9.60 561	28	9.64 415̄	34	0.35 585	9.96 146	6	13
48	9.60 589	29	9.64 449	34	0.35 551	9.96 140	5	12
49	9.60 618	28	9.64 483	34	0.35 517	9.96 135̄	6	11
50	9.60 646	29	9.64 517	35	0.35 483	9.96 129	6	10
51	9.60 675	29	9.64 552	34	0.35 448	9.96 123	5	9
52	9.60 704	28	9.64 586	34	0.35 414	9.96 118	6	8
53	9.60 732	29	9.64 620	34	0.35 380	9.96 112	5	7
54	9.60 761	28	9.64 654	34	0.35 346	9.96 107	6	6
55	9.60 789	29	9.64 688	34	0.35 312	9.96 101	6	5
56	9.60 818	28	9.64 722	34	0.35 278	9.96 095	5	4
57	9.60 846	29	9.64 756	34	0.35 244	9.96 090	6	3
58	9.60 875̄	28	9.64 790	34	0.35 210	9.96 084	5	2
59	9.60 903	28	9.64 824	34	0.35 176	9.96 079	6	1
60	9.60 931		9.64 858		0.35 142	9.96 073		0
	L Cos	d	L Cot	c d	L Tan	L Sin	d	′

P P

	36	35	34
1	0.6	0.6	0.6
2	1.2	1.2	1.1
3	1.8	1.8	1.7
4	2.4	2.3	2.3
5	3.0	2.9	2.8
6	3.6	3.5	3.4
7	4.2	4.1	4.0
8	4.8	4.7	4.5
9	5.4	5.2	5.1
10	6.0	5.8	5.7
20	12.0	11.7	11.3
30	18.0	17.5	17.0
40	24.0	23.3	22.7
50	30.0	29.2	28.3

	30	29	28
1	0.5	0.5̄	0.5̄
2	1.0	1.0	0.9
3	1.5	1.4	1.4
4	2.0	1.9	1.9
5	2.5	2.4	2.3
6	3.0	2.9	2.8
7	3.5	3.4	3.3
8	4.0	3.9	3.7
9	4.5	4.4	4.2
10	5.0	4.8	4.7
20	10.0	9.7	9.3
30	15.0	14.5	14.0
40	20.0	19.3	18.7
50	25.0	24.2	23.3

	6	5
1	0.1	0.1
2	0.2	0.2
3	0.3	0.2
4	0.4	0.3
5	0.5	0.4
6	0.6	0.5
7	0.7	0.6
8	0.8	0.7
9	0.9	0.8
10	1.0	0.8
20	2.0	1.7
30	3.0	2.5
40	4.0	3.3
50	5.0	4.2

	6/36	6/35	6/34
0			
	3.0	2.9	2.8
1			
	9.0	8.8	8.5
2			
	15.0	14.6	14.2
3			
	21.0	20.4	19.8
4			
	27.0	26.2	25.5
5			
	33.0	32.1	31.2
6			

	5/35	5/34
0		
	3.5	3.4
1		
	10.5	10.2
2		
	17.5	17.0
3		
	24.5	23.8
4		
	31.5	30.6
5		

P P

*156° 246° *336°

66°

24°

*114° 204° *294°

′	L Sin	d	L Tan	c d	L Cot	L Cos	d	′
0	9.60 931		9.64 858		0.35 142	9.96 073		60
1	9.60 960	29	9.64 892	34	0.35 108	9.96 067	6	59
2	9.60 988	28	9.64 926	34	0.35 074	9.96 062	5	58
3	9.61 016	28	9.64 960	34	0.35 040	9.96 056	6	57
4	9.61 045̄	29	9.64 994	34	0.35 006	9.96 050	6	56
5	9.61 073	28	9.65 028	34	0.34 972	9.96 045̄	5	55
6	9.61 101	28	9.65 062	34	0.34 938	9.96 039	6	54
7	9.61 129	28	9.65 096	34	0.34 904	9.96 034	5	53
8	9.61 158	29	9.65 130	34	0.34 870	9.96 028	6	52
9	9.61 186	28	9.65 164	34	0.34 836	9.96 022	6	51
10	9.61 214	28	9.65 197	33	0.34 803	9.96 017	5	50
11	9.61 242	28	9.65 231	34	0.34 769	9.96 011	6	49
12	9.61 270	28	9.65 265	34	0.34 735̄	9.96 005	6	48
13	9.61 298	28	9.65 299	34	0.34 701	9.96 000	5	47
14	9.61 326	28	9.65 333	34	0.34 667	9.95 994	6	46
15	9.61 354	28	9.65 366	33	0.34 634	9.95 988	6	45
16	9.61 382	28	9.65 400	34	0.34 600	9.95 982	6	44
17	9.61 411	29	9.65 434	34	0.34 566	9.95 977	5	43
18	9.61 438	27	9.65 467	33	0.34 533	9.95 971	6	42
19	9.61 466	28	9.65 501	34	0.34 499	9.95 965	6	41
20	9.61 494	28	9.65 535̄	34	0.34 465	9.95 960	5	40
21	9.61 522	28	9.65 568	33	0.34 432	9.95 954	6	39
22	9.61 550	28	9.65 602	34	0.34 398	9.95 948	6	38
23	9.61 578	28	9.65 636	34	0.34 364	9.95 942	6	37
24	9.61 606	28	9.65 669	33	0.34 331	9.95 937	5	36
25	9.61 634	28	9.65 703	34	0.34 297	9.95 931	6	35
26	9.61 662	28	9.65 736	33	0.34 264	9.95 925	6	34
27	9.61 689	27	9.65 770	34	0.34 230	9.95 920	5	33
28	9.61 717	28	9.65 803	33	0.34 197	9.95 914	6	32
29	9.61 745̄	28	9.65 837	34	0.34 163	9.95 908	6	31
30	9.61 773	28	9.65 870	33	0.34 130	9.95 902	6	30
31	9.61 800	27	9.65 904	34	0.34 096	9.95 897	5	29
32	9.61 828	28	9.65 937	33	0.34 063	9.95 891	6	28
33	9.61 856	28	9.65 971	34	0.34 029	9.95 885̄	6	27
34	9.61 883	27	9.66 004	33	0.33 996	9.95 879	6	26
35	9.61 911	28	9.66 038	34	0.33 962	9.95 873	6	25
36	9.61 939	28	9.66 071	33	0.33 929	9.95 868	5	24
37	9.61 966	27	9.66 104	33	0.33 896	9.95 862	6	23
38	9.61 994	28	9.66 138	34	0.33 862	9.95 856	6	22
39	9.62 021	27	9.66 171	33	0.33 829	9.95 850	6	21
40	9.62 049	28	9.66 204	33	0.33 796	9.95 844	6	20
41	9.62 076	27	9.66 238	34	0.33 762	9.95 839	5	19
42	9.62 104	28	9.66 271	33	0.33 729	9.95 833	6	18
43	9.62 131	27	9.66 304	33	0.33 696	9.95 827	6	17
44	9.62 159	28	9.66 337	33	0.33 663	9.95 821	6	16
45	9.62 186	27	9.66 371	34	0.33 629	9.95 815	6	15
46	9.62 214	28	9.66 404	33	0.33 596	9.95 810	5	14
47	9.62 241	27	9.66 437	33	0.33 563	9.95 804	6	13
48	9.62 268	27	9.66 470	33	0.33 530	9.95 798	6	12
49	9.62 296	28	9.66 503	33	0.33 497	9.95 792	6	11
50	9.62 323	27	9.66 537	34	0.33 463	9.95 786	6	10
51	9.62 350	27	9.66 570	33	0.33 430	9.95 780	6	9
52	9.62 377	27	9.66 603	33	0.33 397	9.95 775̄	5	8
53	9.62 405̄	28	9.66 636	33	0.33 364	9.95 769	6	7
54	9.62 432	27	9.66 669	33	0.33 331	9.95 763	6	6
55	9.62 459	27	9.66 702	33	0.33 298	9.95 757	6	5
56	9.62 486	27	9.66 735	33	0.33 265̄	9.95 751	6	4
57	9.62 513	27	9.66 768	33	0.33 232	9.95 745	6	3
58	9.62 541	28	9.66 801	33	0.33 199	9.95 739	6	2
59	9.62 568	27	9.66 834	33	0.33 166	9.95 733	6	1
60	9.62 595̄	27	9.66 867	33	0.33 133	9.95 728	5	0
	L Cos	d	L Cot	c d	L Tan	L Sin	d	′

P P

	34	33
1	0.6	0.6
2	1.1	1.1
3	1.7	1.6
4	2.3	2.2
5	2.8	2.8
6	3.4	3.3
7	4.0	3.8
8	4.5	4.4
9	5.1	5̄.0
10	5.7	5.5
20	11.3	11.0
30	17.0	16.5
40	22.7	22.0
50	28.3	27.5

	29	28	27
1	0.5̄	0.5̄	0.4
2	1.0	0.9	0.9
3	1.4	1.4	1.4
4	1.9	1.9	1.8
5	2.4	2.3	2.2
6	2.9	2.8	2.7
7	3.4	3.3	3.2
8	3.9	3.7	3.6
9	4.4	4.2	4.0
10	4.8	4.7	4.5
20	9.7	9.3	9.0
30	14.5	14.0	13.5
40	19.3	18.7	18.0
50	24.2	23.3	22.5

	6	5
1	0.1	0.1
2	0.2	0.2
3	0.3	0.2
4	0.4	0.3
5	0.5	0.4
6	0.6	0.5
7	0.7	0.6
8	0.8	0.7
9	0.9	0.8
10	1.0	0.8
20	2.0	1.7
30	3.0	2.5
40	4.0	3.3
50	5.0	4.2

	6/34	6/33	5/34
0			
	2.8	2.8	3.4
1			
	8.5	8.2	10.2
2			
	14.2	13.8	17.0
3			
	19.8	19.2	23.8
4			
	25.5	24.8	30.6
5			
	31.2	30.2	—
6			

P P

*155° 245° *335°

65°

25° *115° 205° *295°

′	L Sin	d	L Tan	c d	L Cot	L Cos	d	′
0	9.62 595̄	27	9.66 867	33	0.33 133	9.95 728	6	60
1	9.62 622	27	9.66 900	33	0.33 100	9.95 722	6	59
2	9.62 649	27	9.66 933	33	0.33 067	9.95 716	6	58
3	9.62 676	27	9.66 966	33	0.33 034	9.95 710	6	57
4	9.62 703	27	9.66 999	33	0.33 001	9.95 704	6	56
5	9.62 730	27	9.67 032	33	0.32 968	9.95 698	6	55
6	9.62 757	27	9.67 065̄	33	0.32 935	9.95 692	6	54
7	9.62 784	27	9.67 098	33	0.32 902	9.95 686	6	53
8	9.62 811	27	9.67 131	32	0.32 869	9.95 680	6	52
9	9.62 838	27	9.67 163	33	0.32 837	9.95 674	6	51
10	9.62 865̄	27	9.67 196	33	0.32 804	9.95 668	5	50
11	9.62 892	26	9.67 229	33	0.32 771	9.95 663	6	49
12	9.62 918	27	9.67 262	33	0.32 738	9.95 657	6	48
13	9.62 945	27	9.67 295̄	32	0.32 705	9.95 651	6	47
14	9.62 972	27	9.67 327	33	0.32 673	9.95 645̄	6	46
15	9.62 999	27	9.67 360	33	0.32 640	9.95 639	6	45
16	9.63 026	26	9.67 393	33	0.32 607	9.95 633	6	44
17	9.63 052	27	9.67 426	32	0.32 574	9.95 627	6	43
18	9.63 079	27	9.67 458	33	0.32 542	9.95 621	6	42
19	9.63 106	27	9.67 491	33	0.32 509	9.95 615̄	6	41
20	9.63 133	26	9.67 524	32	0.32 476	9.95 609	6	40
21	9.63 159	27	9.67 556	33	0.32 444	9.95 603	6	39
22	9.63 186	27	9.67 589	33	0.32 411	9.95 597	6	38
23	9.63 213	26	9.67 622	32	0.32 378	9.95 591	6	37
24	9.63 239	27	9.67 654	33	0.32 346	9.95 585̄	6	36
25	9.63 266	26	9.67 687	32	0.32 313	9.95 579	6	35
26	9.63 292	27	9.67 719	33	0.32 281	9.95 573	6	34
27	9.63 319	26	9.67 752	33	0.32 248	9.95 567	6	33
28	9.63 345	27	9.67 785̄	32	0.32 215	9.95 561	6	32
29	9.63 372	26	9.67 817	33	0.32 183	9.95 555̄	6	31
30	9.63 398	27	9.67 85̄0	32	0.32 150	9.95 549	6	30
31	9.63 425̄	26	9.67 882	33	0.32 118	9.95 543	6	29
32	9.63 451	27	9.67 915̄	32	0.32 085	9.95 537	6	28
33	9.63 478	26	9.67 947	33	0.32 053	9.95 531	6	27
34	9.63 504	27	9.67 980	32	0.32 020	9.95 525̄	6	26
35	9.63 531	26	9.68 012	32	0.31 988	9.95 519	6	25
36	9.63 557	26	9.68 044	33	0.31 956	9.95 513	6	24
37	9.63 583	27	9.68 077	32	0.31 923	9.95 507	7	23
38	9.63 610	26	9.68 109	33	0.31 891	9.95 500	6	22
39	9.63 636	26	9.68 142	32	0.31 858	9.95 494	6	21
40	9.63 662	27	9.68 174	32	0.31 826	9.95 488	6	20
41	9.63 689	26	9.68 206	33	0.31 794	9.95 482	6	19
42	9.63 715̄	26	9.68 239	32	0.31 761	9.95 476	6	18
43	9.63 741	26	9.68 271	32	0.31 729	9.95 470	6	17
44	9.63 767	27	9.68 303	33	0.31 697	9.95 464	6	16
45	9.63 794	26	9.68 336	32	0.31 664	9.95 458	6	15
46	9.63 820	26	9.68 368	32	0.31 632	9.95 452	6	14
47	9.63 846	26	9.68 400	32	0.31 600	9.95 446	6	13
48	9.63 872	26	9.68 432	33	0.31 568	9.95 440	6	12
49	9.63 898	26	9.68 465̄	32	0.31 535	9.95 434	7	11
50	9.63 924	26	9.68 497	32	0.31 503	9.95 427	6	10
51	9.63 950	26	9.68 529	32	0.31 471	9.95 421	6	9
52	9.63 976	26	9.68 561	32	0.31 439	9.95 415	6	8
53	9.64 002	26	9.68 593	33	0.31 407	9.95 409	6	7
54	9.64 028	26	9.68 626	32	0.31 374	9.95 403	6	6
55	9.64 054	26	9.68 658	32	0.31 342	9.95 397	6	5
56	9.64 080	26	9.68 690	32	0.31 310	9.95 391	7	4
57	9.64 106	26	9.68 722	32	0.31 278	9.95 384	6	3
58	9.64 132	26	9.68 754	32	0.31 246	9.95 378	6	2
59	9.64 158	26	9.68 786	32	0.31 214	9.95 372	6	1
60	9.64 184		9.68 818		0.31 182	9.95 366		0
	L Cos	d	L Cot	c d	L Tan	L Sin	d	′

P P

	33	32
1	0.6	0.5
2	1.1	1.1
3	1.6	1.6
4	2.2	2.1
5	2.8	2.7
6	3.3	3.2
7	3.8	3.7
8	4.4	4.3
9	5̄.0	4.8
10	5.5	5.3
20	11.0	10.7
30	16.5	16.0
40	22.0	21.3
50	27.5	26.7

	27	26
1	0.4	0.4
2	0.9	0.9
3	1.4	1.3
4	1.8	1.7
5	2.2	2.2
6	2.7	2.6
7	3.2	3.0
8	3.6	3.5̄
9	4.0	3.9
10	4.5	4.3
20	9.0	8.7
30	13.5	13.0
40	18.0	17.3
50	22.5	21.7

	7	6	5
1	0.1	0.1	0.1
2	0.2	0.2	0.2
3	0.4	0.3	0.2
4	0.5̄	0.4	0.3
5	0.6	0.5	0.4
6	0.7	0.6	0.5
7	0.8	0.7	0.6
8	0.9	0.8	0.7
9	1.0	0.9	0.8
10	1.2	1.0	0.8
20	2.3	2.0	1.7
30	3.5	3.0	2.5
40	4.7	4.0	3.3
50	5.8	5.0	4.2

	7/32	6/32	5/33
0	2.3	2.7	3.3
1	6.9	8.0	9.9
2	11.4	13.3	16.5
3	16.0	18.7	23.1
4	20.6	24.0	29.7
5	25.1	29.3	—
6	29.7	—	—
7			

P P

*154° 244° *334° 64°

26°

*116° 206° *296°

′	L Sin	d	L Tan	c d	L Cot	L Cos	d	′
0	9.64 184	26	9.68 818	32	0.31 182	9.95 366	6	60
1	9.64 210	26	9.68 850	32	0.31 150	9.95 360	6	59
2	9.64 236	26	9.68 882	32	0.31 118	9.95 354	6	58
3	9.64 262	26	9.68 914	32	0.31 086	9.95 348	7	57
4	9.64 288	25	9.68 946	32	0.31 054	9.95 341	6	56
5	9.64 313	26	9.68 978	32	0.31 022	9.95 335	6	55
6	9.64 339	26	9.69 010	32	0.30 990	9.95 329	6	54
7	9.64 365	26	9.69 042	32	0.30 958	9.95 323	6	53
8	9.64 391	26	9.69 074	32	0.30 926	9.95 317	7	52
9	9.64 417	25	9.69 106	32	0.30 894	9.95 310	6	51
10	9.64 442	26	9.69 138	32	0.30 862	9.95 304	6	50
11	9.64 468	26	9.69 170	32	0.30 830	9.95 298	6	49
12	9.64 494	25	9.69 202	32	0.30 798	9.95 292	6	48
13	9.64 519	26	9.69 234	32	0.30 766	9.95 286	7	47
14	9.64 545̄	26	9.69 266	32	0.30 734	9.95 279	6	46
15	9.64 571	25	9.69 298	31	0.30 702	9.95 273	6	45
16	9.64 596	26	9.69 329	32	0.30 671	9.95 267	6	44
17	9.64 622	25	9.69 361	32	0.30 639	9.95 261	7	43
18	9.64 647	26	9.69 393	32	0.30 607	9.95 254	6	42
19	9.64 673	25	9.69 425̄	32	0.30 575	9.95 248	6	41
20	9.64 698	26	9.69 457	31	0.30 543	9.95 242	6	40
21	9.64 724	25	9.69 488	32	0.30 512	9.95 236	7	39
22	9.64 749	26	9.69 520	32	0.30 480	9.95 229	6	38
23	9.64 775̄	25	9.69 552	32	0.30 448	9.95 223	6	37
24	9.64 800	26	9.69 584	31	0.30 416	9.95 217	6	36
25	9.64 826	25	9.69 615	32	0.30 385̄	9.95 211	7	35
26	9.64 851	26	9.69 647	32	0.30 353	9.95 204	6	34
27	9.64 877	25	9.69 679	31	0.30 321	9.95 198	6	33
28	9.64 902	25	9.69 710	32	0.30 290	9.95 192	7	32
29	9.64 927	26	9.69 742	32	0.30 258	9.95 185	6	31
30	9.64 953	25	9.69 774	31	0.30 226	9.95 179	6	30
31	9.64 978	25	9.69 805	32	0.30 195̄	9.95 173	6	29
32	9.65 003	26	9.69 837	31	0.30 163	9.95 167	7	28
33	9.65 029	25	9.69 868	32	0.30 132	9.95 160	6	27
34	9.65 054	25	9.69 900	32	0.30 100	9.95 154	6	26
35	9.65 079	25	9.69 932	31	0.30 068	9.95 148	7	25
36	9.65 104	26	9.69 963	32	0.30 037	9.95 141	6	24
37	9.65 130	25	9.69 995̄	31	0.30 005	9.95 135̄	6	23
38	9.65 155̄	25	9.70 026	32	0.29 974	9.95 129	7	22
39	9.65 180	25	9.70 058	31	0.29 942	9.95 122	6	21
40	9.65 205	25	9.70 089	32	0.29 911	9.95 116	6	20
41	9.65 230	25	9.70 121	31	0.29 879	9.95 110	7	19
42	9.65 255	26	9.70 152	32	0.29 848	9.95 103	6	18
43	9.65 281	25	9.70 184	31	0.29 816	9.95 097	7	17
44	9.65 306	25	9.70 215	32	0.29 785̄	9.95 090	6	16
45	9.65 331	25	9.70 247	31	0.29 753	9.95 084	6	15
46	9.65 356	25	9.70 278	31	0.29 722	9.95 078	7	14
47	9.65 381	25	9.70 309	32	0.29 691	9.95 071	6	13
48	9.65 406	25	9.70 341	31	0.29 659	9.95 065̄	6	12
49	9.65 431	25	9.70 372	32	0.29 628	9.95 059	7	11
50	9.65 456	25	9.70 404	31	0.29 596	9.95 052	6	10
51	9.65 481	25	9.70 435̄	31	0.29 565	9.95 046	7	9
52	9.65 506	25	9.70 466	32	0.29 534	9.95 039	6	8
53	9.65 531	25	9.70 498	31	0.29 502	9.95 033	6	7
54	9.65 556	24	9.70 529	31	0.29 471	9.95 027	7	6
55	9.65 580	25	9.70 560	32	0.29 440	9.95 020	6	5
56	9.65 605	25	9.70 592	31	0.29 408	9.95 014	7	4
57	9.65 630	25	9.70 623	31	0.29 377	9.95 007	6	3
58	9.65 655	25	9.70 654	31	0.29 346	9.95 001	6	2
59	9.65 680	25	9.70 685	32	0.29 315̄	9.94 995̄	7	1
60	9.65 705̄		9.70 717		0.29 283	9.94 988		0
	L Cos	d	L Cot	c d	L Tan	L Sin	d	′

P P

	32	31
1	0.5	0.5
2	1.1	1.0
3	1.6	1.6
4	2.1	2.1
5	2.7	2.6
6	3.2	3.1
7	3.7	3.6
8	4.3	4.1
9	4.8	4.6
10	5.3	5.2
20	10.7	10.3
30	16.0	15.5
40	21.3	20.7
50	26.7	25.8

	26	25	24
1	0.4	0.4	0.4
2	0.9	0.8	0.8
3	1.3	1.2	1.2
4	1.7	1.7	1.6
5	2.2	2.1	2.0
6	2.6	2.5	2.4
7	3.0	2.9	2.8
8	3.5̄	3.3	3.2
9	3.9	3.8	3.6
10	4.3	4.2	4.0
20	8.7	8.3	8.0
30	13.0	12.5	12.0
40	17.3	16.7	16.0
50	21.7	20.8	20.0

	7	6
1	0.1	0.1
2	0.2	0.2
3	0.4	0.3
4	0.5̄	0.4
5	0.6	0.5
6	0.7	0.6
7	0.8	0.7
8	0.9	0.8
9	1.0	0.9
10	1.2	1.0
20	2.3	2.0
30	3.5	3.0
40	4.7	4.0
50	5.8	5.0

	7/32	7/31	6/32
0			
	2.3	2.2	2.7
1			
	6.9	6.6	8.0
2			
	11.4	11.1	13.3
3			
	16.0	15.5	18.7
4			
	20.6	19.9	24.0
5			
	25.1	24.4	29.3
6			
	29.7	28.8	—
7			

P P

*153° 243° *333° 63°

27° *117° 207° *297°

′	L Sin	d	L Tan	c d	L Cot	L Cos	d	
0	9.65 705̄	24	9.70 717	31	0.29 283	9.94 988	6	60
1	9.65 729	25	9.70 748	31	0.29 252	9.94 982	7	59
2	9.65 754	25	9.70 779	31	0.29 221	9.94 975	6	58
3	9.65 779	25	9.70 810	31	0.29 190	9.94 969	7	57
4	9.65 804	24	9.70 841	32	0.29 159	9.94 962	6	56
5	9.65 828	25	9.70 873	31	0.29 127	9.94 956	7	55
6	9.65 853	25	9.70 904	31	0.29 096	9.94 949	6	54
7	9.65 878	24	9.70 935̄	31	0.29 065	9.94 943	7	53
8	9.65 902	25	9.70 966	31	0.29 034	9.94 936	6	52
9	9.65 927	25	9.70 997	31	0.29 003	9.94 930	7	51
10	9.65 952	24	9.71 028	31	0.28 972	9.94 923	6	50
11	9.65 976	25	9.71 059	31	0.28 941	9.94 917	6	49
12	9.66 001	24	9.71 090	31	0.28 910	9.94 911	7	48
13	9.66 025	25	9.71 121	32	0.28 879	9.94 904	6	47
14	9.66 050	25	9.71 153	31	0.28 847	9.94 898	7	46
15	9.66 075̄	24	9.71 184	31	0.28 816	9.94 891	6	45
16	9.66 099	25	9.71 215̄	31	0.28 785	9.94 885̄	7	44
17	9.66 124	24	9.71 246	31	0.28 754	9.94 878	7	43
18	9.66 148	25	9.71 277	31	0.28 723	9.94 871	6	42
19	9.66 173	24	9.71 308	31	0.28 692	9.94 865̄	7	41
20	9.66 197	24	9.71 339	31	0.28 661	9.94 858	6	40
21	9.66 221	25	9.71 370	31	0.28 630	9.94 852	7	39
22	9.66 246	24	9.71 401	30	0.28 599	9.94 845	6	38
23	9.66 270	25	9.71 431	31	0.28 569	9.94 839	7	37
24	9.66 295̄	24	9.71 462	31	0.28 538	9.94 832	6	36
25	9.66 319	24	9.71 493	31	0.28 507	9.94 826	7	35
26	9.66 343	25	9.71 524	31	0.28 476	9.94 819	6	34
27	9.66 368	24	9.71 555	31	0.28 445̄	9.94 813	7	33
28	9.66 392	24	9.71 586	31	0.28 414	9.94 806	7	32
29	9.66 416	25	9.71 617	31	0.28 383	9.94 799	6	31
30	9.66 441	24	9.71 648	31	0.28 352	9.94 793	7	30
31	9.66 465̄	24	9.71 679	30	0.28 321	9.94 786	6	29
32	9.66 489	24	9.71 709	31	0.28 291	9.94 780	7	28
33	9.66 513	24	9.71 740	31	0.28 260	9.94 773	6	27
34	9.66 537	25	9.71 771	31	0.28 229	9.94 767	7	26
35	9.66 562	24	9.71 802	31	0.28 198	9.94 760	7	25
36	9.66 586	24	9.71 833	30	0.28 167	9.94 753	6	24
37	9.66 610	24	9.71 863	31	0.28 137	9.94 747	7	23
38	9.66 634	24	9.71 894	31	0.28 106	9.94 740	6	22
39	9.66 658	24	9.71 925̄	30	0.28 075	9.94 734	7	21
40	9.66 682	24	9.71 955	31	0.28 045̄	9.94 727	7	20
41	9.66 706	25	9.71 986	31	0.28 014	9.94 720	6	19
42	9.66 731	24	9.72 017	31	0.27 983	9.94 714	7	18
43	9.66 755̄	24	9.72 048	30	0.27 952	9.94 707	7	17
44	9.66 779	24	9.72 078	31	0.27 922	9.94 700	6	16
45	9.66 803	24	9.72 109	31	0.27 891	9.94 694	7	15
46	9.66 827	24	9.72 140	30	0.27 860	9.94 687	7	14
47	9.66 851	24	9.72 170	31	0.27 830	9.94 680	6	13
48	9.66 875̄	24	9.72 201	30	0.27 799	9.94 674	7	12
49	9.66 899	23	9.72 231	31	0.27 769	9.94 667	7	11
50	9.66 922	24	9.72 262	31	0.27 738	9.94 660	6	10
51	9.66 946	24	9.72 293	30	0.27 707	9.94 654	7	9
52	9.66 970	24	9.72 323	31	0.27 677	9.94 647	7	8
53	9.66 994	24	9.72 354	30	0.27 646	9.94 640	6	7
54	9.67 018	24	9.72 384	31	0.27 616	9.94 634	7	6
55	9.67 042	24	9.72 415̄	30	0.27 585	9.94 627	7	5
56	9.67 066	24	9.72 445	31	0.27 555̄	9.94 620	6	4
57	9.67 090	23	9.72 476	30	0.27 524	9.94 614	7	3
58	9.67 113	24	9.72 506	31	0.27 494	9.94 607	7	2
59	9.67 137	24	9.72 537	30	0.27 463	9.94 600	7	1
60	9.67 161		9.72 567		0.27 433	9.94 593		0
	L Cos	d	L Cot	c d	L Tan	L Sin	d	′

P P

	32	31	30
1	0.5	0.5	0.5
2	1.1	1.0	1.0
3	1.6	1.6	1.5
4	2.1	2.1	2.0
5	2.7	2.6	2.5
6	3.2	3.1	3.0
7	3.7	3.6	3.5
8	4.3	4.1	4.0
9	4.8	4.6	4.5
10	5.3	5.2	5.0
20	10.7	10.3	10.0
30	16.0	15.5	15.0
40	21.3	20.7	20.0
50	26.7	25.8	25.0

	25	24	23
1	0.4	0.4	0.4
2	0.8	0.8	0.8
3	1.2	1.2	1.2
4	1.7	1.6	1.5
5	2.1	2.0	1.9
6	2.5	2.4	2.3
7	2.9	2.8	2.7
8	3.3	3.2	3.1
9	3.8	3.6	3.4
10	4.2	4.0	3.8
20	8.3	8.0	7.7
30	12.5	12.0	11.5
40	16.7	16.0	15.3
50	20.8	20.0	19.2

	7	6
1	0.1	0.1
2	0.2	0.2
3	0.4	0.3
4	0.5̄	0.4
5	0.6	0.5
6	0.7	0.6
7	0.8	0.7
8	0.9	0.8
9	1.0	0.9
10	1.2	1.0
20	2.3	2.0
30	3.5	3.0
40	4.7	4.0
50	5.8	5.0

	7/30	6/31	6/30
0	2.1	2.6	2.5
1	6.4	7.8	7.5
2	10.7	12.9	12.5
3	15.0	18.1	17.5
4	19.3	23.2	22.5
5	23.6	28.4	27.5
6	27.9	—	—
7			

P P

*152° 242° *332° 62°

28° *118° 208° *298°

′	L Sin	d	L Tan	c d	L Cot	L Cos	d	
0	9.67 161		9.72 567		0.27 433	9.94 593		60
1	9.67 185̄	24	9.72 598	31	0.27 402	9.94 587	6	59
2	9.67 208	23	9.72 628	30	0.27 372	9.94 580	7	58
3	9.67 232	24	9.72 659	31	0.27 341	9.94 573	7	57
4	9.67 256	24	9.72 689	30	0.27 311	9.94 567	6	56
5	9.67 280	24	9.72 720	31	0.27 280	9.94 560	7	55
6	9.67 303	23	9.72 750	30	0.27 250̄	9.94 553	7	54
7	9.67 327	24	9.72 780	30	0.27 220	9.94 546	7	53
8	9.67 350	23	9.72 811	31	0.27 189	9.94 540	6	52
9	9.67 374	24	9.72 841	30	0.27 159	9.94 533	7	51
10	9.67 398	24	9.72 872	31	0.27 128	9.94 526	7	50
11	9.67 421	23	9.72 902	30	0.27 098	9.94 519	7	49
12	9.67 445̄	24	9.72 932	30	0.27 068	9.94 513	6	48
13	9.67 468	23	9.72 963	31	0.27 037	9.94 506	7	47
14	9.67 492	24	9.72 993	30	0.27 007	9.94 499	7	46
15	9.67 515	23	9.73 023	30	0.26 977	9.94 492	7	45
16	9.67 539	24	9.73 054	31	0.26 946	9.94 485	7	44
17	9.67 562	23	9.73 084	30	0.26 916	9.94 479	6	43
18	9.67 586	24	9.73 114	30	0.26 886	9.94 472	7	42
19	9.67 609	23	9.73 144	30	0.26 856	9.94 465	7	41
20	9.67 633	24	9.73 175̄	31	0.26 825	9.94 458	7	40
21	9.67 656	23	9.73 205̄	30	0.26 795	9.94 451	7	39
22	9.67 680	24	9.73 235	30	0.26 765̄	9.94 445̄	6	38
23	9.67 703	23	9.73 265	30	0.26 735̄	9.94 438	7	37
24	9.67 726	23	9.73 295	30	0.26 705̄	9.94 431	7	36
25	9.67 750̄	24	9.73 326	31	0.26 674	9.94 424	7	35
26	9.67 773	23	9.73 356	30	0.26 644	9.94 417	7	34
27	9.67 796	23	9.73 386	30	0.26 614	9.94 410	7	33
28	9.67 820	24	9.73 416	30	0.26 584	9.94 404	6	32
29	9.67 843	23	9.73 446	30	0.26 554	9.94 397	7	31
30	9.67 866	23	9.73 476	30	0.26 524	9.94 390	7	30
31	9.67 890	24	9.73 507	31	0.26 493	9.94 383	7	29
32	9.67 913	23	9.73 537	30	0.26 463	9.94 376	7	28
33	9.67 936	23	9.73 567	30	0.26 433	9.94 369	7	27
34	9.67 959	23	9.73 597	30	0.26 403	9.94 362	7	26
35	9.67 982	23	9.73 627	30	0.26 373	9.94 355	7	25
36	9.68 006	24	9.73 657	30	0.26 343	9.94 349	6	24
37	9.68 029	23	9.73 687	30	0.26 313	9.94 342	7	23
38	9.68 052	23	9.73 717	30	0.26 283	9.94 335̄	7	22
39	9.68 075	23	9.73 747	30	0.26 253	9.94 328	7	21
40	9.68 098	23	9.73 777	30	0.26 223	9.94 321	7	20
41	9.68 121	23	9.73 807	30	0.26 193	9.94 314	7	19
42	9.68 144	23	9.73 837	30	0.26 163	9.94 307	7	18
43	9.68 167	23	9.73 867	30	0.26 133	9.94 300	7	17
44	9.68 190	23	9.73 897	30	0.26 103	9.94 293	7	16
45	9.68 213	23	9.73 927	30	0.26 073	9.94 286	7	15
46	9.68 237	24	9.73 957	30	0.26 043	9.94 279	7	14
47	9.68 260	23	9.73 987	30	0.26 013	9.94 273	6	13
48	9.68 283	23	9.74 017	30	0.25 983	9.94 266	7	12
49	9.68 305	22	9.74 047	30	0.25 953	9.94 259	7	11
50	9.68 328	23	9.74 077	30	0.25 923	9.94 252	7	10
51	9.68 351	23	9.74 107	30	0.25 893	9.94 245̄	7	9
52	9.68 374	23	9.74 137	30	0.25 863	9.94 238	7	8
53	9.68 397	23	9.74 166	29	0.25 834	9.94 231	7	7
54	9.68 420	23	9.74 196	30	0.25 804	9.94 224	7	6
55	9.68 443	23	9.74 226	30	0.25 774	9.94 217	7	5
56	9.68 466	23	9.74 256	30	0.25 744	9.94 210	7	4
57	9.68 489	23	9.74 286	30	0.25 714	9.94 203	7	3
58	9.68 512	23	9.74 316	30	0.25 684	9.94 196	7	2
59	9.68 534	22	9.74 345	29	0.25 655̄	9.94 189	7	1
60	9.68 557	23	9.74 375	30	0.25 625̄	9.94 182	7	0
	L Cos	d	L Cot	c d	L Tan	L Sin	d	′

P P

	31	30	29
1	0.5	0.5	0.5̄
2	1.0	1.0	1.0
3	1.6	1.5	1.4
4	2.1	2.0	1.9
5	2.6	2.5	2.4
6	3.1	3.0	2.9
7	3.6	3.5	3.4
8	4.1	4.0	3.9
9	4.6	4.5	4.4
10	5.2	5.0	4.8
20	10.3	10.0	9.7
30	15.5	15.0	14.5
40	20.7	20.0	19.3
50	25.8	25.0	24.2

	24	23	22
1	0.4	0.4	0.4
2	0.8	0.8	0.7
3	1.2	1.2	1.1
4	1.6	1.5	1.5̄
5	2.0	1.9	1.8
6	2.4	2.3	2.2
7	2.8	2.7	2.6
8	3.2	3.1	2.9
9	3.6	3.4	3.3
10	4.0	3.8	3.7
20	8.0	7.7	7.3
30	12.0	11.5	11.0
40	16.0	15.3	14.7
50	20.0	19.2	18.3

	7	6
1	0.1	0.1
2	0.2	0.2
3	0.4	0.3
4	0.5̄	0.4
5	0.6	0.5
6	0.7	0.6
7	0.8	0.7
8	0.9	0.8
9	1.0	0.9
10	1.2	1.0
20	2.3	2.0
30	3.5	3.0
40	4.7	4.0
50	5.8	5.0

	7/31	6/31	6/30
0	2.2	2.6	2.5
1	6.6	7.8	7.5
2	11.1	12.9	12.5
3	15.5	18.1	17.5
4	10.9	23.2	22.5
5	24.4	28.4	27.5
6	28.8	—	—
7			

P P

*151° 241° *331° 61°

29°

*119° 209° *299°

′	L Sin	d	L Tan	c d	L Cot	L Cos	d	′
0	9.68 557	23	9.74 375	30	0.25 625̄	9.94 182	7	60
1	9.68 580	23	9.74 405̄	30	0.25 595	9.94 175̄	7	59
2	9.68 603	22	9.74 435̄	30	0.25 565	9.94 168	7	58
3	9.68 625	23	9.74 465̄	29	0.25 535	9.94 161	7	57
4	9.68 648	23	9.74 494	30	0.25 506	9.94 154	7	56
5	9.68 671	23	9.74 524	30	0.25 476	9.94 147	7	55
6	9.68 694	22	9.74 554	29	0.25 446	9.94 140	7	54
7	9.68 716	23	9.74 583	30	0.25 417	9.94 133	7	53
8	9.68 739	23	9.74 613	30	0.25 387	9.94 126	7	52
9	9.68 762	22	9.74 643	30	0.25 357	9.94 119	7	51
10	9.68 784	23	9.74 673	29	0.25 327	9.94 112	7	50
11	9.68 807	22	9.74 702	30	0.25 298	9.94 105̄	7	49
12	9.68 829	23	9.74 732	30	0.25 268	9.94 098	8	48
13	9.68 852	23	9.74 762	29	0.25 238	9.94 090	7	47
14	9.68 875̄	22	9.74 791	30	0.25 209	9.94 083	7	46
15	9.68 897	23	9.74 821	30	0.25 179	9.94 076	7	45
16	9.68 920	22	9.74 851	29	0.25 149	9.94 069	7	44
17	9.68 942	23	9.74 880	30	0.25 120	9.94 062	7	43
18	9.68 965̄	22	9.74 910	29	0.25 090	9.94 055	7	42
19	9.68 987	23	9.74 939	30	0.25 061	9.94 048	7	41
20	9.69 010	22	9.74 969	29	0.25 031	9.94 041	7	40
21	9.69 032	23	9.74 998	30	0.25 002	9.94 034	7	39
22	9.69 055̄	22	9.75 028	30	0.24 972	9.94 027	7	38
23	9.69 077	23	9.75 058	29	0.24 942	9.94 020	8	37
24	9.69 100	22	9.75 087	30	0.24 913	9.94 012	7	36
25	9.69 122	22	9.75 117	29	0.24 883	9.94 005	7	35
26	9.69 144	23	9.75 146	30	0.24 854	9.93 998	7	34
27	9.69 167	22	9.75 176	29	0.24 824	9.93 991	7	33
28	9.69 189	23	9.75 205	30	0.24 795̄	9.93 984	7	32
29	9.69 212	22	9.75 235̄	29	0.24 765	9.93 977	7	31
30	9.69 234	22	9.75 264	30	0.24 736	9.93 970	7	30
31	9.69 256	23	9.75 294	29	0.24 706	9.93 963	8	29
32	9.69 279	22	9.75 323	30	0.24 677	9.93 955	7	28
33	9.69 301	22	9.75 353	29	0.24 647	9.93 948	7	27
34	9.69 323	22	9.75 382	29	0.24 618	9.93 941	7	26
35	9.69 345	23	9.75 411	30	0.24 589	9.93 934	7	25
36	9.69 368	22	9.75 441	29	0.24 559	9.93 927	7	24
37	9.69 390	22	9.75 470	30	0.24 530	9.93 920	8	23
38	9.69 412	22	9.75 5̄00	29	0.24 500	9.93 912	7	22
39	9.69 434	22	9.75 529	29	0.24 471	9.93 905	7	21
40	9.69 456	23	9.75 558	30	0.24 442	9.93 898	7	20
41	9.69 479	22	9.75 588	29	0.24 412	9.93 891	7	19
42	9.69 501	22	9.75 617	30	0.24 383	9.93 884	8	18
43	9.69 523	22	9.75 647	29	0.24 353	9.93 876	7	17
44	9.69 545	22	9.75 676	29	0.24 324	9.93 869	7	16
45	9.69 567	22	9.75 705	30	0.24 295̄	9.93 862	7	15
46	9.69 589	22	9.75 735̄	29	0.24 265	9.93 855̄	8	14
47	9.69 611	22	9.75 764	29	0.24 236	9.93 847	7	13
48	9.69 633	22	9.75 793	29	0.24 207	9.93 840	7	12
49	9.69 655	22	9.75 822	30	0.24 178	9.93 833	7	11
50	9.69 677	22	9.75 852	29	0.24 148	9.93 826	7	10
51	9.69 699	22	9.75 881	29	0.24 119	9.93 819	8	9
52	9.69 721	22	9.75 910	29	0.24 090	9.93 811	7	8
53	9.69 743	22	9.75 939	30	0.24 061	9.93 804	7	7
54	9.69 765	22	9.75 969	29	0.24 031	9.93 797	8	6
55	9.69 787	22	9.75 998	29	0.24 002	9.93 789	7	5
56	9.69 809	22	9.76 027	29	0.23 973	9.93 782	7	4
57	9.69 831	22	9.76 056	30	0.23 944	9.93 775̄	7	3
58	9.69 853	22	9.76 086	29	0.23 914	9.93 768	8	2
59	9.69 875	22	9.76 115̄	29	0.23 885	9.93 760	7	1
60	9.69 897		9.76 144		0.23 856	9.93 753		0
	L Cos	d	L Cot	c d	L Tan	L Sin	d	′

P P

	30	29	23
1	0.5	0.5̄	0.4
2	1.0	1.0	0.8
3	1.5	1.4	1.2
4	2.0	1.9	1.5
5	2.5	2.4	1.9
6	3.0	2.9	2.3
7	3.5	3.4	2.7
8	4.0	3.9	3.1
9	4.5	4.4	3.4
10	5.0	4.8	3.8
20	10.0	9.7	7.7
30	15.0	14.5	11.5
40	20.0	19.3	15.3
50	25.0	24.2	19.2

	22	8	7
1	0.4	0.1	0.1
2	0.7	0.3	0.2
3	1.1	0.4	0.4
4	1.5̄	0.5	0.5̄
5	1.8	0.7	0.6
6	2.2	0.8	0.7
7	2.6	0.9	0.8
8	2.9	1.1	0.9
9	3.3	1.2	1.0
10	3.7	1.3	1.2
20	7.3	2.7	2.3
30	11.0	4.0	3.5
40	14.7	5.3	4.7
50	18.3	6.7	5.8

	8/30	8/29
0		
	1.9	1.8
1		
	5.6	5.4
2		
	9.4	9.1
3		
	13.1	12.7
4		
	16.9	16.3
5		
	20.6	19.9
6		
	24.4	23.6
7		
	28.1	27.2
8		

	7/30	7/29
0		
	2.1	2.1
1		
	6.4	6.2
2		
	10.7	10.4
3		
	15.0	14.5
4		
	19.3	18.6
5		
	23.6	22.8
6		
	27.9	26.9
7		

P P

*150° 240° *330°

60°

′	L Sin	d	L Tan	c d	L Cot	L Cos	d	
0	9.69 897	22	9.76 144	29	0.23 856	9.93 753	7	60
1	9.69 919	22	9.76 173	29	0.23 827	9.93 746	8	59
2	9.69 941	22	9.76 202	29	0.23 798	9.93 738	7	58
3	9.69 963	21	9.76 231	30	0.23 769	9.93 731	7	57
4	9.69 984	22	9.76 261	29	0.23 739	9.93 724	7	56
5	9.70 006	22	9.76 290	29	0.23 710	9.93 717	8	55
6	9.70 028	22	9.76 319	29	0.23 681	9.93 709	7	54
7	9.70 050	22	9.76 348	29	0.23 652	9.93 702	7	53
8	9.70 072	21	9.76 377	29	0.23 623	9.93 695	8	52
9	9.70 093	22	9.76 406	29	0.23 594	9.93 687	7	51
10	9.70 115	22	9.76 435	29	0.23 565̄	9.93 680	7	50
11	9.70 137	22	9.76 464	29	0.23 536	9.93 673	8	49
12	9.70 159	21	9.76 493	29	0.23 507	9.93 665	7	48
13	9.70 180	22	9.76 522	29	0.23 478	9.93 658	8	47
14	9.70 202	22	9.76 551	29	0.23 449	9.93 650	7	46
15	9.70 224	21	9.76 580	29	0.23 420	9.93 643	7	45
16	9.70 245	22	9.76 609	30	0.23 391	9.93 636	8	44
17	9.70 267	21	9.76 639	29	0.23 361	9.93 628	7	43
18	9.70 288	22	9.76 668	29	0.23 332	9.93 621	7	42
19	9.70 310	22	9.76 697	28	0.23 303	9.93 614	8	41
20	9.70 332	21	9.76 725	29	0.23 275̄	9.93 606	7	40
21	9.70 353	22	9.76 754	29	0.23 246	9.93 599	8	39
22	9.70 375̄	21	9.76 783	29	0.23 217	9.93 591	7	38
23	9.70 396	22	9.76 812	29	0.23 188	9.93 584	7	37
24	9.70 418	21	9.76 841	29	0.23 159	9.93 577	8	36
25	9.70 439	22	9.76 870	29	0.23 130	9.93 569	7	35
26	9.70 461	21	9.76 899	29	0.23 101	9.93 562	8	34
27	9.70 482	22	9.76 928	29	0.23 072	9.93 554	7	33
28	9.70 504	21	9.76 957	29	0.23 043	9.93 547	8	32
29	9.70 525	22	9.76 986	29	0.23 014	9.93 539	7	31
30	9.70 547	21	9.77 015̄	29	0.22 985	9.93 532	7	30
31	9.70 568	22	9.77 044	29	0.22 956	9.93 525̄	8	29
32	9.70 590	21	9.77 073	28	0.22 927	9.93 517	7	28
33	9.70 611	22	9.77 101	29	0.22 899	9.93 510	8	27
34	9.70 633	21	9.77 130	29	0.22 870	9.93 502	7	26
35	9.70 654	21	9.77 159	29	0.22 841	9.93 495̄	8	25
36	9.70 675	22	9.77 188	29	0.22 812	9.93 487	7	24
37	9.70 697	21	9.77 217	29	0.22 783	9.93 480	8	23
38	9.70 718	21	9.77 246	28	0.22 754	9.93 472	7	22
39	9.70 739	22	9.77 274	29	0.22 726	9.93 465̄	8	21
40	9.70 761	21	9.77 303	29	0.22 697	9.93 457	7	20
41	9.70 782	21	9.77 332	29	0.22 668	9.93 450̄	8	19
42	9.70 803	21	9.77 361	29	0.22 639	9.93 442	7	18
43	9.70 824	22	9.77 390	28	0.22 610	9.93 435̄	8	17
44	9.70 846	21	9.77 418	29	0.22 582	9.93 427	7	16
45	9.70 867	21	9.77 447	29	0.22 553	9.93 420	8	15
46	9.70 888	21	9.77 476	29	0.22 524	9.93 412	7	14
47	9.70 909	22	9.77 505̄	28	0.22 495	9.93 405̄	8	13
48	9.70 931	21	9.77 533	29	0.22 467	9.93 397	7	12
49	9.70 952	21	9.77 562	29	0.22 438	9.93 390	8	11
50	9.70 973	21	9.77 591	28	0.22 409	9.93 382	7	10
51	9.70 994	21	9.77 619	29	0.22 381	9.93 375̄	8	9
52	9.71 015	21	9.77 648	29	0.22 352	9.93 367	7	8
53	9.71 036	22	9.77 677	29	0.22 323	9.93 360	8	7
54	9.71 058	21	9.77 706	28	0.22 294	9.93 352	8	6
55	9.71 079	21	9.77 734	29	0.22 266	9.93 344	7	5
56	9.71 100	21	9.77 763	28	0.22 237	9.93 337	8	4
57	9.71 121	21	9.77 791	29	0.22 209	9.93 329	7	3
58	9.71 142	21	9.77 820	29	0.22 180	9.93 322	8	2
59	9.71 163	21	9.77 849	28	0.22 151	9.93 314	7	1
60	9.71 184		9.77 877		0.22 123	9.93 307		0
	L Cos	d	L Cot	c d	L Tan	L Sin	d	′

P P

	30	29	28
1	0.5	0.5̄	0.5̄
2	1.0	1.0	0.9
3	1.5	1.4	1.4
4	2.0	1.9	1.9
5	2.5	2.4	2.3
6	3.0	2.9	2.8
7	3.5	3.4	3.3
8	4.0	3.9	3.7
9	4.5	4.4	4.2
10	5.0	4.8	4.7
20	10.0	9.7	9.3
30	15.0	14.5	14.0
40	20.0	19.3	18.7
50	25.0	24.2	23.3

	22	21
1	0.4	0.4
2	0.7	0.7
3	1.1	1.0
4	1.5̄	1.4
5	1.8	1.8
6	2.2	2.1
7	2.6	2.4
8	2.9	2.8
9	3.3	3.2
10	3.7	3.5
20	7.3	7.0
30	11.0	10.5
40	14.7	14.0
50	18.3	17.5

	8	7
1	0.1	0.1
2	0.3	0.2
3	0.4	0.4
4	0.5	0.5̄
5	0.7	0.6
6	0.8	0.7
7	0.9	0.8
8	1.1	0.9
9	1.2	1.0
10	1.3	1.2
20	2.7	2.3
30	4.0	3.5
40	5.3	4.7
50	6.7	5.8

	7/30	7/29	7/28
0	2.1	2.1	2.0
1	6.4	6.2	6.0
2	10.7	10.4	10.0
3	15.0	14.5	14.0
4	19.3	18.6	18.0
5	23.6	22.8	22.0
6	27.9	26.9	26.0
7			

P P

31°

′	L Sin	d	L Tan	c d	L Cot	L Cos	d	
0	9.71 184	21	9.77 877	29	0.22 123	9.93 307	8	60
1	9.71 205̄	21	9.77 906	29	0.22 094	9.93 299	8	59
2	9.71 226	21	9.77 935̄	28	0.22 065	9.93 291	7	58
3	9.71 247	21	9.77 963	29	0.22 037	9.93 284	8	57
4	9.71 268	21	9.77 992	28	0.22 008	9.93 276	7	56
5	9.71 289	21	9.78 020	29	0.21 980	9.93 269	8	55
6	9.71 310	21	9.78 049	28	0.21 951	9.93 261	8	54
7	9.71 331	21	9.78 077	29	0.21 923	9.93 253	7	53
8	9.71 352	21	9.78 106	29	0.21 894	9.93 246	8	52
9	9.71 373	20	9.78 135̄	28	0.21 865	9.93 238	8	51
10	9.71 393	21	9.78 163	29	0.21 837	9.93 230	7	50
11	9.71 414	21	9.78 192	28	0.21 808	9.93 223	8	49
12	9.71 435	21	9.78 220	29	0.21 780	9.93 215	8	48
13	9.71 456	21	9.78 249	28	0.21 751	9.93 207	7	47
14	9.71 477	21	9.78 277	29	0.21 723	9.93 200	8	46
15	9.71 498	21	9.78 306	28	0.21 694	9.93 192	8	45
16	9.71 519	20	9.78 334	29	0.21 666	9.93 184	7	44
17	9.71 539	21	9.78 363	28	0.21 637	9.93 177	8	43
18	9.71 560	21	9.78 391	28	0.21 609	9.93 169	8	42
19	9.71 581	21	9.78 419	29	0.21 581	9.93 161	7	41
20	9.71 602	20	9.78 448	28	0.21 552	9.93 154	8	40
21	9.71 622	21	9.78 476	29	0.21 524	9.93 146	8	39
22	9.71 643	21	9.78 505̄	28	0.21 495	9.93 138	7	38
23	9.71 664	21	9.78 533	29	0.21 467	9.93 131	8	37
24	9.71 685̄	20	9.78 562	28	0.21 438	9.93 123	8	36
25	9.71 705	21	9.78 590	28	0.21 410	9.93 115	7	35
26	9.71 726	21	9.78 618	29	0.21 382	9.93 108	8	34
27	9.71 747	20	9.78 647	28	0.21 353	9.93 100	8	33
28	9.71 767	21	9.78 675	29	0.21 325̄	9.93 092	8	32
29	9.71 788	21	9.78 704	28	0.21 296	9.93 084	7	31
30	9.71 809	20	9.78 732	28	0.21 268	9.93 077	8	30
31	9.71 829	21	9.78 760	29	0.21 240	9.93 069	8	29
32	9.71 85̄0	20	9.78 789	28	0.21 211	9.93 061	8	28
33	9.71 870	21	9.78 817	28	0.21 183	9.93 053	7	27
34	9.71 891	20	9.78 845	29	0.21 155̄	9.93 046	8	26
35	9.71 911	21	9.78 874	28	0.21 126	9.93 038	8	25
36	9.71 932	20	9.78 902	28	0.21 098	9.93 030	8	24
37	9.71 952	21	9.78 930	29	0.21 070	9.93 022	8	23
38	9.71 973	21	9.78 959	28	0.21 041	9.93 014	7	22
39	9.71 994	20	9.78 987	28	0.21 013	9.93 007	8	21
40	9.72 014	20	9.79 015	28	0.20 985̄	9.92 999	8	20
41	9.72 034	21	9.79 043	29	0.20 957	9.92 991	8	19
42	9.72 055̄	20	9.79 072	28	0.20 928	9.92 983	7	18
43	9.72 075	21	9.79 100	28	0.20 900	9.92 976	8	17
44	9.72 096	20	9.79 128	28	0.20 872	9.92 968	8	16
45	9.72 116	21	9.79 156	29	0.20 844	9.92 960	8	15
46	9.72 137	20	9.79 185̄	28	0.20 815	9.92 952	8	14
47	9.72 157	20	9.79 213	28	0.20 787	9.92 944	8	13
48	9.72 177	21	9.79 241	28	0.20 759	9.92 936	7	12
49	9.72 198	20	9.79 269	28	0.20 731	9.92 929	8	11
50	9.72 218	20	9.79 297	29	0.20 703	9.92 921	8	10
51	9.72 238	21	9.79 326	28	0.20 674	9.92 913	8	9
52	9.72 259	20	9.79 354	28	0.20 646	9.92 905	8	8
53	9.72 279	20	9.79 382	28	0.20 618	9.92 897	8	7
54	9.72 299	21	9.79 410	28	0.20 590	9.92 889	8	6
55	9.72 320	20	9.79 438	28	0.20 562	9.92 881	7	5
56	9.72 340	20	9.79 466	29	0.20 534	9.92 874	8	4
57	9.72 360	21	9.79 495̄	28	0.20 505	9.92 866	8	3
58	9.72 381	20	9.79 523	28	0.20 477	9.92 858	8	2
59	9.72 401	20	9.79 551	28	0.20 449	9.92 85̄0	8	1
60	9.72 421		9.79 579		0.20 421	9.92 842		0
	L Cos	d	L Cot	c d	L Tan	L Sin	d	′

P P

	29	28
1	0.5̄	0.5̄
2	1.0	0.9
3	1.4	1.4
4	1.9	1.9
5	2.4	2.3
6	2.9	2.8
7	3.4	3.3
8	3.9	3.7
9	4.4	4.2
10	4.8	4.7
20	9.7	9.3
30	14.5	14.0
40	19.3	18.7
50	24.2	23.3

	21	20
1	0.4	0.3
2	0.7	0.7
3	1.0	1.0
4	1.4	1.3
5	1.8	1.7
6	2.1	2.0
7	2.4	2.3
8	2.8	2.7
9	3.2	3.0
10	3.5	3.3
20	7.0	6.7
30	10.5	10.0
40	14.0	13.3
50	17.5	16.7

	8	7
1	0.1	0.1
2	0.3	0.2
3	0.4	0.4
4	0.5	0.5̄
5	0.7	0.6
6	0.8	0.7
7	0.9	0.8
8	1.1	0.9
9	1.2	1.0
10	1.3	1.2
20	2.7	2.3
30	4.0	3.5
40	5.3	4.7
50	6.7	5.8

	8/30	8/29	8/28
0	1.9	1.8	1.8
1	5.6	5.4	5.2
2	9.4	9.1	8.8
3	13.1	12.7	12.2
4	16.9	16.3	15.8
5	20.6	19.9	19.2
6	24.4	23.6	22.8
7	28.1	27.2	26.2
8			

P P

58°

32°

*122° 212° *302°

′	L Sin	d	L Tan	c d	L Cot	L Cos	d	′
0	9.72 421	20	9.79 579	28	0.20 421	9.92 842	8	60
1	9.72 441	20	9.79 607	28	0.20 393	9.92 834	8	59
2	9.72 461	21	9.79 635	28	0.20 365̄	9.92 826	8	58
3	9.72 482	20	9.79 663	28	0.20 337	9.92 818	8	57
4	9.72 502	20	9.79 691	28	0.20 309	9.92 810	7	56
5	9.72 522	20	9.79 719	28	0.20 281	9.92 803	8	55
6	9.72 542	20	9.79 747	29	0.20 253	9.92 795̄	8	54
7	9.72 562	20	9.79 776	28	0.20 224	9.92 787	8	53
8	9.72 582	20	9.79 804	28	0.20 196	9.92 779	8	52
9	9.72 602	20	9.79 832	28	0.20 168	9.92 771	8	51
10	9.72 622	21	9.79 860	28	0.20 140	9.92 763	8	50
11	9.72 643	20	9.79 888	28	0.20 112	9.92 755̄	8	49
12	9.72 663	20	9.79 916	28	0.20 084	9.92 747	8	48
13	9.72 683	20	9.79 944	28	0.20 056	9.92 739	8	47
14	9.72 703	20	9.79 972	28	0.20 028	9.92 731	8	46
15	9.72 723	20	9.80 000	28	0.20 000	9.92 723	8	45
16	9.72 743	20	9.80 028	28	0.19 972	9.92 715	8	44
17	9.72 763	20	9.80 056	28	0.19 944	9.92 707	8	43
18	9.72 783	20	9.80 084	28	0.19 916	9.92 699	8	42
19	9.72 803	20	9.80 112	28	0.19 888	9.92 691	8	41
20	9.72 823	20	9.80 140	28	0.19 860	9.92 683	8	40
21	9.72 843	20	9.80 168	27	0.19 832	9.92 675	8	39
22	9.72 863	20	9.80 195	28	0.19 805̄	9.92 667	8	38
23	9.72 883	19	9.80 223	28	0.19 777	9.92 659	8	37
24	9.72 902	20	9.80 251	28	0.19 749	9.92 651	8	36
25	9.72 922	20	9.80 279	28	0.19 721	9.92 643	8	35
26	9.72 942	20	9.80 307	28	0.19 693	9.92 635	8	34
27	9.72 962	20	9.80 335	28	0.19 665̄	9.92 627	8	33
28	9.72 982	20	9.80 363	28	0.19 637	9.92 619	8	32
29	9.73 002	20	9.80 391	28	0.19 609	9.92 611	8	31
30	9.73 022	19	9.80 419	28	0.19 581	9.92 603	8	30
31	9.73 041	20	9.80 447	27	0.19 553	9.92 595̄	8	29
32	9.73 061	20	9.80 474	28	0.19 526	9.92 587	8	28
33	9.73 081	20	9.80 502	28	0.19 498	9.92 579	8	27
34	9.73 101	20	9.80 530	28	0.19 470	9.92 571	8	26
35	9.73 121	19	9.80 558	28	0.19 442	9.92 563	8	25
36	9.73 140	20	9.80 586	28	0.19 414	9.92 555̄	9	24
37	9.73 160	20	9.80 614	28	0.19 386	9.92 546	8	23
38	9.73 180	20	9.80 642	27	0.19 358	9.92 538	8	22
39	9.73 200	19	9.80 669	28	0.19 331	9.92 530	8	21
40	9.73 219	20	9.80 697	28	0.19 303	9.92 522	8	20
41	9.73 239	20	9.80 725̄	28	0.19 275	9.92 514	8	19
42	9.73 259	19	9.80 753	28	0.19 247	9.92 506	8	18
43	9.73 278	20	9.80 781	27	0.19 219	9.92 498	8	17
44	9.73 298	20	9.80 808	28	0.19 192	9.92 490	8	16
45	9.73 318	19	9.80 836	28	0.19 164	9.92 482	9	15
46	9.73 337	20	9.80 864	28	0.19 136	9.92 473	8	14
47	9.73 357	20	9.80 892	27	0.19 108	9.92 465	8	13
48	9.73 377	19	9.80 919	28	0.19 081	9.92 457	8	12
49	9.73 396	20	9.80 947	28	0.19 053	9.92 449	8	11
50	9.73 416	19	9.80 975̄	28	0.19 025	9.92 441	8	10
51	9.73 435	20	9.81 003	27	0.18 997	9.92 433	8	9
52	9.73 455̄	19	9.81 030	28	0.18 970	9.92 425̄	9	8
53	9.73 474	20	9.81 058	28	0.18 942	9.92 416	8	7
54	9.73 494	19	9.81 086	27	0.18 914	9.92 408	8	6
55	9.73 513	20	9.81 113	28	0.18 887	9.92 400	8	5
56	9.73 533	19	9.81 141	28	0.18 859	9.92 392	8	4
57	9.73 552	20	9.81 169	27	0.18 831	9.92 384	8	3
58	9.73 572	19	9.81 196	28	0.18 804	9.92 376	9	2
59	9.73 591	20	9.81 224	28	0.18 776	9.92 367	8	1
60	9.73 611		9.81 252		0.18 748	9.92 359		0
	L Cos	d	L Cot	c d	L Tan	L Sin	d	′

P P

	29	28	27
1	0.5̄	0.5̄	0.4
2	1.0	0.9	0.9
3	1.4	1.4	1.4
4	1.9	1.9	1.8
5	2.4	2.3	2.2
6	2.9	2.8	2.7
7	3.4	3.3	3.2
8	3.9	3.7	3.6
9	4.4	4.2	4.0
10	4.8	4.7	4.5
20	9.7	9.3	9.0
30	14.5	14.0	13.5
40	19.3	18.7	18.0
50	24.2	23.3	22.5

	21	20	19
1	0.4	0.3	0.3
2	0.7	0.7	0.6
3	1.0	1.0	1.0
4	1.4	1.3	1.3
5	1.8	1.7	1.6
6	2.1	2.0	1.9
7	2.4	2.3	2.2
8	2.8	2.7	2.5
9	3.2	3.0	2.8
10	3.5	3.3	3.2
20	7.0	6.7	6.3
30	10.5	10.0	9.5
40	14.0	13.3	12.7
50	17.5	16.7	15.8

	9	8	7
1	0.2	0.1	0.1
2	0.3	0.3	0.2
3	0.4	0.4	0.4
4	0.6	0.5	0.5̄
5	0.8	0.7	0.6
6	0.9	0.8	0.7
7	1.0	0.9	0.8
8	1.2	1.1	0.9
9	1.4	1.2	1.0
10	1.5	1.3	1.2
20	3.0	2.7	2.3
30	4.5	4.0	3.5
40	6.0	5.3	4.7
50	7.5	6.7	5.8

	8/29	8/28	7/28
0	1.8	1.8	2.0
1	5.4	5.2	6.0
2	9.1	8.8	10.0
3	12.7	12.2	14.0
4	16.3	15.8	18.0
5	19.9	19.2	22.0
6	23.6	22.8	26.0
7	27.2	26.2	—
8			

P P

*147° 237° *327°

57°

33° *123° 213° *303°

′	L Sin	d	L Tan	c d	L Cot	L Cos	d	
0	9.73 611		9.81 252		0.18 748	9.92 359		60
1	9.73 630	19	9.81 279	27	0.18 721	9.92 351	8	59
2	9.73 650	20	9.81 307	28	0.18 693	9.92 343	8	58
3	9.73 669	19	9.81 335̄	28	0.18 665	9.92 335̄	8	57
4	9.73 689	20	9.81 362	27	0.18 638	9.92 326	9	56
5	9.73 708	19	9.81 390	28	0.18 610	9.92 318	8	55
6	9.73 727	19	9.81 418	28	0.18 582	9.92 310	8	54
7	9.73 747	20	9.81 445	27	0.18 555̄	9.92 302	8	53
8	9.73 766	19	9.81 473	28	0.18 527	9.92 293	9	52
9	9.73 785	19	9.81 500	27	0.18 500	9.92 285	8	51
10	9.73 805̄	20	9.81 528	28	0.18 472	9.92 277	8	50
11	9.73 824	19	9.81 556	28	0.18 444	9.92 269	8	49
12	9.73 843	19	9.81 583	27	0.18 417	9.92 260	9	48
13	9.73 863	20	9.81 611	28	0.18 389	9.92 252	8	47
14	9.73 882	19	9.81 638	27	0.18 362	9.92 244	8	46
15	9.73 901	19	9.81 666	28	0.18 334	9.92 235	9	45
16	9.73 921	20	9.81 693	27	0.18 307	9.92 227	8	44
17	9.73 940	19	9.81 721	28	0.18 279	9.92 219	8	43
18	9.73 959	19	9.81 748	27	0.18 252	9.92 211	8	42
19	9.73 978	19	9.81 776	28	0.18 224	9.92 202	9	41
20	9.73 997	19	9.81 803	27	0.18 197	9.92 194	8	40
21	9.74 017	20	9.81 831	28	0.18 169	9.92 186	8	39
22	9.74 036	19	9.81 858	27	0.18 142	9.92 177	9	38
23	9.74 055	19	9.81 886	28	0.18 114	9.92 169	8	37
24	9.74 074	19	9.81 913	27	0.18 087	9.92 161	8	36
25	9.74 093	19	9.81 941	28	0.18 059	9.92 152	9	35
26	9.74 113	20	9.81 968	27	0.18 032	9.92 144	8	34
27	9.74 132	19	9.81 996	28	0.18 004	9.92 136	8	33
28	9.74 151	19	9.82 023	27	0.17 977	9.92 127	9	32
29	9.74 170	19	9.82 051	28	0.17 949	9.92 119	8	31
30	9.74 189	19	9.82 078	27	0.17 922	9.92 111	8	30
31	9.74 208	19	9.82 106	28	0.17 894	9.92 102	9	29
32	9.74 227	19	9.82 133	27	0.17 867	9.92 094	8	28
33	9.74 246	19	9.82 161	28	0.17 839	9.92 086	8	27
34	9.74 265	19	9.82 188	27	0.17 812	9.92 077	9	26
35	9.74 284	19	9.82 215	27	0.17 785̄	9.92 069	8	25
36	9.74 303	19	9.82 243	28	0.17 757	9.92 060	9	24
37	9.74 322	19	9.82 270	27	0.17 730	9.92 052	8	23
38	9.74 341	19	9.82 298	28	0.17 702	9.92 044	8	22
39	9.74 360	19	9.82 325	27	0.17 675̄	9.92 035	9	21
40	9.74 379	19	9.82 352	27	0.17 648	9.92 027	8	20
41	9.74 398	19	9.82 380	28	0.17 620	9.92 018	9	19
42	9.74 417	19	9.82 407	27	0.17 593	9.92 010	8	18
43	9.74 436	19	9.82 435̄	28	0.17 565	9.92 002	8	17
44	9.74 455̄	19	9.82 462	27	0.17 538	9.91 993	9	16
45	9.74 474	19	9.82 489	27	0.17 511	9.91 985̄	8	15
46	9.74 493	19	9.82 517	28	0.17 483	9.91 976	9	14
47	9.74 512	19	9.82 544	27	0.17 456	9.91 968	8	13
48	9.74 531	19	9.82 571	27	0.17 429	9.91 959	9	12
49	9.74 549	18	9.82 599	28	0.17 401	9.91 951	8	11
50	9.74 568	19	9.82 626	27	0.17 374	9.91 942	9	10
51	9.74 587	19	9.82 653	27	0.17 347	9.91 934	8	9
52	9.74 606	19	9.82 681	28	0.17 319	9.91 925	9	8
53	9.74 625̄	19	9.82 708	27	0.17 292	9.91 917	8	7
54	9.74 644	19	9.82 735	27	0.17 265̄	9.91 908	9	6
55	9.74 662	18	9.82 762	27	0.17 238	9.91 900	8	5
56	9.74 681	19	9.82 790	28	0.17 210	9.91 891	9	4
57	9.74 700	19	9.82 817	27	0.17 183	9.91 883	8	3
58	9.74 719	19	9.82 844	27	0.17 156	9.91 874	9	2
59	9.74 737	18	9.82 871	27	0.17 129	9.91 866	8	1
60	9.74 756	19	9.82 899	28	0.17 101	9.91 857	9	0
	L Cos	d	L Cot	c d	L Tan	L Sin	d	′

P P

	28	27
1	0.5̄	0.4
2	0.9	0.9
3	1.4	1.4
4	1.9	1.8
5	2.3	2.2
6	2.8	2.7
7	3.3	3.2
8	3.7	3.6
9	4.2	4.0
10	4.7	4.5
20	9.3	9.0
30	14.0	13.5
40	18.7	18.0
50	23.3	22.5

	20	19	18
1	0.3	0.3	0.3
2	0.7	0.6	0.6
3	1.0	1.0	0.9
4	1.3	1.3	1.2
5	1.7	1.6	1.5
6	2.0	1.9	1.8
7	2.3	2.2	2.1
8	2.7	2.5	2.4
9	3.0	2.8	2.7
10	3.3	3.2	3.0
20	6.7	6.3	6.0
30	10.0	9.5	9.0
40	13.3	12.7	12.0
50	16.7	15.8	15.0

	9	8
1	0.2	0.1
2	0.3	0.3
3	0.4	0.4
4	0.6	0.5
5	0.8	0.7
6	0.9	0.8
7	1.0	0.9
8	1.2	1.1
9	1.4	1.2
10	1.5	1.3
20	3.0	2.7
30	4.5	4.0
40	6.0	5.3
50	7.5	6.7

	9/28	9/27	8/27
0	1.6	1.5	1.7
1	4.7	4.5	5.1
2	7.8	7.5	8.4
3	10.9	10.5	11.8
4	14.0	13.5	15.2
5	17.1	16.5	18.6
6	20.2	19.5	21.9
7	23.3	22.5	25.3
8	26.4	25.5	—
9			

P P

*146° 236° *326° 56°

34°

*124° 214° *304°

′	L Sin	d	L Tan	c d	L Cot	L Cos	d	
0	9.74 756	19	9.82 899	27	0.17 101	9.91 857	8	60
1	9.74 775̄	19	9.82 926	27	0.17 074	9.91 849	9	59
2	9.74 794	18	9.82 953	27	0.17 047	9.91 840	8	58
3	9.74 812	19	9.82 980	28	0.17 020	9.91 832	9	57
4	9.74 831	19	9.83 008	27	0.16 992	9.91 823	8	56
5	9.74 850	18	9.83 035̄	27	0.16 965	9.91 815̄	9	55
6	9.74 868	19	9.83 062	27	0.16 938	9.91 806	8	54
7	9.74 887	19	9.83 089	28	0.16 911	9.91 798	9	53
8	9.74 906	18	9.83 117	27	0.16 883	9.91 789	8	52
9	9.74 924	19	9.83 144	27	0.16 856	9.91 781	9	51
10	9.74 943	18	9.83 171	27	0.16 829	9.91 772	9	50
11	9.74 961	19	9.83 198	27	0.16 802	9.91 763	8	49
12	9.74 980	19	9.83 225	27	0.16 775̄	9.91 755̄	9	48
13	9.74 999	18	9.83 252	28	0.16 748	9.91 746	8	47
14	9.75 017	19	9.83 280	27	0.16 720	9.91 738	9	46
15	9.75 036	18	9.83 307	27	0.16 693	9.91 729	9	45
16	9.75 054	19	9.83 334	27	0.16 666	9.91 720	8	44
17	9.75 073	18	9.83 361	27	0.16 639	9.91 712	9	43
18	9.75 091	19	9.83 388	27	0.16 612	9.91 703	8	42
19	9.75 110	18	9.83 415	27	0.16 585̄	9.91 695̄	9	41
20	9.75 128	19	9.83 442	28	0.16 558	9.91 686	9	40
21	9.75 147	18	9.83 470	27	0.16 530	9.91 677	8	39
22	9.75 165	19	9.83 497	27	0.16 503	9.91 669	9	38
23	9.75 184	18	9.83 524	27	0.16 476	9.91 660	9	37
24	9.75 202	19	9.83 551	27	0.16 449	9.91 651	8	36
25	9.75 221	18	9.83 578	27	0.16 422	9.91 643	9	35
26	9.75 239	19	9.83 605	27	0.16 395̄	9.91 634	9	34
27	9.75 258	18	9.83 632	27	0.16 368	9.91 625	8	33
28	9.75 276	18	9.83 659	27	0.16 341	9.91 617	9	32
29	9.75 294	19	9.83 686	27	0.16 314	9.91 608	9	31
30	9.75 313	18	9.83 713	27	0.16 287	9.91 599	8	30
31	9.75 331	19	9.83 740	28	0.16 260	9.91 591	9	29
32	9.75 350	18	9.83 768	27	0.16 232	9.91 582	9	28
33	9.75 368	18	9.83 795̄	27	0.16 205	9.91 573	8	27
34	9.75 386	19	9.83 822	27	0.16 178	9.91 565̄	9	26
35	9.75 405̄	18	9.83 849	27	0.16 151	9.91 556	9	25
36	9.75 423	18	9.83 876	27	0.16 124	9.91 547	9	24
37	9.75 441	18	9.83 903	27	0.16 097	9.91 538	8	23
38	9.75 459	19	9.83 930	27	0.16 070	9.91 530	9	22
39	9.75 478	18	9.83 957	27	0.16 043	9.91 521	9	21
40	9.75 496	18	9.83 984	27	0.16 016	9.91 512	8	20
41	9.75 514	19	9.84 011	27	0.15 989	9.91 504	9	19
42	9.75 533	18	9.84 038	27	0.15 962	9.91 495̄	9	18
43	9.75 551	18	9.84 065̄	27	0.15 935	9.91 486	9	17
44	9.75 569	18	9.84 092	27	0.15 908	9.91 477	8	16
45	9.75 587	18	9.84 119	27	0.15 881	9.91 469	9	15
46	9.75 605	19	9.84 146	27	0.15 854	9.91 460	9	14
47	9.75 624	18	9.84 173	27	0.15 827	9.91 451	9	13
48	9.75 642	18	9.84 200	27	0.15 800	9.91 442	9	12
49	9.75 660	18	9.84 227	27	0.15 773	9.91 433	8	11
50	9.75 678	18	9.84 254	26	0.15 746	9.91 425̄	9	10
51	9.75 696	18	9.84 280	27	0.15 720	9.91 416	9	9
52	9.75 714	19	9.84 307	27	0.15 693	9.91 407	9	8
53	9.75 733	18	9.84 334	27	0.15 666	9.91 398	9	7
54	9.75 751	18	9.84 361	27	0.15 639	9.91 389	8	6
55	9.75 769	18	9.84 388	27	0.15 612	9.91 381	9	5
56	9.75 787	18	9.84 415	27	0.15 585̄	9.91 372	9	4
57	9.75 805̄	18	9.84 442	27	0.15 558	9.91 363	9	3
58	9.75 823	18	9.84 469	27	0.15 531	9.91 354	9	2
59	9.75 841	18	9.84 496	27	0.15 504	9.91 345	9	1
60	9.75 859		9.84 523		0.15 477	9.91 336		0
	L Cos	d	L Cot	c d	L Tan	L Sin	d	′

P P

	28	27	26
1	0.5̄	0.4	0.4
2	0.9	0.9	0.9
3	1.4	1.4	1.3
4	1.9	1.8	1.7
5	2.3	2.2	2.2
6	2.8	2.7	2.6
7	3.3	3.2	3.0
8	3.7	3.6	3.5̄
9	4.2	4.0	3.9
10	4.7	4.5	4.3
20	9.3	9.0	8.7
30	14.0	13.5	13.0
40	18.7	18.0	17.3
50	23.3	22.5	21.7

	19	18
1	0.3	0.3
2	0.6	0.6
3	1.0	0.9
4	1.3	1.2
5	1.6	1.5
6	1.9	1.8
7	2.2	2.1
8	2.5	2.4
9	2.8	2.7
10	3.2	3.0
20	6.3	6.0
30	9.5	9.0
40	12.7	12.0
50	15.8	15.0

	9	8
1	0.2	0.1
2	0.3	0.3
3	0.4	0.4
4	0.6	0.5
5	0.8	0.7
6	0.9	0.8
7	1.0	0.9
8	1.2	1.1
9	1.4	1.2
10	1.5	1.3
20	3.0	2.7
30	4.5	4.0
40	6.0	5.3
50	7.5	6.7

	9/28	8/28	8/27
0	1.6	1.8	1.7
1	4.7	5.2	5.1
2	7.8	8.8	8.4
3	10.9	12.2	11.8
4	14.0	15.8	15.2
5	17.1	19.2	18.6
6	20.2	22.8	21.9
7	23.3	26.2	25.3
8	26.4	—	—
9			

P P

*145° 235° *325°

55°

35° *125° 215° *305°

′	L Sin	d	L Tan	cd	L Cot	L Cos	d	
0	9.75 859	18	9.84 523	27	0.15 477	9.91 336	8	60
1	9.75 877	18	9.84 550	26	0.15 450	9.91 328	9	59
2	9.75 895	18	9.84 576	27	0.15 424	9.91 319	9	58
3	9.75 913	18	9.84 603	27	0.15 397	9.91 310	9	57
4	9.75 931	18	9.84 630	27	0.15 370	9.91 301	9	56
5	9.75 949	18	9.84 657	27	0.15 343	9.91 292	9	55
6	9.75 967	18	9.84 684	27	0.15 316	9.91 283	9	54
7	9.75 985	18	9.84 711	27	0.15 289	9.91 274	8	53
8	9.76 003	18	9.84 738	26	0.15 262	9.91 266	9	52
9	9.76 021	18	9.84 764	27	0.15 236	9.91 257	9	51
10	9.76 039	18	9.84 791	27	0.15 209	9.91 248	9	50
11	9.76 057	18	9.84 818	27	0.15 182	9.91 239	9	49
12	9.76 075	18	9.84 845	27	0.15 155	9.91 230	9	48
13	9.76 093	18	9.84 872	27	0.15 128	9.91 221	9	47
14	9.76 111	18	9.84 899	26	0.15 101	9.91 212	9	46
15	9.76 129	17	9.84 925	27	0.15 075	9.91 203	9	45
16	9.76 146	18	9.84 952	27	0.15 048	9.91 194	9	44
17	9.76 164	18	9.84 979	27	0.15 021	9.91 185	9	43
18	9.76 182	18	9.85 006	27	0.14 994	9.91 176	9	42
19	9.76 200	18	9.85 033	26	0.14 967	9.91 167	9	41
20	9.76 218	18	9.85 059	27	0.14 941	9.91 158	9	40
21	9.76 236	17	9.85 086	27	0.14 914	9.91 149	8	39
22	9.76 253	18	9.85 113	27	0.14 887	9.91 141	9	38
23	9.76 271	18	9.85 140	26	0.14 860	9.91 132	9	37
24	9.76 289	18	9.85 166	27	0.14 834	9.91 123	9	36
25	9.76 307	17	9.85 193	27	0.14 807	9.91 114	9	35
26	9.76 324	18	9.85 220	27	0.14 780	9.91 105	9	34
27	9.76 342	18	9.85 247	26	0.14 753	9.91 096	9	33
28	9.76 360	18	9.85 273	27	0.14 727	9.91 087	9	32
29	9.76 378	17	9.85 300	27	0.14 700	9.91 078	9	31
30	9.76 395	18	9.85 327	27	0.14 673	9.91 069	9	30
31	9.76 413	18	9.85 354	26	0.14 646	9.91 060	9	29
32	9.76 431	17	9.85 380	27	0.14 620	9.91 051	9	28
33	9.76 448	18	9.85 407	27	0.14 593	9.91 042	9	27
34	9.76 466	18	9.85 434	26	0.14 566	9.91 033	10	26
35	9.76 484	17	9.85 460	27	0.14 540	9.91 023	9	25
36	9.76 501	18	9.85 487	27	0.14 513	9.91 014	9	24
37	9.76 519	18	9.85 514	26	0.14 486	9.91 005	9	23
38	9.76 537	17	9.85 540	27	0.14 460	9.90 996	9	22
39	9.76 554	18	9.85 567	27	0.14 433	9.90 987	9	21
40	9.76 572	18	9.85 594	26	0.14 406	9.90 978	9	20
41	9.76 590	17	9.85 620	27	0.14 380	9.90 969	9	19
42	9.76 607	18	9.85 647	27	0.14 353	9.90 960	9	18
43	9.76 625	17	9.85 674	26	0.14 326	9.90 951	9	17
44	9.76 642	18	9.85 700	27	0.14 300	9.90 942	9	16
45	9.76 660	17	9.85 727	27	0.14 273	9.90 933	9	15
46	9.76 677	18	9.85 754	26	0.14 246	9.90 924	9	14
47	9.76 695	17	9.85 780	27	0.14 220	9.90 915	9	13
48	9.76 712	18	9.85 807	27	0.14 193	9.90 906	10	12
49	9.76 730	17	9.85 834	26	0.14 166	9.90 896	9	11
50	9.76 747	18	9.85 860	27	0.14 140	9.90 887	9	10
51	9.76 765	17	9.85 887	26	0.14 113	9.90 878	9	9
52	9.76 782	18	9.85 913	27	0.14 087	9.90 869	9	8
53	9.76 800	17	9.85 940	27	0.14 060	9.90 860	9	7
54	9.76 817	18	9.85 967	26	0.14 033	9.90 851	9	6
55	9.76 835	17	9.85 993	27	0.14 007	9.90 842	10	5
56	9.76 852	18	9.86 020	26	0.13 980	9.90 832	9	4
57	9.76 870	17	9.86 046	27	0.13 954	9.90 823	9	3
58	9.76 887	17	9.86 073	27	0.13 927	9.90 814	9	2
59	9.76 904	18	9.86 100	26	0.13 900	9.90 805	9	1
60	9.76 922		9.86 126		0.13 874	9.90 796		0
	L Cos	d	L Cot	cd	L Tan	L Sin	d	′

P P

	27	26	18
1	0.4	0.4	0.3
2	0.9	0.9	0.6
3	1.4	1.3	0.9
4	1.8	1.7	1.2
5	2.2	2.2	1.5
6	2.7	2.6	1.8
7	3.2	3.0	2.1
8	3.6	3.5	2.4
9	4.0	3.9	2.7
10	4.5	4.3	3.0
20	9.0	8.7	6.0
30	13.5	13.0	9.0
40	18.0	17.3	12.0
50	22.5	21.7	15.0

	17	10	9	8
1	0.3	0.2	0.2	0.1
2	0.6	0.3	0.3	0.3
3	0.8	0.5	0.4	0.4
4	1.1	0.7	0.6	0.5
5	1.4	0.8	0.8	0.7
6	1.7	1.0	0.9	0.8
7	2.0	1.2	1.0	0.9
8	2.3	1.3	1.2	1.1
9	2.6	1.5	1.4	1.2
10	2.8	1.7	1.5	1.3
20	5.7	3.3	3.0	2.7
30	8.5	5.0	4.5	4.0
40	11.3	6.7	6.0	5.3
50	14.2	8.3	7.5	6.7

	10/27	10/26
0		
1	1.4	1.3
2	4.1	3.9
3	6.8	6.5
4	9.4	9.1
5	12.2	11.7
6	14.8	14.3
7	17.6	16.9
8	20.2	19.5
9	22.9	22.1
10	25.6	24.7

	9/27	9/26
0		
1	1.5	1.4
2	4.5	4.3
3	7.5	7.2
4	10.5	10.1
5	13.5	13.0
6	16.5	15.9
7	19.5	18.8
8	22.5	21.7
9	25.5	24.6

P P

*144° 234° *324° 54°

′	L Sin	d	L Tan	c d	L Cot	L Cos	d	
0	9.76 922	17	9.86 126	27	0.13 874	9.90 796	9	60
1	9.76 939	18	9.86 153	26	0.13 847	9.90 787	10	59
2	9.76 957	17	9.86 179	27	0.13 821	9.90 777	9	58
3	9.76 974	17	9.86 206	26	0.13 794	9.90 768	9	57
4	9.76 991	18	9.86 232	27	0.13 768	9.90 759	9	56
5	9.77 009	17	9.86 259	26	0.13 741	9.90 750	9	55
6	9.77 026	17	9.86 285	27	0.13 715	9.90 741	10	54
7	9.77 043	18	9.86 312	26	0.13 688	9.90 731	9	53
8	9.77 061	17	9.86 338	27	0.13 662	9.90 722	9	52
9	9.77 078	17	9.86 365	27	0.13 635	9.90 713	9	51
10	9.77 095	17	9.86 392	26	0.13 608	9.90 704	10	50
11	9.77 112	18	9.86 418	27	0.13 582	9.90 694	9	49
12	9.77 130	17	9.86 445	26	0.13 555	9.90 685	9	48
13	9.77 147	17	9.86 471	27	0.13 529	9.90 676	9	47
14	9.77 164	17	9.86 498	26	0.13 502	9.90 667	10	46
15	9.77 181	18	9.86 524	27	0.13 476	9.90 657	9	45
16	9.77 199	17	9.86 551	26	0.13 449	9.90 648	9	44
17	9.77 216	17	9.86 577	26	0.13 423	9.90 639	9	43
18	9.77 233	17	9.86 603	27	0.13 397	9.90 630	10	42
19	9.77 250	18	9.86 630	26	0.13 370	9.90 620	9	41
20	9.77 268	17	9.86 656	27	0.13 344	9.90 611	9	40
21	9.77 285	17	9.86 683	26	0.13 317	9.90 602	10	39
22	9.77 302	17	9.86 709	27	0.13 291	9.90 592	9	38
23	9.77 319	17	9.86 736	26	0.13 264	9.90 583	9	37
24	9.77 336	17	9.86 762	27	0.13 238	9.90 574	9	36
25	9.77 353	17	9.86 789	26	0.13 211	9.90 565	10	35
26	9.77 370	17	9.86 815	27	0.13 185	9.90 555	9	34
27	9.77 387	18	9.86 842	26	0.13 158	9.90 546	9	33
28	9.77 405	17	9.86 868	26	0.13 132	9.90 537	10	32
29	9.77 422	17	9.86 894	27	0.13 106	9.90 527	9	31
30	9.77 439	17	9.86 921	26	0.13 079	9.90 518	9	30
31	9.77 456	17	9.86 947	27	0.13 053	9.90 509	10	29
32	9.77 473	17	9.86 974	26	0.13 026	9.90 499	9	28
33	9.77 490	17	9.87 000	27	0.13 000	9.90 490	10	27
34	9.77 507	17	9.87 027	26	0.12 973	9.90 480	9	26
35	9.77 524	17	9.87 053	26	0.12 947	9.90 471	9	25
36	9.77 541	17	9.87 079	27	0.12 921	9.90 462	10	24
37	9.77 558	17	9.87 106	26	0.12 894	9.90 452	9	23
38	9.77 575	17	9.87 132	26	0.12 868	9.90 443	9	22
39	9.77 592	17	9.87 158	27	0.12 842	9.90 434	10	21
40	9.77 609	17	9.87 185	26	0.12 815	9.90 424	9	20
41	9.77 626	17	9.87 211	27	0.12 789	9.90 415	10	19
42	9.77 643	17	9.87 238	26	0.12 762	9.90 405	9	18
43	9.77 660	17	9.87 264	26	0.12 736	9.90 396	10	17
44	9.77 677	17	9.87 290	27	0.12 710	9.90 386	9	16
45	9.77 694	17	9.87 317	26	0.12 683	9.90 377	9	15
46	9.77 711	17	9.87 343	26	0.12 657	9.90 368	10	14
47	9.77 728	16	9.87 369	27	0.12 631	9.90 358	9	13
48	9.77 744	17	9.87 396	26	0.12 604	9.90 349	10	12
49	9.77 761	17	9.87 422	26	0.12 578	9.90 339	9	11
50	9.77 778	17	9.87 448	27	0.12 552	9.90 330	10	10
51	9.77 795	17	9.87 475	26	0.12 525	9.90 320	9	9
52	9.77 812	17	9.87 501	26	0.12 499	9.90 311	10	8
53	9.77 829	17	9.87 527	27	0.12 473	9.90 301	9	7
54	9.77 846	16	9.87 554	26	0.12 446	9.90 292	10	6
55	9.77 862	17	9.87 580	26	0.12 420	9.90 282	9	5
56	9.77 879	17	9.87 606	27	0.12 394	9.90 273	10	4
57	9.77 896	17	9.87 633	26	0.12 367	9.90 263	9	3
58	9.77 913	17	9.87 659	26	0.12 341	9.90 254	10	2
59	9.77 930	16	9.87 685	26	0.12 315	9.90 244	9	1
60	9.77 946		9.87 711		0.12 289	9.90 235		0
	L Cos	d	L Cot	c d	L Tan	L Sin	d	′

P P

	27	26
1	0.4	0.4
2	0.9	0.9
3	1.4	1.3
4	1.8	1.7
5	2.2	2.2
6	2.7	2.6
7	3.2	3.0
8	3.6	3.5
9	4.0	3.9
10	4.5	4.3
20	9.0	8.7
30	13.5	13.0
40	18.0	17.3
50	22.5	21.7

	18	17	16
1	0.3	0.3	0.3
2	0.6	0.6	0.5
3	0.9	0.8	0.8
4	1.2	1.1	1.1
5	1.5	1.4	1.3
6	1.8	1.7	1.6
7	2.1	2.0	1.9
8	2.4	2.3	2.1
9	2.7	2.6	2.4
10	3.0	2.8	2.7
20	6.0	5.7	5.3
30	9.0	8.5	8.0
40	12.0	11.3	10.7
50	15.0	14.2	13.3

	10	9
1	0.2	0.2
2	0.3	0.3
3	0.5	0.4
4	0.7	0.6
5	0.8	0.8
6	1.0	0.9
7	1.2	1.0
8	1.3	1.2
9	1.5	1.4
10	1.7	1.5
20	3.3	3.0
30	5.0	4.5
40	6.7	6.0
50	8.3	7.5

	9/27	9/26
0		
	1.5	1.4
1		
	4.5	4.3
2		
	7.5	7.2
3		
	10.5	10.1
4		
	13.5	13.0
5		
	16.5	15.9
6		
	19.5	18.8
7		
	22.5	21.7
8		
	25.5	24.6
9		

P P

37° *127° 217° *307°

′	L Sin	d	L Tan	cd	L Cot	L Cos	d	′
0	9.77 946	17	9.87 711	27	0.12 289	9.90 235̄	10	60
1	9.77 963	17	9.87 738	26	0.12 262	9.90 225	9	59
2	9.77 980	17	9.87 764	26	0.12 236	9.90 216	10	58
3	9.77 997	16	9.87 790	27	0.12 210	9.90 206	9	57
4	9.78 013	17	9.87 817	26	0.12 183	9.90 197	10	56
5	9.78 030	17	9.87 843	26	0.12 157	9.90 187	9	55
6	9.78 047	16	9.87 869	26	0.12 131	9.90 178	10	54
7	9.78 063	17	9.87 895	27	0.12 105̄	9.90 168	9	53
8	9.78 080	17	9.87 922	26	0.12 078	9.90 159	10	52
9	9.78 097	16	9.87 948	26	0.12 052	9.90 149	10	51
10	9.78 113	17	9.87 974	26	0.12 026	9.90 139	9	50
11	9.78 130	17	9.88 000	27	0.12 000	9.90 130	10	49
12	9.78 147	16	9.88 027	26	0.11 973	9.90 120	9	48
13	9.78 163	17	9.88 053	26	0.11 947	9.90 111	10	47
14	9.78 180	17	9.88 079	26	0.11 921	9.90 101	10	46
15	9.78 197	16	9.88 105	26	0.11 895̄	9.90 091	9	45
16	9.78 213	17	9.88 131	27	0.11 869	9.90 082	10	44
17	9.78 230	16	9.88 158	26	0.11 842	9.90 072	9	43
18	9.78 246	17	9.88 184	26	0.11 816	9.90 063	10	42
19	9.78 263	17	9.88 210	26	0.11 790	9.90 053	10	41
20	9.78 280	16	9.88 236	26	0.11 764	9.90 043	9	40
21	9.78 296	17	9.88 262	27	0.11 738	9.90 034	10	39
22	9.78 313	16	9.88 289	26	0.11 711	9.90 024	10	38
23	9.78 329	17	9.88 315̄	26	0.11 685	9.90 014	9	37
24	9.78 346	16	9.88 341	26	0.11 659	9.90 005̄	10	36
25	9.78 362	17	9.88 367	26	0.11 633	9.89 995	10	35
26	9.78 379	16	9.88 393	27	0.11 607	9.89 985	9	34
27	9.78 395	17	9.88 420	26	0.11 580	9.89 976	10	33
28	9.78 412	16	9.88 446	26	0.11 554	9.89 966	10	32
29	9.78 428	17	9.88 472	26	0.11 528	9.89 956	9	31
30	9.78 445̄	16	9.88 498	26	0.11 502	9.89 947	10	30
31	9.78 461	17	9.88 524	26	0.11 476	9.89 937	10	29
32	9.78 478	16	9.88 550	27	0.11 450	9.89 927	9	28
33	9.78 494	16	9.88 577	26	0.11 423	9.89 918	10	27
34	9.78 510	17	9.88 603	26	0.11 397	9.89 908	10	26
35	9.78 527	16	9.88 629	26	0.11 371	9.89 898	10	25
36	9.78 543	17	9.88 655̄	26	0.11 345	9.89 888	9	24
37	9.78 560	16	9.88 681	26	0.11 319	9.89 879	10	23
38	9.78 576	16	9.88 707	26	0.11 293	9.89 869	10	22
39	9.78 592	17	9.88 733	26	0.11 267	9.89 859	10	21
40	9.78 609	16	9.88 759	27	0.11 241	9.89 849	9	20
41	9.78 625	17	9.88 786	26	0.11 214	9.89 840	10	19
42	9.78 642	16	9.88 812	26	0.11 188	9.89 830	10	18
43	9.78 658	16	9.88 838	26	0.11 162	9.89 820	10	17
44	9.78 674	17	9.88 864	26	0.11 136	9.89 810	9	16
45	9.78 691	16	9.88 890	26	0.11 110	9.89 801	10	15
46	9.78 707	16	9.88 916	26	0.11 084	9.89 791	10	14
47	9.78 723	16	9.88 942	26	0.11 058	9.89 781	10	13
48	9.78 739	17	9.88 968	26	0.11 032	9.89 771	10	12
49	9.78 756	16	9.88 994	26	0.11 006	9.89 761	9	11
50	9.78 772	16	9.89 020	26	0.10 980	9.89 752	10	10
51	9.78 788	17	9.89 046	27	0.10 954	9.89 742	10	9
52	9.78 805̄	16	9.89 073	26	0.10 927	9.89 732	10	8
53	9.78 821	16	9.89 099	26	0.10 901	9.89 722	10	7
54	9.78 837	16	9.89 125̄	26	0.10 875	9.89 712	10	6
55	9.78 853	16	9.89 151	26	0.10 849	9.89 702	9	5
56	9.78 869	17	9.89 177	26	0.10 823	9.89 693	10	4
57	9.78 886	16	9.89 203	26	0.10 797	9.89 683	10	3
58	9.78 902	16	9.89 229	26	0.10 771	9.89 673	10	2
59	9.78 918	16	9.89 255̄	26	0.10 745	9.89 663	10	1
60	9.78 934		9.89 281		0.10 719	9.89 653		0
	L Cos	d	L Cot	cd	L Tan	L Sin	d	′

P P

	27	26
1	0.4	0.4
2	0.9	0.9
3	1.4	1.3
4	1.8	1.7
5	2.2	2.2
6	2.7	2.6
7	3.2	3.0
8	3.6	3.5
9	4.0	3.9
10	4.5	4.3
20	9.0	8.7
30	13.5	13.0
40	18.0	17.3
50	22.5	21.7

	17	16
1	0.3	0.3
2	0.6	0.5
3	0.8	0.8
4	1.1	1.1
5	1.4	1.3
6	1.7	1.6
7	2.0	1.9
8	2.3	2.1
9	2.6	2.4
10	2.8	2.7
20	5.7	5.3
30	8.5	8.0
40	11.3	10.7
50	14.2	13.3

	10	9
1	0.2	0.2
2	0.3	0.3
3	0.5	0.4
4	0.7	0.6
5	0.8	0.8
6	1.0	0.9
7	1.2	1.0
8	1.3	1.2
9	1.5	1.4
10	1.7	1.5
20	3.3	3.0
30	5.0	4.5
40	6.7	6.0
50	8.3	7.5

	10/27	10/26
0	1.4	1.3
1	4.1	3.9
2	6.8	6.5
3	9.4	9.1
4	12.2	11.7
5	14.8	14.3
6	17.6	16.9
7	20.2	19.5
8	22.9	22.1
9	25.6	24.7
10		

P P

*142° 232° *322° **52°**

′	L Sin	d	L Tan	c d	L Cot	L Cos	d	
0	9.78 934	16	9.89 281	26	0.10 719	9.89 653	10	60
1	9.78 950	17	9.89 307	26	0.10 693	9.89 643	10	59
2	9.78 967	16	9.89 333	26	0.10 667	9.89 633	9	58
3	9.78 983	16	9.89 359	26	0.10 641	9.89 624	10	57
4	9.78 999	16	9.89 385	26	0.10 615̄	9.89 614	10	56
5	9.79 015̄	16	9.89 411	26	0.10 589	9.89 604	10	55
6	9.79 031	16	9.89 437	26	0.10 563	9.89 594	10	54
7	9.79 047	16	9.89 463	26	0.10 537	9.89 584	10	53
8	9.79 063	16	9.89 489	26	0.10 511	9.89 574	10	52
9	9.79 079	16	9.89 515	26	0.10 485̄	9.89 564	10	51
10	9.79 095	16	9.89 541	26	0.10 459	9.89 554	10	50
11	9.79 111	17	9.89 567	26	0.10 433	9.89 544	10	49
12	9.79 128	16	9.89 593	26	0.10 407	9.89 534	10	48
13	9.79 144	16	9.89 619	26	0.10 381	9.89 524	10	47
14	9.79 160	16	9.89 645	26	0.10 355̄	9.89 514	10	46
15	9.79 176	16	9.89 671	26	0.10 329	9.89 504	9	45
16	9.79 192	16	9.89 697	26	0.10 303	9.89 495̄	10	44
17	9.79 208	16	9.89 723	26	0.10 277	9.89 485̄	10	43
18	9.79 224	16	9.89 749	26	0.10 251	9.89 475̄	10	42
19	9.79 240	16	9.89 775	26	0.10 225̄	9.89 465̄	10	41
20	9.79 256	16	9.89 801	26	0.10 199	9.89 455̄	10	40
21	9.79 272	16	9.89 827	26	0.10 173	9.89 445̄	10	39
22	9.79 288	16	9.89 853	26	0.10 147	9.89 435̄	10	38
23	9.79 304	15	9.89 879	26	0.10 121	9.89 425̄	10	37
24	9.79 319	16	9.89 905̄	26	0.10 095	9.89 415̄	10	36
25	9.79 335	16	9.89 931	26	0.10 069	9.89 405̄	10	35
26	9.79 351	16	9.89 957	26	0.10 043	9.89 395̄	10	34
27	9.79 367	16	9.89 983	26	0.10 017	9.89 385̄	10	33
28	9.79 383	16	9.90 009	26	0.09 991	9.89 375̄	11	32
29	9.79 399	16	9.90 035̄	26	0.09 965	9.89 364	10	31
30	9.79 415̄	16	9.90 061	25	0.09 939	9.89 354	10	30
31	9.79 431	16	9.90 086	26	0.09 914	9.89 344	10	29
32	9.79 447	16	9.90 112	26	0.09 888	9.89 334	10	28
33	9.79 463	15	9.90 138	26	0.09 862	9.89 324	10	27
34	9.79 478	16	9.90 164	26	0.09 836	9.89 314	10	26
35	9.79 494	16	9.90 190	26	0.09 810	9.89 304	10	25
36	9.79 510	16	9.90 216	26	0.09 784	9.89 294	10	24
37	9.79 526	16	9.90 242	26	0.09 758	9.89 284	10	23
38	9.79 542	16	9.90 268	26	0.09 732	9.89 274	10	22
39	9.79 558	15	9.90 294	26	0.09 706	9.89 264	10	21
40	9.79 573	16	9.90 320	26	0.09 680	9.89 254	10	20
41	9.79 589	16	9.90 346	25	0.09 654	9.89 244	11	19
42	9.79 605̄	16	9.90 371	26	0.09 629	9.89 233	10	18
43	9.79 621	15	9.90 397	26	0.09 603	9.89 223	10	17
44	9.79 636	16	9.90 423	26	0.09 577	9.89 213	10	16
45	9.79 652	16	9.90 449	26	0.09 551	9.89 203	10	15
46	9.79 668	16	9.90 475̄	26	0.09 525	9.89 193	10	14
47	9.79 684	15	9.90 501	26	0.09 499	9.89 183	10	13
48	9.79 699	16	9.90 527	26	0.09 473	9.89 173	11	12
49	9.79 715	16	9.90 553	25	0.09 447	9.89 162	10	11
50	9.79 731	15	9.90 578	26	0.09 422	9.89 152	10	10
51	9.79 746	16	9.90 604	26	0.09 396	9.89 142	10	9
52	9.79 762	16	9.90 630	26	0.09 370	9.89 132	10	8
53	9.79 778	15	9.90 656	26	0.09 344	9.89 122	10	7
54	9.79 793	16	9.90 682	26	0.09 318	9.89 112	11	6
55	9.79 809	16	9.90 708	26	0.09 292	9.89 101	10	5
56	9.79 825̄	15	9.90 734	25	0.09 266	9.89 091	10	4
57	9.79 840	16	9.90 759	26	0.09 241	9.89 081	10	3
58	9.79 856	16	9.90 785	26	0.09 215̄	9.89 071	11	2
59	9.79 872	15	9.90 811	26	0.09 189	9.89 060	10	1
60	9.79 887		9.90 837		0.09 163	9.89 050		0
	L Cos	d	L Cot	c d	L Tan	L Sin	d	′

P P

	26	25
1	0.4	0.4
2	0.9	0.8
3	1.3	1.2
4	1.7	1.7
5	2.2	2.1
6	2.6	2.5
7	3.0	2.9
8	3.5̄	3.3
9	3.9	3.8
10	4.3	4.2
20	8.7	8.3
30	13.0	12.5
40	17.3	16.7
50	21.7	20.8

	17	16	15
1	0.3	0.3	0.2
2	0.6	0.5	0.5
3	0.8	0.8	0.8
4	1.1	1.1	1.0
5	1.4	1.3	1.2
6	1.7	1.6	1.5
7	2.0	1.9	1.8
8	2.3	2.1	2.0
9	2.6	2.4	2.2
10	2.8	2.7	2.5
20	5.7	5.3	5.0
30	8.5	8.0	7.5
40	11.3	10.7	10.0
50	14.2	13.3	12.5

	11	10	9
1	0.2	0.2	0.2
2	0.4	0.3	0.3
3	0.6	0.5	0.4
4	0.7	0.7	0.6
5	0.9	0.8	0.8
6	1.1	1.0	0.9
7	1.3	1.2	1.0
8	1.5̄	1.3	1.2
9	1.6	1.5	1.4
10	1.8	1.7	1.5
20	3.7	3.3	3.0
30	5.5	5.0	4.5
40	7.3	6.7	6.0
50	9.2	8.3	7.5

	10/26	10/25	9/26
0			
1	1.3	1.2	1.4
2	3.9	3.8	4.3
3	6.5	6.2	7.2
4	9.1	8.8	10.1
5	11.7	11.2	13.0
6	14.3	13.8	15.9
7	16.9	16.2	18.8
8	19.5	18.8	21.7
9	22.1	21.2	24.6
10	24.7	23.8	—

P P

39°

*129° 219° *309°

′	L Sin	d	L Tan	c d	L Cot	L Cos	d	
0	9.79 887	16	9.90 837	26	0.09 163	9.89 050	10	60
1	9.79 903	15	9.90 863	26	0.09 137	9.89 040	10	59
2	9.79 918	16	9.90 889	25	0.09 111	9.89 030	10	58
3	9.79 934	16	9.90 914	26	0.09 086	9.89 020	11	57
4	9.79 950	15	9.90 940	26	0.09 060	9.89 009	10	56
5	9.79 965	16	9.90 966	26	0.09 034	9.88 999	10	55
6	9.79 981	15	9.90 992	26	0.09 008	9.88 989	11	54
7	9.79 996	16	9.91 018	25	0.08 982	9.88 978	10	53
8	9.80 012	15	9.91 043	26	0.08 957	9.88 968	10	52
9	9.80 027	16	9.91 069	26	0.08 931	9.88 958	10	51
10	9.80 043	15	9.91 095	26	0.08 905	9.88 948	11	50
11	9.80 058	16	9.91 121	26	0.08 879	9.88 937	10	49
12	9.80 074	15	9.91 147	25	0.08 853	9.88 927	10	48
13	9.80 089	16	9.91 172	26	0.08 828	9.88 917	11	47
14	9.80 105	15	9.91 198	26	0.08 802	9.88 906	10	46
15	9.80 120	16	9.91 224	26	0.08 776	9.88 896	10	45
16	9.80 136	15	9.91 250	26	0.08 750	9.88 886	11	44
17	9.80 151	15	9.91 276	25	0.08 724	9.88 875	10	43
18	9.80 166	16	9.91 301	26	0.08 699	9.88 865	10	42
19	9.80 182	15	9.91 327	26	0.08 673	9.88 855	11	41
20	9.80 197	16	9.91 353	26	0.08 647	9.88 844	10	40
21	9.80 213	15	9.91 379	25	0.08 621	9.88 834	10	39
22	9.80 228	16	9.91 404	26	0.08 596	9.88 824	11	38
23	9.80 244	15	9.91 430	26	0.08 570	9.88 813	10	37
24	9.80 259	15	9.91 456	26	0.08 544	9.88 803	10	36
25	9.80 274	16	9.91 482	25	0.08 518	9.88 793	11	35
26	9.80 290	15	9.91 507	26	0.08 493	9.88 782	10	34
27	9.80 305	15	9.91 533	26	0.08 467	9.88 772	11	33
28	9.80 320	16	9.91 559	26	0.08 441	9.88 761	10	32
29	9.80 336	15	9.91 585	25	0.08 415	9.88 751	10	31
30	9.80 351	15	9.91 610	26	0.08 390	9.88 741	11	30
31	9.80 366	16	9.91 636	26	0.08 364	9.88 730	10	29
32	9.80 382	15	9.91 662	26	0.08 338	9.88 720	11	28
33	9.80 397	15	9.91 688	25	0.08 312	9.88 709	10	27
34	9.80 412	16	9.91 713	26	0.08 287	9.88 699	11	26
35	9.80 428	15	9.91 739	26	0.08 261	9.88 688	10	25
36	9.80 443	15	9.91 765	26	0.08 235	9.88 678	10	24
37	9.80 458	15	9.91 791	25	0.08 209	9.88 668	11	23
38	9.80 473	16	9.91 816	26	0.08 184	9.88 657	10	22
39	9.80 489	15	9.91 842	26	0.08 158	9.88 647	11	21
40	9.80 504	15	9.91 868	25	0.08 132	9.88 636	10	20
41	9.80 519	15	9.91 893	26	0.08 107	9.88 626	11	19
42	9.80 534	16	9.91 919	26	0.08 081	9.88 615	10	18
43	9.80 550	15	9.91 945	26	0.08 055	9.88 605	11	17
44	9.80 565	15	9.91 971	25	0.08 029	9.88 594	10	16
45	9.80 580	15	9.91 996	26	0.08 004	9.88 584	11	15
46	9.80 595	15	9.92 022	26	0.07 978	9.88 573	10	14
47	9.80 610	15	9.92 048	25	0.07 952	9.88 563	11	13
48	9.80 625	16	9.92 073	26	0.07 927	9.88 552	10	12
49	9.80 641	15	9.92 099	26	0.07 901	9.88 542	11	11
50	9.80 656	15	9.92 125	25	0.07 875	9.88 531	10	10
51	9.80 671	15	9.92 150	26	0.07 850	9.88 521	11	9
52	9.80 686	15	9.92 176	26	0.07 824	9.88 510	11	8
53	9.80 701	15	9.92 202	25	0.07 798	9.88 499	10	7
54	9.80 716	15	9.92 227	26	0.07 773	9.88 489	11	6
55	9.80 731	15	9.92 253	26	0.07 747	9.88 478	10	5
56	9.80 746	16	9.92 279	25	0.07 721	9.88 468	11	4
57	9.80 762	15	9.92 304	26	0.07 696	9.88 457	10	3
58	9.80 777	15	9.92 330	26	0.07 670	9.88 447	11	2
59	9.80 792	15	9.92 356	25	0.07 644	9.88 436	11	1
60	9.80 807		9.92 381		0.07 619	9.88 425		0
	L Cos	d	L Cot	c d	L Tan	L Sin	d	′

P P

	26	25
1	0.4	0.4
2	0.9	0.8
3	1.3	1.2
4	1.7	1.7
5	2.2	2.1
6	2.6	2.5
7	3.0	2.9
8	3.5	3.3
9	3.9	3.8
10	4.3	4.2
20	8.7	8.3
30	13.0	12.5
40	17.3	16.7
50	21.7	20.8

	16	15
1	0.3	0.2
2	0.5	0.5
3	0.8	0.8
4	1.1	1.0
5	1.3	1.2
6	1.6	1.5
7	1.9	1.8
8	2.1	2.0
9	2.4	2.2
10	2.7	2.5
20	5.3	5.0
30	8.0	7.5
40	10.7	10.0
50	13.3	12.5

	11	10
1	0.2	0.2
2	0.4	0.3
3	0.6	0.5
4	0.7	0.7
5	0.9	0.8
6	1.1	1.0
7	1.3	1.2
8	1.5	1.3
9	1.6	1.5
10	1.8	1.7
20	3.7	3.3
30	5.5	5.0
40	7.3	6.7
50	9.2	8.3

	11/26	11/25
0	1.2	1.1
1	3.5	3.4
2	5.9	5.7
3	8.3	7.9
4	10.6	10.2
5	13.0	12.5
6	15.4	14.8
7	17.7	17.1
8	20.1	19.3
9	22.5	21.6
10	24.8	23.9
11		

P P

*140° 230° *320° **50°**

40° *130° 220° *310°

′	L Sin	d	L Tan	c d	L Cot	L Cos	d	
0	9.80 807	15	9.92 381	26	0.07 619	9.88 425	10	60
1	9.80 822	15	9.92 407	26	0.07 593	9.88 415	11	59
2	9.80 837	15	9.92 433	25	0.07 567	9.88 404	10	58
3	9.80 852	15	9.92 458	26	0.07 542	9.88 394	11	57
4	9.80 867	15	9.92 484	26	0.07 516	9.88 383	11	56
5	9.80 882	15	9.92 510	25	0.07 490	9.88 372	10	55
6	9.80 897	15	9.92 535	26	0.07 465	9.88 362	11	54
7	9.80 912	15	9.92 561	26	0.07 439	9.88 351	11	53
8	9.80 927	15	9.92 587	25	0.07 413	9.88 340	10	52
9	9.80 942	15	9.92 612	26	0.07 388	9.88 330	11	51
10	9.80 957	15	9.92 638	25	0.07 362	9.88 319	11	50
11	9.80 972	15	9.92 663	26	0.07 337	9.88 308	10	49
12	9.80 987	15	9.92 689	26	0.07 311	9.88 298	11	48
13	9.81 002	15	9.92 715	25	0.07 285	9.88 287	11	47
14	9.81 017	15	9.92 740	26	0.07 260	9.88 276	10	46
15	9.81 032	15	9.92 766	26	0.07 234	9.88 266	11	45
16	9.81 047	14	9.92 792	25	0.07 208	9.88 255	11	44
17	9.81 061	15	9.92 817	26	0.07 183	9.88 244	10	43
18	9.81 076	15	9.92 843	25	0.07 157	9.88 234	11	42
19	9.81 091	15	9.92 868	26	0.07 132	9.88 223	11	41
20	9.81 106	15	9.92 894	26	0.07 106	9.88 212	11	40
21	9.81 121	15	9.92 920	25	0.07 080	9.88 201	10	39
22	9.81 136	15	9.92 945	26	0.07 055	9.88 191	11	38
23	9.81 151	15	9.92 971	25	0.07 029	9.88 180	11	37
24	9.81 166	14	9.92 996	26	0.07 004	9.88 169	11	36
25	9.81 180	15	9.93 022	26	0.06 978	9.88 158	10	35
26	9.81 195	15	9.93 048	25	0.06 952	9.88 148	11	34
27	9.81 210	15	9.93 073	26	0.06 927	9.88 137	11	33
28	9.81 225	15	9.93 099	25	0.06 901	9.88 126	11	32
29	9.81 240	14	9.93 124	26	0.06 876	9.88 115	10	31
30	9.81 254	15	9.93 150	25	0.06 850	9.88 105	11	30
31	9.81 269	15	9.93 175	26	0.06 825	9.88 094	11	29
32	9.81 284	15	9.93 201	26	0.06 799	9.88 083	11	28
33	9.81 299	15	9.93 227	25	0.06 773	9.88 072	11	27
34	9.81 314	14	9.93 252	26	0.06 748	9.88 061	10	26
35	9.81 328	15	9.93 278	25	0.06 722	9.88 051	11	25
36	9.81 343	15	9.93 303	26	0.06 697	9.88 040	11	24
37	9.81 358	14	9.93 329	25	0.06 671	9.88 029	11	23
38	9.81 372	15	9.93 354	26	0.06 646	9.88 018	11	22
39	9.81 387	15	9.93 380	26	0.06 620	9.88 007	11	21
40	9.81 402	15	9.93 406	25	0.06 594	9.87 996	11	20
41	9.81 417	14	9.93 431	26	0.06 569	9.87 985	10	19
42	9.81 431	15	9.93 457	25	0.06 543	9.87 975	11	18
43	9.81 446	15	9.93 482	26	0.06 518	9.87 964	11	17
44	9.81 461	14	9.93 508	25	0.06 492	9.87 953	11	16
45	9.81 475	15	9.93 533	26	0.06 467	9.87 942	11	15
46	9.81 490	15	9.93 559	25	0.06 441	9.87 931	11	14
47	9.81 505	14	9.93 584	26	0.06 416	9.87 920	11	13
48	9.81 519	15	9.93 610	26	0.06 390	9.87 909	11	12
49	9.81 534	15	9.93 636	25	0.06 364	9.87 898	11	11
50	9.81 549	14	9.93 661	26	0.06 339	9.87 887	10	10
51	9.81 563	15	9.93 687	25	0.06 313	9.87 877	11	9
52	9.81 578	14	9.93 712	26	0.06 288	9.87 866	11	8
53	9.81 592	15	9.93 738	25	0.06 262	9.87 855	11	7
54	9.81 607	15	9.93 763	26	0.06 237	9.87 844	11	6
55	9.81 622	14	9.93 789	25	0.06 211	9.87 833	11	5
56	9.81 636	15	9.93 814	26	0.06 186	9.87 822	11	4
57	9.81 651	14	9.93 840	25	0.06 160	9.87 811	11	3
58	9.81 665	15	9.93 865	26	0.06 135	9.87 800	11	2
59	9.81 680	14	9.93 891	25	0.06 109	9.87 789	11	1
60	9.81 694		9.93 916		0.06 084	9.87 778		0
	L Cos	d	L Cot	c d	L Tan	L Sin	d	′

P P

	26	25
1	0.4	0.4
2	0.9	0.8
3	1.3	1.2
4	1.7	1.7
5	2.2	2.1
6	2.6	2.5
7	3.0	2.9
8	3.5	3.3
9	3.9	3.8
10	4.3	4.2
20	8.7	8.3
30	13.0	12.5
40	17.3	16.7
50	21.7	20.8

	15	14
1	0.2	0.2
2	0.5	0.5
3	0.8	0.7
4	1.0	0.9
5	1.2	1.2
6	1.5	1.4
7	1.8	1.6
8	2.0	1.9
9	2.2	2.1
10	2.5	2.3
20	5.0	4.7
30	7.5	7.0
40	10.0	9.3
50	12.5	11.7

	11	10
1	0.2	0.2
2	0.4	0.3
3	0.6	0.5
4	0.7	0.7
5	0.9	0.8
6	1.1	1.0
7	1.3	1.2
8	1.5	1.3
9	1.6	1.5
10	1.8	1.7
20	3.7	3.3
30	5.5	5.0
40	7.3	6.7
50	9.2	8.3

	11/26	10/26	10/25
0			
1	1.2	1.3	1.2
2	3.5	3.9	3.8
3	5.9	6.5	6.2
4	8.3	9.1	8.8
5	10.6	11.7	11.2
6	13.0	14.3	13.8
7	15.4	16.9	16.2
8	17.7	19.5	18.8
9	20.1	22.1	21.2
10	22.5	24.7	23.8
11	24.8	—	—

P P

*139° 229° *319° 49°

41° *131° 221° *311°

′	L Sin	d	L Tan	c d	L Cot	L Cos	d	′
0	9.81 694	15	9.93 916	26	0.06 084	9.87 778	11	60
1	9.81 709	14	9.93 942	25	0.06 058	9.87 767	11	59
2	9.81 723	15	9.93 967	26	0.06 033	9.87 756	11	58
3	9.81 738	14	9.93 993	25	0.06 007	9.87 745	11	57
4	9.81 752	15	9.94 018	26	0.05 982	9.87 734	11	56
5	9.81 767	14	9.94 044	25	0.05 956	9.87 723	11	55
6	9.81 781	15	9.94 069	26	0.05 931	9.87 712	11	54
7	9.81 796	14	9.94 095̄	25	0.05 905	9.87 701	11	53
8	9.81 810	15	9.94 120	26	0.05 880	9.87 690	11	52
9	9.81 825̄	14	9.94 146	25	0.05 854	9.87 679	11	51
10	9.81 839	15	9.94 171	26	0.05 829	9.87 668	11	50
11	9.81 854	14	9.94 197	25	0.05 803	9.87 657	11	49
12	9.81 868	14	9.94 222	26	0.05 778	9.87 646	11	48
13	9.81 882	15	9.94 248	25	0.05 752	9.87 635̄	11	47
14	9.81 897	14	9.94 273	26	0.05 727	9.87 624	11	46
15	9.81 911	15	9.94 299	25	0.05 701	9.87 613	12	45
16	9.81 926	14	9.94 324	26	0.05 676	9.87 601	11	44
17	9.81 940	15	9.94 350̄	25	0.05 650	9.87 590	11	43
18	9.81 955̄	14	9.94 375	26	0.05 625̄	9.87 579	11	42
19	9.81 969	14	9.94 401	25	0.05 599	9.87 568	11	41
20	9.81 983	15	9.94 426	26	0.05 574	9.87 557	11	40
21	9.81 998	14	9.94 452	25	0.05 548	9.87 546	11	39
22	9.82 012	14	9.94 477	26	0.05 523	9.87 535̄	11	38
23	9.82 026	15	9.94 503	25	0.05 497	9.87 524	11	37
24	9.82 041	14	9.94 528	26	0.05 472	9.87 513	12	36
25	9.82 055̄	14	9.94 554	25	0.05 446	9.87 501	11	35
26	9.82 069	15	9.94 579	25	0.05 421	9.87 490	11	34
27	9.82 084	14	9.94 604	26	0.05 396	9.87 479	11	33
28	9.82 098	14	9.94 630	25	0.05 370	9.87 468	11	32
29	9.82 112	14	9.94 655	26	0.05 345̄	9.87 457	11	31
30	9.82 126	15	9.94 681	25	0.05 319	9.87 446	12	30
31	9.82 141	14	9.94 706	26	0.05 294	9.87 434	11	29
32	9.82 155̄	14	9.94 732	25	0.05 268	9.87 423	11	28
33	9.82 169	15	9.94 757	26	0.05 243	9.87 412	11	27
34	9.82 184	14	9.94 783	25	0.05 217	9.87 401	11	26
35	9.82 198	14	9.94 808	26	0.05 192	9.87 390	12	25
36	9.82 212	14	9.94 834	25	0.05 166	9.87 378	11	24
37	9.82 226	14	9.94 859	25	0.05 141	9.87 367	11	23
38	9.82 240	15	9.94 884	26	0.05 116	9.87 356	11	22
39	9.82 255̄	14	9.94 910	25	0.05 090	9.87 345̄	11	21
40	9.82 269	14	9.94 935	26	0.05 065̄	9.87 334	12	20
41	9.82 283	14	9.94 961	25	0.05 039	9.87 322	11	19
42	9.82 297	14	9.94 986	26	0.05 014	9.87 311	11	18
43	9.82 311	15	9.95 012	25	0.04 988	9.87 300	12	17
44	9.82 326	14	9.95 037	25	0.04 963	9.87 288	11	16
45	9.82 340	14	9.95 062	26	0.04 938	9.87 277	11	15
46	9.82 354	14	9.95 088	25	0.04 912	9.87 266	11	14
47	9.82 368	14	9.95 113	26	0.04 887	9.87 255̄	12	13
48	9.82 382	14	9.95 139	25	0.04 861	9.87 243	11	12
49	9.82 396	14	9.95 164	26	0.04 836	9.87 232	11	11
50	9.82 410	14	9.95 190	25	0.04 810	9.87 221	12	10
51	9.82 424	15	9.95 215	25	0.04 785̄	9.87 209	11	9
52	9.82 439	14	9.95 240	26	0.04 760	9.87 198	11	8
53	9.82 453	14	9.95 266	25	0.04 734	9.87 187	12	7
54	9.82 467	14	9.95 291	26	0.04 709	9.87 175	11	6
55	9.82 481	14	9.95 317	25	0.04 683	9.87 164	11	5
56	9.82 495̄	14	9.95 342	26	0.04 658	9.87 153	12	4
57	9.82 509	14	9.95 368	25	0.04 632	9.87 141	11	3
58	9.82 523	14	9.95 393	25	0.04 607	9.87 130	11	2
59	9.82 537	14	9.95 418	26	0.04 582	9.87 119	12	1
60	9.82 551		9.95 444		0.04 556	9.87 107		0
	L Cos	d	L Cot	c d	L Tan	L Sin	d	′

P P

	26	25
1	0.4	0.4
2	0.9	0.8
3	1.3	1.2
4	1.7	1.7
5	2.2	2.1
6	2.6	2.5
7	3.0	2.9
8	3.5̄	3.3
9	3.9	3.8
10	4.3	4.2
20	8.7	8.3
30	13.0	12.5
40	17.3	16.7
50	21.7	20.8

	15	14
1	0.2	0.2
2	0.5	0.5̄
3	0.8	0.7
4	1.0	0.9
5	1.2	1.2
6	1.5	1.4
7	1.8	1.6
8	2.0	1.9
9	2.2	2.1
10	2.5	2.3
20	5.0	4.7
30	7.5	7.0
40	10.0	9.3
50	12.5	11.7

	12	11
1	0.2	0.2
2	0.4	0.4
3	0.6	0.6
4	0.8	0.7
5	1.0	0.9
6	1.2	1.1
7	1.4	1.3
8	1.6	1.5̄
9	1.8	1.6
10	2.0	1.8
20	4.0	3.7
30	6.0	5.5
40	8.0	7.3
50	10.0	9.2

	12/26	12/25	11/25
0			
1	1.1	1.1	1.1
2	3.2	3.1	3.4
3	5.4	5.2	5.7
4	7.6	7.3	7.9
5	9.8	9.4	10.2
6	11.9	11.5̄	12.5
7	14.1	13.5	14.8
8	16.2	15.6	17.1
9	18.4	17.7	19.3
10	20.6	19.8	21.6
11	22.8	21.9	23.9
12	24.9	23.9	—

P P

*138° 228° *318° **48°**

42° *132° 222° *312°

′	L Sin	d	L Tan	c d	L Cot	L Cos	d	
0	9.82 551	14	9.95 444	25	0.04 556	9.87 107	11	60
1	9.82 565	14	9.95 469	26	0.04 531	9.87 096	11	59
2	9.82 579	14	9.95 495̄	25	0.04 505	9.87 085̄	12	58
3	9.82 593	14	9.95 520	25	0.04 480	9.87 073	11	57
4	9.82 607	14	9.95 545	26	0.04 455̄	9.87 062	12	56
5	9.82 621	14	9.95 571	25	0.04 429	9.87 050	11	55
6	9.82 635	14	9.95 596	26	0.04 404	9.87 039	11	54
7	9.82 649	14	9.95 622	25	0.04 378	9.87 028	12	53
8	9.82 663	14	9.95 647	25	0.04 353	9.87 016	11	52
9	9.82 677	14	9.95 672	26	0.04 328	9.87 005̄	12	51
10	9.82 691	14	9.95 698	25	0.04 302	9.86 993	11	50
11	9.82 705̄	14	9.95 723	25	0.04 277	9.86 982	12	49
12	9.82 719	14	9.95 748	26	0.04 252	9.86 970	11	48
13	9.82 733	14	9.95 774	25	0.04 226	9.86 959	12	47
14	9.82 747	14	9.95 799	26	0.04 201	9.86 947	11	46
15	9.82 761	14	9.95 825̄	25	0.04 175	9.86 936	12	45
16	9.82 775̄	13	9.95 850	25	0.04 150	9.86 924	11	44
17	9.82 788	14	9.95 875	26	0.04 125̄	9.86 913	11	43
18	9.82 802	14	9.95 901	25	0.04 099	9.86 902	12	42
19	9.82 816	14	9.95 926	26	0.04 074	9.86 890	11	41
20	9.82 830	14	9.95 952	25	0.04 048	9.86 879	12	40
21	9.82 844	14	9.95 977	25	0.04 023	9.86 867	12	39
22	9.82 858	14	9.96 002	26	0.03 998	9.86 855	11	38
23	9.82 872	13	9.96 028	25	0.03 972	9.86 844	12	37
24	9.82 885	14	9.96 053	25	0.03 947	9.86 832	11	36
25	9.82 899	14	9.96 078	26	0.03 922	9.86 821	12	35
26	9.82 913	14	9.96 104	25	0.03 896	9.86 809	11	34
27	9.82 927	14	9.96 129	26	0.03 871	9.86 798	12	33
28	9.82 941	14	9.96 155̄	25	0.03 845	9.86 786	11	32
29	9.82 955̄	13	9.96 180	25	0.03 820	9.86 775̄	12	31
30	9.82 968	14	9.96 205	26	0.03 795̄	9.86 763	11	30
31	9.82 982	14	9.96 231	25	0.03 769	9.86 752	12	29
32	9.82 996	14	9.96 256	25	0.03 744	9.86 740	12	28
33	9.83 010	13	9.96 281	26	0.03 719	9.86 728	11	27
34	9.83 023	14	9.96 307	25	0.03 693	9.86 717	12	26
35	9.83 037	14	9.96 332	25	0.03 668	9.86 705	11	25
36	9.83 051	14	9.96 357	26	0.03 643	9.86 694	12	24
37	9.83 065̄	13	9.96 383	25	0.03 617	9.86 682	12	23
38	9.83 078	14	9.96 408	25	0.03 592	9.86 670	11	22
39	9.83 092	14	9.96 433	26	0.03 567	9.86 659	12	21
40	9.83 106	14	9.96 459	25	0.03 541	9.86 647	12	20
41	9.83 120	13	9.96 484	26	0.03 516	9.86 635	11	19
42	9.83 133	14	9.96 510	25	0.03 490	9.86 624	12	18
43	9.83 147	14	9.96 535̄	25	0.03 465	9.86 612	12	17
44	9.83 161	13	9.96 560	26	0.03 440	9.86 600	11	16
45	9.83 174	14	9.96 586	25	0.03 414	9.86 589	12	15
46	9.83 188	14	9.96 611	25	0.03 389	9.86 577	12	14
47	9.83 202	13	9.96 636	26	0.03 364	9.86 565	11	13
48	9.83 215	14	9.96 662	25	0.03 338	9.86 554	12	12
49	9.83 229	13	9.96 687	25	0.03 313	9.86 542	12	11
50	9.83 242	14	9.96 712	26	0.03 288	9.86 530	12	10
51	9.83 256	14	9.96 738	25	0.03 262	9.86 518	11	9
52	9.83 270	13	9.96 763	25	0.03 237	9.86 507	12	8
53	9.83 283	14	9.96 788	26	0.03 212	9.86 495	12	7
54	9.83 297	13	9.96 814	25	0.03 186	9.86 483	11	6
55	9.83 310	14	9.96 839	25	0.03 161	9.86 472	12	5
56	9.83 324	14	9.96 864	26	0.03 136	9.86 460	12	4
57	9.83 338	13	9.96 890	25	0.03 110	9.86 448	12	3
58	9.83 351	14	9.96 915̄	25	0.03 085	9.86 436	11	2
59	9.83 365̄	13	9.96 940	26	0.03 060	9.86 425̄	12	1
60	9.83 378		9.96 966		0.03 034	9.86 413		0
	L Cos	d	L Cot	c d	L Tan	L Sin	d	′

P P

	26	25
1	0.4	0.4
2	0.9	0.8
3	1.3	1.2
4	1.7	1.7
5	2.2	2.1
6	2.6	2.5
7	3.0	2.9
8	3.5̄	3.3
9	3.9	3.8
10	4.3	4.2
20	8.7	8.3
30	13.0	12.5
40	17.3	16.7
50	21.7	20.8

	14	13
1	0.2	0.2
2	0.5̄	0.4
3	0.7	0.6
4	0.9	0.9
5	1.2	1.1
6	1.4	1.3
7	1.6	1.5
8	1.9	1.7
9	2.1	2.0
10	2.3	2.2
20	4.7	4.3
30	7.0	6.5
40	9.3	8.7
50	11.7	10.8

	12	11
1	0.2	0.2
2	0.4	0.4
3	0.6	0.6
4	0.8	0.7
5	1.0	0.9
6	1.2	1.1
7	1.4	1.3
8	1.6	1.5̄
9	1.8	1.6
10	2.0	1.8
20	4.0	3.7
30	6.0	5.5
40	8.0	7.3
50	10.0	9.2

	12/26	11/26	11/25
0	1.1	1.2	1.1
1	3.2	3.5	3.4
2	5.4	5.9	5.7
3	7.6	8.3	7.9
4	9.8	10.6	10.2
5	11.9	13.0	12.5
6	14.1	15.4	14.8
7	16.2	17.7	17.1
8	18.4	20.1	19.3
9	20.6	22.5̄	21.6
10	22.8	24.8	23.9
11	24.9	—	—
12			

P P

*137° 227° *317° 47°

43° *133° 223° *313°

′	L Sin	d	L Tan	c d	L Cot	L Cos	d	′
0	9.83 378	14	9.96 966	25	0.03 034	9.86 413	12	60
1	9.83 392	13	9.96 991	25	0.03 009	9.86 401	12	59
2	9.83 405	14	9.97 016	26	0.02 984	9.86 389	12	58
3	9.83 419	13	9.97 042	25	0.02 958	9.86 377	11	57
4	9.83 432	14	9.97 067	25	0.02 933	9.86 366	12	56
5	9.83 446	13	9.97 092	26	0.02 908	9.86 354	12	55
6	9.83 459	14	9.97 118	25	0.02 882	9.86 342	12	54
7	9.83 473	13	9.97 143	25	0.02 857	9.86 330	12	53
8	9.83 486	14	9.97 168	25	0.02 832	9.86 318	12	52
9	9.83 5̄00	13	9.97 193	26	0.02 807	9.86 306	11	51
10	9.83 513	14	9.97 219	25	0.02 781	9.86 295̄	12	50
11	9.83 527	13	9.97 244	25	0.02 756	9.86 283	12	49
12	9.83 540	14	9.97 269	26	0.02 731	9.86 271	12	48
13	9.83 554	13	9.97 295̄	25	0.02 705	9.86 259	12	47
14	9.83 567	14	9.97 320	25	0.02 680	9.86 247	12	46
15	9.83 581	13	9.97 345	26	0.02 655̄	9.86 235	12	45
16	9.83 594	14	9.97 371	25	0.02 629	9.86 223	12	44
17	9.83 608	13	9.97 396	25	0.02 604	9.86 211	11	43
18	9.83 621	13	9.97 421	26	0.02 579	9.86 200	12	42
19	9.83 634	14	9.97 447	25	0.02 553	9.86 188	12	41
20	9.83 648	13	9.97 472	25	0.02 528	9.86 176	12	40
21	9.83 661	13	9.97 497	26	0.02 503	9.86 164	12	39
22	9.83 674	14	9.97 523	25	0.02 477	9.86 152	12	38
23	9.83 688	13	9.97 548	25	0.02 452	9.86 140	12	37
24	9.83 701	14	9.97 573	25	0.02 427	9.86 128	12	36
25	9.83 715̄	13	9.97 598	26	0.02 402	9.86 116	12	35
26	9.83 728	13	9.97 624	25	0.02 376	9.86 104	12	34
27	9.83 741	14	9.97 649	25	0.02 351	9.86 092	12	33
28	9.83 755̄	13	9.97 674	26	0.02 326	9.86 080	12	32
29	9.83 768	13	9.97 700	25	0.02 300	9.86 068	12	31
30	9.83 781	14	9.97 725̄	25	0.02 275	9.86 056	12	30
31	9.83 795̄	13	9.97 750	26	0.02 250̄	9.86 044	12	29
32	9.83 808	13	9.97 776	25	0.02 224	9.86 032	12	28
33	9.83 821	13	9.97 801	25	0.02 199	9.86 020	12	27
34	9.83 834	14	9.97 826	25	0.02 174	9.86 008	12	26
35	9.83 848	13	9.97 851	26	0.02 149	9.85 996	12	25
36	9.83 861	13	9.97 877	25	0.02 123	9.85 984	12	24
37	9.83 874	13	9.97 902	25	0.02 098	9.85 972	12	23
38	9.83 887	14	9.97 927	26	0.02 073	9.85 960	12	22
39	9.83 901	13	9.97 953	25	0.02 047	9.85 948	12	21
40	9.83 914	13	9.97 978	25	0.02 022	9.85 936	12	20
41	9.83 927	13	9.98 003	26	0.01 997	9.85 924	12	19
42	9.83 940	14	9.98 029	25	0.01 971	9.85 912	12	18
43	9.83 954	13	9.98 054	25	0.01 946	9.85 900	12	17
44	9.83 967	13	9.98 079	25	0.01 921	9.85 888	12	16
45	9.83 980	13	9.98 104	26	0.01 896	9.85 876	12	15
46	9.83 993	13	9.98 130	25	0.01 870	9.85 864	13	14
47	9.84 006	14	9.98 155	25	0.01 845̄	9.85 851	12	13
48	9.84 020	13	9.98 180	26	0.01 820	9.85 839	12	12
49	9.84 033	13	9.98 206	25	0.01 794	9.85 827	12	11
50	9.84 046	13	9.98 231	25	0.01 769	9.85 815	12	10
51	9.84 059	13	9.98 256	25	0.01 744	9.85 803	12	9
52	9.84 072	13	9.98 281	26	0.01 719	9.85 791	12	8
53	9.84 085	13	9.98 307	25	0.01 693	9.85 779	13	7
54	9.84 098	14	9.98 332	25	0.01 668	9.85 766	12	6
55	9.84 112	13	9.98 357	26	0.01 643	9.85 754	12	5
56	9.84 125̄	13	9.98 383	25	0.01 617	9.85 742	12	4
57	9.84 138	13	9.98 408	25	0.01 592	9.85 730	12	3
58	9.84 151	13	9.98 433	25	0.01 567	9.85 718	12	2
59	9.84 164	13	9.98 458	26	0.01 542	9.85 706	13	1
60	9.84 177		9.98 484		0.01 516	9.85 693		0
	L Cos	d	L Cot	c d	L Tan	L Sin	d	′

P P

	26	25
1	0.4	0.4
2	0.9	0.8
3	1.3	1.2
4	1.7	1.7
5	2.2	2.1
6	2.6	2.5
7	3.0	2.9
8	3.5̄	3.3
9	3.9	3.8
10	4.3	4.2
20	8.7	8.3
30	13.0	12.5
40	17.3	16.7
50	21.7	20.8

	14	13
1	0.2	0.2
2	0.5̄	0.4
3	0.7	0.6
4	0.9	0.9
5	1.2	1.1
6	1.4	1.3
7	1.6	1.5
8	1.9	1.7
9	2.1	2.0
10	2.3	2.2
20	4.7	4.3
30	7.0	6.5
40	9.3	8.7
50	11.7	10.8

	12	11
1	0.2	0.2
2	0.4	0.4
3	0.6	0.6
4	0.8	0.7
5	1.0	0.9
6	1.2	1.1
7	1.4	1.3
8	1.6	1.5̄
9	1.8	1.6
10	2.0	1.8
20	4.0	3.7
30	6.0	5.5
40	8.0	7.3
50	10.0	9.2

	13/26	13/25	12/25
0	1.0	0.9	1.1
1	3.0	2.9	3.1
2	5.0	4.8	5.2
3	7.0	6.7	7.3
4	9.0	8.7	9.4
5	11.0	10.6	11.5̄
6	13.0	12.5	13.5
7	15.0	14.4	15.6
8	17.0	16.3	17.7
9	19.0	18.3	19.8
10	21.0	20.2	21.9
11	23.0	22.1	23.9
12	25.0	24.1	—
13			

P P

*136° 226° *316° 46°

44°

*134° 224° *314°

′	L Sin	d	L Tan	c d	L Cot	L Cos	d	′
0	9.84 177		9.98 484		0.01 516	9.85 693		60
1	9.84 190	13	9.98 509	25	0.01 491	9.85 681	12	59
2	9.84 203	13	9.98 534	25	0.01 466	9.85 669	12	58
3	9.84 216	13	9.98 560	26	0.01 440	9.85 657	12	57
4	9.84 229	13	9.98 585̄	25	0.01 415	9.85 645̄	12	56
5	9.84 242	13	9.98 610	25	0.01 390	9.85 632	13	55
6	9.84 255	13	9.98 635	25	0.01 365̄	9.85 620	12	54
7	9.84 269	14	9.98 661	26	0.01 339	9.85 608	12	53
8	9.84 282	13	9.98 686	25	0.01 314	9.85 596	12	52
9	9.84 295̄	13	9.98 711	25	0.01 289	9.85 583	13	51
10	9.84 308	13	9.98 737	26	0.01 263	9.85 571	12	50
11	9.84 321	13	9.98 762	25	0.01 238	9.85 559	12	49
12	9.84 334	13	9.98 787	25	0.01 213	9.85 547	12	48
13	9.84 347	13	9.98 812	25	0.01 188	9.85 534	13	47
14	9.84 360	13	9.98 838	26	0.01 162	9.85 522	12	46
15	9.84 373	13	9.98 863	25	0.01 137	9.85 510	12	45
16	9.84 385	12	9.98 888	25	0.01 112	9.85 497	13	44
17	9.84 398	13	9.98 913	25	0.01 087	9.85 485̄	12	43
18	9.84 411	13	9.98 939	26	0.01 061	9.85 473	12	42
19	9.84 424	13	9.98 964	25	0.01 036	9.85 460	13	41
20	9.84 437	13	9.98 989	25	0.01 011	9.85 448	12	40
21	9.84 450	13	9.99 015̄	26	0.00 985	9.85 436	12	39
22	9.84 463	13	9.99 040	25	0.00 960	9.85 423	13	38
23	9.84 476	13	9.99 065	25	0.00 935̄	9.85 411	12	37
24	9.84 489	13	9.99 090	25	0.00 910	9.85 399	12	36
25	9.84 502	13	9.99 116	26	0.00 884	9.85 386	13	35
26	9.84 515̄	13	9.99 141	25	0.00 859	9.85 374	12	34
27	9.84 528	13	9.99 166	25	0.00 834	9.85 361	13	33
28	9.84 540	12	9.99 191	25	0.00 809	9.85 349	12	32
29	9.84 553	13	9.99 217	26	0.00 783	9.85 337	12	31
30	9.84 566	13	9.99 242	25	0.00 758	9.85 324	13	30
31	9.84 579	13	9.99 267	25	0.00 733	9.85 312	12	29
32	9.84 592	13	9.99 293	26	0.00 707	9.85 299	13	28
33	9.84 605̄	13	9.99 318	25	0.00 682	9.85 287	12	27
34	9.84 618	13	9.99 343	25	0.00 657	9.85 274	13	26
35	9.84 630	12	9.99 368	25	0.00 632	9.85 262	12	25
36	9.84 643	13	9.99 394	26	0.00 606	9.85 250	12	24
37	9.84 656	13	9.99 419	25	0.00 581	9.85 237	13	23
38	9.84 669	13	9.99 444	25	0.00 556	9.85 225̄	12	22
39	9.84 682	13	9.99 469	25	0.00 531	9.85 212	13	21
40	9.84 694	12	9.99 495̄	26	0.00 505	9.85 200	12	20
41	9.84 707	13	9.99 520	25	0.00 480	9.85 187	13	19
42	9.84 720	13	9.99 545	25	0.00 455̄	9.85 175̄	12	18
43	9.84 733	13	9.99 570	25	0.00 430	9.85 162	13	17
44	9.84 745	12	9.99 596	26	0.00 404	9.85 150	12	16
45	9.84 758	13	9.99 621	25	0.00 379	9.85 137	13	15
46	9.84 771	13	9.99 646	25	0.00 354	9.85 125̄	12	14
47	9.84 784	13	9.99 672	26	0.00 328	9.85 112	13	13
48	9.84 796	12	9.99 697	25	0.00 303	9.85 100	12	12
49	9.84 809	13	9.99 722	25	0.00 278	9.85 087	13	11
50	9.84 822	13	9.99 747	25	0.00 253	9.85 074	13	10
51	9.84 835̄	13	9.99 773	26	0.00 227	9.85 062	12	9
52	9.84 847	12	9.99 798	25	0.00 202	9.85 049	13	8
53	9.84 860	13	9.99 823	25	0.00 177	9.85 037	12	7
54	9.84 873	13	9.99 848	25	0.00 152	9.85 024	13	6
55	9.84 885	12	9.99 874	26	0.00 126	9.85 012	12	5
56	9.84 898	13	9.99 899	25	0.00 101	9.84 999	13	4
57	9.84 911	13	9.99 924	25	0.00 076	9.84 986	13	3
58	9.84 923	12	9.99 949	25	0.00 051	9.84 974	12	2
59	9.84 936	13	9.99 975̄	26	0.00 025	9.84 961	13	1
60	9.84 949	13	0.00 000	25	0.00 000	9.84 949	12	0
	L Cos	d	L Cot	c d	L Tan	L Sin	d	′

P P

	26	25
1	0.4	0.4
2	0.9	0.8
3	1.3	1.2
4	1.7	1.7
5	2.2	2.1
6	2.6	2.5
7	3.0	2.9
8	3.5̄	3.3
9	3.9	3.8
10	4.3	4.2
20	8.7	8.3
30	13.0	12.5
40	17.3	16.7
50	21.7	20.8

	14	13	12
1	0.2	0.2	0.2
2	0.5̄	0.4	0.4
3	0.7	0.6	0.6
4	0.9	0.9	0.8
5	1.2	1.1	1.0
6	1.4	1.3	1.2
7	1.6	1.5	1.4
8	1.9	1.7	1.6
9	2.1	2.0	1.8
10	2.3	2.2	2.0
20	4.7	4.3	4.0
30	7.0	6.5	6.0
40	9.3	8.7	8.0
50	11.7	10.8	10.0

	13/26	13/25
0		
1	1.0	0.9
2	3.0	2.9
3	5.0	4.8
4	7.0	6.7
5	9.0	8.7
6	11.0	10.6
7	13.0	12.5
8	15.0	14.4
9	17.0	16.3
10	19.0	18.3
11	21.0	20.2
12	23.0	22.1
13	25.0	24.1

	12/26	12/25
0		
1	1.1	1.1
2	3.2	3.1
3	5.4	5.2
4	7.6	7.3
5	9.8	9.4
6	11.9	11.5̄
7	14.1	13.5
8	16.2	15.6
9	18.4	17.7
10	20.6	19.8
11	22.8	21.9
12	24.9	23.9

*135° 225° *315° **45°**

TABLE OF THE NATURAL

TRIGONOMETRIC FUNCTIONS

FROM MINUTE TO MINUTE.

*90° 180° *270° **0°** NATURAL

′	Sin	Tan	Cot	Cos	
0	0.0000	0.0000	∞	1.0000	60
1	0.0003	0.0003	3437.75̄	1.0000	59
2	0.0006	0.0006	1718.87	1.0000	58
3	0.0009	0.0009	1145.92	1.0000	57
4	0.0012	0.0012	859.436	1.0000	56
5	0.0015̄	0.0015̄	687.549	1.0000	55
6	0.0017	0.0017	572.957	1.0000	54
7	0.0020	0.0020	491.106	1.0000	53
8	0.0023	0.0023	429.718	1.0000	52
9	0.0026	0.0026	381.971	1.0000	51
10	0.0029	0.0029	343.774	1.0000	50
11	0.0032	0.0032	312.521	1.0000	49
12	0.0035̄	0.0035̄	286.478	1.0000	48
13	0.0038	0.0038	264.441	1.0000	47
14	0.0041	0.0041	245.552	1.0000	46
15	0.0044	0.0044	229.182	1.0000	45
16	0.0047	0.0047	214.858	1.0000	44
17	0.0049	0.0049	202.219	1.0000	43
18	0.0052	0.0052	190.984	1.0000	42
19	0.0055	0.0055	180.932	1.0000	41
20	0.0058	0.0058	171.885	1.0000	40
21	0.0061	0.0061	163.700	1.0000	39
22	0.0064	0.0064	156.259	1.0000	38
23	0.0067	0.0067	149.465	1.0000	37
24	0.0070	0.0070	143.237	1.0000	36
25	0.0073	0.0073	137.507	1.0000	35
26	0.0076	0.0076	132.219	1.0000	34
27	0.0079	0.0079	127.321	1.0000	33
28	0.0081	0.0081	122.774	1.0000	32
29	0.0084	0.0084	118.540	1.0000	31
30	0.0087	0.0087	114.589	1.0000	30
31	0.0090	0.0090	110.892	1.0000	29
32	0.0093	0.0093	107.426	1.0000	28
33	0.0096	0.0096	104.171	1.0000	27
34	0.0099	0.0099	101.107	1.0000	26
35	0.0102	0.0102	98.2179	0.9999	25
36	0.0105̄	0.0105̄	95.4895̄	0.9999	24
37	0.0108	0.0108	92.9085̄	0.9999	23
38	0.0111	0.0111	90.4633	0.9999	22
39	0.0113	0.0113	88.1436	0.9999	21
40	0.0116	0.0116	85.9398	0.9999	20
41	0.0119	0.0119	83.8435	0.9999	19
42	0.0122	0.0122	81.8470	0.9999	18
43	0.0125	0.0125	79.9434	0.9999	17
44	0.0128	0.0128	78.1263	0.9999	16
45	0.0131	0.0131	76.3900	0.9999	15
46	0.0134	0.0134	74.7292	0.9999	14
47	0.0137	0.0137	73.1390	0.9999	13
48	0.0140	0.0140	71.6151	0.9999	12
49	0.0143	0.0143	70.1533	0.9999	11
50	0.0145	0.0145	68.7501	0.9999	10
51	0.0148	0.0148	67.4019	0.9999	9
52	0.0151	0.0151	66.1055̄	0.9999	8
53	0.0154	0.0154	64.8580	0.9999	7
54	0.0157	0.0157	63.6567	0.9999	6
55	0.0160	0.0160	62.4992	0.9999	5
56	0.0163	0.0163	61.3829	0.9999	4
57	0.0166	0.0166	60.3058	0.9999	3
58	0.0169	0.0169	59.2659	0.9999	2
59	0.0172	0.0172	58.2612	0.9999	1
60	0.0175̄	0.0175̄	57.2900	0.9998	0
	Cos	Cot	Tan	Sin	′

179° 269° *359° **89°** NATURAL

1° *91° 181° *271°

′	Sin	Tan	Cot	Cos	
0	0.0175̄	0.0175̄	57.2900	0.9998	60
1	0.0177	0.0177	56.3506	0.9998	59
2	0.0180	0.0180	55.4415	0.9998	58
3	0.0183	0.0183	54.5613	0.9998	57
4	0.0186	0.0186	53.7086	0.9998	56
5	0.0189	0.0189	52.8821	0.9998	55
6	0.0192	0.0192	52.0807	0.9998	54
7	0.0195̄	0.0195̄	51.3032	0.9998	53
8	0.0198	0.0198	50.5485	0.9998	52
9	0.0201	0.0201	49.8157	0.9998	51
10	0.0204	0.0204	49.1039	0.9998	50
11	0.0207	0.0207	48.4121	0.9998	49
12	0.0209	0.0209	47.7395	0.9998	48
13	0.0212	0.0212	47.0853	0.9998	47
14	0.0215	0.0215	46.4489	0.9998	46
15	0.0218	0.0218	45.8294	0.9998	45
16	0.0221	0.0221	45.2261	0.9998	44
17	0.0224	0.0224	44.6386	0.9997	43
18	0.0227	0.0227	44.0661	0.9997	42
19	0.0230	0.0230	43.5081	0.9997	41
20	0.0233	0.0233	42.9641	0.9997	40
21	0.0236	0.0236	42.4335̄	0.9997	39
22	0.0239	0.0239	41.9158	0.9997	38
23	0.0241	0.0241	41.4106	0.9997	37
24	0.0244	0.0244	40.9174	0.9997	36
25	0.0247	0.0247	40.4358	0.9997	35
26	0.0250	0.0250	39.9655̄	0.9997	34
27	0.0253	0.0253	39.5059	0.9997	33
28	0.0256	0.0256	39.0568	0.9997	32
29	0.0259	0.0259	38.6177	0.9997	31
30	0.0262	0.0262	38.1885̄	0.9997	30
31	0.0265̄	0.0265̄	37.7686	0.9996	29
32	0.0268	0.0268	37.3579	0.9996	28
33	0.0270	0.0271	36.9560	0.9996	27
34	0.0273	0.0274	36.5627	0.9996	26
35	0.0276	0.0276	36.1776	0.9996	25
36	0.0279	0.0279	35.8006	0.9996	24
37	0.0282	0.0282	35.4313	0.9996	23
38	0.0285	0.0285	35.0695	0.9996	22
	0.0288	0.0288	34.7151	0.9996	21
40	0.0291	0.0291	34.3678	0.9996	20
41	0.0294	0.0294	34.0273	0.9996	19
42	0.0297	0.0297	33.6935	0.9996	18
43	0.0300	0.0300	33.3662	0.9996	17
44	0.0302	0.0303	33.0452	0.9995	16
45	0.0305	0.0306	32.7303	0.9995	15
46	0.0308	0.0308	32.4213	0.9995	14
47	0.0311	0.0311	32.1181	0.9995	13
48	0.0314	0.0314	31.8205	0.9995	12
49	0.0317	0.0317	31.5284	0.9995̄	11
50	0.0320	0.0320	31.2416	0.9995̄	10
51	0.0323	0.0323	30.9599	0.9995̄	9
52	0.0326	0.0326	30.6833	0.9995̄	8
53	0.0329	0.0329	30.4116	0.9995̄	7
54	0.0332	0.0332	30.1446	0.9995̄	6
55	0.0334	0.0335̄	29.8823	0.9994	5
56	0.0337	0.0338	29.6245̄	0.9994	4
57	0.0340	0.0340	29.3711	0.9994	3
58	0.0343	0.0343	29.1220	0.9994	2
59	0.0346	0.0346	28.8771	0.9994	1
60	0.0349	0.0349	28.6363	0.9994	0
	Cos	Cot	Tan	Sin	′

88° *178° 268° *358°

*92° 182° *272° **2°** NATURAL

′	Sin	Tan	Cot	Cos	
0	0.0349	0.0349	28.6363	0.9994	60
1	0.0352	0.0352	28.3994	0.9994	59
2	0.0355	0.0355	28.1664	0.9994	58
3	0.0358	0.0358	27.9372	0.9994	57
4	0.0361	0.0361	27.7117	0.9993	56
5	0.0364	0.0364	27.4899	0.9993	55
6	0.0366	0.0367	27.2715	0.9993	54
7	0.0369	0.0370	27.0566	0.9993	53
8	0.0372	0.0373	26.8450	0.9993	52
9	0.0375	0.0375	26.6367	0.9993	51
10	0.0378	0.0378	26.4316	0.9993	50
11	0.0381	0.0381	26.2296	0.9993	49
12	0.0384	0.0384	26.0307	0.9993	48
13	0.0387	0.0387	25.8348	0.9993	47
14	0.0390	0.0390	25.6418	0.9992	46
15	0.0393	0.0393	25.4517	0.9992	45
16	0.0396	0.0396	25.2644	0.9992	44
17	0.0398	0.0399	25.0798	0.9992	43
18	0.0401	0.0402	24.8978	0.9992	42
19	0.0404	0.0405	24.7185	0.9992	41
20	0.0407	0.0407	24.5418	0.9992	40
21	0.0410	0.0410	24.3675	0.9992	39
22	0.0413	0.0413	24.1957	0.9991	38
23	0.0416	0.0416	24.0263	0.9991	37
24	0.0419	0.0419	23.8593	0.9991	36
25	0.0422	0.0422	23.6945	0.9991	35
26	0.0425	0.0425	23.5321	0.9991	34
27	0.0427	0.0428	23.3718	0.9991	33
28	0.0430	0.0431	23.2137	0.9991	32
29	0.0433	0.0434	23.0577	0.9991	31
30	0.0436	0.0437	22.9038	0.9990	30
31	0.0439	0.0440	22.7519	0.9990	29
32	0.0442	0.0442	22.6020	0.9990	28
33	0.0445	0.0445	22.4541	0.9990	27
34	0.0448	0.0448	22.3081	0.9990	26
35	0.0451	0.0451	22.1640	0.9990	25
36	0.0454	0.0454	22.0217	0.9990	24
37	0.0457	0.0457	21.8813	0.9990	23
38	0.0459	0.0460	21.7426	0.9989	22
39	0.0462	0.0463	21.6056	0.9989	21
40	0.0465	0.0466	21.4704	0.9989	20
41	0.0468	0.0469	21.3369	0.9989	19
42	0.0471	0.0472	21.2049	0.9989	18
43	0.0474	0.0475	21.0747	0.9989	17
44	0.0477	0.0477	20.9460	0.9989	16
45	0.0480	0.0480	20.8188	0.9988	15
46	0.0483	0.0483	20.6932	0.9988	14
47	0.0486	0.0486	20.5691	0.9988	13
48	0.0488	0.0489	20.4465	0.9988	12
49	0.0491	0.0492	20.3253	0.9988	11
50	0.0494	0.0495	20.2056	0.9988	10
51	0.0497	0.0498	20.0872	0.9988	9
52	0.0500	0.0501	19.9702	0.9987	8
53	0.0503	0.0504	19.8546	0.9987	7
54	0.0506	0.0507	19.7403	0.9987	6
55	0.0509	0.0509	19.6273	0.9987	5
56	0.0512	0.0512	19.5156	0.9987	4
57	0.0515	0.0515	19.4051	0.9987	3
58	0.0518	0.0518	19.2959	0.9987	2
59	0.0520	0.0521	19.1879	0.9986	1
60	0.0523	0.0524	19.0811	0.9986	0
	Cos	Cot	Tan	Sin	′

*177° 267° *357° **87°** NATURAL

3° *93° 183° *273°

′	Sin	Tan	Cot	Cos	
0	0.0523	0.0524	19.0811	0.9986	60
1	0.0526	0.0527	18.9755	0.9986	59
2	0.0529	0.0530	18.8711	0.9986	58
3	0.0532	0.0533	18.7678	0.9986	57
4	0.0535	0.0536	18.6656	0.9986	56
5	0.0538	0.0539	18.5645	0.9986	55
6	0.0541	0.0542	18.4645	0.9985	54
7	0.0544	0.0544	18.3655	0.9985	53
8	0.0547	0.0547	18.2677	0.9985	52
9	0.0550	0.0550	18.1708	0.9985	51
10	0.0552	0.0553	18.0750	0.9985	50
11	0.0555	0.0556	17.9802	0.9985	49
12	0.0558	0.0559	17.8863	0.9984	48
13	0.0561	0.0562	17.7934	0.9984	47
14	0.0564	0.0565	17.7015	0.9984	46
15	0.0567	0.0568	17.6106	0.9984	45
16	0.0570	0.0571	17.5205	0.9984	44
17	0.0573	0.0574	17.4314	0.9984	43
18	0.0576	0.0577	17.3432	0.9983	42
19	0.0579	0.0580	17.2558	0.9983	41
20	0.0581	0.0582	17.1693	0.9983	40
21	0.0584	0.0585	17.0837	0.9983	39
22	0.0587	0.0588	16.9990	0.9983	38
23	0.0590	0.0591	16.9150	0.9983	37
24	0.0593	0.0594	16.8319	0.9982	36
25	0.0596	0.0597	16.7496	0.9982	35
26	0.0599	0.0600	16.6681	0.9982	34
27	0.0602	0.0603	16.5874	0.9982	33
28	0.0605	0.0606	16.5075	0.9982	32
29	0.0608	0.0609	16.4283	0.9982	31
30	0.0610	0.0612	16.3499	0.9981	30
31	0.0613	0.0615	16.2722	0.9981	29
32	0.0616	0.0617	16.1952	0.9981	28
33	0.0619	0.0620	16.1190	0.9981	27
34	0.0622	0.0623	16.0435	0.9981	26
35	0.0625	0.0626	15.9687	0.9980	25
36	0.0628	0.0629	15.8945	0.9980	24
37	0.0631	0.0632	15.8211	0.9980	23
38	0.0634	0.0635	15.7483	0.9980	22
39	0.0637	0.0638	15.6762	0.9980	21
40	0.0640	0.0641	15.6048	0.9980	20
41	0.0642	0.0644	15.5340	0.9979	19
42	0.0645	0.0647	15.4638	0.9979	18
43	0.0648	0.0650	15.3943	0.9979	17
44	0.0651	0.0653	15.3254	0.9979	16
45	0.0654	0.0655	15.2571	0.9979	15
46	0.0657	0.0658	15.1893	0.9978	14
47	0.0660	0.0661	15.1222	0.9978	13
48	0.0663	0.0664	15.0557	0.9978	12
49	0.0666	0.0667	14.9898	0.9978	11
50	0.0669	0.0670	14.9244	0.9978	10
51	0.0671	0.0673	14.8596	0.9977	9
52	0.0674	0.0676	14.7954	0.9977	8
53	0.0677	0.0679	14.7317	0.9977	7
54	0.0680	0.0682	14.6685	0.9977	6
55	0.0683	0.0685	14.6059	0.9977	5
56	0.0686	0.0688	14.5438	0.9976	4
57	0.0689	0.0690	14.4823	0.9976	3
58	0.0692	0.0693	14.4212	0.9976	2
59	0.0695	0.0696	14.3607	0.9976	1
60	0.0698	0.0699	14.3007	0.9976	0
	Cos	Cot	Tan	Sin	′

86° *176° 266° *356°

*94° 184° *274° 4° — NATURAL

′	Sin	Tan	Cot	Cos	
0	0.0698	0.0699	14.3007	0.9976	60
1	0.0700	0.0702	14.2411	0.9975	59
2	0.0703	0.0705	14.1821	0.9975	58
3	0.0706	0.0708	14.1235	0.9975	57
4	0.0709	0.0711	14.0655	0.9975	56
5	0.0712	0.0714	14.0079	0.9975	55
6	0.0715	0.0717	13.9507	0.9974	54
7	0.0718	0.0720	13.8940	0.9974	53
8	0.0721	0.0723	13.8378	0.9974	52
9	0.0724	0.0726	13.7821	0.9974	51
10	0.0727	0.0729	13.7267	0.9974	50
11	0.0729	0.0731	13.6719	0.9973	49
12	0.0732	0.0734	13.6174	0.9973	48
13	0.0735	0.0737	13.5634	0.9973	47
14	0.0738	0.0740	13.5098	0.9973	46
15	0.0741	0.0743	13.4566	0.9973	45
16	0.0744	0.0746	13.4039	0.9972	44
17	0.0747	0.0749	13.3515	0.9972	43
18	0.0750	0.0752	13.2996	0.9972	42
19	0.0753	0.0755	13.2480	0.9972	41
20	0.0756	0.0758	13.1969	0.9971	40
21	0.0758	0.0761	13.1461	0.9971	39
22	0.0761	0.0764	13.0958	0.9971	38
23	0.0764	0.0767	13.0458	0.9971	37
24	0.0767	0.0769	12.9962	0.9971	36
25	0.0770	0.0772	12.9469	0.9970	35
26	0.0773	0.0775	12.8981	0.9970	34
27	0.0776	0.0778	12.8496	0.9970	33
28	0.0779	0.0781	12.8014	0.9970	32
29	0.0782	0.0784	12.7536	0.9969	31
30	0.0785	0.0787	12.7062	0.9969	30
31	0.0787	0.0790	12.6591	0.9969	29
32	0.0790	0.0793	12.6124	0.9969	28
33	0.0793	0.0796	12.5660	0.9968	27
34	0.0796	0.0799	12.5199	0.9968	26
35	0.0799	0.0802	12.4742	0.9968	25
36	0.0802	0.0805	12.4288	0.9968	24
37	0.0805	0.0808	12.3838	0.9968	23
38	0.0808	0.0810	12.3390	0.9967	22
39	0.0811	0.0813	12.2946	0.9967	21
40	0.0814	0.0816	12.2505	0.9967	20
41	0.0816	0.0819	12.2067	0.9967	19
42	0.0819	0.0822	12.1632	0.9966	18
43	0.0822	0.0825	12.1201	0.9966	17
44	0.0825	0.0828	12.0772	0.9966	16
45	0.0828	0.0831	12.0346	0.9966	15
46	0.0831	0.0834	11.9923	0.9965	14
47	0.0834	0.0837	11.9504	0.9965	13
48	0.0837	0.0840	11.9087	0.9965	12
49	0.0840	0.0843	11.8673	0.9965	11
50	0.0843	0.0846	11.8262	0.9964	10
51	0.0845	0.0849	11.7853	0.9964	9
52	0.0848	0.0851	11.7448	0.9964	8
53	0.0851	0.0854	11.7045	0.9964	7
54	0.0854	0.0857	11.6645	0.9963	6
55	0.0857	0.0860	11.6248	0.9963	5
56	0.0860	0.0863	11.5853	0.9963	4
57	0.0863	0.0866	11.5461	0.9963	3
58	0.0866	0.0869	11.5072	0.9962	2
59	0.0869	0.0872	11.4685	0.9962	1
60	0.0872	0.0875	11.4301	0.9962	0
	Cos	Cot	Tan	Sin	′

*175° 265° *355° 85° — NATURAL

5° *95° 185° *275°

′	Sin	Tan	Cot	Cos	
0	0.0872	0.0875	11.4301	0.9962	60
1	0.0874	0.0878	11.3919	0.9962	59
2	0.0877	0.0881	11.3540	0.9961	58
3	0.0880	0.0884	11.3163	0.9961	57
4	0.0883	0.0887	11.2789	0.9961	56
5	0.0886	0.0890	11.2417	0.9961	55
6	0.0889	0.0892	11.2048	0.9960	54
7	0.0892	0.0895	11.1681	0.9960	53
8	0.0895	0.0898	11.1316	0.9960	52
9	0.0898	0.0901	11.0954	0.9960	51
10	0.0901	0.0904	11.0594	0.9959	50
11	0.0903	0.0907	11.0237	0.9959	49
12	0.0906	0.0910	10.9882	0.9959	48
13	0.0909	0.0913	10.9529	0.9959	47
14	0.0912	0.0916	10.9178	0.9958	46
15	0.0915	0.0919	10.8829	0.9958	45
16	0.0918	0.0922	10.8483	0.9958	44
17	0.0921	0.0925	10.8139	0.9958	43
18	0.0924	0.0928	10.7797	0.9957	42
19	0.0927	0.0931	10.7457	0.9957	41
20	0.0929	0.0934	10.7119	0.9957	40
21	0.0932	0.0936	10.6783	0.9956	39
22	0.0935	0.0939	10.6450	0.9956	38
23	0.0938	0.0942	10.6118	0.9956	37
24	0.0941	0.0945	10.5789	0.9956	36
25	0.0944	0.0948	10.5462	0.9955	35
26	0.0947	0.0951	10.5136	0.9955	34
27	0.0950	0.0954	10.4813	0.9955	33
28	0.0953	0.0957	10.4491	0.9955	32
29	0.0956	0.0960	10.4172	0.9954	31
30	0.0958	0.0963	10.3854	0.9954	30
31	0.0961	0.0966	10.3538	0.9954	29
32	0.0964	0.0969	10.3224	0.9953	28
33	0.0967	0.0972	10.2913	0.9953	27
34	0.0970	0.0975	10.2602	0.9953	26
35	0.0973	0.0978	10.2294	0.9953	25
36	0.0976	0.0981	10.1988	0.9952	24
37	0.0979	0.0983	10.1683	0.9952	23
38	0.0982	0.0986	10.1381	0.9952	22
39	0.0985	0.0989	10.1080	0.9951	21
40	0.0987	0.0992	10.0780	0.9951	20
41	0.0990	0.0995	10.0483	0.9951	19
42	0.0993	0.0998	10.0187	0.9951	18
43	0.0996	0.1001	9.9893	0.9950	17
44	0.0999	0.1004	9.9601	0.9950	16
45	0.1002	0.1007	9.9310	0.9950	15
46	0.1005	0.1010	9.9021	0.9949	14
47	0.1008	0.1013	9.8734	0.9949	13
48	0.1011	0.1016	9.8448	0.9949	12
49	0.1013	0.1019	9.8164	0.9949	11
50	0.1016	0.1022	9.7882	0.9948	10
51	0.1019	0.1025	9.7601	0.9948	9
52	0.1022	0.1028	9.7322	0.9948	8
53	0.1025	0.1030	9.7044	0.9947	7
54	0.1028	0.1033	9.6768	0.9947	6
55	0.1031	0.1036	9.6493	0.9947	5
56	0.1034	0.1039	9.6220	0.9946	4
57	0.1037	0.1042	9.5949	0.9946	3
58	0.1039	0.1045	9.5679	0.9946	2
59	0.1042	0.1048	9.5411	0.9946	1
60	0.1045	0.1051	9.5144	0.9945	0
	Cos	Cot	Tan	Sin	′

84° *174° 264° *354

′	Sin	Tan	Cot	Cos	
0	0.1045	0.1051	9.5144	0.9945	60
1	0.1048	0.1054	9.4878	0.9945	59
2	0.1051	0.1057	9.4614	0.9945	58
3	0.1054	0.1060	9.4352	0.9944	57
4	0.1057	0.1063	9.4090	0.9944	56
5	0.1060	0.1066	9.3831	0.9944	55
6	0.1063	0.1069	9.3572	0.9943	54
7	0.1066	0.1072	9.3315	0.9943	53
8	0.1068	0.1075	9.3060	0.9943	52
9	0.1071	0.1078	9.2806	0.9942	51
10	0.1074	0.1080	9.2553	0.9942	50
11	0.1077	0.1083	9.2302	0.9942	49
12	0.1080	0.1086	9.2052	0.9942	48
13	0.1083	0.1089	9.1803	0.9941	47
14	0.1086	0.1092	9.1555	0.9941	46
15	0.1089	0.1095	9.1309	0.9941	45
16	0.1092	0.1098	9.1065	0.9940	44
17	0.1094	0.1101	9.0821	0.9940	43
18	0.1097	0.1104	9.0579	0.9940	42
19	0.1100	0.1107	9.0338	0.9939	41
20	0.1103	0.1110	9.0098	0.9939	40
21	0.1106	0.1113	8.9860	0.9939	39
22	0.1109	0.1116	8.9623	0.9938	38
23	0.1112	0.1119	8.9387	0.9938	37
24	0.1115	0.1122	8.9152	0.9938	36
25	0.1118	0.1125	8.8919	0.9937	35
26	0.1120	0.1128	8.8686	0.9937	34
27	0.1123	0.1131	8.8455	0.9937	33
28	0.1126	0.1133	8.8225	0.9936	32
29	0.1129	0.1136	8.7996	0.9936	31
30	0.1132	0.1139	8.7769	0.9936	30
31	0.1135	0.1142	8.7542	0.9935	29
32	0.1138	0.1145	8.7317	0.9935	28
33	0.1141	0.1148	8.7093	0.9935	27
34	0.1144	0.1151	8.6870	0.9934	26
35	0.1146	0.1154	8.6648	0.9934	25
36	0.1149	0.1157	8.6427	0.9934	24
37	0.1152	0.1160	8.6208	0.9933	23
38	0.1155	0.1163	8.5989	0.9933	22
39	0.1158	0.1166	8.5772	0.9933	21
40	0.1161	0.1169	8.5555	0.9932	20
41	0.1164	0.1172	8.5340	0.9932	19
42	0.1167	0.1175	8.5126	0.9932	18
43	0.1170	0.1178	8.4913	0.9931	17
44	0.1172	0.1181	8.4701	0.9931	16
45	0.1175	0.1184	8.4490	0.9931	15
46	0.1178	0.1187	8.4280	0.9930	14
47	0.1181	0.1189	8.4071	0.9930	13
48	0.1184	0.1192	8.3863	0.9930	12
49	0.1187	0.1195	8.3656	0.9929	11
50	0.1190	0.1198	8.3450	0.9929	10
51	0.1193	0.1201	8.3245	0.9929	9
52	0.1196	0.1204	8.3041	0.9928	8
53	0.1198	0.1207	8.2838	0.9928	7
54	0.1201	0.1210	8.2636	0.9928	6
55	0.1204	0.1213	8.2434	0.9927	5
56	0.1207	0.1216	8.2234	0.9927	4
57	0.1210	0.1219	8.2035	0.9927	3
58	0.1213	0.1222	8.1837	0.9926	2
59	0.1216	0.1225	8.1640	0.9926	1
60	0.1219	0.1228	8.1443	0.9925	0
	Cos	Cot	Tan	Sin	′

′	Sin	Tan	Cot	Cos	
0	0.1219	0.1228	8.1443	0.9925	60
1	0.1222	0.1231	8.1248	0.9925	59
2	0.1224	0.1234	8.1054	0.9925	58
3	0.1227	0.1237	8.0860	0.9924	57
4	0.1230	0.1240	8.0667	0.9924	56
5	0.1233	0.1243	8.0476	0.9924	55
6	0.1236	0.1246	8.0285	0.9923	54
7	0.1239	0.1249	8.0095	0.9923	53
8	0.1242	0.1251	7.9906	0.9923	52
9	0.1245	0.1254	7.9718	0.9922	51
10	0.1248	0.1257	7.9530	0.9922	50
11	0.1250	0.1260	7.9344	0.9922	49
12	0.1253	0.1263	7.9158	0.9921	48
13	0.1256	0.1266	7.8973	0.9921	47
14	0.1259	0.1269	7.8789	0.9920	46
15	0.1262	0.1272	7.8606	0.9920	45
16	0.1265	0.1275	7.8424	0.9920	44
17	0.1268	0.1278	7.8243	0.9919	43
18	0.1271	0.1281	7.8062	0.9919	42
19	0.1274	0.1284	7.7882	0.9919	41
20	0.1276	0.1287	7.7704	0.9918	40
21	0.1279	0.1290	7.7525	0.9918	39
22	0.1282	0.1293	7.7348	0.9917	38
23	0.1285	0.1296	7.7171	0.9917	37
24	0.1288	0.1299	7.6996	0.9917	36
25	0.1291	0.1302	7.6821	0.9916	35
26	0.1294	0.1305	7.6647	0.9916	34
27	0.1297	0.1308	7.6473	0.9916	33
28	0.1299	0.1311	7.6301	0.9915	32
29	0.1302	0.1314	7.6129	0.9915	31
30	0.1305	0.1317	7.5958	0.9914	30
31	0.1308	0.1319	7.5787	0.9914	29
32	0.1311	0.1322	7.5618	0.9914	28
33	0.1314	0.1325	7.5449	0.9913	27
34	0.1317	0.1328	7.5281	0.9913	26
35	0.1320	0.1331	7.5113	0.9913	25
36	0.1323	0.1334	7.4947	0.9912	24
37	0.1325	0.1337	7.4781	0.9912	23
38	0.1328	0.1340	7.4615	0.9911	22
39	0.1331	0.1343	7.4451	0.9911	21
40	0.1334	0.1346	7.4287	0.9911	20
41	0.1337	0.1349	7.4124	0.9910	19
42	0.1340	0.1352	7.3962	0.9910	18
43	0.1343	0.1355	7.3800	0.9909	17
44	0.1346	0.1358	7.3639	0.9909	16
45	0.1349	0.1361	7.3479	0.9909	15
46	0.1351	0.1364	7.3319	0.9908	14
47	0.1354	0.1367	7.3160	0.9908	13
48	0.1357	0.1370	7.3002	0.9907	12
49	0.1360	0.1373	7.2844	0.9907	11
50	0.1363	0.1376	7.2687	0.9907	10
51	0.1366	0.1379	7.2531	0.9906	9
52	0.1369	0.1382	7.2375	0.9906	8
53	0.1372	0.1385	7.2220	0.9905	7
54	0.1374	0.1388	7.2066	0.9905	6
55	0.1377	0.1391	7.1912	0.9905	5
56	0.1380	0.1394	7.1759	0.9904	4
57	0.1383	0.1397	7.1607	0.9904	3
58	0.1386	0.1399	7.1455	0.9903	2
59	0.1389	0.1402	7.1304	0.9903	1
60	0.1392	0.1405	7.1154	0.9903	0
	Cos	Cot	Tan	Sin	′

*98° 188° *278° **8°** NATURAL **9°** *99° 189° *279°

′	Sin	Tan	Cot	Cos	
0	0.1392	0.1405	7.1154	0.9903	60
1	0.1395	0.1408	7.1004	0.9902	59
2	0.1397	0.1411	7.0855	0.9902	58
3	0.1400	0.1414	7.0706	0.9901	57
4	0.1403	0.1417	7.0558	0.9901	56
5	0.1406	0.1420	7.0410	0.9901	55
6	0.1409	0.1423	7.0264	0.9900	54
7	0.1412	0.1426	7.0117	0.9900	53
8	0.1415	0.1429	6.9972	0.9899	52
9	0.1418	0.1432	6.9827	0.9899	51
10	0.1421	0.1435	6.9682	0.9899	50
11	0.1423	0.1438	6.9538	0.9898	49
12	0.1426	0.1441	6.9395	0.9898	48
13	0.1429	0.1444	6.9252	0.9897	47
14	0.1432	0.1447	6.9110	0.9897	46
15	0.1435	0.1450	6.8969	0.9897	45
16	0.1438	0.1453	6.8828	0.9896	44
17	0.1441	0.1456	6.8687	0.9896	43
18	0.1444	0.1459	6.8548	0.9895	42
19	0.1446	0.1462	6.8408	0.9895	41
20	0.1449	0.1465	6.8269	0.9894	40
21	0.1452	0.1468	6.8131	0.9894	39
22	0.1455	0.1471	6.7994	0.9894	38
23	0.1458	0.1474	6.7856	0.9893	37
24	0.1461	0.1477	6.7720	0.9893	36
25	0.1464	0.1480	6.7584	0.9892	35
26	0.1467	0.1483	6.7448	0.9892	34
27	0.1469	0.1486	6.7313	0.9891	33
28	0.1472	0.1489	6.7179	0.9891	32
29	0.1475	0.1492	6.7045	0.9891	31
30	0.1478	0.1495	6.6912	0.9890	30
31	0.1481	0.1497	6.6779	0.9890	29
32	0.1484	0.1500	6.6646	0.9889	28
33	0.1487	0.1503	6.6514	0.9889	27
34	0.1490	0.1506	6.6383	0.9888	26
35	0.1492	0.1509	6.6252	0.9888	25
36	0.1495	0.1512	6.6122	0.9888	24
37	0.1498	0.1515	6.5992	0.9887	23
38	0.1501	0.1518	6.5863	0.9887	22
39	0.1504	0.1521	6.5734	0.9886	21
40	0.1507	0.1524	6.5606	0.9886	20
41	0.1510	0.1527	6.5478	0.9885	19
42	0.1513	0.1530	6.5350	0.9885	18
43	0.1515	0.1533	6.5223	0.9884	17
44	0.1518	0.1536	6.5097	0.9884	16
45	0.1521	0.1539	6.4971	0.9884	15
46	0.1524	0.1542	6.4846	0.9883	14
47	0.1527	0.1545	6.4721	0.9883	13
48	0.1530	0.1548	6.4596	0.9882	12
49	0.1533	0.1551	6.4472	0.9882	11
50	0.1536	0.1554	6.4348	0.9881	10
51	0.1538	0.1557	6.4225	0.9881	9
52	0.1541	0.1560	6.4103	0.9880	8
53	0.1544	0.1563	6.3980	0.9880	7
54	0.1547	0.1566	6.3859	0.9880	6
55	0.1550	0.1569	6.3737	0.9879	5
56	0.1553	0.1572	6.3617	0.9879	4
57	0.1556	0.1575	6.3496	0.9878	3
58	0.1559	0.1578	6.3376	0.9878	2
59	0.1561	0.1581	6.3257	0.9877	1
60	0.1564	0.1584	6.3138	0.9877	0
	Cos	Cot	Tan	Sin	′

′	Sin	Tan	Cot	Cos	
0	0.1564	0.1584	6.3138	0.9877	60
1	0.1567	0.1587	6.3019	0.9876	59
2	0.1570	0.1590	6.2901	0.9876	58
3	0.1573	0.1593	6.2783	0.9876	57
4	0.1576	0.1596	6.2666	0.9875	56
5	0.1579	0.1599	6.2549	0.9875	55
6	0.1582	0.1602	6.2432	0.9874	54
7	0.1584	0.1605	6.2316	0.9874	53
8	0.1587	0.1608	6.2200	0.9873	52
9	0.1590	0.1611	6.2085	0.9873	51
10	0.1593	0.1614	6.1970	0.9872	50
11	0.1596	0.1617	6.1856	0.9872	49
12	0.1599	0.1620	6.1742	0.9871	48
13	0.1602	0.1623	6.1628	0.9871	47
14	0.1605	0.1626	6.1515	0.9870	46
15	0.1607	0.1629	6.1402	0.9870	45
16	0.1610	0.1632	6.1290	0.9869	44
17	0.1613	0.1635	6.1178	0.9869	43
18	0.1616	0.1638	6.1066	0.9869	42
19	0.1619	0.1641	6.0955	0.9868	41
20	0.1622	0.1644	6.0844	0.9868	40
21	0.1625	0.1647	6.0734	0.9867	39
22	0.1628	0.1650	6.0624	0.9867	38
23	0.1630	0.1653	6.0514	0.9866	37
24	0.1633	0.1655	6.0405	0.9866	36
25	0.1636	0.1658	6.0296	0.9865	35
26	0.1639	0.1661	6.0188	0.9865	34
27	0.1642	0.1664	6.0080	0.9864	33
28	0.1645	0.1667	5.9972	0.9864	32
29	0.1648	0.1670	5.9865	0.9863	31
30	0.1650	0.1673	5.9758	0.9863	30
31	0.1653	0.1676	5.9651	0.9862	29
32	0.1656	0.1679	5.9545	0.9862	28
33	0.1659	0.1682	5.9439	0.9861	27
34	0.1662	0.1685	5.9333	0.9861	26
35	0.1665	0.1688	5.9228	0.9860	25
36	0.1668	0.1691	5.9124	0.9860	24
37	0.1671	0.1694	5.9019	0.9859	23
38	0.1673	0.1697	5.8915	0.9859	22
39	0.1676	0.1700	5.8811	0.9859	21
40	0.1679	0.1703	5.8708	0.9858	20
41	0.1682	0.1706	5.8605	0.9858	19
42	0.1685	0.1709	5.8502	0.9857	18
43	0.1688	0.1712	5.8400	0.9857	17
44	0.1691	0.1715	5.8298	0.9856	16
45	0.1693	0.1718	5.8197	0.9856	15
46	0.1696	0.1721	5.8095	0.9855	14
47	0.1699	0.1724	5.7994	0.9855	13
48	0.1702	0.1727	5.7894	0.9854	12
49	0.1705	0.1730	5.7794	0.9854	11
50	0.1708	0.1733	5.7694	0.9853	10
51	0.1711	0.1736	5.7594	0.9853	9
52	0.1714	0.1739	5.7495	0.9852	8
53	0.1716	0.1742	5.7396	0.9852	7
54	0.1719	0.1745	5.7297	0.9851	6
55	0.1722	0.1748	5.7199	0.9851	5
56	0.1725	0.1751	5.7101	0.9850	4
57	0.1728	0.1754	5.7004	0.9850	3
58	0.1731	0.1757	5.6906	0.9849	2
59	0.1734	0.1760	5.6809	0.9849	1
60	0.1736	0.1763	5.6713	0.9848	0
	Cos	Cot	Tan	Sin	′

*171° 261° *351° **81°** NATURAL **80°** *170° 260° *350°

′	Sin	Tan	Cot	Cos	
0	0.1736	0.1763	5.6713	0.9848	60
1	0.1739	0.1766	5.6617	0.9848	59
2	0.1742	0.1769	5.6521	0.9847	58
3	0.1745	0.1772	5.6425	0.9847	57
4	0.1748	0.1775	5.6329	0.9846	56
5	0.1751	0.1778	5.6234	0.9846	55
6	0.1754	0.1781	5.6140	0.9845	54
7	0.1757	0.1784	5.6045	0.9845	53
8	0.1759	0.1787	5.5951	0.9844	52
9	0.1762	0.1790	5.5857	0.9843	51
10	0.1765	0.1793	5.5764	0.9843	50
11	0.1768	0.1796	5.5671	0.9842	49
12	0.1771	0.1799	5.5578	0.9842	48
13	0.1774	0.1802	5.5485	0.9841	47
14	0.1777	0.1805	5.5393	0.9841	46
15	0.1779	0.1808	5.5301	0.9840	45
16	0.1782	0.1811	5.5209	0.9840	44
17	0.1785	0.1814	5.5118	0.9839	43
18	0.1788	0.1817	5.5026	0.9839	42
19	0.1791	0.1820	5.4936	0.9838	41
20	0.1794	0.1823	5.4845	0.9838	40
21	0.1797	0.1826	5.4755	0.9837	39
22	0.1799	0.1829	5.4665	0.9837	38
23	0.1802	0.1832	5.4575	0.9836	37
24	0.1805	0.1835	5.4486	0.9836	36
25	0.1808	0.1838	5.4397	0.9835	35
26	0.1811	0.1841	5.4308	0.9835	34
27	0.1814	0.1844	5.4219	0.9834	33
28	0.1817	0.1847	5.4131	0.9834	32
29	0.1819	0.1850	5.4043	0.9833	31
30	0.1822	0.1853	5.3955	0.9833	30
31	0.1825	0.1856	5.3868	0.9832	29
32	0.1828	0.1859	5.3781	0.9831	28
33	0.1831	0.1862	5.3694	0.9831	27
34	0.1834	0.1865	5.3607	0.9830	26
35	0.1837	0.1868	5.3521	0.9830	25
36	0.1840	0.1871	5.3435	0.9829	24
37	0.1842	0.1874	5.3349	0.9829	23
38	0.1845	0.1877	5.3263	0.9828	22
39	0.1848	0.1880	5.3178	0.9828	21
40	0.1851	0.1883	5.3093	0.9827	20
41	0.1854	0.1887	5.3008	0.9827	19
42	0.1857	0.1890	5.2924	0.9826	18
43	0.1860	0.1893	5.2839	0.9826	17
44	0.1862	0.1896	5.2755	0.9825	16
45	0.1865	0.1899	5.2672	0.9825	15
46	0.1868	0.1902	5.2588	0.9824	14
47	0.1871	0.1905	5.2505	0.9823	13
48	0.1874	0.1908	5.2422	0.9823	12
49	0.1877	0.1911	5.2339	0.9822	11
50	0.1880	0.1914	5.2257	0.9822	10
51	0.1882	0.1917	5.2174	0.9821	9
52	0.1885	0.1920	5.2092	0.9821	8
53	0.1888	0.1923	5.2011	0.9820	7
54	0.1891	0.1926	5.1929	0.9820	6
55	0.1894	0.1929	5.1848	0.9819	5
56	0.1897	0.1932	5.1767	0.9818	4
57	0.1900	0.1935	5.1686	0.9818	3
58	0.1902	0.1938	5.1606	0.9817	2
59	0.1905	0.1941	5.1526	0.9817	1
60	0.1908	0.1944	5.1446	0.9816	0
	Cos	Cot	Tan	Sin	′

′	Sin	Tan	Cot	Cos	
0	0.1908	0.1944	5.1446	0.9816	60
1	0.1911	0.1947	5.1366	0.9816	59
2	0.1914	0.1950	5.1286	0.9815	58
3	0.1917	0.1953	5.1207	0.9815	57
4	0.1920	0.1956	5.1128	0.9814	56
5	0.1922	0.1959	5.1049	0.9813	55
6	0.1925	0.1962	5.0970	0.9813	54
7	0.1928	0.1965	5.0892	0.9812	53
8	0.1931	0.1968	5.0814	0.9812	52
9	0.1934	0.1971	5.0736	0.9811	51
10	0.1937	0.1974	5.0658	0.9811	50
11	0.1939	0.1977	5.0581	0.9810	49
12	0.1942	0.1980	5.0504	0.9810	48
13	0.1945	0.1983	5.0427	0.9809	47
14	0.1948	0.1986	5.0350	0.9808	46
15	0.1951	0.1989	5.0273	0.9808	45
16	0.1954	0.1992	5.0197	0.9807	44
17	0.1957	0.1995	5.0121	0.9807	43
18	0.1959	0.1998	5.0045	0.9806	42
19	0.1962	0.2001	4.9969	0.9806	41
20	0.1965	0.2004	4.9894	0.9805	40
21	0.1968	0.2007	4.9819	0.9804	39
22	0.1971	0.2010	4.9744	0.9804	38
23	0.1974	0.2013	4.9669	0.9803	37
24	0.1977	0.2016	4.9594	0.9803	36
25	0.1979	0.2019	4.9520	0.9802	35
26	0.1982	0.2022	4.9446	0.9802	34
27	0.1985	0.2025	4.9372	0.9801	33
28	0.1988	0.2028	4.9298	0.9800	32
29	0.1991	0.2031	4.9225	0.9800	31
30	0.1994	0.2035	4.9152	0.9799	30
31	0.1997	0.2038	4.9078	0.9799	29
32	0.1999	0.2041	4.9006	0.9798	28
33	0.2002	0.2044	4.8933	0.9798	27
34	0.2005	0.2047	4.8860	0.9797	26
35	0.2008	0.2050	4.8788	0.9796	25
36	0.2011	0.2053	4.8716	0.9796	24
37	0.2014	0.2056	4.8644	0.9795	23
38	0.2016	0.2059	4.8573	0.9795	22
39	0.2019	0.2062	4.8501	0.9794	21
40	0.2022	0.2065	4.8430	0.9793	20
41	0.2025	0.2068	4.8359	0.9793	19
42	0.2028	0.2071	4.8288	0.9792	18
43	0.2031	0.2074	4.8218	0.9792	17
44	0.2034	0.2077	4.8147	0.9791	16
45	0.2036	0.2080	4.8077	0.9790	15
46	0.2039	0.2083	4.8007	0.9790	14
47	0.2042	0.2086	4.7937	0.9789	13
48	0.2045	0.2089	4.7867	0.9789	12
49	0.2048	0.2092	4.7798	0.9788	11
50	0.2051	0.2095	4.7729	0.9787	10
51	0.2054	0.2098	4.7659	0.9787	9
52	0.2056	0.2101	4.7591	0.9786	8
53	0.2059	0.2104	4.7522	0.9786	7
54	0.2062	0.2107	4.7453	0.9785	6
55	0.2065	0.2110	4.7385	0.9784	5
56	0.2068	0.2113	4.7317	0.9784	4
57	0.2071	0.2116	4.7249	0.9783	3
58	0.2073	0.2119	4.7181	0.9783	2
59	0.2076	0.2123	4.7114	0.9782	1
60	0.2079	0.2126	4.7046	0.9781	0
	Cos	Cot	Tan	Sin	′

*102° 192° *282° **12°** Natural

′	Sin	Tan	Cot	Cos	
0	0.2079	0.2126	4.7046	0.9781	60
1	0.2082	0.2129	4.6979	0.9781	59
2	0.2085̄	0.2132	4.6912	0.9780	58
3	0.2088	0.2135̄	4.6845	0.9780	57
4	0.2090	0.2138	4.6779	0.9779	56
5	0.2093	0.2141	4.6712	0.9778	55
6	0.2096	0.2144	4.6646	0.9778	54
7	0.2099	0.2147	4.6580	0.9777	53
8	0.2102	0.2150	4.6514	0.9777	52
9	0.2105̄	0.2153	4.6448	0.9776	51
10	0.2108	0.2156	4.6382	0.9775	50
11	0.2110	0.2159	4.6317	0.9775̄	49
12	0.2113	0.2162	4.6252	0.9774	48
13	0.2116	0.2165	4.6187	0.9774	47
14	0.2119	0.2168	4.6122	0.9773	46
15	0.2122	0.2171	4.6057	0.9772	45
16	0.2125̄	0.2174	4.5993	0.9772	44
17	0.2127	0.2177	4.5928	0.9771	43
18	0.2130	0.2180	4.5864	0.9770	42
19	0.2133	0.2183	4.5800	0.9770	41
20	0.2136	0.2186	4.5736	0.9769	40
21	0.2139	0.2189	4.5673	0.9769	39
22	0.2142	0.2193	4.5609	0.9768	38
23	0.2145̄	0.2196	4.5546	0.9767	37
24	0.2147	0.2199	4.5483	0.9767	36
25	0.2150	0.2202	4.5420	0.9766	35
26	0.2153	0.2205̄	4.5357	0.9765	34
27	0.2156	0.2208	4.5294	0.9765̄	33
28	0.2159	0.2211	4.5232	0.9764	32
29	0.2162	0.2214	4.5169	0.9764	31
30	0.2164	0.2217	4.5107	0.9763	30
31	0.2167	0.2220	4.5045	0.9762	29
32	0.2170	0.2223	4.4983	0.9762	28
33	0.2173	0.2226	4.4922	0.9761	27
34	0.2176	0.2229	4.4860	0.9760	26
35	0.2179	0.2232	4.4799	0.9760	25
36	0.2181	0.2235	4.4737	0.9759	24
37	0.2184	0.2238	4.4676	0.9759	23
38	0.2187	0.2241	4.4615	0.9758	22
39	0.2190	0.2244	4.4555̄	0.9757	21
40	0.2193	0.2247	4.4494	0.9757	20
41	0.2196	0.2251	4.4434	0.9756	19
42	0.2198	0.2254	4.4373	0.9755	18
43	0.2201	0.2257	4.4313	0.9755̄	17
44	0.2204	0.2260	4.4253	0.9754	16
45	0.2207	0.2263	4.4194	0.9753	15
46	0.2210	0.2266	4.4134	0.9753	14
47	0.2213	0.2269	4.4075̄	0.9752	13
48	0.2215	0.2272	4.4015	0.9751	12
49	0.2218	0.2275	4.3956	0.9751	11
50	0.2221	0.2278	4.3897	0.9750	10
51	0.2224	0.2281	4.3838	0.9750	9
52	0.2227	0.2284	4.3779	0.9749	8
53	0.2230	0.2287	4.3721	0.9748	7
54	0.2233	0.2290	4.3662	0.9748	6
55	0.2235	0.2293	4.3604	0.9747	5
56	0.2238	0.2296	4.3546	0.9746	4
57	0.2241	0.2299	4.3488	0.9746	3
58	0.2244	0.2303	4.3430	0.9745	2
59	0.2247	0.2306	4.3372	0.9744	1
60	0.2250	0.2309	4.3315̄	0.9744	0
	Cos	Cot	Tan	Sin	′

*167° 257° *347° **77°** Natural

*103° 193° *283° **13°** Natural

′	Sin	Tan	Cot	Cos	
0	0.2250	0.2309	4.3315̄	0.9744	60
1	0.2252	0.2312	4.3257	0.9743	59
2	0.2255	0.2315̄	4.3200	0.9742	58
3	0.2258	0.2318	4.3143	0.9742	57
4	0.2261	0.2321	4.3086	0.9741	56
5	0.2264	0.2324	4.3029	0.9740	55
6	0.2267	0.2327	4.2972	0.9740	54
7	0.2269	0.2330	4.2916	0.9739	53
8	0.2272	0.2333	4.2859	0.9738	52
9	0.2275	0.2336	4.2803	0.9738	51
10	0.2278	0.2339	4.2747	0.9737	50
11	0.2281	0.2342	4.2691	0.9736	49
12	0.2284	0.2345	4.2635	0.9736	48
13	0.2286	0.2349	4.2580	0.9735	47
14	0.2289	0.2352	4.2524	0.9734	46
15	0.2292	0.2355̄	4.2468	0.9734	45
16	0.2295̄	0.2358	4.2413	0.9733	44
17	0.2298	0.2361	4.2358	0.9732	43
18	0.2300	0.2364	4.2303	0.9732	42
19	0.2303	0.2367	4.2248	0.9731	41
20	0.2306	0.2370	4.2193	0.9730	40
21	0.2309	0.2373	4.2139	0.9730	39
22	0.2312	0.2376	4.2084	0.9729	38
23	0.2315̄	0.2379	4.2030	0.9728	37
24	0.2317	0.2382	4.1976	0.9728	36
25	0.2320	0.2385	4.1922	0.9727	35
26	0.2323	0.2388	4.1868	0.9726	34
27	0.2326	0.2392	4.1814	0.9726	33
28	0.2329	0.2395̄	4.1760	0.9725	32
29	0.2332	0.2398	4.1706	0.9724	31
30	0.2334	0.2401	4.1653	0.9724	30
31	0.2337	0.2404	4.1600	0.9723	29
32	0.2340	0.2407	4.1547	0.9722	28
33	0.2343	0.2410	4.1493	0.9722	27
34	0.2346	0.2413	4.1441	0.9721	26
35	0.2349	0.2416	4.1388	0.9720	25
36	0.2351	0.2419	4.1335	0.9720	24
37	0.2354	0.2422	4.1282	0.9719	23
38	0.2357	0.2425	4.1230	0.9718	22
39	0.2360	0.2428	4.1178	0.9718	21
40	0.2363	0.2432	4.1126	0.9717	20
41	0.2366	0.2435̄	4.1074	0.9716	19
42	0.2368	0.2438	4.1022	0.9715	18
43	0.2371	0.2441	4.0970	0.9715̄	17
44	0.2374	0.2444	4.0918	0.9714	16
45	0.2377	0.2447	4.0867	0.9713	15
46	0.2380	0.2450	4.0815	0.9713	14
47	0.2383	0.2453	4.0764	0.9712	13
48	0.2385	0.2456	4.0713	0.9711	12
49	0.2388	0.2459	4.0662	0.9711	11
50	0.2391	0.2462	4.0611	0.9710	10
51	0.2394	0.2465	4.0560	0.9709	9
52	0.2397	0.2469	4.0509	0.9709	8
53	0.2399	0.2472	4.0459	0.9708	7
54	0.2402	0.2475̄	4.0408	0.9707	6
55	0.2405	0.2478	4.0358	0.9706	5
56	0.2408	0.2481	4.0308	0.9706	4
57	0.2411	0.2484	4.0257	0.9705	3
58	0.2414	0.2487	4.0207	0.9704	2
59	0.2416	0.2490	4.0158	0.9704	1
60	0.2419	0.2493	4.0108	0.9703	0
	Cos	Cot	Tan	Sin	′

*166° 256° *346° **76°** Natural

*104° 194° *284° **14°** NATURAL

′	Sin	Tan	Cot	Cos	
0	0.2419	0.2493	4.0108	0.9703	60
1	0.2422	0.2496	4.0058	0.9702	59
2	0.2425̄	0.2499	4.0009	0.9702	58
3	0.2428	0.2503	3.9959	0.9701	57
4	0.2431	0.2506	3.9910	0.9700	56
5	0.2433	0.2509	3.9861	0.9699	55
6	0.2436	0.2512	3.9812	0.9699	54
7	0.2439	0.2515̄	3.9763	0.9698	53
8	0.2442	0.2518	3.9714	0.9697	52
9	0.2445̄	0.2521	3.9665	0.9697	51
10	0.2447	0.2524	3.9617	0.9696	50
11	0.2450	0.2527	3.9568	0.9695	49
12	0.2453	0.2530	3.9520	0.9694	48
13	0.2456	0.2533	3.9471	0.9694	47
14	0.2459	0.2537	3.9423	0.9693	46
15	0.2462	0.2540	3.9375	0.9692	45
16	0.2464	0.2543	3.9327	0.9692	44
17	0.2467	0.2546	3.9279	0.9691	43
18	0.2470	0.2549	3.9232	0.9690	42
19	0.2473	0.2552	3.9184	0.9689	41
20	0.2476	0.2555	3.9136	0.9689	40
21	0.2478	0.2558	3.9089	0.9688	39
22	0.2481	0.2561	3.9042	0.9687	38
23	0.2484	0.2564	3.8995̄	0.9687	37
24	0.2487	0.2568	3.8947	0.9686	36
25	0.2490	0.2571	3.8900	0.9685	35
26	0.2493	0.2574	3.8854	0.9684	34
27	0.2495	0.2577	3.8807	0.9684	33
28	0.2498	0.2580	3.8760	0.9683	32
29	0.2501	0.2583	3.8714	0.9682	31
30	0.2504	0.2586	3.8667	0.9681	30
31	0.2507	0.2589	3.8621	0.9681	29
32	0.2509	0.2592	3.8575̄	0.9680	28
33	0.2512	0.2595	3.8528	0.9679	27
34	0.2515	0.2599	3.8482	0.9679	26
35	0.2518	0.2602	3.8436	0.9678	25
36	0.2521	0.2605̄	3.8391	0.9677	24
37	0.2524	0.2608	3.8345̄	0.9676	23
38	0.2526	0.2611	3.8299	0.9676	22
39	0.2529	0.2614	3.8254	0.9675̄	21
40	0.2532	0.2617	3.8208	0.9674	20
41	0.2535̄	0.2620	3.8163	0.9673	19
42	0.2538	0.2623	3.8118	0.9673	18
43	0.2540	0.2627	3.8073	0.9672	17
44	0.2543	0.2630	3.8028	0.9671	16
45	0.2546	0.2633	3.7983	0.9670	15
46	0.2549	0.2636	3.7938	0.9670	14
47	0.2552	0.2639	3.7893	0.9669	13
48	0.2554	0.2642	3.7848	0.9668	12
49	0.2557	0.2645	3.7804	0.9667	11
50	0.2560	0.2648	3.7760	0.9667	10
51	0.2563	0.2651	3.7715	0.9666	9
52	0.2566	0.2655̄	3.7671	0.9665	8
53	0.2569	0.2658	3.7627	0.9665̄	7
54	0.2571	0.2661	3.7583	0.9664	6
55	0.2574	0.2664	3.7539	0.9663	5
56	0.2577	0.2667	3.7495̄	0.9662	4
57	0.2580	0.2670	3.7451	0.9662	3
58	0.2583	0.2673	3.7408	0.9661	2
59	0.2585	0.2676	3.7364	0.9660	1
60	0.2588	0.2679	3.7321	0.9659	0
	Cos	Cot	Tan	Sin	′

*165° 255° *345° **75°** NATURAL

15° *105° 195° *285°

′	Sin	Tan	Cot	Cos	
0	0.2588	0.2679	3.7321	0.9659	60
1	0.2591	0.2683	3.7277	0.9659	59
2	0.2594	0.2686	3.7234	0.9658	58
3	0.2597	0.2689	3.7191	0.9657	57
4	0.2599	0.2692	3.7148	0.9656	56
5	0.2602	0.2695	3.7105̄	0.9655	55
6	0.2605	0.2698	3.7062	0.9655̄	54
7	0.2608	0.2701	3.7019	0.9654	53
8	0.2611	0.2704	3.6976	0.9653	52
9	0.2613	0.2708	3.6933	0.9652	51
10	0.2616	0.2711	3.6891	0.9652	50
11	0.2619	0.2714	3.6848	0.9651	49
12	0.2622	0.2717	3.6806	0.9650	48
13	0.2625̄	0.2720	3.6764	0.9649	47
14	0.2628	0.2723	3.6722	0.9649	46
15	0.2630	0.2726	3.6680	0.9648	45
16	0.2633	0.2729	3.6638	0.9647	44
17	0.2636	0.2733	3.6596	0.9646	43
18	0.2639	0.2736	3.6554	0.9646	42
19	0.2642	0.2739	3.6512	0.9645̄	41
20	0.2644	0.2742	3.6470	0.9644	40
21	0.2647	0.2745	3.6429	0.9643	39
22	0.2650̄	0.2748	3.6387	0.9642	38
23	0.2653	0.2751	3.6346	0.9642	37
24	0.2656	0.2754	3.6305̄	0.9641	36
25	0.2658	0.2758	3.6264	0.9640	35
26	0.2661	0.2761	3.6222	0.9639	34
27	0.2664	0.2764	3.6181	0.9639	33
28	0.2667	0.2767	3.6140	0.9638	32
29	0.2670	0.2770	3.6100	0.9637	31
30	0.2672	0.2773	3.6059	0.9636	30
31	0.2675	0.2776	3.6018	0.9636	29
32	0.2678	0.2780	3.5978	0.9635̄	28
33	0.2681	0.2783	3.5937	0.9634	27
34	0.2684	0.2786	3.5897	0.9633	26
35	0.2686	0.2789	3.5856	0.9632	25
36	0.2689	0.2792	3.5816	0.9632	24
37	0.2692	0.2795	3.5776	0.9631	23
38	0.2695̄	0.2798	3.5736	0.9630	22
39	0.2698	0.2801	3.5696	0.9629	21
40	0.2700	0.2805̄	3.5656	0.9628	20
41	0.2703	0.2808	3.5616	0.9628	19
42	0.2706	0.2811	3.5576	0.9627	18
43	0.2709	0.2814	3.5536	0.9626	17
44	0.2712	0.2817	3.5497	0.9625	16
45	0.2714	0.2820	3.5457	0.9625̄	15
46	0.2717	0.2823	3.5418	0.9624	14
47	0.2720	0.2827	3.5379	0.9623	13
48	0.2723	0.2830	3.5339	0.9622	12
49	0.2726	0.2833	3.5300	0.9621	11
50	0.2728	0.2836	3.5261	0.9621	10
51	0.2731	0.2839	3.5222	0.9620	9
52	0.2734	0.2842	3.5183	0.9619	8
53	0.2737	0.2845	3.5144	0.9618	7
54	0.2740	0.2849	3.5105	0.9617	6
55	0.2742	0.2852	3.5067	0.9617	5
56	0.2745	0.2855̄	3.5028	0.9616	4
57	0.2748	0.2858	3.4989	0.9615	3
58	0.2751	0.2861	3.4951	0.9614	2
59	0.2754	0.2864	3.4912	0.9613	1
60	0.2756	0.2867	3.4874	0.9613	0
	Cos	Cot	Tan	Sin	′

74° *164° 254° *344°

*106° 196° *286° **16°** NATURAL

′	Sin	Tan	Cot	Cos	
0	0.2756	0.2867	3.4874	0.9613	60
1	0.2759	0.2871	3.4836	0.9612	59
2	0.2762	0.2874	3.4798	0.9611	58
3	0.2765	0.2877	3.4760	0.9610	57
4	0.2768	0.2880	3.4722	0.9609	56
5	0.2770	0.2883	3.4684	0.9609	55
6	0.2773	0.2886	3.4646	0.9608	54
7	0.2776	0.2890	3.4608	0.9607	53
8	0.2779	0.2893	3.4570	0.9606	52
9	0.2782	0.2896	3.4533	0.9605	51
10	0.2784	0.2899	3.4495	0.9605	50
11	0.2787	0.2902	3.4458	0.9604	49
12	0.2790	0.2905	3.4420	0.9603	48
13	0.2793	0.2908	3.4383	0.9602	47
14	0.2795	0.2912	3.4346	0.9601	46
15	0.2798	0.2915	3.4308	0.9600	45
16	0.2801	0.2918	3.4271	0.9600	44
17	0.2804	0.2921	3.4234	0.9599	43
18	0.2807	0.2924	3.4197	0.9598	42
19	0.2809	0.2927	3.4160	0.9597	41
20	0.2812	0.2931	3.4124	0.9596	40
21	0.2815	0.2934	3.4087	0.9596	39
22	0.2818	0.2937	3.4050	0.9595	38
23	0.2821	0.2940	3.4014	0.9594	37
24	0.2823	0.2943	3.3977	0.9593	36
25	0.2826	0.2946	3.3941	0.9592	35
26	0.2829	0.2949	3.3904	0.9591	34
27	0.2832	0.2953	3.3868	0.9591	33
28	0.2835	0.2956	3.3832	0.9590	32
29	0.2837	0.2959	3.3796	0.9589	31
30	0.2840	0.2962	3.3759	0.9588	30
31	0.2843	0.2965	3.3723	0.9587	29
32	0.2846	0.2968	3.3687	0.9587	28
33	0.2849	0.2972	3.3652	0.9586	27
34	0.2851	0.2975	3.3616	0.9585	26
35	0.2854	0.2978	3.3580	0.9584	25
36	0.2857	0.2981	3.3544	0.9583	24
37	0.2860	0.2984	3.3509	0.9582	23
38	0.2862	0.2987	3.3473	0.9582	22
39	0.2865	0.2991	3.3438	0.9581	21
40	0.2868	0.2994	3.3402	0.9580	20
41	0.2871	0.2997	3.3367	0.9579	19
42	0.2874	0.3000	3.3332	0.9578	18
43	0.2876	0.3003	3.3297	0.9577	17
44	0.2879	0.3006	3.3261	0.9577	16
45	0.2882	0.3010	3.3226	0.9576	15
46	0.2885	0.3013	3.3191	0.9575	14
47	0.2888	0.3016	3.3156	0.9574	13
48	0.2890	0.3019	3.3122	0.9573	12
49	0.2893	0.3022	3.3087	0.9572	11
50	0.2896	0.3026	3.3052	0.9572	10
51	0.2899	0.3029	3.3017	0.9571	9
52	0.2901	0.3032	3.2983	0.9570	8
53	0.2904	0.3035	3.2948	0.9569	7
54	0.2907	0.3038	3.2914	0.9568	6
55	0.2910	0.3041	3.2879	0.9567	5
56	0.2913	0.3045	3.2845	0.9566	4
57	0.2915	0.3048	3.2811	0.9566	3
58	0.2918	0.3051	3.2777	0.9565	2
59	0.2921	0.3054	3.2743	0.9564	1
60	0.2924	0.3057	3.2709	0.9563	0
	Cos	Cot	Tan	Sin	′

163° 253° *343° **73°** NATURAL

17° *107° 197° *287°

′	Sin	Tan	Cot	Cos	
0	0.2924	0.3057	3.2709	0.9563	60
1	0.2926	0.3060	3.2675	0.9562	59
2	0.2929	0.3064	3.2641	0.9561	58
3	0.2932	0.3067	3.2607	0.9560	57
4	0.2935	0.3070	3.2573	0.9560	56
5	0.2938	0.3073	3.2539	0.9559	55
6	0.2940	0.3076	3.2506	0.9558	54
7	0.2943	0.3080	3.2472	0.9557	53
8	0.2946	0.3083	3.2438	0.9556	52
9	0.2949	0.3086	3.2405	0.9555	51
10	0.2952	0.3089	3.2371	0.9555	50
11	0.2954	0.3092	3.2338	0.9554	49
12	0.2957	0.3096	3.2305	0.9553	48
13	0.2960	0.3099	3.2272	0.9552	47
14	0.2963	0.3102	3.2238	0.9551	46
15	0.2965	0.3105	3.2205	0.9550	45
16	0.2968	0.3108	3.2172	0.9549	44
17	0.2971	0.3111	3.2139	0.9548	43
18	0.2974	0.3115	3.2106	0.9548	42
19	0.2977	0.3118	3.2073	0.9547	41
20	0.2979	0.3121	3.2041	0.9546	40
21	0.2982	0.3124	3.2008	0.9545	39
22	0.2985	0.3127	3.1975	0.9544	38
23	0.2988	0.3131	3.1943	0.9543	37
24	0.2990	0.3134	3.1910	0.9542	36
25	0.2993	0.3137	3.1878	0.9542	35
26	0.2996	0.3140	3.1845	0.9541	34
27	0.2999	0.3143	3.1813	0.9540	33
28	0.3002	0.3147	3.1780	0.9539	32
29	0.3004	0.3150	3.1748	0.9538	31
30	0.3007	0.3153	3.1716	0.9537	30
31	0.3010	0.3156	3.1684	0.9536	29
32	0.3013	0.3159	3.1652	0.9535	28
33	0.3015	0.3163	3.1620	0.9535	27
34	0.3018	0.3166	3.1588	0.9534	26
35	0.3021	0.3169	3.1556	0.9533	25
36	0.3024	0.3172	3.1524	0.9532	24
37	0.3026	0.3175	3.1492	0.9531	23
38	0.3029	0.3179	3.1460	0.9530	22
39	0.3032	0.3182	3.1429	0.9529	21
40	0.3035	0.3185	3.1397	0.9528	20
41	0.3038	0.3188	3.1366	0.9527	19
42	0.3040	0.3191	3.1334	0.9527	18
43	0.3043	0.3195	3.1303	0.9526	17
44	0.3046	0.3198	3.1271	0.9525	16
45	0.3049	0.3201	3.1240	0.9524	15
46	0.3051	0.3204	3.1209	0.9523	14
47	0.3054	0.3207	3.1178	0.9522	13
48	0.3057	0.3211	3.1146	0.9521	12
49	0.3060	0.3214	3.1115	0.9520	11
50	0.3062	0.3217	3.1084	0.9520	10
51	0.3065	0.3220	3.1053	0.9519	9
52	0.3068	0.3223	3.1022	0.9518	8
53	0.3071	0.3227	3.0991	0.9517	7
54	0.3074	0.3230	3.0961	0.9516	6
55	0.3076	0.3233	3.0930	0.9515	5
56	0.3079	0.3236	3.0899	0.9514	4
57	0.3082	0.3240	3.0868	0.9513	3
58	0.3085	0.3243	3.0838	0.9512	2
59	0.3087	0.3246	3.0807	0.9511	1
60	0.3090	0.3249	3.0777	0.9511	0
	Cos	Cot	Tan	Sin	′

72° *162° 252° *342°

18°

′	Sin	Tan	Cot	Cos	
0	0.3090	0.3249	3.0777	0.9511	60
1	0.3093	0.3252	3.0746	0.9510	59
2	0.3096	0.3256	3.0716	0.9509	58
3	0.3098	0.3259	3.0686	0.9508	57
4	0.3101	0.3262	3.0655	0.9507	56
5	0.3104	0.3265	3.0625	0.9506	55
6	0.3107	0.3269	3.0595	0.9505	54
7	0.3110	0.3272	3.0565	0.9504	53
8	0.3112	0.3275	3.0535	0.9503	52
9	0.3115	0.3278	3.0505	0.9502	51
10	0.3118	0.3281	3.0475	0.9502	50
11	0.3121	0.3285	3.0445	0.9501	49
12	0.3123	0.3288	3.0415	0.9500	48
13	0.3126	0.3291	3.0385	0.9499	47
14	0.3129	0.3294	3.0356	0.9498	46
15	0.3132	0.3298	3.0326	0.9497	45
16	0.3134	0.3301	3.0296	0.9496	44
17	0.3137	0.3304	3.0267	0.9495	43
18	0.3140	0.3307	3.0237	0.9494	42
19	0.3143	0.3310	3.0208	0.9493	41
20	0.3145	0.3314	3.0178	0.9492	40
21	0.3148	0.3317	3.0149	0.9492	39
22	0.3151	0.3320	3.0120	0.9491	38
23	0.3154	0.3323	3.0090	0.9490	37
24	0.3156	0.3327	3.0061	0.9489	36
25	0.3159	0.3330	3.0032	0.9488	35
26	0.3162	0.3333	3.0003	0.9487	34
27	0.3165	0.3336	2.9974	0.9486	33
28	0.3168	0.3339	2.9945	0.9485	32
29	0.3170	0.3343	2.9916	0.9484	31
30	0.3173	0.3346	2.9887	0.9483	30
31	0.3176	0.3349	2.9858	0.9482	29
32	0.3179	0.3352	2.9829	0.9481	28
33	0.3181	0.3356	2.9800	0.9480	27
34	0.3184	0.3359	2.9772	0.9480	26
35	0.3187	0.3362	2.9743	0.9479	25
36	0.3190	0.3365	2.9714	0.9478	24
37	0.3192	0.3369	2.9686	0.9477	23
38	0.3195	0.3372	2.9657	0.9476	22
39	0.3198	0.3375	2.9629	0.9475	21
40	0.3201	0.3378	2.9600	0.9474	20
41	0.3203	0.3382	2.9572	0.9473	19
42	0.3206	0.3385	2.9544	0.9472	18
43	0.3209	0.3388	2.9515	0.9471	17
44	0.3212	0.3391	2.9487	0.9470	16
45	0.3214	0.3395	2.9459	0.9469	15
46	0.3217	0.3398	2.9431	0.9468	14
47	0.3220	0.3401	2.9403	0.9467	13
48	0.3223	0.3404	2.9375	0.9466	12
49	0.3225	0.3408	2.9347	0.9466	11
50	0.3228	0.3411	2.9319	0.9465	10
51	0.3231	0.3414	2.9291	0.9464	9
52	0.3234	0.3417	2.9263	0.9463	8
53	0.3236	0.3421	2.9235	0.9462	7
54	0.3239	0.3424	2.9208	0.9461	6
55	0.3242	0.3427	2.9180	0.9460	5
56	0.3245	0.3430	2.9152	0.9459	4
57	0.3247	0.3434	2.9125	0.9458	3
58	0.3250	0.3437	2.9097	0.9457	2
59	0.3253	0.3440	2.9070	0.9456	1
60	0.3256	0.3443	2.9042	0.9455	0
	Cos	Cot	Tan	Sin	′

71°

19°

′	Sin	Tan	Cot	Cos	
0	0.3256	0.3443	2.9042	0.9455	60
1	0.3258	0.3447	2.9015	0.9454	59
2	0.3261	0.3450	2.8987	0.9453	58
3	0.3264	0.3453	2.8960	0.9452	57
4	0.3267	0.3456	2.8933	0.9451	56
5	0.3269	0.3460	2.8905	0.9450	55
6	0.3272	0.3463	2.8878	0.9449	54
7	0.3275	0.3466	2.8851	0.9449	53
8	0.3278	0.3469	2.8824	0.9448	52
9	0.3280	0.3473	2.8797	0.9447	51
10	0.3283	0.3476	2.8770	0.9446	50
11	0.3286	0.3479	2.8743	0.9445	49
12	0.3289	0.3482	2.8716	0.9444	48
13	0.3291	0.3486	2.8689	0.9443	47
14	0.3294	0.3489	2.8662	0.9442	46
15	0.3297	0.3492	2.8636	0.9441	45
16	0.3300	0.3495	2.8609	0.9440	44
17	0.3302	0.3499	2.8582	0.9439	43
18	0.3305	0.3502	2.8556	0.9438	42
19	0.3308	0.3505	2.8529	0.9437	41
20	0.3311	0.3508	2.8502	0.9436	40
21	0.3313	0.3512	2.8476	0.9435	39
22	0.3316	0.3515	2.8449	0.9434	38
23	0.3319	0.3518	2.8423	0.9433	37
24	0.3322	0.3522	2.8397	0.9432	36
25	0.3324	0.3525	2.8370	0.9431	35
26	0.3327	0.3528	2.8344	0.9430	34
27	0.3330	0.3531	2.8318	0.9429	33
28	0.3333	0.3535	2.8291	0.9428	32
29	0.3335	0.3538	2.8265	0.9427	31
30	0.3338	0.3541	2.8239	0.9426	30
31	0.3341	0.3544	2.8213	0.9425	29
32	0.3344	0.3548	2.8187	0.9424	28
33	0.3346	0.3551	2.8161	0.9423	27
34	0.3349	0.3554	2.8135	0.9423	26
35	0.3352	0.3558	2.8109	0.9422	25
36	0.3355	0.3561	2.8083	0.9421	24
37	0.3357	0.3564	2.8057	0.9420	23
38	0.3360	0.3567	2.8032	0.9419	22
39	0.3363	0.3571	2.8006	0.9418	21
40	0.3365	0.3574	2.7980	0.9417	20
41	0.3368	0.3577	2.7955	0.9416	19
42	0.3371	0.3581	2.7929	0.9415	18
43	0.3374	0.3584	2.7903	0.9414	17
44	0.3376	0.3587	2.7878	0.9413	16
45	0.3379	0.3590	2.7852	0.9412	15
46	0.3382	0.3594	2.7827	0.9411	14
47	0.3385	0.3597	2.7801	0.9410	13
48	0.3387	0.3600	2.7776	0.9409	12
49	0.3390	0.3604	2.7751	0.9408	11
50	0.3393	0.3607	2.7725	0.9407	10
51	0.3396	0.3610	2.7700	0.9406	9
52	0.3398	0.3613	2.7675	0.9405	8
53	0.3401	0.3617	2.7650	0.9404	7
54	0.3404	0.3620	2.7625	0.9403	6
55	0.3407	0.3623	2.7600	0.9402	5
56	0.3409	0.3627	2.7575	0.9401	4
57	0.3412	0.3630	2.7550	0.9400	3
58	0.3415	0.3633	2.7525	0.9399	2
59	0.3417	0.3636	2.7500	0.9398	1
60	0.3420	0.3640	2.7475	0.9397	0
	Cos	Cot	Tan	Sin	′

70°

*110° 200° *290° **20°** NATURAL

′	Sin	Tan	Cot	Cos	
0	0.3420	0.3640	2.7475	0.9397	60
1	0.3423	0.3643	2.7450	0.9396	59
2	0.3426	0.3646	2.7425	0.9395	58
3	0.3428	0.3650	2.7400	0.9394	57
4	0.3431	0.3653	2.7376	0.9393	56
5	0.3434	0.3656	2.7351	0.9392	55
6	0.3437	0.3659	2.7326	0.9391	54
7	0.3439	0.3663	2.7302	0.9390	53
8	0.3442	0.3666	2.7277	0.9389	52
9	0.3445	0.3669	2.7253	0.9388	51
10	0.3448	0.3673	2.7228	0.9387	50
11	0.3450	0.3676	2.7204	0.9386	49
12	0.3453	0.3679	2.7179	0.9385	48
13	0.3456	0.3683	2.7155	0.9384	47
14	0.3458	0.3686	2.7130	0.9383	46
15	0.3461	0.3689	2.7106	0.9382	45
16	0.3464	0.3693	2.7082	0.9381	44
17	0.3467	0.3696	2.7058	0.9380	43
18	0.3469	0.3699	2.7034	0.9379	42
19	0.3472	0.3702	2.7009	0.9378	41
20	0.3475	0.3706	2.6985	0.9377	40
21	0.3478	0.3709	2.6961	0.9376	39
22	0.3480	0.3712	2.6937	0.9375	38
23	0.3483	0.3716	2.6913	0.9374	37
24	0.3486	0.3719	2.6889	0.9373	36
25	0.3488	0.3722	2.6865	0.9372	35
26	0.3491	0.3726	2.6841	0.9371	34
27	0.3494	0.3729	2.6818	0.9370	33
28	0.3497	0.3732	2.6794	0.9369	32
29	0.3499	0.3736	2.6770	0.9368	31
30	0.3502	0.3739	2.6746	0.9367	30
31	0.3505	0.3742	2.6723	0.9366	29
32	0.3508	0.3745	2.6699	0.9365	28
33	0.3510	0.3749	2.6675	0.9364	27
34	0.3513	0.3752	2.6652	0.9363	26
35	0.3516	0.3755	2.6628	0.9362	25
36	0.3518	0.3759	2.6605	0.9361	24
37	0.3521	0.3762	2.6581	0.9360	23
38	0.3524	0.3765	2.6558	0.9359	22
39	0.3527	0.3769	2.6534	0.9358	21
40	0.3529	0.3772	2.6511	0.9356	20
41	0.3532	0.3775	2.6488	0.9355	19
42	0.3535	0.3779	2.6464	0.9354	18
43	0.3537	0.3782	2.6441	0.9353	17
44	0.3540	0.3785	2.6418	0.9352	16
45	0.3543	0.3789	2.6395	0.9351	15
46	0.3546	0.3792	2.6371	0.9350	14
47	0.3548	0.3795	2.6348	0.9349	13
48	0.3551	0.3799	2.6325	0.9348	12
49	0.3554	0.3802	2.6302	0.9347	11
50	0.3557	0.3805	2.6279	0.9346	10
51	0.3559	0.3809	2.6256	0.9345	9
52	0.3562	0.3812	2.6233	0.9344	8
53	0.3565	0.3815	2.6210	0.9343	7
54	0.3567	0.3819	2.6187	0.9342	6
55	0.3570	0.3822	2.6165	0.9341	5
56	0.3573	0.3825	2.6142	0.9340	4
57	0.3576	0.3829	2.6119	0.9339	3
58	0.3578	0.3832	2.6096	0.9338	2
59	0.3581	0.3835	2.6074	0.9337	1
60	0.3584	0.3839	2.6051	0.9336	0
	Cos	Cot	Tan	Sin	′

*159° 249° *339° **69°** NATURAL

21° *111° 201° *291°

′	Sin	Tan	Cot	Cos	
0	0.3584	0.3839	2.6051	0.9336	60
1	0.3586	0.3842	2.6028	0.9335	59
2	0.3589	0.3845	2.6006	0.9334	58
3	0.3592	0.3849	2.5983	0.9333	57
4	0.3595	0.3852	2.5961	0.9332	56
5	0.3597	0.3855	2.5938	0.9331	55
6	0.3600	0.3859	2.5916	0.9330	54
7	0.3603	0.3862	2.5893	0.9328	53
8	0.3605	0.3865	2.5871	0.9327	52
9	0.3608	0.3869	2.5848	0.9326	51
10	0.3611	0.3872	2.5826	0.9325	50
11	0.3614	0.3875	2.5804	0.9324	49
12	0.3616	0.3879	2.5782	0.9323	48
13	0.3619	0.3882	2.5759	0.9322	47
14	0.3622	0.3885	2.5737	0.9321	46
15	0.3624	0.3889	2.5715	0.9320	45
16	0.3627	0.3892	2.5693	0.9319	44
17	0.3630	0.3895	2.5671	0.9318	43
18	0.3633	0.3899	2.5649	0.9317	42
19	0.3635	0.3902	2.5627	0.9316	41
20	0.3638	0.3906	2.5605	0.9315	40
21	0.3641	0.3909	2.5583	0.9314	39
22	0.3643	0.3912	2.5561	0.9313	38
23	0.3646	0.3916	2.5539	0.9312	37
24	0.3649	0.3919	2.5517	0.9311	36
25	0.3651	0.3922	2.5495	0.9309	35
26	0.3654	0.3926	2.5473	0.9308	34
27	0.3657	0.3929	2.5452	0.9307	33
28	0.3660	0.3932	2.5430	0.9306	32
29	0.3662	0.3936	2.5408	0.9305	31
30	0.3665	0.3939	2.5386	0.9304	30
31	0.3668	0.3942	2.5365	0.9303	29
32	0.3670	0.3946	2.5343	0.9302	28
33	0.3673	0.3949	2.5322	0.9301	27
34	0.3676	0.3953	2.5300	0.9300	26
35	0.3679	0.3956	2.5279	0.9299	25
36	0.3681	0.3959	2.5257	0.9298	24
37	0.3684	0.3963	2.5236	0.9297	23
38	0.3687	0.3966	2.5214	0.9296	22
39	0.3689	0.3969	2.5193	0.9295	21
40	0.3692	0.3973	2.5172	0.9293	20
41	0.3695	0.3976	2.5150	0.9292	19
42	0.3697	0.3979	2.5129	0.9291	18
43	0.3700	0.3983	2.5108	0.9290	17
44	0.3703	0.3986	2.5086	0.9289	16
45	0.3706	0.3990	2.5065	0.9288	15
46	0.3708	0.3993	2.5044	0.9287	14
47	0.3711	0.3996	2.5023	0.9286	13
48	0.3714	0.4000	2.5002	0.9285	12
49	0.3716	0.4003	2.4981	0.9284	11
50	0.3719	0.4006	2.4960	0.9283	10
51	0.3722	0.4010	2.4939	0.9282	9
52	0.3724	0.4013	2.4918	0.9281	8
53	0.3727	0.4017	2.4897	0.9279	7
54	0.3730	0.4020	2.4876	0.9278	6
55	0.3733	0.4023	2.4855	0.9277	5
56	0.3735	0.4027	2.4834	0.9276	4
57	0.3738	0.4030	2.4813	0.9275	3
58	0.3741	0.4033	2.4792	0.9274	2
59	0.3743	0.4037	2.4772	0.9273	1
60	0.3746	0.4040	2.4751	0.9272	0
	Cos	Cot	Tan	Sin	′

68° *158° 248° *338°

′	Sin	Tan	Cot	Cos	
0	0.3746	0.4040	2.4751	0.9272	60
1	0.3749	0.4044	2.4730	0.9271	59
2	0.3751	0.4047	2.4709	0.9270	58
3	0.3754	0.4050	2.4689	0.9269	57
4	0.3757	0.4054	2.4668	0.9267	56
5	0.3760	0.4057	2.4648	0.9266	55
6	0.3762	0.4061	2.4627	0.9265	54
7	0.3765	0.4064	2.4606	0.9264	53
8	0.3768	0.4067	2.4586	0.9263	52
9	0.3770	0.4071	2.4566	0.9262	51
10	0.3773	0.4074	2.4545	0.9261	50
11	0.3776	0.4078	2.4525	0.9260	49
12	0.3778	0.4081	2.4504	0.9259	48
13	0.3781	0.4084	2.4484	0.9258	47
14	0.3784	0.4088	2.4464	0.9257	46
15	0.3786	0.4091	2.4443	0.9255	45
16	0.3789	0.4095	2.4423	0.9254	44
17	0.3792	0.4098	2.4403	0.9253	43
18	0.3795	0.4101	2.4383	0.9252	42
19	0.3797	0.4105	2.4362	0.9251	41
20	0.3800	0.4108	2.4342	0.9250	40
21	0.3803	0.4111	2.4322	0.9249	39
22	0.3805	0.4115	2.4302	0.9248	38
23	0.3808	0.4118	2.4282	0.9247	37
24	0.3811	0.4122	2.4262	0.9245	36
25	0.3813	0.4125	2.4242	0.9244	35
26	0.3816	0.4129	2.4222	0.9243	34
27	0.3819	0.4132	2.4202	0.9242	33
28	0.3821	0.4135	2.4182	0.9241	32
29	0.3824	0.4139	2.4162	0.9240	31
30	0.3827	0.4142	2.4142	0.9239	30
31	0.3830	0.4146	2.4122	0.9238	29
32	0.3832	0.4149	2.4102	0.9237	28
33	0.3835	0.4152	2.4083	0.9235	27
34	0.3838	0.4156	2.4063	0.9234	26
35	0.3840	0.4159	2.4043	0.9233	25
36	0.3843	0.4163	2.4023	0.9232	24
37	0.3846	0.4166	2.4004	0.9231	23
38	0.3848	0.4169	2.3984	0.9230	22
39	0.3851	0.4173	2.3964	0.9229	21
40	0.3854	0.4176	2.3945	0.9228	20
41	0.3856	0.4180	2.3925	0.9227	19
42	0.3859	0.4183	2.3906	0.9225	18
43	0.3862	0.4187	2.3886	0.9224	17
44	0.3864	0.4190	2.3867	0.9223	16
45	0.3867	0.4193	2.3847	0.9222	15
46	0.3870	0.4197	2.3828	0.9221	14
47	0.3872	0.4200	2.3808	0.9220	13
48	0.3875	0.4204	2.3789	0.9219	12
49	0.3878	0.4207	2.3770	0.9218	11
50	0.3881	0.4210	2.3750	0.9216	10
51	0.3883	0.4214	2.3731	0.9215	9
52	0.3886	0.4217	2.3712	0.9214	8
53	0.3889	0.4221	2.3693	0.9213	7
54	0.3891	0.4224	2.3673	0.9212	6
55	0.3894	0.4228	2.3654	0.9211	5
56	0.3897	0.4231	2.3635	0.9210	4
57	0.3899	0.4234	2.3616	0.9208	3
58	0.3902	0.4238	2.3597	0.9207	2
59	0.3905	0.4241	2.3578	0.9206	1
60	0.3907	0.4245	2.3559	0.9205	0
	Cos	Cot	Tan	Sin	′

′	Sin	Tan	Cot	Cos	
0	0.3907	0.4245	2.3559	0.9205	60
1	0.3910	0.4248	2.3539	0.9204	59
2	0.3913	0.4252	2.3520	0.9203	58
3	0.3915	0.4255	2.3501	0.9202	57
4	0.3918	0.4258	2.3483	0.9200	56
5	0.3921	0.4262	2.3464	0.9199	55
6	0.3923	0.4265	2.3445	0.9198	54
7	0.3926	0.4269	2.3426	0.9197	53
8	0.3929	0.4272	2.3407	0.9196	52
9	0.3931	0.4276	2.3388	0.9195	51
10	0.3934	0.4279	2.3369	0.9194	50
11	0.3937	0.4283	2.3351	0.9192	49
12	0.3939	0.4286	2.3332	0.9191	48
13	0.3942	0.4289	2.3313	0.9190	47
14	0.3945	0.4293	2.3294	0.9189	46
15	0.3947	0.4296	2.3276	0.9188	45
16	0.3950	0.4300	2.3257	0.9187	44
17	0.3953	0.4303	2.3238	0.9186	43
18	0.3955	0.4307	2.3220	0.9184	42
19	0.3958	0.4310	2.3201	0.9183	41
20	0.3961	0.4314	2.3183	0.9182	40
21	0.3963	0.4317	2.3164	0.9181	39
22	0.3966	0.4320	2.3146	0.9180	38
23	0.3969	0.4324	2.3127	0.9179	37
24	0.3971	0.4327	2.3109	0.9178	36
25	0.3974	0.4331	2.3090	0.9176	35
26	0.3977	0.4334	2.3072	0.9175	34
27	0.3979	0.4338	2.3053	0.9174	33
28	0.3982	0.4341	2.3035	0.9173	32
29	0.3985	0.4345	2.3017	0.9172	31
30	0.3987	0.4348	2.2998	0.9171	30
31	0.3990	0.4352	2.2980	0.9169	29
32	0.3993	0.4355	2.2962	0.9168	28
33	0.3995	0.4359	2.2944	0.9167	27
34	0.3998	0.4362	2.2925	0.9166	26
35	0.4001	0.4365	2.2907	0.9165	25
36	0.4003	0.4369	2.2889	0.9164	24
37	0.4006	0.4372	2.2871	0.9162	23
38	0.4009	0.4376	2.2853	0.9161	22
39	0.4011	0.4379	2.2835	0.9160	21
40	0.4014	0.4383	2.2817	0.9159	20
41	0.4017	0.4386	2.2799	0.9158	19
42	0.4019	0.4390	2.2781	0.9157	18
43	0.4022	0.4393	2.2763	0.9155	17
44	0.4025	0.4397	2.2745	0.9154	16
45	0.4027	0.4400	2.2727	0.9153	15
46	0.4030	0.4404	2.2709	0.9152	14
47	0.4033	0.4407	2.2691	0.9151	13
48	0.4035	0.4411	2.2673	0.9150	12
49	0.4038	0.4414	2.2655	0.9148	11
50	0.4041	0.4417	2.2637	0.9147	10
51	0.4043	0.4421	2.2620	0.9146	9
52	0.4046	0.4424	2.2602	0.9145	8
53	0.4049	0.4428	2.2584	0.9144	7
54	0.4051	0.4431	2.2566	0.9143	6
55	0.4054	0.4435	2.2549	0.9141	5
56	0.4057	0.4438	2.2531	0.9140	4
57	0.4059	0.4442	2.2513	0.9139	3
58	0.4062	0.4445	2.2496	0.9138	2
59	0.4065	0.4449	2.2478	0.9137	1
60	0.4067	0.4452	2.2460	0.9135	0
	Cos	Cot	Tan	Sin	′

′	Sin	Tan	Cot	Cos	
0	0.4067	0.4452	2.2460	0.9135	60
1	0.4070	0.4456	2.2443	0.9134	59
2	0.4073	0.4459	2.2425	0.9133	58
3	0.4075	0.4463	2.2408	0.9132	57
4	0.4078	0.4466	2.2390	0.9131	56
5	0.4081	0.4470	2.2373	0.9130	55
6	0.4083	0.4473	2.2355	0.9128	54
7	0.4086	0.4477	2.2338	0.9127	53
8	0.4089	0.4480	2.2320	0.9126	52
9	0.4091	0.4484	2.2303	$0.912\bar{5}$	51
10	0.4094	0.4487	2.2286	0.9124	50
11	0.4097	0.4491	2.2268	0.9122	49
12	0.4099	0.4494	2.2251	0.9121	48
13	0.4102	0.4498	2.2234	0.9120	47
14	$0.410\bar{5}$	0.4501	2.2216	0.9119	46
15	0.4107	$0.450\bar{5}$	2.2199	0.9118	45
16	0.4110	0.4508	2.2182	0.9116	44
17	0.4112	0.4512	$2.216\bar{5}$	0.9115	43
18	0.4115	0.4515	2.2148	0.9114	42
19	0.4118	0.4519	2.2130	0.9113	41
20	0.4120	0.4522	2.2113	0.9112	40
21	0.4123	0.4526	2.2096	0.9110	39
22	0.4126	0.4529	2.2079	0.9109	38
23	0.4128	0.4533	2.2062	0.9108	37
24	0.4131	0.4536	$2.204\bar{5}$	0.9107	36
25	0.4134	0.4540	2.2028	0.9106	35
26	0.4136	0.4543	2.2011	0.9104	34
27	0.4139	0.4547	2.1994	0.9103	33
28	0.4142	0.4550	2.1977	0.9102	32
29	0.4144	0.4554	2.1960	0.9101	31
30	0.4147	0.4557	2.1943	0.9100	30
31	$0.41\bar{5}0$	0.4561	2.1926	0.9098	29
32	0.4152	0.4564	2.1909	0.9097	28
33	$0.415\bar{5}$	0.4568	2.1892	0.9096	27
34	0.4158	0.4571	2.1876	$0.909\bar{5}$	26
35	0.4160	$0.457\bar{5}$	2.1859	0.9094	25
36	0.4163	0.4578	2.1842	0.9092	24
37	0.4165	0.4582	2.1825	0.9091	23
38	0.4168	0.4585	2.1808	0.9090	22
39	0.4171	0.4589	2.1792	0.9089	21
40	0.4173	0.4592	$2.177\bar{5}$	0.9088	20
41	0.4176	0.4596	2.1758	0.9086	19
42	0.4179	0.4599	2.1742	0.9085	18
43	0.4181	0.4603	$2.172\bar{5}$	0.9084	17
44	0.4184	0.4607	2.1708	0.9083	16
45	0.4187	0.4610	2.1692	0.9081	15
46	0.4189	0.4614	2.1675	0.9080	14
47	0.4192	0.4617	2.1659	0.9079	13
48	$0.419\bar{5}$	0.4621	2.1642	0.9078	12
49	0.4197	0.4624	2.1625	0.9077	11
50	0.4200	0.4628	2.1609	0.9075	10
51	0.4202	0.4631	2.1592	0.9074	9
52	0.4205	$0.463\bar{5}$	2.1576	0.9073	8
53	0.4208	0.4638	2.1560	0.9072	7
54	0.4210	0.4642	2.1543	0.9070	6
55	0.4213	0.4645	2.1527	0.9069	5
56	0.4216	0.4649	2.1510	0.9068	4
57	0.4218	0.4652	2.1494	0.9067	3
58	0.4221	0.4656	2.1478	0.9066	2
59	0.4224	0.4660	2.1461	0.9064	1
60	0.4226	0.4663	2.1445	0.9063	0
	Cos	Cot	Tan	Sin	′

′	Sin	Tan	Cot	Cos	
0	0.4226	0.4663	2.1445	0.9063	60
1	0.4229	0.4667	2.1429	0.9062	59
2	0.4231	0.4670	2.1413	0.9061	58
3	0.4234	0.4674	2.1396	0.9059	57
4	0.4237	0.4677	2.1380	0.9058	56
5	0.4239	0.4681	2.1364	0.9057	55
6	0.4242	0.4684	2.1348	0.9056	54
7	$0.424\bar{5}$	0.4688	2.1332	0.9054	53
8	0.4247	0.4691	2.1315	0.9053	52
9	$0.42\bar{5}0$	$0.469\bar{5}$	2.1299	0.9052	51
10	0.4253	0.4699	2.1283	0.9051	50
11	0.4255	0.4702	2.1267	$0.90\bar{5}0$	49
12	0.4258	0.4706	2.1251	0.9048	48
13	0.4260	0.4709	2.1235	0.9047	47
14	0.4263	0.4713	2.1219	0.9046	46
15	0.4266	0.4716	2.1203	$0.904\bar{5}$	45
16	0.4268	0.4720	2.1187	0.9043	44
17	0.4271	0.4723	2.1171	0.9042	43
18	0.4274	0.4727	2.1155	0.9041	42
19	0.4276	0.4731	2.1139	0.9040	41
20	0.4279	0.4734	2.1123	0.9038	40
21	0.4281	0.4738	2.1107	0.9037	39
22	0.4284	0.4741	2.1092	0.9036	38
23	0.4287	$0.474\bar{5}$	2.1076	$0.903\bar{5}$	37
24	0.4289	0.4748	2.1060	0.9033	36
25	0.4292	0.4752	2.1044	0.9032	35
26	$0.429\bar{5}$	0.4755	2.1028	0.9031	34
27	0.4297	0.4759	2.1013	0.9030	33
28	0.4300	0.4763	2.0997	0.9028	32
29	0.4302	0.4766	2.0981	0.9027	31
30	0.4305	0.4770	2.0965	0.9026	30
31	0.4308	0.4773	$2.09\bar{5}0$	$0.902\bar{5}$	29
32	0.4310	0.4777	2.0934	0.9023	28
33	0.4313	0.4780	2.0918	0.9022	27
34	0.4316	0.4784	2.0903	0.9021	26
35	0.4318	0.4788	2.0887	0.9020	25
36	0.4321	0.4791	2.0872	0.9018	24
37	0.4323	$0.479\bar{5}$	2.0856	0.9017	23
38	0.4326	0.4798	2.0840	0.9016	22
39	0.4329	0.4802	$2.082\bar{5}$	$0.901\bar{5}$	21
40	0.4331	0.4806	2.0809	0.9013	20
41	0.4334	0.4809	2.0794	0.9012	19
42	0.4337	0.4813	2.0778	0.9011	18
43	0.4339	0.4816	2.0763	0.9010	17
44	0.4342	0.4820	2.0748	0.9008	16
45	0.4344	0.4823	2.0732	0.9007	15
46	0.4347	0.4827	2.0717	0.9006	14
47	$0.43\bar{5}0$	0.4831	2.0701	0.9004	13
48	0.4352	0.4834	2.0686	0.9003	12
49	$0.435\bar{5}$	0.4838	2.0671	0.9002	11
50	0.4358	0.4841	2.0655	0.9001	10
51	0.4360	$0.484\bar{5}$	2.0640	0.8999	9
52	0.4363	0.4849	$2.062\bar{5}$	0.8998	8
53	0.4365	0.4852	2.0609	0.8997	7
54	0.4368	0.4856	2.0594	0.8996	6
55	0.4371	0.4859	2.0579	0.8994	5
56	0.4373	0.4863	2.0564	0.8993	4
57	0.4376	0.4867	2.0549	0.8992	3
58	0.4378	0.4870	2.0533	0.8990	2
59	0.4381	0.4874	2.0518	0.8989	1
60	0.4384	0.4877	2.0503	0.8988	0
	Cos	Cot	Tan	Sin	′

*116° 206° *296° **26°** NATURAL

′	Sin	Tan	Cot	Cos	
0	0.4384	0.4877	2.0503	0.8988	60
1	0.4386	0.4881	2.0488	0.8987	59
2	0.4389	0.4885̄	2.0473	0.8985	58
3	0.4392	0.4888	2.0458	0.8984	57
4	0.4394	0.4892	2.0443	0.8983	56
5	0.4397	0.4895	2.0428	0.8982	55
6	0.4399	0.4899	2.0413	0.8980	54
7	0.4402	0.4903	2.0398	0.8979	53
8	0.4405̄	0.4906	2.0383	0.8978	52
9	0.4407	0.4910	2.0368	0.8976	51
10	0.4410	0.4913	2.0353	0.8975	50
11	0.4412	0.4917	2.0338	0.8974	49
12	0.4415	0.4921	2.0323	0.8973	48
13	0.4418	0.4924	2.0308	0.8971	47
14	0.4420	0.4928	2.0293	0.8970	46
15	0.4423	0.4931	2.0278	0.8969	45
16	0.4425	0.4935	2.0263	0.8967	44
17	0.4428	0.4939	2.0248	0.8966	43
18	0.4431	0.4942	2.0233	0.8965̄	42
19	0.4433	0.4946	2.0219	0.8964	41
20	0.4436	0.4950̄	2.0204	0.8962	40
21	0.4439	0.4953	2.0189	0.8961	39
22	0.4441	0.4957	2.0174	0.8960	38
23	0.4444	0.4960	2.0160	0.8958	37
24	0.4446	0.4964	2.0145̄	0.8957	36
25	0.4449	0.4968	2.0130	0.8956	35
26	0.4452	0.4971	2.0115	0.8955̄	34
27	0.4454	0.4975̄	2.0101	0.8953	33
28	0.4457	0.4979	2.0086	0.8952	32
29	0.4459	0.4982	2.0072	0.8951	31
30	0.4462	0.4986	2.0057	0.8949	30
31	0.4465̄	0.4989	2.0042	0.8948	29
32	0.4467	0.4993	2.0028	0.8947	28
33	0.4470	0.4997	2.0013	0.8945	27
34	0.4472	0.5000	1.9999	0.8944	26
35	0.4475̄	0.5004	1.9984	0.8943	25
36	0.4478	0.5008	1.9970	0.8942	24
37	0.4480	0.5011	1.9955	0.8940	23
38	0.4483	0.5015̄	1.9941	0.8939	22
39	0.4485	0.5019	1.9926	0.8938	21
40	0.4488	0.5022	1.9912	0.8936	20
41	0.4491	0.5026	1.9897	0.8935	19
42	0.4493	0.5029	1.9883	0.8934	18
43	0.4496	0.5033	1.9868	0.8932	17
44	0.4498	0.5037	1.9854	0.8931	16
45	0.4501	0.5040	1.9840	0.8930	15
46	0.4504	0.5044	1.9825	0.8928	14
47	0.4506	0.5048	1.9811	0.8927	13
48	0.4509	0.5051	1.9797	0.8926	12
49	0.4511	0.5055	1.9782	0.8925̄	11
50	0.4514	0.5059	1.9768	0.8923	10
51	0.4517	0.5062	1.9754	0.8922	9
52	0.4519	0.5066	1.9740	0.8921	8
53	0.4522	0.5070	1.9725	0.8919	7
54	0.4524	0.5073	1.9711	0.8918	6
55	0.4527	0.5077	1.9697	0.8917	5
56	0.4530	0.5081	1.9683	0.8915	4
57	0.4532	0.5084	1.9669	0.8914	3
58	0.4535̄	0.5088	1.9654	0.8913	2
59	0.4537	0.5092	1.9640	0.8911	1
60	0.4540	0.5095	1.9626	0.8910	0
	Cos	Cot	Tan	Sin	′

*153° 243° *333° **63°** NATURAL

27° *117° 207° *297°

′	Sin	Tan	Cot	Cos	
0	0.4540	0.5095	1.9626	0.8910	60
1	0.4542	0.5099	1.9612	0.8909	59
2	0.4545	0.5103	1.9598	0.8907	58
3	0.4548	0.5106	1.9584	0.8906	57
4	0.4550	0.5110	1.9570	0.8905̄	56
5	0.4553	0.5114	1.9556	0.8903	55
6	0.4555	0.5117	1.9542	0.8902	54
7	0.4558	0.5121	1.9528	0.8901	53
8	0.4561	0.5125̄	1.9514	0.8899	52
9	0.4563	0.5128	1.95̄00	0.8898	51
10	0.4566	0.5132	1.9486	0.8897	50
11	0.4568	0.5136	1.9472	0.8895	49
12	0.4571	0.5139	1.9458	0.8894	48
13	0.4574	0.5143	1.9444	0.8893	47
14	0.4576	0.5147	1.9430	0.8892	46
15	0.4579	0.5150	1.9416	0.8890	45
16	0.4581	0.5154	1.9402	0.8889	44
17	0.4584	0.5158	1.9388	0.8888	43
18	0.4586	0.5161	1.9375̄	0.8886	42
19	0.4589	0.5165	1.9361	0.8885̄	41
20	0.4592	0.5169	1.9347	0.8884	40
21	0.4594	0.5172	1.9333	0.8882	39
22	0.4597	0.5176	1.9319	0.8881	38
23	0.4599	0.5180	1.9306	0.8879	37
24	0.4602	0.5184	1.9292	0.8878	36
25	0.4605̄	0.5187	1.9278	0.8877	35
26	0.4607	0.5191	1.9265̄	0.8875	34
27	0.4610	0.5195̄	1.9251	0.8874	33
28	0.4612	0.5198	1.9237	0.8873	32
29	0.4615̄	0.5202	1.9223	0.8871	31
30	0.4617	0.5206	1.9210	0.8870	30
31	0.4620	0.5209	1.9196	0.8869	29
32	0.4623	0.5213	1.9183	0.8867	28
33	0.4625	0.5217	1.9169	0.8866	27
34	0.4628	0.5220	1.9155	0.8865̄	26
35	0.4630	0.5224	1.9142	0.8863	25
36	0.4633	0.5228	1.9128	0.8862	24
37	0.4636	0.5232	1.9115̄	0.8861	23
38	0.4638	0.5235	1.9101	0.8859	22
39	0.4641	0.5239	1.9088	0.8858	21
40	0.4643	0.5243	1.9074	0.8857	20
41	0.4646	0.5246	1.9061	0.8855	19
42	0.4648	0.5250	1.9047	0.8854	18
43	0.4651	0.5254	1.9034	0.8853	17
44	0.4654	0.5258	1.9020	0.8851	16
45	0.4656	0.5261	1.9007	0.8850̄	15
46	0.4659	0.5265̄	1.8993	0.8849	14
47	0.4661	0.5269	1.8980	0.8847	13
48	0.4664	0.5272	1.8967	0.8846	12
49	0.4666	0.5276	1.8953	0.8844	11
50	0.4669	0.5280	1.8940	0.8843	10
51	0.4672	0.5284	1.8927	0.8842	9
52	0.4674	0.5287	1.8913	0.8840	8
53	0.4677	0.5291	1.8900	0.8839	7
54	0.4679	0.5295̄	1.8887	0.8838	6
55	0.4682	0.5298	1.8873	0.8836	5
56	0.4684	0.5302	1.8860	0.8835̄	4
57	0.4687	0.5306	1.8847	0.8834	3
58	0.4690	0.5310	1.8834	0.8832	2
59	0.4692	0.5313	1.8820	0.8831	1
60	0.4695̄	0.5317	1.8807	0.8829	0
	Cos	Cot	Tan	Sin	′

62° *152° 242° *332°

*118° 208° *298° **28°** NATURAL

′	Sin	Tan	Cot	Cos	
0	0.4695	0.5317	1.8807	0.8829	60
1	0.4697	0.5321	1.8794	0.8828	59
2	0.4700	0.5325	1.8781	0.8827	58
3	0.4702	0.5328	1.8768	0.8825	57
4	0.4705	0.5332	1.8755	0.8824	56
5	0.4708	0.5336	1.8741	0.8823	55
6	0.4710	0.5340	1.8728	0.8821	54
7	0.4713	0.5343	1.8715	0.8820	53
8	0.4715	0.5347	1.8702	0.8819	52
9	0.4718	0.5351	1.8689	0.8817	51
10	0.4720	0.5354	1.8676	0.8816	50
11	0.4723	0.5358	1.8663	0.8814	49
12	0.4726	0.5362	1.8650	0.8813	48
13	0.4728	0.5366	1.8637	0.8812	47
14	0.4731	0.5369	1.8624	0.8810	46
15	0.4733	0.5373	1.8611	0.8809	45
16	0.4736	0.5377	1.8598	0.8808	44
17	0.4738	0.5381	1.8585	0.8806	43
18	0.4741	0.5384	1.8572	0.8805	42
19	0.4743	0.5388	1.8559	0.8803	41
20	0.4746	0.5392	1.8546	0.8802	40
21	0.4749	0.5396	1.8533	0.8801	39
22	0.4751	0.5399	1.8520	0.8799	38
23	0.4754	0.5403	1.8507	0.8798	37
24	0.4756	0.5407	1.8495	0.8796	36
25	0.4759	0.5411	1.8482	0.8795	35
26	0.4761	0.5415	1.8469	0.8794	34
27	0.4764	0.5418	1.8456	0.8792	33
28	0.4766	0.5422	1.8443	0.8791	32
29	0.4769	0.5426	1.8430	0.8790	31
30	0.4772	0.5430	1.8418	0.8788	30
31	0.4774	0.5433	1.8405	0.8787	29
32	0.4777	0.5437	1.8392	0.8785	28
33	0.4779	0.5441	1.8379	0.8784	27
34	0.4782	0.5445	1.8367	0.8783	26
35	0.4784	0.5448	1.8354	0.8781	25
36	0.4787	0.5452	1.8341	0.8780	24
37	0.4789	0.5456	1.8329	0.8778	23
38	0.4792	0.5460	1.8316	0.8777	22
39	0.4795	0.5464	1.8303	0.8776	21
40	0.4797	0.5467	1.8291	0.8774	20
41	0.4800	0.5471	1.8278	0.8773	19
42	0.4802	0.5475	1.8265	0.8771	18
43	0.4805	0.5479	1.8253	0.8770	17
44	0.4807	0.5482	1.8240	0.8769	16
45	0.4810	0.5486	1.8228	0.8767	15
46	0.4812	0.5490	1.8215	0.8766	14
47	0.4815	0.5494	1.8202	0.8764	13
48	0.4818	0.5498	1.8190	0.8763	12
49	0.4820	0.5501	1.8177	0.8762	11
50	0.4823	0.5505	1.8165	0.8760	10
51	0.4825	0.5509	1.8152	0.8759	9
52	0.4828	0.5513	1.8140	0.8757	8
53	0.4830	0.5517	1.8127	0.8756	7
54	0.4833	0.5520	1.8115	0.8755	6
55	0.4835	0.5524	1.8103	0.8753	5
56	0.4838	0.5528	1.8090	0.8752	4
57	0.4840	0.5532	1.8078	0.8750	3
58	0.4843	0.5535	1.8065	0.8749	2
59	0.4846	0.5539	1.8053	0.8748	1
60	0.4848	0.5543	1.8040	0.8746	0
	Cos	Cot	Tan	Sin	′

*151° 241° *331° **61°** NATURAL

29° *119° 209° *299°

′	Sin	Tan	Cot	Cos	
0	0.4848	0.5543	1.8040	0.8746	60
1	0.4851	0.5547	1.8028	0.8745	59
2	0.4853	0.5551	1.8016	0.8743	58
3	0.4856	0.5555	1.8003	0.8742	57
4	0.4858	0.5558	1.7991	0.8741	56
5	0.4861	0.5562	1.7979	0.8739	55
6	0.4863	0.5566	1.7966	0.8738	54
7	0.4866	0.5570	1.7954	0.8736	53
8	0.4868	0.5574	1.7942	0.8735	52
9	0.4871	0.5577	1.7930	0.8733	51
10	0.4874	0.5581	1.7917	0.8732	50
11	0.4876	0.5585	1.7905	0.8731	49
12	0.4879	0.5589	1.7893	0.8729	48
13	0.4881	0.5593	1.7881	0.8728	47
14	0.4884	0.5596	1.7868	0.8726	46
15	0.4886	0.5600	1.7856	0.8725	45
16	0.4889	0.5604	1.7844	0.8724	44
17	0.4891	0.5608	1.7832	0.8722	43
18	0.4894	0.5612	1.7820	0.8721	42
19	0.4896	0.5616	1.7808	0.8719	41
20	0.4899	0.5619	1.7796	0.8718	40
21	0.4901	0.5623	1.7783	0.8716	39
22	0.4904	0.5627	1.7771	0.8715	38
23	0.4907	0.5631	1.7759	0.8714	37
24	0.4909	0.5635	1.7747	0.8712	36
25	0.4912	0.5639	1.7735	0.8711	35
26	0.4914	0.5642	1.7723	0.8709	34
27	0.4917	0.5646	1.7711	0.8708	33
28	0.4919	0.5650	1.7699	0.8706	32
29	0.4922	0.5654	1.7687	0.8705	31
30	0.4924	0.5658	1.7675	0.8704	30
31	0.4927	0.5662	1.7663	0.8702	29
32	0.4929	0.5665	1.7651	0.8701	28
33	0.4932	0.5669	1.7639	0.8699	27
34	0.4934	0.5673	1.7627	0.8698	26
35	0.4937	0.5677	1.7615	0.8696	25
36	0.4939	0.5681	1.7603	0.8695	24
37	0.4942	0.5685	1.7591	0.8694	23
38	0.4944	0.5688	1.7579	0.8692	22
39	0.4947	0.5692	1.7567	0.8691	21
40	0.4950	0.5696	1.7556	0.8689	20
41	0.4952	0.5700	1.7544	0.8688	19
42	0.4955	0.5704	1.7532	0.8686	18
43	0.4957	0.5708	1.7520	0.8685	17
44	0.4960	0.5712	1.7508	0.8683	16
45	0.4962	0.5715	1.7496	0.8682	15
46	0.4965	0.5719	1.7485	0.8681	14
47	0.4967	0.5723	1.7473	0.8679	13
48	0.4970	0.5727	1.7461	0.8678	12
49	0.4972	0.5731	1.7449	0.8676	11
50	0.4975	0.5735	1.7437	0.8675	10
51	0.4977	0.5739	1.7426	0.8673	9
52	0.4980	0.5743	1.7414	0.8672	8
53	0.4982	0.5746	1.7402	0.8670	7
54	0.4985	0.5750	1.7391	0.8669	6
55	0.4987	0.5754	1.7379	0.8668	5
56	0.4990	0.5758	1.7367	0.8666	4
57	0.4992	0.5762	1.7355	0.8665	3
58	0.4995	0.5766	1.7344	0.8663	2
59	0.4997	0.5770	1.7332	0.8662	1
60	0.5000	0.5774	1.7321	0.8660	0
	Cos	Cot	Tan	Sin	′

60° *150° 240° *330°

*120° 210° *300° **30°** NATURAL

′	Sin	Tan	Cot	Cos	
0	0.5000	0.5774	1.7321	0.8660	60
1	0.5003	0.5777	1.7309	0.8659	59
2	0.5005	0.5781	1.7297	0.8657	58
3	0.5008	0.5785	1.7286	0.8656	57
4	0.5010	0.5789	1.7274	0.8654	56
5	0.5013	0.5793	1.7262	0.8653	55
6	0.5015	0.5797	1.7251	0.8652	54
7	0.5018	0.5801	1.7239	0.8650	53
8	0.5020	0.5805	1.7228	0.8649	52
9	0.5023	0.5808	1.7216	0.8647	51
10	0.5025	0.5812	1.7205	0.8646	50
11	0.5028	0.5816	1.7193	0.8644	49
12	0.5030	0.5820	1.7182	0.8643	48
13	0.5033	0.5824	1.7170	0.8641	47
14	0.5035	0.5828	1.7159	0.8640	46
15	0.5038	0.5832	1.7147	0.8638	45
16	0.5040	0.5836	1.7136	0.8637	44
17	0.5043	0.5840	1.7124	0.8635	43
18	0.5045	0.5844	1.7113	0.8634	42
19	0.5048	0.5847	1.7102	0.8632	41
20	0.5050	0.5851	1.7090	0.8631	40
21	0.5053	0.5855	1.7079	0.8630	39
22	0.5055	0.5859	1.7067	0.8628	38
23	0.5058	0.5863	1.7056	0.8627	37
24	0.5060	0.5867	1.7045	0.8625	36
25	0.5063	0.5871	1.7033	0.8624	35
26	0.5065	0.5875	1.7022	0.8622	34
27	0.5068	0.5879	1.7011	0.8621	33
28	0.5070	0.5883	1.6999	0.8619	32
29	0.5073	0.5887	1.6988	0.8618	31
30	0.5075	0.5890	1.6977	0.8616	30
31	0.5078	0.5894	1.6965	0.8615	29
32	0.5080	0.5898	1.6954	0.8613	28
33	0.5083	0.5902	1.6943	0.8612	27
34	0.5085	0.5906	1.6932	0.8610	26
35	0.5088	0.5910	1.6920	0.8609	25
36	0.5090	0.5914	1.6909	0.8607	24
37	0.5093	0.5918	1.6898	0.8606	23
38	0.5095	0.5922	1.6887	0.8604	22
39	0.5098	0.5926	1.6875	0.8603	21
40	0.5100	0.5930	1.6864	0.8601	20
41	0.5103	0.5934	1.6853	0.8600	19
42	0.5105	0.5938	1.6842	0.8599	18
43	0.5108	0.5942	1.6831	0.8597	17
44	0.5110	0.5945	1.6820	0.8596	16
45	0.5113	0.5949	1.6808	0.8594	15
46	0.5115	0.5953	1.6797	0.8593	14
47	0.5118	0.5957	1.6786	0.8591	13
48	0.5120	0.5961	1.6775	0.8590	12
49	0.5123	0.5965	1.6764	0.8588	11
50	0.5125	0.5969	1.6753	0.8587	10
51	0.5128	0.5973	1.6742	0.8585	9
52	0.5130	0.5977	1.6731	0.8584	8
53	0.5133	0.5981	1.6720	0.8582	7
54	0.5135	0.5985	1.6709	0.8581	6
55	0.5138	0.5989	1.6698	0.8579	5
56	0.5140	0.5993	1.6687	0.8578	4
57	0.5143	0.5997	1.6676	0.8576	3
58	0.5145	0.6001	1.6665	0.8575	2
59	0.5148	0.6005	1.6654	0.8573	1
60	0.5150	0.6009	1.6643	0.8572	0
	Cos	Cot	Tan	Sin	′

*149° 239° *329° **59°** NATURAL

31° *121° 211° *301°

′	Sin	Tan	Cot	Cos	
0	0.5150	0.6009	1.6643	0.8572	60
1	0.5153	0.6013	1.6632	0.8570	59
2	0.5155	0.6017	1.6621	0.8569	58
3	0.5158	0.6020	1.6610	0.8567	57
4	0.5160	0.6024	1.6599	0.8566	56
5	0.5163	0.6028	1.6588	0.8564	55
6	0.5165	0.6032	1.6577	0.8563	54
7	0.5168	0.6036	1.6566	0.8561	53
8	0.5170	0.6040	1.6555	0.8560	52
9	0.5173	0.6044	1.6545	0.8558	51
10	0.5175	0.6048	1.6534	0.8557	50
11	0.5178	0.6052	1.6523	0.8555	49
12	0.5180	0.6056	1.6512	0.8554	48
13	0.5183	0.6060	1.6501	0.8552	47
14	0.5185	0.6064	1.6490	0.8551	46
15	0.5188	0.6068	1.6479	0.8549	45
16	0.5190	0.6072	1.6469	0.8548	44
17	0.5193	0.6076	1.6458	0.8546	43
18	0.5195	0.6080	1.6447	0.8545	42
19	0.5198	0.6084	1.6436	0.8543	41
20	0.5200	0.6088	1.6426	0.8542	40
21	0.5203	0.6092	1.6415	0.8540	39
22	0.5205	0.6096	1.6404	0.8539	38
23	0.5208	0.6100	1.6393	0.8537	37
24	0.5210	0.6104	1.6383	0.8536	36
25	0.5213	0.6108	1.6372	0.8534	35
26	0.5215	0.6112	1.6361	0.8532	34
27	0.5218	0.6116	1.6351	0.8531	33
28	0.5220	0.6120	1.6340	0.8529	32
29	0.5223	0.6124	1.6329	0.8528	31
30	0.5225	0.6128	1.6319	0.8526	30
31	0.5227	0.6132	1.6308	0.8525	29
32	0.5230	0.6136	1.6297	0.8523	28
33	0.5232	0.6140	1.6287	0.8522	27
34	0.5235	0.6144	1.6276	0.8520	26
35	0.5237	0.6148	1.6265	0.8519	25
36	0.5240	0.6152	1.6255	0.8517	24
37	0.5242	0.6156	1.6244	0.8516	23
38	0.5245	0.6160	1.6234	0.8514	22
39	0.5247	0.6164	1.6223	0.8513	21
40	0.5250	0.6168	1.6212	0.8511	20
41	0.5252	0.6172	1.6202	0.8510	19
42	0.5255	0.6176	1.6191	0.8508	18
43	0.5257	0.6180	1.6181	0.8507	17
44	0.5260	0.6184	1.6170	0.8505	16
45	0.5262	0.6188	1.6160	0.8504	15
46	0.5265	0.6192	1.6149	0.8502	14
47	0.5267	0.6196	1.6139	0.8500	13
48	0.5270	0.6200	1.6128	0.8499	12
49	0.5272	0.6204	1.6118	0.8497	11
50	0.5275	0.6208	1.6107	0.8496	10
51	0.5277	0.6212	1.6097	0.8494	9
52	0.5279	0.6216	1.6087	0.8493	8
53	0.5282	0.6220	1.6076	0.8491	7
54	0.5284	0.6224	1.6066	0.8490	6
55	0.5287	0.6228	1.6055	0.8488	5
56	0.5289	0.6233	1.6045	0.8487	4
57	0.5292	0.6237	1.6034	0.8485	3
58	0.5294	0.6241	1.6024	0.8484	2
59	0.5297	0.6245	1.6014	0.8482	1
60	0.5299	0.6249	1.6003	0.8480	0
	Cos	Cot	Tan	Sin	′

58° *148° 238° *328°

*122° 212° *302° **32°** NATURAL

′	Sin	Tan	Cot	Cos	
0	0.5299	0.6249	1.6003	0.8480	60
1	0.5302	0.6253	1.5993	0.8479	59
2	0.5304	0.6257	1.5983	0.8477	58
3	0.5307	0.6261	1.5972	0.8476	57
4	0.5309	0.6265	1.5962	0.8474	56
5	0.5312	0.6269	1.5952	0.8473	55
6	0.5314	0.6273	1.5941	0.8471	54
7	0.5316	0.6277	1.5931	0.8470	53
8	0.5319	0.6281	1.5921	0.8468	52
9	0.5321	0.6285	1.5911	0.8467	51
10	0.5324	0.6289	1.5900	0.8465	50
11	0.5326	0.6293	1.5890	0.8463	49
12	0.5329	0.6297	1.5880	0.8462	48
13	0.5331	0.6301	1.5869	0.8460	47
14	0.5334	0.6305	1.5859	0.8459	46
15	0.5336	0.6310	1.5849	0.8457	45
16	0.5339	0.6314	1.5839	0.8456	44
17	0.5341	0.6318	1.5829	0.8454	43
18	0.5344	0.6322	1.5818	0.8453	42
19	0.5346	0.6326	1.5808	0.8451	41
20	0.5348	0.6330	1.5798	0.8450	40
21	0.5351	0.6334	1.5788	0.8448	39
22	0.5353	0.6338	1.5778	0.8446	38
23	0.5356	0.6342	1.5768	0.8445	37
24	0.5358	0.6346	1.5757	0.8443	36
25	0.5361	0.6350	1.5747	0.8442	35
26	0.5363	0.6354	1.5737	0.8440	34
27	0.5366	0.6358	1.5727	0.8439	33
28	0.5368	0.6363	1.5717	0.8437	32
29	0.5371	0.6367	1.5707	0.8435	31
30	0.5373	0.6371	1.5697	0.8434	30
31	0.5375	0.6375	1.5687	0.8432	29
32	0.5378	0.6379	1.5677	0.8431	28
33	0.5380	0.6383	1.5667	0.8429	27
34	0.5383	0.6387	1.5657	0.8428	26
35	0.5385	0.6391	1.5647	0.8426	25
36	0.5388	0.6395	1.5637	0.8425	24
37	0.5390	0.6399	1.5627	0.8423	23
38	0.5393	0.6403	1.5617	0.8421	22
39	0.5395	0.6408	1.5607	0.8420	21
40	0.5398	0.6412	1.5597	0.8418	20
41	0.5400	0.6416	1.5587	0.8417	19
42	0.5402	0.6420	1.5577	0.8415	18
43	0.5405	0.6424	1.5567	0.8414	17
44	0.5407	0.6428	1.5557	0.8412	16
45	0.5410	0.6432	1.5547	0.8410	15
46	0.5412	0.6436	1.5537	0.8409	14
47	0.5415	0.6440	1.5527	0.8407	13
48	0.5417	0.6445	1.5517	0.8406	12
49	0.5420	0.6449	1.5507	0.8404	11
50	0.5422	0.6453	1.5497	0.8403	10
51	0.5424	0.6457	1.5487	0.8401	9
52	0.5427	0.6461	1.5477	0.8399	8
53	0.5429	0.6465	1.5468	0.8398	7
54	0.5432	0.6469	1.5458	0.8396	6
55	0.5434	0.6473	1.5448	0.8395	5
56	0.5437	0.6478	1.5438	0.8393	4
57	0.5439	0.6482	1.5428	0.8391	3
58	0.5442	0.6486	1.5418	0.8390	2
59	0.5444	0.6490	1.5408	0.8388	1
60	0.5446	0.6494	1.5399	0.8387	0
	Cos	Cot	Tan	Sin	′

*147° 237° *327° **57°** NATURAL

33° *123° 213° *303°

′	Sin	Tan	Cot	Cos	
0	0.5446	0.6494	1.5399	0.8387	60
1	0.5449	0.6498	1.5389	0.8385	59
2	0.5451	0.6502	1.5379	0.8384	58
3	0.5454	0.6506	1.5369	0.8382	57
4	0.5456	0.6511	1.5359	0.8380	56
5	0.5459	0.6515	1.5350	0.8379	55
6	0.5461	0.6519	1.5340	0.8377	54
7	0.5463	0.6523	1.5330	0.8376	53
8	0.5466	0.6527	1.5320	0.8374	52
9	0.5468	0.6531	1.5311	0.8372	51
10	0.5471	0.6536	1.5301	0.8371	50
11	0.5473	0.6540	1.5291	0.8369	49
12	0.5476	0.6544	1.5282	0.8368	48
13	0.5478	0.6548	1.5272	0.8366	47
14	0.5480	0.6552	1.5262	0.8364	46
15	0.5483	0.6556	1.5253	0.8363	45
16	0.5485	0.6560	1.5243	0.8361	44
17	0.5488	0.6565	1.5233	0.8360	43
18	0.5490	0.6569	1.5224	0.8358	42
19	0.5493	0.6573	1.5214	0.8356	41
20	0.5495	0.6577	1.5204	0.8355	40
21	0.5498	0.6581	1.5195	0.8353	39
22	0.5500	0.6585	1.5185	0.8352	38
23	0.5502	0.6590	1.5175	0.8350	37
24	0.5505	0.6594	1.5166	0.8348	36
25	0.5507	0.6598	1.5156	0.8347	35
26	0.5510	0.6602	1.5147	0.8345	34
27	0.5512	0.6606	1.5137	0.8344	33
28	0.5515	0.6610	1.5127	0.8342	32
29	0.5517	0.6615	1.5118	0.8340	31
30	0.5519	0.6619	1.5108	0.8339	30
31	0.5522	0.6623	1.5099	0.8337	29
32	0.5524	0.6627	1.5089	0.8336	28
33	0.5527	0.6631	1.5080	0.8334	27
34	0.5529	0.6636	1.5070	0.8332	26
35	0.5531	0.6640	1.5061	0.8331	25
36	0.5534	0.6644	1.5051	0.8329	24
37	0.5536	0.6648	1.5042	0.8328	23
38	0.5539	0.6652	1.5032	0.8326	22
39	0.5541	0.6657	1.5023	0.8324	21
40	0.5544	0.6661	1.5013	0.8323	20
41	0.5546	0.6665	1.5004	0.8321	19
42	0.5548	0.6669	1.4994	0.8320	18
43	0.5551	0.6673	1.4985	0.8318	17
44	0.5553	0.6678	1.4975	0.8316	16
45	0.5556	0.6682	1.4966	0.8315	15
46	0.5558	0.6686	1.4957	0.8313	14
47	0.5561	0.6690	1.4947	0.8311	13
48	0.5563	0.6694	1.4938	0.8310	12
49	0.5565	0.6699	1.4928	0.8308	11
50	0.5568	0.6703	1.4919	0.8307	10
51	0.5570	0.6707	1.4910	0.8305	9
52	0.5573	0.6711	1.4900	0.8303	8
53	0.5575	0.6715	1.4891	0.8302	7
54	0.5577	0.6720	1.4882	0.8300	6
55	0.5580	0.6724	1.4872	0.8298	5
56	0.5582	0.6728	1.4863	0.8297	4
57	0.5585	0.6732	1.4854	0.8295	3
58	0.5587	0.6737	1.4844	0.8294	2
59	0.5590	0.6741	1.4835	0.8292	1
60	0.5592	0.6745	1.4826	0.8290	0
	Cos	Cot	Tan	Sin	′

56° *146° 236° *326°

*124° 214° *304° **34°** NATURAL

'	Sin	Tan	Cot	Cos	
0	0.5592	0.6745	1.4826	0.8290	60
1	0.5594	0.6749	1.4816	0.8289	59
2	0.5597	0.6754	1.4807	0.8287	58
3	0.5599	0.6758	1.4798	0.8285	57
4	0.5602	0.6762	1.4788	0.8284	56
5	0.5604	0.6766	1.4779	0.8282	55
6	0.5606	0.6771	1.4770	0.8281	54
7	0.5609	0.6775	1.4761	0.8279	53
8	0.5611	0.6779	1.4751	0.8277	52
9	0.5614	0.6783	1.4742	0.8276	51
10	0.5616	0.6787	1.4733	0.8274	50
11	0.5618	0.6792	1.4724	0.8272	49
12	0.5621	0.6796	1.4715	0.8271	48
13	0.5623	0.6800	1.4705	0.8269	47
14	0.5626	0.6805	1.4696	0.8268	46
15	0.5628	0.6809	1.4687	0.8266	45
16	0.5630	0.6813	1.4678	0.8264	44
17	0.5633	0.6817	1.4669	0.8263	43
18	0.5635	0.6822	1.4659	0.8261	42
19	0.5638	0.6826	1.4650	0.8259	41
20	0.5640	0.6830	1.4641	0.8258	40
21	0.5642	0.6834	1.4632	0.8256	39
22	0.5645	0.6839	1.4623	0.8254	38
23	0.5647	0.6843	1.4614	0.8253	37
24	0.5650	0.6847	1.4605	0.8251	36
25	0.5652	0.6851	1.4596	0.8249	35
26	0.5654	0.6856	1.4586	0.8248	34
27	0.5657	0.6860	1.4577	0.8246	33
28	0.5659	0.6864	1.4568	0.8245	32
29	0.5662	0.6869	1.4559	0.8243	31
30	0.5664	0.6873	1.4550	0.8241	30
31	0.5666	0.6877	1.4541	0.8240	29
32	0.5669	0.6881	1.4532	0.8238	28
33	0.5671	0.6886	1.4523	0.8236	27
34	0.5674	0.6890	1.4514	0.8235	26
35	0.5676	0.6894	1.4505	0.8233	25
36	0.5678	0.6899	1.4496	0.8231	24
37	0.5681	0.6903	1.4487	0.8230	23
38	0.5683	0.6907	1.4478	0.8228	22
39	0.5686	0.6911	1.4469	0.8226	21
40	0.5688	0.6916	1.4460	0.8225	20
41	0.5690	0.6920	1.4451	0.8223	19
42	0.5693	0.6924	1.4442	0.8221	18
43	0.5695	0.6929	1.4433	0.8220	17
44	0.5698	0.6933	1.4424	0.8218	16
45	0.5700	0.6937	1.4415	0.8216	15
46	0.5702	0.6942	1.4406	0.8215	14
47	0.5705	0.6946	1.4397	0.8213	13
48	0.5707	0.6950	1.4388	0.8211	12
49	0.5710	0.6954	1.4379	0.8210	11
50	0.5712	0.6959	1.4370	0.8208	10
51	0.5714	0.6963	1.4361	0.8207	9
52	0.5717	0.6967	1.4352	0.8205	8
53	0.5719	0.6972	1.4344	0.8203	7
54	0.5721	0.6976	1.4335	0.8202	6
55	0.5724	0.6980	1.4326	0.8200	5
56	0.5726	0.6985	1.4317	0.8198	4
57	0.5729	0.6989	1.4308	0.8197	3
58	0.5731	0.6993	1.4299	0.8195	2
59	0.5733	0.6998	1.4290	0.8193	1
60	0.5736	0.7002	1.4281	0.8192	0
	Cos	Cot	Tan	Sin	'

*145° 235° *325° **55°** NATURAL

35° *125° 215° *305°

'	Sin	Tan	Cot	Cos	
0	0.5736	0.7002	1.4281	0.8192	60
1	0.5738	0.7006	1.4273	0.8190	59
2	0.5741	0.7011	1.4264	0.8188	58
3	0.5743	0.7015	1.4255	0.8187	57
4	0.5745	0.7019	1.4246	0.8185	56
5	0.5748	0.7024	1.4237	0.8183	55
6	0.5750	0.7028	1.4229	0.8181	54
7	0.5752	0.7032	1.4220	0.8180	53
8	0.5755	0.7037	1.4211	0.8178	52
9	0.5757	0.7041	1.4202	0.8176	51
10	0.5760	0.7046	1.4193	0.8175	50
11	0.5762	0.7050	1.4185	0.8173	49
12	0.5764	0.7054	1.4176	0.8171	48
13	0.5767	0.7059	1.4167	0.8170	47
14	0.5769	0.7063	1.4158	0.8168	46
15	0.5771	0.7067	1.4150	0.8166	45
16	0.5774	0.7072	1.4141	0.8165	44
17	0.5776	0.7076	1.4132	0.8163	43
18	0.5779	0.7080	1.4124	0.8161	42
19	0.5781	0.7085	1.4115	0.8160	41
20	0.5783	0.7089	1.4106	0.8158	40
21	0.5786	0.7094	1.4097	0.8156	39
22	0.5788	0.7098	1.4089	0.8155	38
23	0.5790	0.7102	1.4080	0.8153	37
24	0.5793	0.7107	1.4071	0.8151	36
25	0.5795	0.7111	1.4063	0.8150	35
26	0.5798	0.7115	1.4054	0.8148	34
27	0.5800	0.7120	1.4045	0.8146	33
28	0.5802	0.7124	1.4037	0.8145	32
29	0.5805	0.7129	1.4028	0.8143	31
30	0.5807	0.7133	1.4019	0.8141	30
31	0.5809	0.7137	1.4011	0.8139	29
32	0.5812	0.7142	1.4002	0.8138	28
33	0.5814	0.7146	1.3994	0.8136	27
34	0.5816	0.7151	1.3985	0.8134	26
35	0.5819	0.7155	1.3976	0.8133	25
36	0.5821	0.7159	1.3968	0.8131	24
37	0.5824	0.7164	1.3959	0.8129	23
38	0.5826	0.7168	1.3951	0.8128	22
39	0.5828	0.7173	1.3942	0.8126	21
40	0.5831	0.7177	1.3934	0.8124	20
41	0.5833	0.7181	1.3925	0.8123	19
42	0.5835	0.7186	1.3916	0.8121	18
43	0.5838	0.7190	1.3908	0.8119	17
44	0.5840	0.7195	1.3899	0.8117	16
45	0.5842	0.7199	1.3891	0.8116	15
46	0.5845	0.7203	1.3882	0.8114	14
47	0.5847	0.7208	1.3874	0.8112	13
48	0.5850	0.7212	1.3865	0.8111	12
49	0.5852	0.7217	1.3857	0.8109	11
50	0.5854	0.7221	1.3848	0.8107	10
51	0.5857	0.7226	1.3840	0.8106	9
52	0.5859	0.7230	1.3831	0.8104	8
53	0.5861	0.7234	1.3823	0.8102	7
54	0.5864	0.7239	1.3814	0.8100	6
55	0.5866	0.7243	1.3806	0.8099	5
56	0.5868	0.7248	1.3798	0.8097	4
57	0.5871	0.7252	1.3789	0.8095	3
58	0.5873	0.7257	1.3781	0.8094	2
59	0.5875	0.7261	1.3772	0.8092	1
60	0.5878	0.7265	1.3764	0.8090	0
	Cos	Cot	Tan	Sin	'

54° *144° 234° *324°

*126° 216° *306° **36°** NATURAL

′	Sin	Tan	Cot	Cos	
0	0.5878	0.7265	1.3764	0.8090	60
1	0.5880	0.7270	1.3755	0.8088	59
2	0.5883	0.7274	1.3747	0.8087	58
3	0.5885	0.7279	1.3739	0.8085	57
4	0.5887	0.7283	1.3730	0.8083	56
5	0.5890	0.7288	1.3722	0.8082	55
6	0.5892	0.7292	1.3713	0.8080	54
7	0.5894	0.7297	1.3705	0.8078	53
8	0.5897	0.7301	1.3697	0.8076	52
9	0.5899	0.7306	1.3688	0.8075	51
10	0.5901	0.7310	1.3680	0.8073	50
11	0.5904	0.7314	1.3672	0.8071	49
12	0.5906	0.7319	1.3663	0.8070	48
13	0.5908	0.7323	1.3655	0.8068	47
14	0.5911	0.7328	1.3647	0.8066	46
15	0.5913	0.7332	1.3638	0.8064	45
16	0.5915	0.7337	1.3630	0.8063	44
17	0.5918	0.7341	1.3622	0.8061	43
18	0.5920	0.7346	1.3613	0.8059	42
19	0.5922	0.7350	1.3605	0.8058	41
20	0.5925	0.7355	1.3597	0.8056	40
21	0.5927	0.7359	1.3588	0.8054	39
22	0.5930	0.7364	1.3580	0.8052	38
23	0.5932	0.7368	1.3572	0.8051	37
24	0.5934	0.7373	1.3564	0.8049	36
25	0.5937	0.7377	1.3555	0.8047	35
26	0.5939	0.7382	1.3547	0.8045	34
27	0.5941	0.7386	1.3539	0.8044	33
28	0.5944	0.7391	1.3531	0.8042	32
29	0.5946	0.7395	1.3522	0.8040	31
30	0.5948	0.7400	1.3514	0.8039	30
31	0.5951	0.7404	1.3506	0.8037	29
32	0.5953	0.7409	1.3498	0.8035	28
33	0.5955	0.7413	1.3490	0.8033	27
34	0.5958	0.7418	1.3481	0.8032	26
35	0.5960	0.7422	1.3473	0.8030	25
36	0.5962	0.7427	1.3465	0.8028	24
37	0.5965	0.7431	1.3457	0.8026	23
38	0.5967	0.7436	1.3449	0.8025	22
39	0.5969	0.7440	1.3440	0.8023	21
40	0.5972	0.7445	1.3432	0.8021	20
41	0.5974	0.7449	1.3424	0.8019	19
42	0.5976	0.7454	1.3416	0.8018	18
43	0.5979	0.7458	1.3408	0.8016	17
44	0.5981	0.7463	1.3400	0.8014	16
45	0.5983	0.7467	1.3392	0.8013	15
46	0.5986	0.7472	1.3384	0.8011	14
47	0.5988	0.7476	1.3375	0.8009	13
48	0.5990	0.7481	1.3367	0.8007	12
49	0.5993	0.7485	1.3359	0.8006	11
50	0.5995	0.7490	1.3351	0.8004	10
51	0.5997	0.7495	1.3343	0.8002	9
52	0.6000	0.7499	1.3335	0.8000	8
53	0.6002	0.7504	1.3327	0.7999	7
54	0.6004	0.7508	1.3319	0.7997	6
55	0.6007	0.7513	1.3311	0.7995	5
56	0.6009	0.7517	1.3303	0.7993	4
57	0.6011	0.7522	1.3295	0.7992	3
58	0.6014	0.7526	1.3287	0.7990	2
59	0.6016	0.7531	1.3278	0.7988	1
60	0.6018	0.7536	1.3270	0.7986	0
	Cos	Cot	Tan	Sin	′

*143° 233° *323° **53°** NATURAL

37° *127° 217° *307°

′	Sin	Tan	Cot	Cos	
0	0.6018	0.7536	1.3270	0.7986	60
1	0.6020	0.7540	1.3262	0.7985	59
2	0.6023	0.7545	1.3254	0.7983	58
3	0.6025	0.7549	1.3246	0.7981	57
4	0.6027	0.7554	1.3238	0.7979	56
5	0.6030	0.7558	1.3230	0.7978	55
6	0.6032	0.7563	1.3222	0.7976	54
7	0.6034	0.7568	1.3214	0.7974	53
8	0.6037	0.7572	1.3206	0.7972	52
9	0.6039	0.7577	1.3198	0.7971	51
10	0.6041	0.7581	1.3190	0.7969	50
11	0.6044	0.7586	1.3182	0.7967	49
12	0.6046	0.7590	1.3175	0.7965	48
13	0.6048	0.7595	1.3167	0.7964	47
14	0.6051	0.7600	1.3159	0.7962	46
15	0.6053	0.7604	1.3151	0.7960	45
16	0.6055	0.7609	1.3143	0.7958	44
17	0.6058	0.7613	1.3135	0.7956	43
18	0.6060	0.7618	1.3127	0.7955	42
19	0.6062	0.7623	1.3119	0.7953	41
20	0.6065	0.7627	1.3111	0.7951	40
21	0.6067	0.7632	1.3103	0.7949	39
22	0.6069	0.7636	1.3095	0.7948	38
23	0.6071	0.7641	1.3087	0.7946	37
24	0.6074	0.7646	1.3079	0.7944	36
25	0.6076	0.7650	1.3072	0.7942	35
26	0.6078	0.7655	1.3064	0.7941	34
27	0.6081	0.7659	1.3056	0.7939	33
28	0.6083	0.7664	1.3048	0.7937	32
29	0.6085	0.7669	1.3040	0.7935	31
30	0.6088	0.7673	1.3032	0.7934	30
31	0.6090	0.7678	1.3024	0.7932	29
32	0.6092	0.7683	1.3017	0.7930	28
33	0.6095	0.7687	1.3009	0.7928	27
34	0.6097	0.7692	1.3001	0.7926	26
35	0.6099	0.7696	1.2993	0.7925	25
36	0.6101	0.7701	1.2985	0.7923	24
37	0.6104	0.7706	1.2977	0.7921	23
38	0.6106	0.7710	1.2970	0.7919	22
39	0.6108	0.7715	1.2962	0.7918	21
40	0.6111	0.7720	1.2954	0.7916	20
41	0.6113	0.7724	1.2946	0.7914	19
42	0.6115	0.7729	1.2938	0.7912	18
43	0.6118	0.7734	1.2931	0.7910	17
44	0.6120	0.7738	1.2923	0.7909	16
45	0.6122	0.7743	1.2915	0.7907	15
46	0.6124	0.7747	1.2907	0.7905	14
47	0.6127	0.7752	1.2900	0.7903	13
48	0.6129	0.7757	1.2892	0.7902	12
49	0.6131	0.7761	1.2884	0.7900	11
50	0.6134	0.7766	1.2876	0.7898	10
51	0.6136	0.7771	1.2869	0.7896	9
52	0.6138	0.7775	1.2861	0.7894	8
53	0.6141	0.7780	1.2853	0.7893	7
54	0.6143	0.7785	1.2846	0.7891	6
55	0.6145	0.7789	1.2838	0.7889	5
56	0.6147	0.7794	1.2830	0.7887	4
57	0.6150	0.7799	1.2822	0.7885	3
58	0.6152	0.7803	1.2815	0.7884	2
59	0.6154	0.7808	1.2807	0.7882	1
60	0.6157	0.7813	1.2799	0.7880	0
	Cos	Cot	Tan	Sin	′

52° *142° 232° *322°

38°

′	Sin	Tan	Cot	Cos	
0	0.6157	0.7813	1.2799	0.7880	60
1	0.6159	0.7818	1.2792	0.7878	59
2	0.6161	0.7822	1.2784	0.7877	58
3	0.6163	0.7827	1.2776	0.7875	57
4	0.6166	0.7832	1.2769	0.7873	56
5	0.6168	0.7836	1.2761	0.7871	55
6	0.6170	0.7841	1.2753	0.7869	54
7	0.6173	0.7846	1.2746	0.7868	53
8	0.6175	0.7850	1.2738	0.7866	52
9	0.6177	0.7855	1.2731	0.7864	51
10	0.6180	0.7860	1.2723	0.7862	50
11	0.6182	0.7865	1.2715	0.7860	49
12	0.6184	0.7869	1.2708	0.7859	48
13	0.6186	0.7874	1.2700	0.7857	47
14	0.6189	0.7879	1.2693	0.7855	46
15	0.6191	0.7883	1.2685	0.7853	45
16	0.6193	0.7888	1.2677	0.7851	44
17	0.6196	0.7893	1.2670	0.7850	43
18	0.6198	0.7898	1.2662	0.7848	42
19	0.6200	0.7902	1.2655	0.7846	41
20	0.6202	0.7907	1.2647	0.7844	40
21	0.6205	0.7912	1.2640	0.7842	39
22	0.6207	0.7916	1.2632	0.7841	38
23	0.6209	0.7921	1.2624	0.7839	37
24	0.6211	0.7926	1.2617	0.7837	36
25	0.6214	0.7931	1.2609	0.7835	35
26	0.6216	0.7935	1.2602	0.7833	34
27	0.6218	0.7940	1.2594	0.7832	33
28	0.6221	0.7945	1.2587	0.7830	32
29	0.6223	0.7950	1.2579	0.7828	31
30	0.6225	0.7954	1.2572	0.7826	30
31	0.6227	0.7959	1.2564	0.7824	29
32	0.6230	0.7964	1.2557	0.7822	28
33	0.6232	0.7969	1.2549	0.7821	27
34	0.6234	0.7973	1.2542	0.7819	26
35	0.6237	0.7978	1.2534	0.7817	25
36	0.6239	0.7983	1.2527	0.7815	24
37	0.6241	0.7988	1.2519	0.7813	23
38	0.6243	0.7992	1.2512	0.7812	22
39	0.6246	0.7997	1.2504	0.7810	21
40	0.6248	0.8002	1.2497	0.7808	20
41	0.6250	0.8007	1.2489	0.7806	19
42	0.6252	0.8012	1.2482	0.7804	18
43	0.6255	0.8016	1.2475	0.7802	17
44	0.6257	0.8021	1.2467	0.7801	16
45	0.6259	0.8026	1.2460	0.7799	15
46	0.6262	0.8031	1.2452	0.7797	14
47	0.6264	0.8035	1.2445	0.7795	13
48	0.6266	0.8040	1.2437	0.7793	12
49	0.6268	0.8045	1.2430	0.7792	11
50	0.6271	0.8050	1.2423	0.7790	10
51	0.6273	0.8055	1.2415	0.7788	9
52	0.6275	0.8059	1.2408	0.7786	8
53	0.6277	0.8064	1.2401	0.7784	7
54	0.6280	0.8069	1.2393	0.7782	6
55	0.6282	0.8074	1.2386	0.7781	5
56	0.6284	0.8079	1.2378	0.7779	4
57	0.6286	0.8083	1.2371	0.7777	3
58	0.6289	0.8088	1.2364	0.7775	2
59	0.6291	0.8093	1.2356	0.7773	1
60	0.6293	0.8098	1.2349	0.7771	0
	Cos	Cot	Tan	Sin	′

39°

′	Sin	Tan	Cot	Cos	
0	0.6293	0.8098	1.2349	0.7771	60
1	0.6295	0.8103	1.2342	0.7770	59
2	0.6298	0.8107	1.2334	0.7768	58
3	0.6300	0.8112	1.2327	0.7766	57
4	0.6302	0.8117	1.2320	0.7764	56
5	0.6305	0.8122	1.2312	0.7762	55
6	0.6307	0.8127	1.2305	0.7760	54
7	0.6309	0.8132	1.2298	0.7759	53
8	0.6311	0.8136	1.2290	0.7757	52
9	0.6314	0.8141	1.2283	0.7755	51
10	0.6316	0.8146	1.2276	0.7753	50
11	0.6318	0.8151	1.2268	0.7751	49
12	0.6320	0.8156	1.2261	0.7749	48
13	0.6323	0.8161	1.2254	0.7748	47
14	0.6325	0.8165	1.2247	0.7746	46
15	0.6327	0.8170	1.2239	0.7744	45
16	0.6329	0.8175	1.2232	0.7742	44
17	0.6332	0.8180	1.2225	0.7740	43
18	0.6334	0.8185	1.2218	0.7738	42
19	0.6336	0.8190	1.2210	0.7737	41
20	0.6338	0.8195	1.2203	0.7735	40
21	0.6341	0.8199	1.2196	0.7733	39
22	0.6343	0.8204	1.2189	0.7731	38
23	0.6345	0.8209	1.2181	0.7729	37
24	0.6347	0.8214	1.2174	0.7727	36
25	0.6350	0.8219	1.2167	0.7725	35
26	0.6352	0.8224	1.2160	0.7724	34
27	0.6354	0.8229	1.2153	0.7722	33
28	0.6356	0.8234	1.2145	0.7720	32
29	0.6359	0.8238	1.2138	0.7718	31
30	0.6361	0.8243	1.2131	0.7716	30
31	0.6363	0.8248	1.2124	0.7714	29
32	0.6365	0.8253	1.2117	0.7713	28
33	0.6368	0.8258	1.2109	0.7711	27
34	0.6370	0.8263	1.2102	0.7709	26
35	0.6372	0.8268	1.2095	0.7707	25
36	0.6374	0.8273	1.2088	0.7705	24
37	0.6376	0.8278	1.2081	0.7703	23
38	0.6379	0.8283	1.2074	0.7701	22
39	0.6381	0.8287	1.2066	0.7700	21
40	0.6383	0.8292	1.2059	0.7698	20
41	0.6385	0.8297	1.2052	0.7696	19
42	0.6388	0.8302	1.2045	0.7694	18
43	0.6390	0.8307	1.2038	0.7692	17
44	0.6392	0.8312	1.2031	0.7690	16
45	0.6394	0.8317	1.2024	0.7688	15
46	0.6397	0.8322	1.2017	0.7687	14
47	0.6399	0.8327	1.2009	0.7685	13
48	0.6401	0.8332	1.2002	0.7683	12
49	0.6403	0.8337	1.1995	0.7681	11
50	0.6406	0.8342	1.1988	0.7679	10
51	0.6408	0.8346	1.1981	0.7677	9
52	0.6410	0.8351	1.1974	0.7675	8
53	0.6412	0.8356	1.1967	0.7674	7
54	0.6414	0.8361	1.1960	0.7672	6
55	0.6417	0.8366	1.1953	0.7670	5
56	0.6419	0.8371	1.1946	0.7668	4
57	0.6421	0.8376	1.1939	0.7666	3
58	0.6423	0.8381	1.1932	0.7664	2
59	0.6426	0.8386	1.1925	0.7662	1
60	0.6428	0.8391	1.1918	0.7660	0
	Cos	Cot	Tan	Sin	′

*130° 220° *310° **40°** NATURAL

′	Sin	Tan	Cot	Cos	
0	0.6428	0.8391	1.1918	0.7660	60
1	0.6430	0.8396	1.1910	0.7659	59
2	0.6432	0.8401	1.1903	0.7657	58
3	0.6435	0.8406	1.1896	0.7655	57
4	0.6437	0.8411	1.1889	0.7653	56
5	0.6439	0.8416	1.1882	0.7651	55
6	0.6441	0.8421	1.1875	0.7649	54
7	0.6443	0.8426	1.1868	0.7647	53
8	0.6446	0.8431	1.1861	0.7645	52
9	0.6448	0.8436	1.1854	0.7644	51
10	0.6450	0.8441	1.1847	0.7642	50
11	0.6452	0.8446	1.1840	0.7640	49
12	0.6455	0.8451	1.1833	0.7638	48
13	0.6457	0.8456	1.1826	0.7636	47
14	0.6459	0.8461	1.1819	0.7634	46
15	0.6461	0.8466	1.1812	0.7632	45
16	0.6463	0.8471	1.1806	0.7630	44
17	0.6466	0.8476	1.1799	0.7629	43
18	0.6468	0.8481	1.1792	0.7627	42
19	0.6470	0.8486	1.1785	0.7625	41
20	0.6472	0.8491	1.1778	0.7623	40
21	0.6475	0.8496	1.1771	0.7621	39
22	0.6477	0.8501	1.1764	0.7619	38
23	0.6479	0.8506	1.1757	0.7617	37
24	0.6481	0.8511	1.1750	0.7615	36
25	0.6483	0.8516	1.1743	0.7613	35
26	0.6486	0.8521	1.1736	0.7612	34
27	0.6488	0.8526	1.1729	0.7610	33
28	0.6490	0.8531	1.1722	0.7608	32
29	0.6492	0.8536	1.1715	0.7606	31
30	0.6494	0.8541	1.1708	0.7604	30
31	0.6497	0.8546	1.1702	0.7602	29
32	0.6499	0.8551	1.1695	0.7600	28
33	0.6501	0.8556	1.1688	0.7598	27
34	0.6503	0.8561	1.1681	0.7596	26
35	0.6506	0.8566	1.1674	0.7595	25
36	0.6508	0.8571	1.1667	0.7593	24
37	0.6510	0.8576	1.1660	0.7591	23
38	0.6512	0.8581	1.1653	0.7589	22
39	0.6514	0.8586	1.1647	0.7587	21
40	0.6517	0.8591	1.1640	0.7585	20
41	0.6519	0.8596	1.1633	0.7583	19
42	0.6521	0.8601	1.1626	0.7581	18
43	0.6523	0.8606	1.1619	0.7579	17
44	0.6525	0.8611	1.1612	0.7578	16
45	0.6528	0.8617	1.1606	0.7576	15
46	0.6530	0.8622	1.1599	0.7574	14
47	0.6532	0.8627	1.1592	0.7572	13
48	0.6534	0.8632	1.1585	0.7570	12
49	0.6536	0.8637	1.1578	0.7568	11
50	0.6539	0.8642	1.1571	0.7566	10
51	0.6541	0.8647	1.1565	0.7564	9
52	0.6543	0.8652	1.1558	0.7562	8
53	0.6545	0.8657	1.1551	0.7560	7
54	0.6547	0.8662	1.1544	0.7559	6
55	0.6550	0.8667	1.1538	0.7557	5
56	0.6552	0.8672	1.1531	0.7555	4
57	0.6554	0.8678	1.1524	0.7553	3
58	0.6556	0.8683	1.1517	0.7551	2
59	0.6558	0.8688	1.1510	0.7549	1
60	0.6561	0.8693	1.1504	0.7547	0
	Cos	Cot	Tan	Sin	′

*139° 229° *319° **49°** NATURAL

41° *131° 221° *311°

′	Sin	Tan	Cot	Cos	
0	0.6561	0.8693	1.1504	0.7547	60
1	0.6563	0.8698	1.1497	0.7545	59
2	0.6565	0.8703	1.1490	0.7543	58
3	0.6567	0.8708	1.1483	0.7541	57
4	0.6569	0.8713	1.1477	0.7539	56
5	0.6572	0.8718	1.1470	0.7538	55
6	0.6574	0.8724	1.1463	0.7536	54
7	0.6576	0.8729	1.1456	0.7534	53
8	0.6578	0.8734	1.1450	0.7532	52
9	0.6580	0.8739	1.1443	0.7530	51
10	0.6583	0.8744	1.1436	0.7528	50
11	0.6585	0.8749	1.1430	0.7526	49
12	0.6587	0.8754	1.1423	0.7524	48
13	0.6589	0.8759	1.1416	0.7522	47
14	0.6591	0.8765	1.1410	0.7520	46
15	0.6593	0.8770	1.1403	0.7518	45
16	0.6596	0.8775	1.1396	0.7516	44
17	0.6598	0.8780	1.1389	0.7515	43
18	0.6600	0.8785	1.1383	0.7513	42
19	0.6602	0.8790	1.1376	0.7511	41
20	0.6604	0.8796	1.1369	0.7509	40
21	0.6607	0.8801	1.1363	0.7507	39
22	0.6609	0.8806	1.1356	0.7505	38
23	0.6611	0.8811	1.1349	0.7503	37
24	0.6613	0.8816	1.1343	0.7501	36
25	0.6615	0.8821	1.1336	0.7499	35
26	0.6617	0.8827	1.1329	0.7497	34
27	0.6620	0.8832	1.1323	0.7495	33
28	0.6622	0.8837	1.1316	0.7493	32
29	0.6624	0.8842	1.1310	0.7491	31
30	0.6626	0.8847	1.1303	0.7490	30
31	0.6628	0.8852	1.1296	0.7488	29
32	0.6631	0.8858	1.1290	0.7486	28
33	0.6633	0.8863	1.1283	0.7484	27
34	0.6635	0.8868	1.1276	0.7482	26
35	0.6637	0.8873	1.1270	0.7480	25
36	0.6639	0.8878	1.1263	0.7478	24
37	0.6641	0.8884	1.1257	0.7476	23
38	0.6644	0.8889	1.1250	0.7474	22
39	0.6646	0.8894	1.1243	0.7472	21
40	0.6648	0.8899	1.1237	0.7470	20
41	0.6650	0.8904	1.1230	0.7468	19
42	0.6652	0.8910	1.1224	0.7466	18
43	0.6654	0.8915	1.1217	0.7464	17
44	0.6657	0.8920	1.1211	0.7463	16
45	0.6659	0.8925	1.1204	0.7461	15
46	0.6661	0.8931	1.1197	0.7459	14
47	0.6663	0.8936	1.1191	0.7457	13
48	0.6665	0.8941	1.1184	0.7455	12
49	0.6667	0.8946	1.1178	0.7453	11
50	0.6670	0.8952	1.1171	0.7451	10
51	0.6672	0.8957	1.1165	0.7449	9
52	0.6674	0.8962	1.1158	0.7447	8
53	0.6676	0.8967	1.1152	0.7445	7
54	0.6678	0.8972	1.1145	0.7443	6
55	0.6680	0.8978	1.1139	0.7441	5
56	0.6683	0.8983	1.1132	0.7439	4
57	0.6685	0.8988	1.1126	0.7437	3
58	0.6687	0.8994	1.1119	0.7435	2
59	0.6689	0.8999	1.1113	0.7433	1
60	0.6691	0.9004	1.1106	0.7431	0
	Cos	Cot	Tan	Sin	′

48° *138° 228° *318°

*132° 222° *312° **42°** NATURAL

′	Sin	Tan	Cot	Cos	
0	0.6691	0.9004	1.1106	0.7431	60
1	0.6693	0.9009	1.1100	0.7430	59
2	0.6696	0.9015	1.1093	0.7428	58
3	0.6698	0.9020	1.1087	0.7426	57
4	0.6700	0.9025	1.1080	0.7424	56
5	0.6702	0.9030	1.1074	0.7422	55
6	0.6704	0.9036	1.1067	0.7420	54
7	0.6706	0.9041	1.1061	0.7418	53
8	0.6709	0.9046	1.1054	0.7416	52
9	0.6711	0.9052	1.1048	0.7414	51
10	0.6713	0.9057	1.1041	0.7412	50
11	0.6715	0.9062	1.1035	0.7410	49
12	0.6717	0.9067	1.1028	0.7408	48
13	0.6719	0.9073	1.1022	0.7406	47
14	0.6722	0.9078	1.1016	0.7404	46
15	0.6724	0.9083	1.1009	0.7402	45
16	0.6726	0.9089	1.1003	0.7400	44
17	0.6728	0.9094	1.0996	0.7398	43
18	0.6730	0.9099	1.0990	0.7396	42
19	0.6732	0.9105	1.0983	0.7394	41
20	0.6734	0.9110	1.0977	0.7392	40
21	0.6737	0.9115	1.0971	0.7390	39
22	0.6739	0.9121	1.0964	0.7388	38
23	0.6741	0.9126	1.0958	0.7387	37
24	0.6743	0.9131	1.0951	0.7385	36
25	0.6745	0.9137	1.0945	0.7383	35
26	0.6747	0.9142	1.0939	0.7381	34
27	0.6749	0.9147	1.0932	0.7379	33
28	0.6752	0.9153	1.0926	0.7377	32
29	0.6754	0.9158	1.0919	0.7375	31
30	0.6756	0.9163	1.0913	0.7373	30
31	0.6758	0.9169	1.0907	0.7371	29
32	0.6760	0.9174	1.0900	0.7369	28
33	0.6762	0.9179	1.0894	0.7367	27
34	0.6764	0.9185	1.0888	0.7365	26
35	0.6767	0.9190	1.0881	0.7363	25
36	0.6769	0.9195	1.0875	0.7361	24
37	0.6771	0.9201	1.0869	0.7359	23
38	0.6773	0.9206	1.0862	0.7357	22
39	0.6775	0.9212	1.0856	0.7355	21
40	0.6777	0.9217	1.0850	0.7353	20
41	0.6779	0.9222	1.0843	0.7351	19
42	0.6782	0.9228	1.0837	0.7349	18
43	0.6784	0.9233	1.0831	0.7347	17
44	0.6786	0.9239	1.0824	0.7345	16
45	0.6788	0.9244	1.0818	0.7343	15
46	0.6790	0.9249	1.0812	0.7341	14
47	0.6792	0.9255	1.0805	0.7339	13
48	0.6794	0.9260	1.0799	0.7337	12
49	0.6797	0.9266	1.0793	0.7335	11
50	0.6799	0.9271	1.0786	0.7333	10
51	0.6801	0.9276	1.0780	0.7331	9
52	0.6803	0.9282	1.0774	0.7329	8
53	0.6805	0.9287	1.0768	0.7327	7
54	0.6807	0.9293	1.0761	0.7325	6
55	0.6809	0.9298	1.0755	0.7323	5
56	0.6811	0.9303	1.0749	0.7321	4
57	0.6814	0.9309	1.0742	0.7319	3
58	0.6816	0.9314	1.0736	0.7318	2
59	0.6818	0.9320	1.0730	0.7316	1
60	0.6820	0.9325	1.0724	0.7314	0
	Cos	Cot	Tan	Sin	′

*137° 227° *317° **47°** NATURAL

43° *133° 223° *313°

′	Sin	Tan	Cot	Cos	
0	0.6820	0.9325	1.0724	0.7314	60
1	0.6822	0.9331	1.0717	0.7312	59
2	0.6824	0.9336	1.0711	0.7310	58
3	0.6826	0.9341	1.0705	0.7308	57
4	0.6828	0.9347	1.0699	0.7306	56
5	0.6831	0.9352	1.0692	0.7304	55
6	0.6833	0.9358	1.0686	0.7302	54
7	0.6835	0.9363	1.0680	0.7300	53
8	0.6837	0.9369	1.0674	0.7298	52
9	0.6839	0.9374	1.0668	0.7296	51
10	0.6841	0.9380	1.0661	0.7294	50
11	0.6843	0.9385	1.0655	0.7292	49
12	0.6845	0.9391	1.0649	0.7290	48
13	0.6848	0.9396	1.0643	0.7288	47
14	0.6850	0.9402	1.0637	0.7286	46
15	0.6852	0.9407	1.0630	0.7284	45
16	0.6854	0.9413	1.0624	0.7282	44
17	0.6856	0.9418	1.0618	0.7280	43
18	0.6858	0.9424	1.0612	0.7278	42
19	0.6860	0.9429	1.0606	0.7276	41
20	0.6862	0.9435	1.0599	0.7274	40
21	0.6865	0.9440	1.0593	0.7272	39
22	0.6867	0.9446	1.0587	0.7270	38
23	0.6869	0.9451	1.0581	0.7268	37
24	0.6871	0.9457	1.0575	0.7266	36
25	0.6873	0.9462	1.0569	0.7264	35
26	0.6875	0.9468	1.0562	0.7262	34
27	0.6877	0.9473	1.0556	0.7260	33
28	0.6879	0.9479	1.0550	0.7258	32
29	0.6881	0.9484	1.0544	0.7256	31
30	0.6884	0.9490	1.0538	0.7254	30
31	0.6886	0.9495	1.0532	0.7252	29
32	0.6888	0.9501	1.0526	0.7250	28
33	0.6890	0.9506	1.0519	0.7248	27
34	0.6892	0.9512	1.0513	0.7246	26
35	0.6894	0.9517	1.0507	0.7244	25
36	0.6896	0.9523	1.0501	0.7242	24
37	0.6898	0.9528	1.0495	0.7240	23
38	0.6900	0.9534	1.0489	0.7238	22
39	0.6903	0.9540	1.0483	0.7236	21
40	0.6905	0.9545	1.0477	0.7234	20
41	0.6907	0.9551	1.0470	0.7232	19
42	0.6909	0.9556	1.0464	0.7230	18
43	0.6911	0.9562	1.0458	0.7228	17
44	0.6913	0.9567	1.0452	0.7226	16
45	0.6915	0.9573	1.0446	0.7224	15
46	0.6917	0.9578	1.0440	0.7222	14
47	0.6919	0.9584	1.0434	0.7220	13
48	0.6921	0.9590	1.0428	0.7218	12
49	0.6924	0.9595	1.0422	0.7216	11
50	0.6926	0.9601	1.0416	0.7214	10
51	0.6928	0.9606	1.0410	0.7212	9
52	0.6930	0.9612	1.0404	0.7210	8
53	0.6932	0.9618	1.0398	0.7208	7
54	0.6934	0.9623	1.0392	0.7206	6
55	0.6936	0.9629	1.0385	0.7203	5
56	0.6938	0.9634	1.0379	0.7201	4
57	0.6940	0.9640	1.0373	0.7199	3
58	0.6942	0.9646	1.0367	0.7197	2
59	0.6944	0.9651	1.0361	0.7195	1
60	0.6947	0.9657	1.0355	0.7193	0
	Cos	Cot	Tan	Sin	′

46° *136° 226° *316°

NATURAL **44°** *134° 224° *314°

′	Sin	Tan	Cot	Cos	
0	0.6947	0.9657	1.0355	0.7193	60
1	0.6949	0.9663	1.0349	0.7191	59
2	0.6951	0.9668	1.0343	0.7189	58
3	0.6953	0.9674	1.0337	0.7187	57
4	0.6955	0.9679	1.0331	0.7185	56
5	0.6957	0.9685	1.0325	0.7183	55
6	0.6959	0.9691	1.0319	0.7181	54
7	0.6961	0.9696	1.0313	0.7179	53
8	0.6963	0.9702	1.0307	0.7177	52
9	0.6965	0.9708	1.0301	0.7175	51
10	0.6967	0.9713	1.0295	0.7173	50
11	0.6970	0.9719	1.0289	0.7171	49
12	0.6972	0.9725	1.0283	0.7169	48
13	0.6974	0.9730	1.0277	0.7167	47
14	0.6976	0.9736	1.0271	0.7165	46
15	0.6978	0.9742	1.0265	0.7163	45
16	0.6980	0.9747	1.0259	0.7161	44
17	0.6982	0.9753	1.0253	0.7159	43
18	0.6984	0.9759	1.0247	0.7157	42
19	0.6986	0.9764	1.0241	0.7155	41
20	0.6988	0.9770	1.0235	0.7153	40
21	0.6990	0.9776	1.0230	0.7151	39
22	0.6992	0.9781	1.0224	0.7149	38
23	0.6995	0.9787	1.0218	0.7147	37
24	0.6997	0.9793	1.0212	0.7145	36
25	0.6999	0.9798	1.0206	0.7143	35
26	0.7001	0.9804	1.0200	0.7141	34
27	0.7003	0.9810	1.0194	0.7139	33
28	0.7005	0.9816	1.0188	0.7137	32
29	0.7007	0.9821	1.0182	0.7135	31
30	0.7009	0.9827	1.0176	0.7133	30
31	0.7011	0.9833	1.0170	0.7130	29
32	0.7013	0.9838	1.0104	0.7128	28
33	0.7015	0.9844	1.0158	0.7126	27
34	0.7017	0.9850	1.0152	0.7124	26
35	0.7019	0.9856	1.0147	0.7122	25
36	0.7022	0.9861	1.0141	0.7120	24
37	0.7024	0.9867	1.0135	0.7118	23
38	0.7026	0.9873	1.0129	0.7116	22
39	0.7028	0.9879	1.0123	0.7114	21
40	0.7030	0.9884	1.0117	0.7112	20
41	0.7032	0.9890	1.0111	0.7110	19
42	0.7034	0.9896	1.0105	0.7108	18
43	0.7036	0.9902	1.0099	0.7106	17
44	0.7038	0.9907	1.0094	0.7104	16
45	0.7040	0.9913	1.0088	0.7102	15
46	0.7042	0.9919	1.0082	0.7100	14
47	0.7044	0.9925	1.0076	0.7098	13
48	0.7046	0.9930	1.0070	0.7096	12
49	0.7048	0.9936	1.0064	0.7094	11
50	0.7050	0.9942	1.0058	0.7092	10
51	0.7053	0.9948	1.0052	0.7090	9
52	0.7055	0.9954	1.0047	0.7088	8
53	0.7057	0.9959	1.0041	0.7085	7
54	0.7059	0.9965	1.0035	0.7083	6
55	0.7061	0.9971	1.0029	0.7081	5
56	0.7063	0.9977	1.0023	0.7079	4
57	0.7065	0.9983	1.0017	0.7077	3
58	0.7067	0.9988	1.0012	0.7075	2
59	0.7069	0.9994	1.0006	0.7073	1
60	0.7071	1.0000	1.0000	0.7071	0
	Cos	Cot	Tan	Sin	′

*135° 225° *315° **45°** NATURAL

VALUES AND LOGARITHMS OF HAVERSINES

[Characteristics of Logarithms omitted — determine by rule from the value]

°	0′ Value	0′ Log10	10′ Value	10′ Log10	20′ Value	20′ Log10	30′ Value	30′ Log10	40′ Value	40′ Log10	50′ Value	50′ Log10
0	.0000		.0000	4.3254	.0000	4.9275	.0000	5.2796	.0000	5.5295	.0001	5.7223
1	.0001	5.8817	.0001	6.0156	.0001	6.1315	.0002	.2338	.0002	.3254	.0003	.4081
2	.0003	.4837	.0004	.5532	.0004	.6176	.0005	.6775	.0005	.7336	.0006	.7862
3	.0007	.8358	.0008	.8828	.0008	.9273	.0009	.9697	.0010	.0101	.0011	.0487
4	.0012	.0856	.0013	.1211	.0014	.1551	.0015	.1879	.0017	.2195	.0018	.2499
5	.0019	.2793	.0020	.3078	.0022	.3354	.0023	.3621	.0024	.3880	.0026	.4132
6	.0027	.4376	.0029	.4614	.0031	.4845	.0032	.5071	.0034	.5290	.0036	.5504
7	.0037	.5713	.0039	.5918	.0041	.6117	.0043	.6312	.0045	.6503	.0047	.6689
8	.0049	.6872	.0051	.7051	.0053	.7226	.0055	.7397	.0057	.7566	.0059	.7731
9	.0062	.7893	.0064	.8052	.0066	.8208	.0069	.8361	.0071	.8512	.0073	.8660
10	.0076	.8806	.0079	.8949	.0081	.9090	.0084	.9229	.0086	.9365	.0089	.9499
11	.0092	.9631	.0095	.9762	.0097	.9890	.0100	.0016	.0103	.0141	.0106	.0264
12	.0109	.0385	.0112	.0504	.0115	.0622	.0119	.0738	.0122	.0853	.0125	.0966
13	.0128	.1077	.0131	.1187	.0135	.1296	.0138	.1404	.0142	.1510	.0145	.1614
14	.0149	.1718	.0152	.1820	.0156	.1921	.0159	.2021	.0163	.2120	.0167	.2218
15	.0170	.2314	.0174	.2409	.0178	.2504	.0182	.2597	.0186	.2689	.0190	.2781
16	.0194	.2871	.0198	.2961	.0202	.3049	.0206	.3137	.0210	.3223	.0214	.3309
17	.0218	.3394	.0223	.3478	.0227	.3561	.0231	.3644	.0236	.3726	.0240	.3806
18	.0245	.3887	.0249	.3966	.0254	.4045	.0258	.4123	.0263	.4200	.0268	.4276
19	.0272	.4352	.0277	.4427	.0282	.4502	.0287	.4576	.0292	.4649	.0297	.4721
20	.0302	.4793	.0307	.4865	.0312	.4936	.0317	.5006	.0322	.5075	.0327	.5144
21	.0332	.5213	.0337	.5281	.0343	.5348	.0348	.5415	.0353	.5481	.0359	.5547
22	.0364	.5612	.0370	.5677	.0375	.5741	.0381	.5805	.0386	.5868	.0392	.5931
23	.0397	.5993	.0403	.6055	.0409	.6116	.0415	.6177	.0421	.6238	.0426	.6298
24	.0432	.6357	.0438	.6417	.0444	.6476	.0450	.6534	.0456	.6592	.0462	.6650
25	.0468	.6707	.0475	.6764	.0481	.6820	.0487	.6876	.0493	.6932	.0500	.6987
26	.0506	.7042	.0512	.7096	.0519	.7151	.0525	.7204	.0532	.7258	.0538	.7311
27	.0545	.7364	.0552	.7416	.0558	.7468	.0565	.7520	.0572	.7572	.0578	.7623
28	.0585	.7673	.0592	.7724	.0599	.7774	.0606	.7824	.0613	.7874	.0620	.7923
29	.0627	.7972	.0634	.8020	.0641	.8069	.0648	.8117	.0655	.8165	.0663	.8213
30	.0670	.8260	.0677	.8307	.0684	.8354	.0692	.8400	.0699	.8446	.0707	.8492
31	.0714	.8538	.0722	.8583	.0729	.8629	.0737	.8673	.0744	.8718	.0752	.8763
32	.0760	.8807	.0767	.8851	.0775	.8894	.0783	.8938	.0791	.8981	.0799	.9024
33	.0807	.9067	.0815	.9109	.0823	.9152	.0831	.9194	.0839	.9236	.0847	.9277
34	.0855	.9319	.0863	.9360	.0871	.9401	.0879	.9442	.0888	.9482	.0896	.9523
35	.0904	.9563	.0913	.9603	.0921	.9643	.0929	.9682	.0938	.9722	.0946	.9761
36	.0955	.9800	.0963	.9838	.0972	.9877	.0981	.9915	.0989	.9954	.0998	.9992
37	.1007	.0030	.1016	.0067	.1024	.0105	.1033	.0142	.1042	.0179	.1051	.0216
38	.1060	.0253	.1069	.0289	.1078	.0326	.1087	.0362	.1096	.0398	.1105	.0434
39	.1114	.0470	.1123	.0505	.1133	.0541	.1142	.0576	.1151	.0611	.1160	.0646
40	.1170	.0681	.1179	.0716	.1189	.0750	.1198	.0784	.1207	.0817	.1217	.0853
41	.1226	.0887	.1236	.0920	.1246	.0954	.1255	.0987	.1265	.1021	.1275	.1054
42	.1284	.1087	.1294	.1119	.1304	.1152	.1314	.1185	.1323	.1217	.1333	.1249
43	.1343	.1282	.1353	.1314	.1363	.1345	.1373	.1377	.1383	.1409	.1393	.1440
44	.1403	.1472	.1413	.1503	.1424	.1534	.1434	.1565	.1444	.1596	.1454	.1626
45	.1464	.1657	.1475	.1687	.1485	.1718	.1495	.1748	.1506	.1778	.1516	.1808
46	.1527	.1838	.1538	.1867	.1548	.1897	.1558	.1926	.1569	.1956	.1579	.1985
47	.1590	.2014	.1600	.2043	.1611	.2072	.1622	.2101	.1633	.2129	.1644	.2158
48	.1654	.2186	.1665	.2215	.1676	.2243	.1687	.2271	.1698	.2299	.1709	.2327
49	.1720	.2355	.1731	.2382	.1742	.2410	.1753	.2437	.1764	.2465	.1775	.2492
50	.1786	.2519	.1797	.2546	.1808	.2573	.1820	.2600	.1831	.2627	.1842	.2653
51	.1853	.2680	.1865	.2706	.1876	.2732	.1887	.2759	.1899	.2785	.1910	.2811
52	.1922	.2837	.1933	.2863	.1945	.2888	.1956	.2914	.1968	.2940	.1979	.2965
53	.1991	.2991	.2003	.3016	.2014	.3041	.2026	.3066	.2038	.3091	.2049	.3116
54	.2061	.3141	.2073	.3166	.2085	.3190	.2096	.3215	.2108	.3239	.2120	.3264
55	.2132	.3288	.2144	.3312	.2156	.3336	.2168	.3361	.2180	.3384	.2192	.3408
56	.2204	.3432	.2216	.3456	.2228	.3480	.2240	.3503	.2252	.3527	.2265	.3550
57	.2277	.3573	.2289	.3596	.2301	.3620	.2314	.3643	.2326	.3666	.2338	.3689
58	.2350	.3711	.2363	.3734	.2375	.3757	.2388	.3779	.2400	.3802	.2412	.3824
59	.2425	.3847	.2437	.3869	.2450	.3891	.2462	.3913	.2475	.3935	.2487	.3957

°	0′ Value	0′ Log₁₀	10′ Value	10′ Log₁₀	20′ Value	20′ Log₁₀	30′ Value	30′ Log₁₀	40′ Value	40′ Log₁₀	50′ Value	50′ Log₁₀
60	.2500	.3979	.2513	.4001	.2525	.4023	.2538	.4045	.2551	.4066	.2563	.4088
61	.2576	.4109	.2589	.4131	.2601	.4152	.2614	.4173	.2627	.4195	.2640	.4216
62	.2653	.4237	.2665	.4258	.2678	.4279	.2691	.4300	.2704	.4320	.2717	.4341
63	.2730	.4362	.2743	.4382	.2756	.4403	.2769	.4423	.2782	.4444	.2795	.4464
64	.2808	.4484	.2821	.4504	.2834	.4524	.2847	.4545	.2861	.4565	.2874	.4584
65	.2887	.4604	.2900	.4624	.2913	.4644	.2927	.4664	.2940	.4683	.2953	.4703
66	.2966	.4722	.2980	.4742	.2993	.4761	.3006	.4780	.3020	.4799	.3033	.4819
67	.3046	.4838	.3060	.4857	.3073	.4876	.3087	.4895	.3100	.4914	.3113	.4932
68	.3127	.4951	.3140	.4970	.3154	.4989	.3167	.5007	.3181	.5026	.3195	.5044
69	.3208	.5063	.3222	.5081	.3235	.5099	.3249	.5117	.3263	.5136	.3276	.5154
70	.3290	.5172	.3304	.5190	.3317	.5208	.3331	.5226	.3345	.5244	.3358	.5261
71	.3372	.5279	.3386	.5297	.3400	.5314	.3413	.5332	.3427	.5349	.3441	.5367
72	.3455	.5384	.3469	.5402	.3483	.5419	.3496	.5436	.3510	.5454	.3524	.5471
73	.3538	.5488	.3552	.5505	.3566	.5522	.3580	.5539	.3594	.5556	.3608	.5572
74	.3622	.5589	.3636	.5606	.3650	.5623	.3664	.5639	.3678	.5656	.3692	.5672
75	.3706	.5689	.3720	.5705	.3734	.5722	.3748	.5738	.3762	.5754	.3776	.5771
76	.3790	.5787	.3805	.5803	.3819	.5819	.3833	.5835	.3847	.5851	.3861	.5867
77	.3875	.5883	.3889	.5899	.3904	.5915	.3918	.5930	.3932	.5946	.3946	.5962
78	.3960	.5977	.3975	.5993	.3989	.6009	.4003	.6024	.4017	.6039	.4032	.6055
79	.4046	.6070	.4060	.6085	.4075	.6101	.4089	.6116	.4103	.6131	.4117	.6146
80	.4132	.6161	.4146	.6176	.4160	.6191	.4175	.6206	.4189	.6221	.4203	.6236
81	.4218	.6251	.4232	.6266	.4247	.6280	.4261	.6295	.4275	.6310	.4290	.6324
82	.4304	.6339	.4319	.6353	.4333	.6368	.4347	.6382	.4362	.6397	.4376	.6411
83	.4391	.6425	.4405	.6440	.4420	.6454	.4434	.6468	.4448	.6482	.4463	.6496
84	.4477	.6510	.4492	.6524	.4506	.6538	.4521	.6552	.4535	.6566	.4550	.6580
85	.4564	.6594	.4579	.6607	.4593	.6621	.4608	.6635	.4622	.6649	.4637	.6663
86	.4651	.6676	.4666	.6689	.4680	.6703	.4695	.6716	.4709	.6730	.4724	.6743
87	.4738	.6756	.4753	.6770	.4767	.6783	.4782	.6796	.4796	.6809	.4811	.6822
88	.4826	.6835	.4840	.6848	.4855	.6862	.4869	.6875	.4884	.6887	.4898	.6900
89	.4913	.6913	.4937	.6926	.4942	.6939	.4956	.6952	.4971	.6964	.4985	.6977
90	.5000	.6990	.5015	.7002	.5029	.7015	.5044	.7027	.5058	.7040	.5073	.7052
91	.5087	.7065	.5102	.7077	.5116	.7090	.5131	.7102	.5145	.7114	.5160	.7126
92	.5174	.7139	.5189	.7151	.5204	.7163	.5218	.7175	.5233	.7187	.5247	.7199
93	.5262	.7211	.5276	.7223	.5291	.7235	.5305	.7247	.5320	.7259	.5334	.7271
94	.5349	.7283	.5363	.7294	.5378	.7306	.5392	.7318	.5407	.7329	.5421	.7341
95	.5436	.7353	.5450	.7364	.5465	.7376	.5479	.7387	.5494	.7399	.5508	.7410
96	.5523	.7421	.5537	.7433	.5552	.7444	.5566	.7455	.5580	.7467	.5595	.7478
97	.5609	.7489	.5624	.7500	.5638	.7511	.5653	.7523	.5667	.7534	.5682	.7545
98	.5696	.7556	.5710	.7567	.5725	.7577	.5739	.7588	.5753	.7599	.5768	.7610
99	.5782	.7621	.5797	.7632	.5811	.7642	.5825	.7653	.5840	.7664	.5854	.7674
100	.5868	.7685	.5883	.7696	.5897	.7706	.5911	.7717	.5925	.7727	.5940	.7738
101	.5954	.7748	.5968	.7759	.5983	.7769	.5997	.7779	.6011	.7790	.6025	.7800
102	.6040	.7810	.6054	.7820	.6068	.7830	.6082	.7841	.6096	.7851	.6111	.7861
103	.6125	.7871	.6139	.7881	.6153	.7891	.6167	.7901	.6181	.7911	.6195	.7921
104	.6210	.7931	.6224	.7940	.6238	.7950	.6252	.7960	.6266	.7970	.6280	.7980
105	.6294	.7989	.6308	.7999	.6322	.8009	.6336	.8018	.6350	.8028	.6364	.8037
106	.6378	.8047	.6392	.8056	.6406	.8066	.6420	.8075	.6434	.8085	.6448	.8094
107	.6462	.8104	.6476	.8113	.6490	.8122	.6504	.8131	.6517	.8141	.6531	.8150
108	.6545	.8159	.6559	.8168	.6573	.8177	.6587	.8187	.6600	.8196	.6614	.8205
109	.6628	.8214	.6642	.8223	.6655	.8232	.6669	.8241	.6683	.8250	.6696	.8258
110	.6710	.8267	.6724	.8276	.6737	.8285	.6751	.8294	.6765	.8302	.6778	.8311
111	.6792	.8320	.6805	.8329	.6819	.8337	.6833	.8346	.6846	.8354	.6860	.8363
112	.6873	.8371	.6887	.8380	.6900	.8388	.6913	.8397	.6927	.8405	.6940	.8414
113	.6954	.8422	.6967	.8430	.6980	.8439	.6994	.8447	.7007	.8455	.7020	.8464
114	.7034	.8472	.7047	.8480	.7060	.8488	.7073	.8496	.7087	.8504	.7100	.8513
115	.7113	.8521	.7126	.8529	.7139	.8537	.7153	.8545	.7166	.8553	.7179	.8561
116	.7192	.8568	.7205	.8576	.7218	.8584	.7231	.8592	.7244	.8600	.7257	.8608
117	.7270	.8615	.7283	.8623	.7296	.8631	.7309	.8638	.7322	.8646	.7335	.8654
118	.7347	.8661	.7360	.8669	.7373	.8676	.7386	.8684	.7399	.8691	.7411	.8699
119	.7424	.8706	.7437	.8714	.7449	.8721	.7462	.8729	.7475	.8736	.7487	.8743

°	0′ Value	0′ Log₁₀	10′ Value	10′ Log₁₀	20′ Value	20′ Log₁₀	30′ Value	30′ Log₁₀	40′ Value	40′ Log₁₀	50′ Value	50′ Log₁₀
120	.7500	.8751	.7513	.8758	.7525	.8765	.7538	.8772	.7550	.8780	.7563	.8787
121	.7575	.8794	.7588	.8801	.7600	.8808	.7612	.8815	.7625	.8822	.7637	.8829
122	.7650	.8836	.7662	.8843	.7674	.8850	.7686	.8857	.7699	.8864	.7711	.8871
123	.7723	.8878	.7735	.8885	.7748	.8892	.7760	.8898	.7772	.8905	.7784	.8912
124	.7796	.8919	.7808	.8925	.7820	.8932	.7832	.8939	.7844	.8945	.7856	.8952
125	.7868	.8959	.7880	.8965	.7892	.8972	.7904	.8978	.7915	.8985	.7927	.8991
126	.7939	.8998	.7951	.9004	.7962	.9010	.7974	.9017	.7986	.9023	.7997	.9030
127	.8009	.9036	.8021	.9042	.8032	.9048	.8044	.9055	.8055	.9061	.8067	.9067
128	.8078	.9073	.8090	.9079	.8101	.9085	.8113	.9092	.8124	.9098	.8135	.9104
129	.8147	.9110	.8158	.9116	.8169	.9122	.8180	.9128	.8192	.9134	.8203	.9140
130	.8214	.9146	.8225	.9151	.8236	.9157	.8247	.9163	.8258	.9169	.8269	.9175
131	.8280	.9180	.8291	.9186	.8302	.9192	.8313	.9198	.8324	.9203	.8335	.9209
132	.8346	.9215	.8356	.9220	.8367	.9226	.8378	.9231	.8389	.9237	.8399	.9242
133	.8410	.9248	.8421	.9253	.8431	.9259	.8442	.9264	.8452	.9270	.8463	.9275
134	.8473	.9281	.8484	.9286	.8494	.9291	.8501	.9297	.8515	.9302	.8525	.9307
135	.8536	.9312	.8546	.9318	.8556	.9323	.8566	.9328	.8576	.9333	.8587	.9338
136	.8597	.9343	.8607	.9348	.8617	.9353	.8627	.9359	.8637	.9364	.8647	.9369
137	.8657	.9374	.8667	.9379	.8677	.9383	.8686	.9388	.8696	.9393	.8706	.9398
138	.8716	.9403	.8725	.9408	.8735	.9413	.8745	.9417	.8754	.9422	.8764	.9427
139	.8774	.9432	.8783	.9436	.8793	.9441	.8802	.9446	.8811	.9450	.8821	.9455
140	.8830	.9460	.8840	.9464	.8849	.9469	.8858	.9473	.8867	.9478	.8877	.9482
141	.8886	.9487	.8895	.9491	.8904	.9496	.8913	.9500	.8922	.9505	.8931	.9509
142	.8940	.9513	.8949	.9518	.8958	.9522	.8967	.9526	.8976	.9531	.8984	.9535
143	.8993	.9539	.9002	.9543	.9011	.9548	.9019	.9552	.9028	.9556	.9037	.9560
144	.9045	.9564	.9054	.9568	.9062	.9572	.9071	.9576	.9079	.9580	.9087	.9584
145	.9096	.9588	.9104	.9592	.9112	.9596	.9121	.9600	.9129	.9604	.9137	.9608
146	.9145	.9612	.9153	.9616	.9161	.9620	.9169	.9623	.9177	.9627	.9185	.9631
147	.9193	.9635	.9201	.9638	.9209	.9642	.9217	.9646	.9225	.9650	.9233	.9653
148	.9240	.9657	.9248	.9660	.9256	.9664	.9263	.9668	.9271	.9671	.9278	.9675
149	.9286	.9678	.9293	.9682	.9301	.9685	.9308	.9689	.9316	.9692	.9323	.9695
150	.9330	.9699	.9337	.9702	.9345	.9706	.9352	.9709	.9359	.9712	.9366	.9716
151	.9373	.9719	.9380	.9722	.9387	.9725	.9394	.9729	.9401	.9732	.9408	.9735
152	.9415	.9738	.9422	.9741	.9428	.9744	.9435	.9747	.9442	.9751	.9448	.9754
153	.9455	.9757	.9462	.9760	.9468	.9763	.9475	.9766	.9481	.9769	.9488	.9772
154	.9494	.9774	.9500	.9777	.9507	.9780	.9513	.9783	.9519	.9786	.9525	.9789
155	.9532	.9792	.9538	.9794	.9544	.9797	.9550	.9800	.9556	.9803	.9562	.9805
156	.9568	.9808	.9574	.9811	.9579	.9813	.9585	.9816	.9591	.9819	.9597	.9821
157	.9603	.9824	.9608	.9826	.9614	.9829	.9619	.9831	.9625	.9834	.9630	.9836
158	.9636	.9839	.9641	.9841	.9647	.9844	.9652	.9846	.9657	.9849	.9663	.9851
159	.9668	.9853	.9673	.9856	.9678	.9858	.9683	.9860	.9688	.9863	.9693	.9865
160	.9698	.9867	.9703	.9869	.9708	.9871	.9713	.9874	.9718	.9876	.9723	.9878
161	.9728	.9880	.9732	.9882	.9737	.9884	.9742	.9886	.9746	.9888	.9751	.9890
162	.9755	.9892	.9760	.9894	.9764	.9896	.9769	.9898	.9773	.9900	.9777	.9902
163	.9782	.9904	.9786	.9906	.9790	.9908	.9794	.9910	.9798	.9911	.9802	.9913
164	.9806	.9915	.9810	.9917	.9814	.9919	.9818	.9920	.9822	.9922	.9826	.9923
165	.9830	.9925	.9833	.9927	.9837	.9929	.9841	.9930	.9844	.9932	.9848	.9933
166	.9851	.9935	.9855	.9937	.9858	.9938	.9862	.9940	.9865	.9941	.9869	.9943
167	.9872	.9944	.9875	.9945	.9878	.9947	.9881	.9948	.9885	.9950	.9888	.9951
168	.9891	.9952	.9894	.9954	.9897	.9955	.9900	.9956	.9903	.9957	.9905	.9959
169	.9908	.9960	.9911	.9961	.9914	.9962	.9916	.9963	.9919	.9965	.9921	.9966
170	.9924	.9967	.9927	.9968	.9929	.9969	.9931	.9970	.9934	.9971	.9936	.9972
171	.9938	.9973	.9941	.9974	.9943	.9975	.9945	.9976	.9947	.9977	.9949	.9978
172	.9951	.9979	.9953	.9980	.9955	.9981	.9957	.9981	.9959	.9982	.9961	.9983
173	.9963	.9984	.9964	.9984	.9966	.9985	.9968	.9986	.9969	.9987	.9971	.9987
174	.9973	.9988	.9974	.9988	.9976	.9989	.9977	.9990	.9978	.9991	.9980	.9991
175	.9981	.9992	.9982	.9992	.9983	.9993	.9985	.9993	.9986	.9994	.9987	.9994
176	.9988	.9995	.9989	.9995	.9990	.9996	.9991	.9996	.9992	.9996	.9992	.9997
177	.9993	.9997	.9994	.9997	.9995	.9998	.9995	.9998	.9996	.9998	.9996	.9998
178	.9997	.9999	.9997	.9999	.9998	.9999	.9998	.9999	.9999	.9999	.9999	.9999
179	.9999	.9999	.9999	.9999	.9999	.9999	.9999	.9999	.9999	.0000	1.0000	.0000

POWERS AND ROOTS

No.	Square	Cube	Square Root	Cube Root	No.	Square	Cube	Square Root	Cube Root
1	1	1	1.000	1.000	**51**	2 601	132 651	7.141	3.708
2	4	8	1.414	1.260	**52**	2 704	140 608	7.211	3.733
3	9	27	1.732	1.442	**53**	2 809	148 877	7.280	3.756
4	16	64	2.000	1.587	**54**	2 916	157 464	7.348	3.780
5	25	125	2.236	1.710	**55**	3 025	166 375	7.416	3.803
6	36	216	2.449	1.817	**56**	3 136	175 616	7.483	3.826
7	49	343	2.646	1.913	**57**	3 249	185 193	7.550	3.849
8	64	512	2.828	2.000	**58**	3 364	195 112	7.616	3.871
9	81	729	3.000	2.080	**59**	3 481	205 379	7.681	3.893
10	100	1 000	3.162	2.154	**60**	3 600	216 000	7.746	3.915
11	121	1 331	3.317	2.224	**61**	3 721	226 981	7.810	3.936
12	144	1 728	3.464	2.289	**62**	3 844	238 328	7.874	3.958
13	169	2 197	3.606	2.351	**63**	3 969	250 047	7.937	3.979
14	196	2 744	3.742	2.410	**64**	4 096	262 144	8.000	4.000
15	225	3 375	3.873	2.466	**65**	4 225	274 625	8.062	4.021
16	256	4 096	4.000	2.520	**66**	4 356	287 496	8.124	4.041
17	289	4 913	4.123	2.571	**67**	4 489	300 763	8.185	4.062
18	324	5 832	4.243	2.621	**68**	4 624	314 432	8.246	4.082
19	361	6 859	4.359	2.668	**69**	4 761	328 509	8.307	4.102
20	400	8 000	4.472	2.714	**70**	4 900	343 000	8.367	4.121
21	441	9 261	4.583	2.759	**71**	5 041	357 911	8.426	4.141
22	484	10 648	4.690	2.802	**72**	5 184	373 248	8.485	4.160
23	529	12 167	4.796	2.844	**73**	5 329	389 017	8.544	4.179
24	576	13 824	4.899	2.884	**74**	5 476	405 224	8.602	4.198
25	625	15 625	5.000	2.924	**75**	5 625	421 875	8.660	4.217
26	676	17 576	5.099	2.962	**76**	5 776	438 976	8.718	4.236
27	729	19 683	5.196	3.000	**77**	5 929	456 533	8.775	4.254
28	784	21 952	5.292	3.037	**78**	6 084	474 552	8.832	4.273
29	841	24 389	5.385	3.072	**79**	6 241	493 039	8.888	4.291
30	900	27 000	5.477	3.107	**80**	6 400	512 000	8.944	4.309
31	961	29 791	5.568	3.141	**81**	6 561	531 441	9.000	4.327
32	1 024	32 768	5.657	3.175	**82**	6 724	551 368	9.055	4.344
33	1 089	35 937	5.745	3.208	**83**	6 889	571 787	9.110	4.362
34	1 156	39 304	5.831	3.240	**84**	7 056	592 704	9.165	4.380
35	1 225	42 875	5.916	3.271	**85**	7 225	614 125	9.220	4.397
36	1 296	46 656	6.000	3.302	**86**	7 396	636 056	9.274	4.414
37	1 369	50 653	6.083	3.332	**87**	7 569	658 503	9.327	4.431
38	1 444	54 872	6.164	3.362	**88**	7 744	681 472	9.381	4.448
39	1 521	59 319	6.245	3.391	**89**	7 921	704 969	9.434	4.465
40	1 600	64 000	6.325	3.420	**90**	8 100	729 000	9.487	4.481
41	1 681	68 921	6.403	3.448	**91**	8 281	753 571	9.539	4.498
42	1 764	74 088	6.481	3.476	**92**	8 464	778 688	9.592	4.514
43	1 849	79 507	6.557	3.503	**93**	8 649	804 357	9.644	4.531
44	1 936	85 184	6.633	3.530	**94**	8 836	830 584	9.695	4.547
45	2 025	91 125	6.708	3.557	**95**	9 025	857 375	9.747	4.563
46	2 116	97 336	6.782	3.583	**96**	9 216	884 736	9.798	4.579
47	2 209	103 823	6.856	3.609	**97**	9 409	912 673	9.849	4.595
48	2 304	110 592	6.928	3.634	**98**	9 604	941 192	9.899	4.610
49	2 401	117 649	7.000	3.659	**99**	9 801	970 299	9.950	4.626
50	2 500	125 000	7.071	3.684	**100**	10 000	1 000 000	10.000	4.642

DEGREES, MINUTES AND SECONDS TO RADIANS

1 degree = 0.01745 32925 19943 radians

Degrees						Minutes		Seconds	
0°	0.00000 00	60°	1.04719 76	120°	2.09439 51	0	0.00000 00	0	0.00000 00
1	0.01745 33	61	1.06465 08	121	2.11184 84	1	0.00029 09	1	0.00000 48
2	0.03490 66	62	1.08210 41	122	2.12930 17	2	0.00058 18	2	0.00000 97
3	0.05235 99	63	1.09955 74	123	2.14675 50	3	0.00087 27	3	0.00001 45
4	0.06981 32	64	1.11701 07	124	2.16420 83	4	0.00116 36	4	0.00001 94
5	0.08726 65	65	1.13446 40	125	2.18166 16	5	0.00145 44	5	0.00002 42
6	0.10471 98	66	1.15191 73	126	2.19911 49	6	0.00174 53	6	0.00002 91
7	0.12217 30	67	1.16937 06	127	2.21656 82	7	0.00203 62	7	0.00003 39
8	0.13962 63	68	1.18682 39	128	2.23402 14	8	0.00232 71	8	0.00003 88
9	0.15707 96	69	1.20427 72	129	2.25147 47	9	0.00261 80	9	0.00004 36
10	0.17453 29	70	1.22173 05	130	2.26892 80	10	0.00290 89	10	0.00004 85
11	0.19198 62	71	1.23918 38	131	2.28638 13	11	0.00319 98	11	0.00005 33
12	0.20943 95	72	1.25663 71	132	2.30383 46	12	0.00349 07	12	0.00005 82
13	0.22689 28	73	1.27409 04	133	2.32128 79	13	0.00378 15	13	0.00006 30
14	0.24434 61	74	1.29154 36	134	2.33874 12	14	0.00407 24	14	0.00006 79
15	0.26179 94	75	1.30899 69	135	2.35619 45	15	0.00436 33	15	0.00007 27
16	0.27925 27	76	1.32645 02	136	2.37364 78	16	0.00465 42	16	0.00007 76
17	0.29670 60	77	1.34390 35	137	2.39110 11	17	0.00494 51	17	0.00008 24
18	0.31415 93	78	1.36135 68	138	2.40855 44	18	0.00523 60	18	0.00008 73
19	0.33161 26	79	1.37881 01	139	2.42600 77	19	0.00552 69	19	0.00009 21
20	0.34906 59	80	1.39626 34	140	2.44346 10	20	0.00581 78	20	0.00009 70
21	0.36651 91	81	1.41371 67	141	2.46091 42	21	0.00610 87	21	0.00010 18
22	0.38397 24	82	1.43117 00	142	2.47836 75	22	0.00639 95	22	0.00010 67
23	0.40142 57	83	1.44862 33	143	2.49582 08	23	0.00669 04	23	0.00011 15
24	0.41887 90	84	1.46607 66	144	2.51327 41	24	0.00698 13	24	0.00011 64
25	0.43633 23	85	1.48352 99	145	2.53072 74	25	0.00727 22	25	0.00012 12
26	0.45378 56	86	1.50098 32	146	2.54818 07	26	0.00756 31	26	0.00012 61
27	0.47123 89	87	1.51843 64	147	2.56563 40	27	0.00785 40	27	0.00013 09
28	0.48869 22	88	1.53588 97	148	2.58308 73	28	0.00814 49	28	0.00013 57
29	0.50614 55	89	1.55334 30	149	2.60054 06	29	0.00843 58	29	0.00014 06
30	0.52359 88	90	1.57079 63	150	2.61799 39	30	0.00872 66	30	0.00014 54
31	0.54105 21	91	1.58824 96	151	2.63544 72	31	0.00901 75	31	0.00015 03
32	0.55850 54	92	1.60570 29	152	2.65290 05	32	0.00930 84	32	0.00015 51
33	0.57595 87	93	1.62315 62	153	2.67035 38	33	0.00959 93	33	0.00016 00
34	0.59341 19	94	1.64060 95	154	2.68780 70	34	0.00989 02	34	0.00016 48
35	0.61086 52	95	1.65806 28	155	2.70526 03	35	0.01018 11	35	0.00016 97
36	0.62831 85	96	1.67551 61	156	2.72271 36	36	0.01047 20	36	0.00017 45
37	0.64577 18	97	1.69296 94	157	2.74016 69	37	0.01076 29	37	0.00017 94
38	0.66322 51	98	1.71042 27	158	2.75762 02	38	0.01105 38	38	0.00018 42
39	0.68067 84	99	1.72787 60	159	2.77507 35	39	0.01134 46	39	0.00018 91
40	0.69813 17	100	1.74532 93	160	2.79252 68	40	0.01163 55	40	0.00019 39
41	0.71558 50	101	1.76278 25	161	2.80998 01	41	0.01192 64	41	0.00019 88
42	0.73303 83	102	1.78023 58	162	2.82743 34	42	0.01221 73	42	0.00020 36
43	0.75049 16	103	1.79768 91	163	2.84488 67	43	0.01250 82	43	0.00020 85
44	0.76794 49	104	1.81514 24	164	2.86234 00	44	0.01279 91	44	0.00021 33
45	0.78539 82	105	1.83259 57	165	2.87979 33	45	0.01309 00	45	0.00021 82
46	0.80285 15	106	1.85004 90	166	2.89724 66	46	0.01338 09	46	0.00022 30
47	0.82030 47	107	1.86750 23	167	2.91469 99	47	0.01367 17	47	0.00022 79
48	0.83775 80	108	1.88495 56	168	2.93215 31	48	0.01396 26	48	0.00023 27
49	0.85521 13	109	1.90240 89	169	2.94960 64	49	0.01425 35	49	0.00023 76
50	0.87266 46	110	1.91986 22	170	2.96705 97	50	0.01454 44	50	0.00024 24
51	0.89011 79	111	1.93731 55	171	2.98451 30	51	0.01483 53	51	0.00024 73
52	0.90757 12	112	1.95476 88	172	3.00196 63	52	0.01512 62	52	0.00025 21
53	0.92502 45	113	1.97222 21	173	3.01941 96	53	0.01541 71	53	0.00025 70
54	0.94247 78	114	1.98967 53	174	3.03687 29	54	0.01570 80	54	0.00026 18
55	0.95993 11	115	2.00712 86	175	3.05432 62	55	0.01599 89	55	0.00026 66
56	0.97738 44	116	2.02458 19	176	3.07177 95	56	0.01628 97	56	0.00027 15
57	0.99483 77	117	2.04203 52	177	3.08923 28	57	0.01658 06	57	0.00027 63
58	1.01229 10	118	2.05948 85	178	3.10668 61	58	0.01687 15	58	0.00028 12
59	1.02974 43	119	2.07694 18	179	3.12413 94	59	0.01716 24	59	0.00028 60
60	1.04719 76	120	2.09439 51	180	3.14159 27	60	0.01745 33	60	0.00029 09

1 radian = 57.29577 95130 82321 degrees

RADIANS TO DEGREES

	Radians	Tenths	Hundredths	Thousandths	Ten thousandths
1	57°17′44″.8	5°43′46″.5	0°34′22″.6	0° 3′26″.3	0° 0′20″.6
2	114°35′29″.6	11°27′33″.0	1° 8′45″.3	0° 6′52″.5	0° 0′41″.3
3	171°53′14″.4	17°11′19″.4	1°43′07″.9	0°10′18″.8	0° 1′01″.9
4	229°10′59″.2	22°55′05″.9	2°17′30″.6	0°13′45″.1	0° 1′22″.5
5	286°28′44″.0	28°38′52″.4	2°51′53″.2	0°17′11″.3	0° 1′43″.1
6	343°46′28″.8	34°22′38″.9	3°26′15″.9	0°20′37″.6	0° 2′03″.8
7	401° 4′13″.6	40° 6′25″.4	4° 0′38″.5	0°24′03″.9	0° 2′24″.4
8	458°21′58″.4	45°50′11″.8	4°35′01″.2	0°27′30″.1	0° 2′45″.0
9	515°39′43″.3	51°33′58″.3	5° 9′23″.8	0°30′56″.4	0° 3′05″.6

TABLE OF CONSTANTS

VALUES		RECIPROCALS	
π	3.14159 26535 89793	$\frac{1}{\pi}$	0.31830 98861 83791
$\frac{\pi}{2}$	1.57079 63267 94897	$\frac{2}{\pi}$	0.63661 97723 67582
2π	6.28318 53071 79586	$\frac{1}{2\pi}$	0.15915 49430 91895
π^2	9.86960 44010 89359	$\frac{1}{\pi^2}$	0.10132 11836 42338
$\sqrt{\pi}$	1.77245 38509 05516	$\frac{1}{\sqrt{\pi}}$	0.56418 95835 47756
$\sqrt{\frac{\pi}{2}}$	1.25331 41373 15500	$\sqrt{\frac{2}{\pi}}$	0.79788 45608 02865
$\sqrt{2\pi}$	2.50622 82746 31001	$\frac{1}{\sqrt{2\pi}}$	0.39894 22804 01433
e	2.71828 18284 59045	$\frac{1}{e}$	0.36787 94411 71442
e^2	7.38905 60989 30650	$\frac{1}{e^2}$	0.13533 52832 36613
$\sqrt{e}$	1.64872 12707 00128	$\frac{1}{\sqrt{e}}$	0.60653 06597 36613
$\log 10^e$	0.43429 44819 03252	$\log_e 10$	2.30258 50929 94046